Physics for Realists

Electricity and Magnetism

Modern Physics with a Common Sense Grounding

Physics for Realists:

Electricity and Magnetism

Physics with a Common Sense Grounding

Ravi,
   Thanks for all your support through the years. Please know how much it is appreciated.   Your friend,
                                    Anth

Anthony Rizzi

Press of the Institute for Advanced Physics

Published by IAP Press, Baton Rouge, LA (July 14, 2011)
IAP Press: IAPpress@iapweb.org

ISBN: 978-0-9816470-2-9
Library of Congress Control Number: 2011929560
Includes appendices, index.
1. Physics-textbooks. I. Rizzi, Anthony II. Title.

QC522   2011
537-dc22

Printed in the United States of America
Rochester, New York

<u>Author</u>

Anthony Rizzi, Ph.D., Director, Institute for Advanced Physics, LA

<u>Reviewers & Contributors</u>

Benedict Ashley, O.P., Ph.D., philosopher, St. Louis University, MO
(special role in the reviewing process)

Murray Daw, Ph.D., Professor of Physics, Clemson University, SC

Joseph Haller, Ph.D., physicist, Institute for Advanced Physics

Clifton Hill, C.S.Sp., physicist, Institute for Advanced Physics,
Prof. of Physics (retired), Duquesne University, PA

Kenneth Klenk, Ph.D., Director, Science Systems Consulting, Inc., SD

Joseph Martin, Ph.D., physicist, Lockheed Martin Planetary
Science Lab (retired), CO

Daniel Welch, Ph.D, Professor of Physics & Department Chairman.
Wofford College, SC

<u>Cover Final Drawing</u>

Jeffrey Hayden, Space Communications Architect, NASA Glenn Research
Center (GRC), Cleveland, OH

<u>Special thanks to:</u>

IAP's *Foundational Funders* each of whom made this work possible.
Giuseppe Rizzi for reviewing many parts of the manuscript
for clarity and common sense approachableness.

<u>Heartfelt thanks to:</u>

Mrs. Susan Rizzi and Giuseppe Rizzi for much work, and Kateri,
Nicolo and Thomasina Rizzi for their special contributions during
the long process of the creating and writing of this book.

Preface

In this second volume of the *Physics for Realists* series, we cover electricity and magnetism (E&M) from a point of view that keeps contact with one's pre-scientific experience. This is what we mean by *Physics for Realists*. It brings out and emphasizes that physics is about the world we see around us. It starts with what we see in front of us and leads us to study things too small (in size or influence) or too far away to see. It applies to the entire physical universe.

This textbook assumes that you have completed the first course in the series: *Physics for Realists: Mechanics*. Having completed that course on Newtonian mechanics, you have some real understanding of the causes and nature of locomotion, including (analogically) quantitative specifics, and the generic nature of the massive bodies. These latter are bodies that we can move and see the result directly or indirectly from baseballs and apples to rockets and planes to molecules and amoebae.

Whereas the first course could rightly be called the "Book of Impetus," this text might be called "Impetus Meets Space" or, using a new word from the text, "Impetus meets Plana (vacuum)," or, more descriptively, "How Certain Massive Bodies Act over a Distance." Even this last phrase needs much more explanation to fill out what it means, and that, of course, is the task of the text. Here we want only to give you a flavor for the significant difference in depth of content and approach of this series in general and in this text in particular.

Physics for Realists (PFR) is for people who want to be firmly grounded in the world of things around them. There is a second related philosophical sense, where *realism* is opposed to *philosophical idealism*, not against wanting to dream great dreams but against the idea that our minds know only themselves and ideas, not things. Of course, science is about real things, so we cannot accept such an opinion, but oftentimes this is not made clear to students by connecting everything back to their immediate sensorial contact with real physical things.

This starting point makes this text unique. Indeed, this text distinguishes itself in at least six major ways:

1. Common sense starting point, building on the platform established in the first volume. Everything in standard E&M, from the deepest (empiriometric) physics to the most esoteric mathematical formalism, is grounded in the things that we see, hear or feel.

2. Unifying practical theme in addition to theoretical theme. The student learns the principles behind an RF transmitter that he can actually build in the final chapter. The larger theme, initiated in volume one, of a manned trip to Mars is continued in this volume as well.

3. History boxes that continue the history sequence established in the first volume which underscores the continuity of physics over the centuries (and in so doing cements the concepts and gives students an important pathway by which to remember them).

4. The ϕ-field and the A-field are treated as fundamental, which, in light of the fundamental principles introduced in the first volume, simplify the exposition and give new insights.

5. Radiation is treated as a pinnacle of the understanding of the subject.
6. Building on the impetus-based development of special relativity given in the first volume, the relativistic underpinnings of E&M that give deep insight into the interrelation between the magnetic and electric field are given.

The wonder and awe that nature elicits through her deep beauty is hampered when textbooks and teachers leap-frog over the student's common sense. Indeed, it is precisely the contact with reality given in everyday life that is the starting point of science. Simple concepts, such as substance and property, which are at the root of common sense, can be understood at a very young age. Indeed, they are understood at a very young age, at a pre-scientific (or infra-scientific) level, though they are seldom explicitly raised by educators, let alone given a clear definition. Science depends on having experienced much, and, especially as children, we do experience much before thinking deeply about it. In this series, we make explicit use of this experience and critically analyze it to bring out the essential truth common sense contains while purging it of spurious error that attaches to it without such rigorous analysis. For instance, understanding how a charged body can cause motion over a distance can be hard for students, because it appears to contradict their common sense. In addition, the idea that uniform motion needs an external cause, as discussed in the first text, is a block that can again arise in understanding such motion, because each of the areas affected by the false digestion of common sense principles needs to be addressed to overcome thinking that has been habituated for so long. The explicit analysis of the electric field given in this text shows that common sense is right in its deepest insight. However, it also shows that the confused thinking that can exist in "common sense" needs to be subjected to thoughtful analysis and critique. It is important to know where "common sense" went wrong, but it is also important to see the fundamental truths with which it begins and from which it gets its strength. Starting with common sense manifests the firm ground upon which modern science, and all of our knowledge, stands as well as reveals the nature of modern science, what it includes and what it leaves out, showing what we know at the current level, e.g. the physics of classical electricity and magnetism in this text, and what we do not.

This is also true of mathematics. The starting point for mathematics is the physical world. The student will meet mathematical formalisms called divergence and curl and theorems that go along with them. Each will be grounded directly in the things that we see, hear and touch so that he can see clearly the realities that those formalisms encapsulate and thus better understand, remember and use them.

The second item, i.e. the unifying practical theme, is also not seen in standard physics texts. As every experimental physicist knows, nature is always much more complex than appears at first glance. Every experimentalist has to learn by experience how to make his experiment work. This is because when we return from theory to experiment we move from abstraction to the specific reality, which means we have to confront all those things we have not incorporated into our theories. A practical theme, which involves, at least, having to think how one might do something, forces one back into this specificity. It forces one to think more deeply about the meaning of the theory.

The transmitter is chosen for the principled reason that electromagnetic radiation[1] reveals much of the fundamentals of electromagnetism. Every time we see an object, electromagnetic radiation is carrying to us something deeply rooted, though generic, in the

---

[1] Radio waves are part of the spectrum of electromagnetic radiation that includes visible light.

nature of that object. Indeed, every aspect of E&M is, in principle, directly or indirectly involved in radiation. The transmitter is also chosen for the practical reason that it makes use of every aspect of E&M in its components. Also, in this practical light, the transmitter is essential to modern life; for example, it is at the core of cell phones, and even land line phones with their cordless handsets. Lastly, it is also just plain fascinating to be able to communicate over a distance, through walls and other obstacles. Since, we cannot see, hear or touch these waves, it is, indeed, an unexpected marvel to see such transmission occur. It shows the power of the human intellect; starting with sensorial particulars, it reaches things that we *cannot* sense.

The manned Mars mission, with its imagination-firing vision of travel to other planets, inspires us to consider how electricity and magnetism help us understand and deal further with the problems of making the trip to Mars as well as living there after we arrive. The textbook cover shows a few of the E&M phenomena of interest in our space travel. It shows the aurora on the Earth's poles, which are a sign of the Earth's magnetic field that directly protects us from deadly radiation as well as helps protect our atmosphere from being stripped away by that ionizing radiation, reminding us of the problems we face on the way to Mars and even on Mars. The cover also shows lightning which is known to occur within the dust devils. It also shows an electronic compass as well as implies the existence of a host of other electromagnetic devices such as electric motors. Lastly, it shows the central practical theme, the transmitter, which, because communications is crucial, is crucial to any space mission.

Third, the history written in physics texts almost universally begins with Galileo, giving little credit to key predecessors of Galileo, often maligning both the medieval contribution and the contribution of Aristotle. Though historians who study the medieval period now know that this is incorrect at a very important level, few physicists know it, and it has not managed, until now, to reach the physics texts. Knowing this history helps students see the continuity in the evolution of ideas and the dependency that even great thinkers like Newton have on those that came before them, as Newton himself was not hesitant to say. The missing link is the medieval period in which modern science was conceived, finally being born at the end of that period. Seeing this dependency helps students realize their own intellectual dependency and thus to take seriously the need for study. Further, having such knowledge helps the pedagogy, because the history mirrors the path the student must take from common sense grounding to the highly mathematical modern science. Thus, it helps cement his understanding while giving him another pathway by which to remember the material. Skipping over history leads, as is famously said, to repeating its mistakes. In this case, the central mistake is not properly understanding what we include and what we leave out in our thinking in modern physics. We need to know what it is we mean and do not mean by our theories in modern science; we should clearly sort out what we *actually* know about nature from the equations that describe it, and thus also be able to point out what we still seek to know. In this volume, we continue the historical thread, building on the background established in the first volume, pointing out what we know and what we still need to study or even still need to go learn by experiment and analysis.

Next, the fourth numbered item, focusing on first principles leads us to emphasize parts that will be fundamental in more advanced courses, in this case the potentials, the fields called ϕ and A. Without these fields in the principled foreground, it is much harder to avoid confusing truly electric and magnetic phenomena. This starting point allows us to properly distinguish them and thereby show the complementarity that unites them, that is brought home by our later relativistic analysis. This fundamental viewpoint is reached by analyzing experiments to find causes in the light of the first principles given in the first volume including nine fundamental categories of properties of physical things.

The fifth item, radiation, having already been addressed, we turn to the sixth item, the special relativistic underpinnings of Maxwell's theory of E&M. The approach centered on impetus (momentum) allows a clear understanding of the nature of Maxwell's theory that shows exactly what is meant when we say in special relativity that, in a uniformly moving frame, a magnetic field can appear as an electric field and vice versa. Whereas in the first volume the emphasis was on unchanging impetus, with the further specification given by E&M, we are able to dive deeply into the forces,[2] showing how they appear differently when viewed in different frames. We also show, for example, the role that the change that these forces undergo (when the bodies that are responsible for them are in uniform motion) plays in time dilation.

Although the sixth is the last item listed, there is one other important aspect about the book that bears mentioning. As in the first volume, a concentrated effort has been made to include many problems about everyday life, so that students may realize that physics starts with our senses and comes back to them. Again, everything we see around us is the business of physics. Furthermore, physics in the wide sense of the study of the physical world, though it is not all we know, is the *basis* for all we know. It is in this sense that physics *is* the base science, and so essential.

We, the author and contributors at the Institute for Advanced Physics, welcome your comments, as we continually strive to improve the usefulness of the text. In particular, send your suggestions or any corrections you may find to us at iappress@iapweb.org. Though much effort has been put into proofing the text, no doubt errata remain. Hence, for your convenience, we will keep a running compilation of errata at iapweb.org/pfrem/errata. These will be incorporated into the next printing of the book.

Also, there is a solutions manual available to instructors. Please email us at iappress@iapweb.org.

Lastly, we plan on writing teacher's manuals for various levels. Please check IAP's website at iapweb.org/pfrem/teachers_manuals

---

[2] Recall that a force is the ability of a body to cause impetus. In Newtonian mechanics, the word "force" is usually only used for bodies that are *actually* causing impetus or *trying* to cause impetus. We will see in this text that the word "field" is used for the power (ability) more generally, i.e. whether it is actually causing or trying to cause impetus or not. Hence, for the case of a field, we say it causes a force when there is something to receive its action.

Contents

A detailed outline of each chapter is given on the first page of each chapter. Appendices on: i) the concepts of form and abstraction, ii) the proportionality of active and passive charge, iii) the mathematics of conservative forces, iv) "Why do we need plana?", v) Mars, vi) various physical constants and conversions of interest, including some relations between *cgs* and *SI* units, and vii) useful grad, div and curl formulas and vector relations are found in the back of the text along with some flip books and an index.

vi

Chapter I

Physics and Fields Generally

Introduction

Electricity and magnetism are now part of common experience for most people (in the developed world), though this was probably not the case before modern times. In the past, lightning, the only form of electricity manifest to all, could easily be seen as only an extraordinary phenomenon, largely unrelated to the nature of ordinary objects. Today, thanks to easy access to all kinds of materials, people play with magnets from childhood, and everyone has experienced static cling and even heard of products to eliminate it. It is this basic understanding that will serve as our starting point. Of course, we must also make use of other things we have seen and learned from our experience and from our study of mechanics. Since to progress in understanding necessarily means starting from where we are, we should make sure that we have firmly understood the principles up to that point, especially the foundational ones. Thus, we will start with a summary review of the foundation, but a more in-depth review can be gained by studying Chapter 1 of *Physics for Realists: Mechanics* (*PFR: Mechanics* or *PFR-M*) again. Recall also that Appendix I of that textbook goes into further depth; it is highly recommended that you read this multiple times. A review of the entire text, especially Chapters 2,3,4,5 and 8 might also be helpful.

Brief Review of Foundational Principles

In understanding something, we should always start with what our senses reveal to us, for they are our portal to the world. As we saw in *PFR: Mechanics,* we have two types of knowledge, intellectual and sensorial knowledge. Our intellectual knowledge, our knowledge of general things, is abstracted from our sensorial contact with particular physical things. For example, we get the general idea of two, for instance, by seeing two halves of a whole pie. Our *intellectual* knowledge must move from the general to the specific; for example, one could not understand the concept of an *equilateral* triangle unless one understood what a triangle was, for an equilateral triangle is a specific type of triangle. At the most general level, we learned five fundamental principles:

1) *Things exist.*
2) *Things change.*
   More explicitly: Things are something, but can become something else. They are actually something (act), but are potentially something else (potency).
   > **Example***:* An apple is one thing before it is eaten and is another thing after it is digested and part of one's body.
3) *Principle of Non-Contradiction:*
   Something cannot be and not be at the same time and in the same way.
   > **Example:** A shirt cannot be red and green at the same time and in the same way. It could be red and then green *or* it could be part red and part green, but not simultaneously red and green in the same place.
   The principle of logic that we either have X or not X derives from this fact. The daisy is blooming. The daisy is not blooming.
4) *Principle of Causality:*
   Something cannot change itself. Something cannot give itself something it doesn't have. It can, however, act on (change) other things according to what it is.
5) *Substances and their properties*:
   a. Some things (substances) exist of themselves, while others can only exist as aspects of a substance (call them properties).
   b. Obviously, the first (substances) are referred to using nouns and the second (properties) using adjectives.
   > **Example:** "The daisy is yellow." The daisy is the substance, and yellow is a property of that substance. Yellow cannot exist by itself, but only as a property of something(s).

We also learned that each physical substance has nine categories of properties (see also Figure 1-1). Thus, we have a total of 10 categories of physical existence: substance which exists of itself and nine different ways things can exist in that substance.

The first two types of properties are *intrinsic* to the substance:
1) **Quantity** (extension): plane, circle (disc), number, …3-dimensions
2) **Quality:** shape, color, hardness, pressure, hot, cold, tones, smells, tastes. A quality determines a thing to be a certain way, determines its nature in some way. Shape, for example, determines the limits of a substance's extension. Quantity is related to how much of a quality there is, e.g., a more extended circle or a more extended patch of green.

The remaining categories of properties are *relative* to another substance in some way:
3) **Relation**: equality; similarity: two discs of the same *size (extension)*, two shades of the same color; cause and effect
4) **Action**: heating (something), moving (something)
5) **Reception**: being heated, being moved
6) **Place**: a fish in a certain region of water in the tank
7) **Orientation:** the cylinder is rotated by 90 degrees
8) **Environment**: the water around a fish
9) **Time:** now, today, yesterday, a second from now

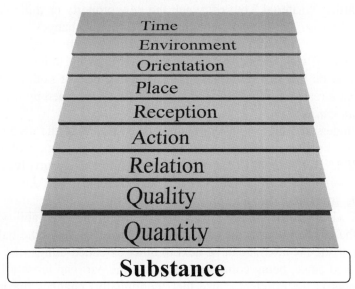

Time
Environment
Orientation
Place
Reception
Action
Relation
Quality
Quantity
Substance

Figure 1-1: Physical substances have nine categories of properties. In gray are the nine categories of properties that are interrelated or co-relative one with the other. Quantity is the base property, upon which all the others stand. By abstracting, we can separate out quantity and consider it isolated from the rest excepting a thin presence of quality (shape, for instance in geometry); this is what we do in mathematics. The importance of this type of abstraction is indicated in the figure by a large separation between the quantity layer and the quality layer. Of course, the properties presuppose a substance, i.e., that which has properties.

We pointed out that *mathematics* is the study of quantity, the first property of physical things. And, we saw that modern physics utilizes a powerful central tool, which we called the empiriometric method. This method looks at the physical world (i.e. the *empirical*, hence the first part of the word) as mathematical (i.e. *metric*, for it is by measurement that we get numbers). The mathematics used in this modern physical method is, in turn, almost always captured by the modern symbolized and highly axiomatized method.

Within the text of *PFR: Mechanics*, we also touched lightly on the four "causes." (This topic is spread through that textbook, not discussed completely in any one place.) We noted that the defining characteristic of physical things was their ability to change, i.e. they are one thing and can change to another.[1] There are two types of changes, changes of substance and changes of property.[2] In both types, there are four key ingredients, or, in extended use of the word, we could say there are four key "causes."[3] The first type of

---

[1] In this way, each physical thing can be said to have two "components;" what it is and what it can be. We call what it is its *form* and what it can become its *matter*.

[2] There are two types of properties, those that belong to a thing essentially, which is the primary meaning of the word, and those that do not.

[3] Since a change of substance is fundamentally different from a change of property, it is clear that we do not mean exactly the same thing by "change" in both cases, but only that they are in some proportional sense the same. This is what we mean by analogy; so, we say this use of "change" is an analogically general meaning

"cause" is the starting "material," out of which the new property or thing is made; this is called the *material cause*. The second is the thing into which it changes, which is called the *formal cause*.[4] The third is the thing that causes the change, which is called the *efficient cause*. Lastly, each efficient cause, which is the everyday meaning of "cause," has a specific tendency to do this or that rather than something else or nothing at all and thus there is behind that a reason for that tendency; the tendency specified by the reason behind it is called the *final cause*.

To fully understand these four ingredients or causes, take the case of a change of property, which is the first meaning of change, rather than a change of substance (see Appendix I in *PFR: Mechanics*). Concretely, consider the change involved in a baseball moving from the point of release at the pitcher's hand. Roughly speaking, the ball and the things around it, including the air and ground are the material causes, i.e. starting "material" out of which the new place is made. More accurately, the ability of the ball to have a place in the environment is the material cause; namely, since the ball is extended, having one part outside the next, it can be next to other extended things (the environment), and it can be moved since, being composed of parts, one part can cross a line before the others. The impetus is the efficient cause that (continually) changes the initial place to another (see below for review on the basics of impetus). The "formal cause" is the new place. That is, for example, the old place above the pitcher's mound must yield to the new place just in front of it. A generic "final cause" or reason for motion, seen in the tendency of the impetus to move the body ever "forward," is the generic need for bodies to move next to each other in order for them to interact and their need to prompt activity by banging into others.[5] This is seen in the fact that impetus continues to move a body until another body de-activates its impetus. Thus, for example, two bodies in a collision will press into each other, distorting each surface of contact, because it will take a definite time before each body can change the impetus of the other enough to stop their relative motion. The "final cause" of the particular motion over home plate is the intention of the pitcher to strike out the batter; this is called an *extrinsic* final cause because it does not belong to the body (ball) itself, while the first is called the *intrinsic* final cause.

Brief Review of the Fundamental Principles of Newtonian Mechanics

We proceeded in *PFR: Mechanics* to further specify these very general principles in our study of Newtonian mechanics. We found that measurement was a relation, for example, between an intensity or extension of a standard and that of the body of interest. We found that bodies[6] move under that action of quality of impetus. *Impetus* is defined as the power (category of quality) activated in a body that moves it at a constant speed in a particular linear direction. It is *second* nature to the body, because it can be gained and lost

---

of "change." Thus, when we talk of the four causes applied to both these analogical types of changes we are also using the four causes in an analogical way.

[4] This can also be simply referred to as the "form," i.e. what the thing it changes into actually is. For more discussion on the word "form," see Appendix I.

[5] Or, at other times the reason may be so that things are kept separate, so that there can be different substances in different places, giving them space for their proper behaviors. If everything was bunched in one place, we could not, for example, have a stable solar system.

[6] The first definition of body was any physical substance, but we later settled into a somewhat indeterminate, vague definition of *body*, i.e. a substance, group of substances or sometimes even a part of a substance. This analogical general definition allowed us to talk about what was analogically common to all these.

without the body changing its fundamental nature. *Mass* is a measure of the receptivity of a body to the action of its impetus, such that the larger the mass, the less speed a given impetus is able to cause. We noted a body with an intensity p of impetus will move at a speed ($v = p/m$, to be discussed below) because of the way the action of the impetus (category of action) is received by the body (category of reception). More precisely, we learned that *momentum* (symbol, p) is the measure of the (intensity and direction of the) impetus or more generally (analogically) the sum of all the measures of the impetus of all relevant bodies. It is the mass times the velocity, $\vec{p} = m\vec{v}$. In this text, as in the first volume, the unit of measure of momentum is the Buridan. (We call the unit a "Buridan" because of John Buridan's important contribution. 1 buridan $= 1\,B = 1\,kg\,m/s$.)

Force is the ability (quality) of one body to change the impetus of another. The measure of the intensity and direction of a force is related to other relevant measures by Newton's second law: $\vec{F} = \dfrac{d\vec{p}}{dt}$. (We call the unit of force a "Newton" because of Isaac Newton's important contribution. *1 newton = 1 N = 1 kg m/s²*.) Newton's first law simply points to a key part of the nature of impetus implicit in the second law, i.e. impetus does not change without the action of a force, so that a body at rest stays at rest and a body in uniform motion stays in uniform motion unless acted on by an outside force. Until **all** of a body's impetus is gone it continues to move. We also learned Newton's third law, which explains that every force acting from one body on another is accompanied by a reaction force of equal strength but oppositely directed from the second body on the first.

Lastly, we defined *kinetic energy* as the rate of transfer of the intensity of impetus of a body (or bodies) across a distance (i.e. rate of transfer of momentum). These basics set the stage for many other discoveries and principles, including Newton's law of gravity.

Through this study of Newtonian mechanics, we discovered two primal types of physical things, massive and non-massive. Massive things are the type we can handle and move around because they can be given impetus. The term "massive," at our level of abstraction, indicates something of the essence of this primal type of physical thing. It is *primarily* called "massive" to indicate its ability to receive impetus reflective of its *inertial* mass, but this ability implies that there must be at least one type of force associated with massive bodies that can activate impetus in them. That key force that is intimately linked with inertial mass is *gravity*. In fact, we saw that the greater a body's inertial mass, the stronger the gravitational field that it can produce at a given distance from it, and the more receptive it is to the action of a given gravitational field. Or, more precisely, the *active* and *passive* mass are each proportional to the *inertial* mass. Thus, the appellation "massive" referring to the inertial mass also brings out indirectly this type of body's essentially gravitational nature.

We also discussed a non-massive type of physical reality, what is often called "space" or "vacuum." Since what we call "space" or "vacuum" is receptive of *fields*, we call it plana, which derives from the Latin word for field. Remember what we call "space" or "vacuum" is not bare extension, which is after all a property, but is a *physical* thing and thus has all nine categories of properties including, of course, extension. This is not to say much particular about plana, but only to emphasize that between two planets or between two atoms is not nothing ("vacuum"), but something; we give that something a name, and in this text, we will begin to learn a little more about it. Recall that in our study of special relativity, we found that the natural empiriometric mode of that formalism left out most aspects of the plana, and indeed it was easier to work in that mode without thinking about plana at all; we will find at some level this is also true of electricity and magnetism, but as we have just noted above, leaving something out does not thereby negate its reality.[7]

We also categorized the types of substances based on what we learned. That categorization is summarized in the chart below.

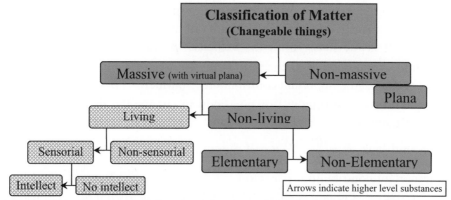

Figure 1-2: Summary of the classification of physical substances. The arrows indicate a transition from lower to higher level substances.

The Transition to Electricity and Magnetism

In moving from these principles to electricity and magnetism (E&M), we are further specifying our understanding. We are "switching on" more things, filling out what we have left out of our (analogical) abstractions.[8] In *PFR: Mechanics*, we left behind or "switched off" even the distinctions between substances, groups of substances and parts of substances, and we only considered the qualities (in some generic sense) of impetus, mass and force. Furthermore, except for gravity and a few atypical cases, we treated forces abstractly as if they could take any form as long as they obeyed Newton's laws. In fact, of course, in reality forces are properties of specific bodies and thus have a specific nature, *not* an *abstract* nature that leaves the form of the force undetermined except for the fact that when it acts on a body that body generates an equal and opposite back reaction. We will see that E&M makes us further fill out our concept of force, specifying our (analogically) abstract understanding in many ways.

This is typical of the progress of our understanding. We first leave many aspects out, and then when, for example, a certain insight arises after sufficient digestion of the empirical realities,[9] we are able to include more of reality in our understanding. If we were not careful to remain sufficiently general in our statements, we would then have to revise those statements to account for their premature over-specification. If we were careful, we would only have to show how to incorporate the new specific findings, and we would then naturally see that what was previously left deliberately vague[10] is, after such specification, less so and thus clearer. In regard to the (analogical) abstractions we make that leave

---

[7] Indeed, in quantum field theory, as noted in the body of the text below, we will see it is natural to bring it back again.

[8] See Appendix I for more discussion of the meaning of the term analogical abstraction.

[9] Or the precision of our measurements increases or as we become able to do certain experiments.

[10] Again, technically, we say they are analogically general.

certain things out (or in the other metaphor, switches things off) in order to more easily focus on the subject at hand, we should always remember that just because something can be fruitfully ignored in this way, even to such a degree that we get used to (i.e., habituated to) treating it as if it did not exist, doesn't make it not exist.

As this course brings to the fore the nature and operation of two fundamental forces: *electricity and magnetism*,[11] it will also bring back important considerations about "space" that are ignored in Newtonian mechanics. In particular, in Newtonian mechanics, we often considered so-called "action at a distance," so that we would not have to consider "space" itself. In particular, we considered gravity to act instantaneously through space so that we could avoid the apparently thorny question of violations of Newton's third law if we allowed any time for propagation; again, this had the side benefit of allowing us, at some level, to ignore most aspects of "space." However, even there we found the need to talk about a power (a field) at every "point" of space, and a power is a property of something, so we still had to deal with "space" or what we call "plana," as that which is receptive to the field. In an advanced course, you can study quantum field theory where it is found that plana ("space") is active in ways few would have expected. For now, following the above prescription of moving from the general to the specific (which is, as we've noted, the way our minds work), we will leave the further specifics given by quantum field theory (and all of quantum theory and many other subfields), so as to focus on understanding all the many specifics that classical E&M itself can give. Indeed, at first, we will even exclude special relativistic considerations, though we will eventually see that the very nature of classical *E&M* calls for the inclusion of special relativity, (which will, in turn, as discussed above and in the special relativity chapter in *PFR: Mechanics*, make it possible for us, at some level, to again leave aside many aspects of the plana). Even with these limits, which are imposed by the proper boundary of the subject as well as by secondary considerations of the amount of space and time available, you will not only find that a large quantity (first category) of intellectual food remains, but that that food consists of a significant variety (category of quality) and is quite delectable (category of quality) so as to satisfy the most discriminating palate.

General Introduction to Electricity and Magnetism

To truly understand electricity and magnetism, as with any subject of study, we should start with the things we see directly. As we indicated, most ancients had little contact with electrical and magnetic phenomena. However, some people in some civilizations did have such contact. The history is summarized in the chart below. To understand how this history fits into the whole of the history of science refer to the chart of that larger history in chapter 1 of *PFR: Mechanics*.

---

[11] Because of the deep relation between these two forces, they are often counted as one. As we study these forces, especially in the context of special relativity, we will see many of the deep relations that reveal their deep interconnection. However, such unity of interrelation does not obliterate the distinction between them as we will also see as we uncover their nature.

History Timeline for E&M

Greek, Chinese and other cultures find magnetite and amber and make some practical use of magnetite but no real advances in its use and no scientific study of either phenomenon. That is, neither pure nor applied science, in the fundamental senses, were at work.

Greeks: <u>Thales</u>, per Aristotle, discusses magnets as having souls...no true physical theory, <u>Aristotle</u>: Torpedo fish, discusses amber only as covering animals. Seems Greeks and Romans thought amber and loadstone powers linked (cf. P. Benjamin, Intellectual Rise of Electricity, hereafter IRE)

Romans: Lucretius (c.99-55BC): talks about loadstone attracting iron; no experimentation, gives guesses based on unsupported ideas such as his idea that everything is made of unbreakable particles. First documentation of some noting repulsion of magnets (see IRE)

Early Middle Ages (c.400-570)   St. Augustine notes properties of electricity (amber) and magnetism (lodestone), especially astonished by his brother bishop moving iron around through a silver plate, so noted in a confused way that magnetic was not shielded like the electric was. It is said he is first to note difference between lodestone's action and that of amber, i.e. between electricity and magnetism but no further study (see IRE). (St. Hippolytus of Rome notes this in a less clear way in c 200AD)

Islamic Period (c.700-1250)**:** Seems to be little or no progress in E&M

High & Late Middle Ages (c.1000-1500)**:** *Petrus Peregrinus de Maricourt: first scientific study of magnetism: Epistola de Magnete* (1269)**,** John of St. Amand, Alexander Neckam, Bartholomew the Englishman, Albert the Great, Gilbert credits St. Thomas Aquinas with noting iron is changed by a magnet.

Gilbert (1544-1603), a follower of Aristotle,[1] codifies medieval E&M work up to that time and adds detail, but no new essential information except perhaps his correct identification of earth as a magnet**.**

Age of Coulomb (1700-1800s)

Coulomb, Ampere, Faraday

Galvani, Volta, Franklin, Priestly, Cavendish, Ohm, Biot, Savart, Orstead, G. Stoney, Fr. Nollet

Age of *Maxwell* (1800's)

Henry, Hertz, Heaviside, Ludwig Lorenz, Lorentz

Quantum...**a new type of magnetic moment**. Stern and Gerlach proved atomic magnetic moments had a quantized character. Compton had suggested in 1921 that the electron also possessed a magnetic moment associated with an intrinsic spin angular momentum which was then discovered by Goudsmit and Uhlenbeck in 1925. The magnetic moment was, relative to the angular momentum, twice the orbital value. In 1928, the Dirac equation that describes spin was discovered.

---

[1] In the primary sense. He also gives explicit credit "To those early forefathers of philosophy, Aristotle, Theophrastus…, let due honor be ever paid; for by them wisdom has been diffused to posterity." (See last page of "To the reader" section of his *De Magnete*)

The earliest record of human contact with inanimate electrical and magnetic phenomena begins with stories about discoveries in China early as c. 2500 BC.[12] However, these "discoveries" are not well established and, more importantly in the end, there is no evidence that the phenomena were treated scientifically, that is analyzed to find causes and reasons for the behaviors within the things themselves. This latter distinction is very important and deserves emphasis. It's one thing to notice a phenomenon, which is indeed an important aspect of science, namely being observant and sensitive to details. However, without a reasoned understanding and approach to the study of nature, opportunities for observing important details or even new effects are reduced largely to chance, being dependent, for example, on accidental accessibility of the requisite resources. For instance, if magnetic material happens to be near the earth's surface in one civilization but not another, it will be more likely for the first to come across the material and note one magnet's effect on another and on certain metals (assuming they are also available), but this doesn't mean a science of magnetic material has begun.

Again, science seeks to understand nature, to trace effects to causes and underlying reasons for the effects, not simply to catalog phenomena. Such scientific analysis begins with the Greeks. Finally, the Greek civilization can be said to have reached, for the first time in human history, a deliberate, grounded and conscious science with Aristotle (384-322 BC)[13], who, as we saw in *PFR: Mechanics*, coined the very word physics and, despite the modern lore to the contrary, highly valued detailed sense observation.

As might be expected given that their distinctive effects were not readily accessible, electricity and magnetism did not get the same attention and thus did not undergo the development that statics or even dynamics did. There is a tradition going back to ancient Egypt circa 2700 BC[14] which discusses electric fish. We, as physicists, however, are looking for primary generic principles, not particularly specified ones as found in an animal such as a fish. Indeed, Aristotle, himself, discusses the abilities of the torpedo fish (which is now called an electric ray) to cause electric shock in other fish, causing a torpor (hence the name of the fish). The principles of electricity (and magnetism) that we seek will apply to fish as well as inanimate objects. However, inanimate objects will more quickly reveal the generic aspects of electricity (or magnetism), and more closely reveal the fundamental powers at work. After all, recall that inanimate bodies, in the form of elements and compounds, will of necessity enter the composition of a higher substance such as an animal, and, through this assimilation, their native powers will become powers of the animal. (At a simple level, for example, it's the inanimate ability of the compounds to resist being broken apart, for instance, that also appears, though specified in new way, in the animal.)

The first mention in Greece of E&M effects in inanimate objects is from Thales of Miletus (c.620-c.546 BC), whom Aristotle quotes and criticizes for thinking magnets have souls.[15] Thales also thinks that amber turns magnetic upon rubbing, thus making one of the first links, though a wholly false one, between the two. Theophrastus, a follower of Aristotle who came immediately after him, points out that both amber and magnetite have

---

[12] History of Philosophy and Sociology of Science, Paul Mottelay, Arno Press, NY, 1975, page 1
[13] See history box on Aristotle in Chapter 1 of *PFR: Mechanics*
[14] See http://en.wikipedia.org/wiki/Electricity
[15] Aristotle *De Anima* 405a:19

"the power of attraction."[16] The very word **magnet** comes from the Greek, *magnetis lithos* (*μαγνήτης λίθος*), which means stone (*lithos*) of Magnesia, an area of Greece. The Greek word for *amber* is *ηλεκτρον,* which in Latin is rendered *electrum* and in English is **electron**. Hence, the two key words of our subject come from amber and magnetite (also known as lodestone). These two are shown in the figures below.

Figure 1-3: *Left:* **a)** Amber (*Gr., ηλεκτρον, L., electrum*), Ancient Greeks and Romans investigated static electricity using amber, fossilized tree resin. By rubbing it against fur, for example, it can pick up pieces of hair and wool. Top shows a spider encased in amber. *Right:* **b)** "Magnetite" or Lodestone (naturally magnetized pieces of Fe_3O_4) attracts metals and attracts or repels other magnetite depending on the orientation.

It was thus known in Greece that magnetite attracts metals, and amber, upon being rubbed with certain materials, attracts certain small lightweight objects such as hair or wool.[17] Amber is a fossilized tree resin (which is distinct from sap) commonly used in ornamental objects in history; because tree resin is sticky, it sometimes traps and encases small animals (or plants) such as the spider shown in Figure 1-3a. Those who read or saw *Jurassic Park* will recall that it was from DNA trapped in amber that the fictional scientists retrieved the genetic code they needed to make their dinosaurs. Magnetite is the most magnetic of all the naturally occurring minerals on Earth.[18] Scientists have found evidence that bees use magnetite as part of a sensory organ that allows them to sense the Earth's magnetic field and thus to navigate. Indeed, many organisms including man

---

[16] *Theophrastus On Stones*, Greek original with English translation and commentary by Caley, Earle R. and Richards, John F. C., Ohio State Univ. 1956

[17] Apparently, also the Chinese and Persians knew about the property of certain rubbed bodies to pick small things up; see *Electricity and Magnetism: a Historical Perspective* by Brian Scott Baigrie, Greenwood Press, CT, 2007, pg1.

[18] "Direct imaging of nanoscale magnetic interactions in minerals" Richard J. Harrison, Rafal E. Dunin-Borkowski, and Andrew Putnis, 16556–16561 PNAS, December 24, 2002, vol. 99, no. 26.

"biochemically precipitate the ferrimagnetic mineral magnetite (Fe_3O_4)."[19] These materials are thus interesting in themselves. However, we want more generic information about electricity and magnetism; we want to know something about the nature of these powers. It's not until the advent of the Christian High Middle Ages, with its belief in the deep rationality of the physical world, that the first real work is done in this direction by Peter the Pilgrim of Maricourt (13th century, in Latin his name is Petrus Peregrinus de Maricourt, see history box). He is, in key ways, the John Buridan (cf. *PFR: Mechanics*, Chapter 3) of E&M. Others further his work, though mostly in non-essential ways, most importantly Gilbert (1544–1603 AD). Still, though 300 years pass with Peter the Pilgrim's work in hand, Galileo (1564-1642 AD) notes and is critical of the lack of quantitative study in electricity and magnetism, though Galileo himself was not able to further this quantitative work. Moreover, it is also true that much of the qualitative work is not done either; for example, even the relationship between moving charge and magnetism is not known until much later. Surprisingly, substantial progress only begins with Coulomb (1736-1806 AD) and the others listed in the box labeled *The Age of Coulomb* in the "History Timeline for E&M" above. E&M reaches its pinnacle with Maxwell (1831-1879 AD). Later generations verified that there were many examples of the electromagnetic waves that his theory described, including: light, radio waves, x-rays, and the whole spectrum that we will discuss in time. Indeed, because of the fundamental nature of electromagnetic waves and the fundamental practical importance of electronics to modern life, we will introduce, as we move through the text, the requisite concepts and you will solve the requisite problems to enable you to understand and build *your own FM band radio transmitter* in the last chapter of the book.

There are many other truly electromagnetic phenomena as well as apparently-related phenomena that can be investigated; these vary from the "sparks" generated by flint and steel to ball lightning. Ball lightning, for example, which was documented, like so many other things, by Aristotle, is still not wholly understood, having only recently been accepted as a real phenomenon. How is it formed, what makes the light, what sustains its glow even while in motion, and so on--perhaps you can contribute to understanding it. More importantly in principle, we now know that electromagnetic activity plays a central role in the chemical properties of the elements and of their compounds, but in the past someone had to wonder. Someone had to ask, for example: "What is the connection between readily visible and apparently simple effects like lightning and static electric pull in pith ball experiments (see next section) and complex chemistry such as the Krebs cycle in a cell?" All science starts with questions, thinking about the intelligibility in nature. Einstein, pondering these starting points of science, said, "Whoever… can no longer wonder, no longer marvel, is as good as dead." The question we will start with is simple. What is static electricity?

---

[19] *Magnetite in human tissues: a mechanism for the biological effects of weak ELF magnetic fields.* Kirschvink JL, Kobayashi-Kirschvink A, Diaz-Ricci JC, Kirschvink SJ. PMID: 1285705.

Peter the Pilgrim of Maricourt (Latin, *Petrus Peregrinus de Maricourt*) was born in France, in c. 1220, probably in Picardy. He can be called the John Buridan of E&M in that he did pivotal first-of-its-kind work in E&M, (though unlike Buridan he did not predict quantitative relations, this did not happen for some time even after Newton and Galileo). In particular, in 1269 he wrote, *Epistola de Magnete*.

He was the first to assign a definite position to the poles of a lodestone and to give directions for determining which is north and south. He mapped the lines of force of a spherical lodestone and noted their similarity to meridians of longitude. He called the ends "poles" in analogy to the Earth's poles, and even called the spherical lodestone a <u>terrella,</u> 'little Earth!"[1] (Despite this pregnant appellation, he erroneously thought that the celestial sphere was the source of the magnetism that drives compasses.) He established that every fragment of a lodestone, however small, is a complete magnet. He invented a new type of compass that used a magnetized needle on a pivot surrounded by a graduated circle.

According to an old manuscript of the *Epistola*, that letter was "Done in camp during the siege of Lucera, August 8, in the year of our Lord 1269." Peregrinus was thus a crusader, probably an engineer, in the army of Charles of Anjou, (brother of Saint Louis IX, King of France) who, in that year, laid siege to the Italian city of Lucera to win it back from Islam.[2,3]

There were no studies of the magnitude of *Epistola* until 1600 (by Wm. Gilbert, who, except for the crucial work on the earth as a magnet, seems to have only added detail). Apparently, the Franciscan Roger Bacon, himself renowned for his respect for experiment, praised Peregrinus as a "master of experiments." From medieval times, *Epistola* was extremely popular; in 1326 Thomas Bradwardine quotes it in his *Tractatus de proportionibus*, and, after his time, the masters of Oxford University make frequent use of it (cf. the *Oxford Calculators* history box in Chap. 2 of *PFR: Mechanics.*)

[1]*page 25 of his Epistola magnete*
[2] *http://www.britannica.com/EBchecked/topic/451293/Peter-Peregrinus-of-Maricourt* **and** http://www.encyclopedia.com/doc/1G2-2830903367.html
[3] *His participation in this part of the crusades is the likely origin of the honorific title "Pilgrim," which was commonly used in this way in the Middle Ages.*

Static Electricity and the Electric Field

With the above overall picture in hand, we can, indeed, now begin to focus on static electricity. To understand it, we begin with what is evident to the senses by simple observation and experiment. The average person has seen the first two of the manifestations of electricity shown in Figure 1-4a,b below and many more. Ordinary lightning, at the simplest level, is very similar to the spark one gets from having walked across a rug and touching a door knob. Friction with the rug builds up a charge on your body as it does on a comb wiped through your hair. To see this, take that comb and touch it close to a large piece of metal and you will get a spark.

Simple experiments can reveal much of the nature of static electricity. Try, for example, rubbing various pairs of bodies together; in particular, consider those bodies shown in the so-called tribo-electric ("tribo" comes from the Greek word to rub) series chart in Figure 1-5a. Note that the order shown in the chart is dependent on many factors and when those factors change, the ordering will change; the chart should thus only be used to give a general idea of order and point to the fact that given a closely controlled experiment there will be a definite order. If you put these near an electroscope such as shown in Figure 1-5b, you will quickly learn that there are two types of charge, one capable of nullifying the effect of the other. Benjamin Franklin (1706-1790AD), the great American experimenter who did much to further knowledge of electricity, unfortunately thought the positive charges (which he assigned by experiment) moved, while negative ones did not. It was, of course, discovered later that negative charges (electrons) are the ones that move in solids, such as wires, while the positive ones, the atomic nuclei, stay (relatively) put.

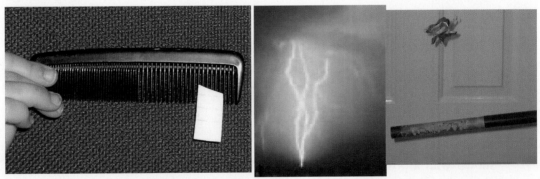

Figure 1-4: Examples of static electricity: (a) *(left)* comb picking up paper (b) *(middle)* Lightning, from discharge of electricity (c) *(right)* Electrostatically charged stick holds thin strips of connected tinsel above it.

The electroscope picture shown in Figure 1-5b shows the effects of a positively charged rod causing some electrons from the leaves to move through the metal connecting-rod up into the metal disc. Clearly, a force must act on the electrons. Furthermore, since the two leaves are both positively charged, they move apart. Note that the rod is not touching the disc; yet it causes electrons and the leaves to move. Or consider the even simpler case of the pith ball (i.e. a small lightweight ball[20]) setup shown in Figure 1-6. If you charge both balls with the same charge, say both with your comb, the balls will move apart. The pith balls are not touching yet they cause impetus in each other. How do the balls act where they are not?

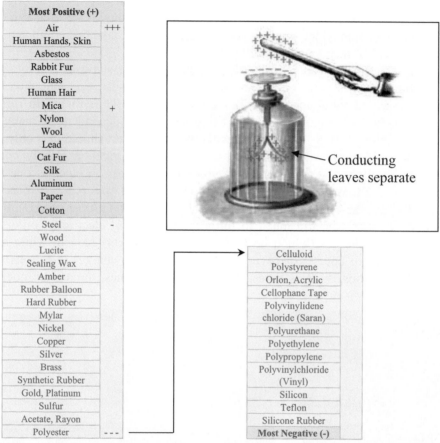

Most Positive (+)	
Air	+++
Human Hands, Skin	
Asbestos	
Rabbit Fur	
Glass	
Human Hair	
Mica	+
Nylon	
Wool	
Lead	
Cat Fur	
Silk	
Aluminum	
Paper	
Cotton	
Steel	-
Wood	
Lucite	
Sealing Wax	
Amber	
Rubber Balloon	
Hard Rubber	
Mylar	
Nickel	
Copper	
Silver	
Brass	
Synthetic Rubber	
Gold, Platinum	
Sulfur	
Acetate, Rayon	
Polyester	- - -

Celluloid
Polystyrene
Orlon, Acrylic
Cellophane Tape
Polyvinylidene chloride (Saran)
Polyurethane
Polyethylene
Polypropylene
Polyvinylchloride (Vinyl)
Silicon
Teflon
Silicone Rubber
Most Negative (-)

Conducting leaves separate

Figure 1-5: a. *(table)* Tribo-electric series **b.***(top right)* An electroscope shown with a positively charged rod near its metal disc. The positively charged rod attracts some electrons from the conducting "leaves" up the metal rod into the metal disk. Thus, the "leaves," previously having a balance of positive and negative charge, are left with a net positive charge; this gives the leaves impetus that moves them apart as shown.

What is actually touching the balls? The air is, but, the role of the air here is minimal, for the repulsion remains even if the air is pumped out. This is analogous to what happens with gravity; a ball dropped inside of a vacuum chamber, or on the moon where there is already a very good vacuum, still falls. As in the case of gravity, it is the plana that

---

[20] Apparently, the name "pith" arises because an 18th century experimenter invented a "pith-ball electroscope" that used a ball that was actually made of pith.

is acting on the ball. It has been given a new power, a new ability, the electric field, that can cause impetus in suitably disposed bodies.

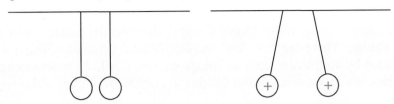

Figure 1-6: **a.** *(left)* Two uncharged pith balls hang vertically. **b.** *(right)* The two positively charged pith balls repel each other, moving them off their vertically hanging positions.

Let's slow down a bit and analyze how this happens. The charged ball has the ability to cause a new quality, a new power, the electric field, to exist in the plana and obviously the plana has the ability to receive this new power. This field then "travels" from one part of the plana to the next until it reaches a part in contact with a charged body that can receive the action of the field-activated plana.[21] In particular, the plana part so activated causes impetus in that second body. The field actually takes some time to "travel" from one ball to the other. It actually travels at what we call "the speed of light"! So, it is so fast that most of the time we cannot notice it. Thus, as we saw for gravity, our infinite speed of travel idealization is justified under many circumstances as an approximation.

A similar analysis can be made of magnetic fields, which also act on objects through a distance, i.e. through the plana. We will soon see the close relationship between electricity and magnetism, but for now we are ready to analyze the electric field in detail, including introducing the needed mathematics.

---

[21] "Activated plana" simply means plana that has this new power we call the electric field. "Activated" here does not imply that the electric field already exists in the plana "waiting" only to be somehow switched on, but only that the plana is receptive to having the electric field, but does not have the field until the field is *caused* in it by a charged body or another part of the plana.

Summary

Physics is the rigorous study of the physical world, the essential characteristic of which is its ability to change. There are first fundamental physical principles which we get from direct observation by abstraction from the things we sense. All of our reasoning makes use of these generics, which we discussed in Chapter 1 of *PFR-Mechanics* and reviewed in this chapter.

In our study of the mathematics in Chapter 2 *PFR-Mechanics*, we reviewed basic realities about the first category of property (quantity or extension) and saw some of how basic facts about quantity are incorporated into an axiomatic symbolized system. In the rest of the text in which we studied Newtonian mechanics (and for some students, special relativity), we specified our generic knowledge, i.e. those fundamental principles mentioned above, to include deep aspects of reality related to locomotion, including impetus, force, mass, plana and even a little on fields.

"Vacuum" or "space" is not nothing but something. Though it does not have mass, it is a physical reality, having extension and all the nine categories of properties. Because it has the ability (power), which is in the category of quality, to have the qualities of gravitational, electric or magnetic fields as well as other fields, we call it plana, which derives from the Latin word for field. Chapter 7 of this text shows how we can, making use of deep aspects of nature, avoid *direct* consideration of much of the nature of the plana through the use of the empiriometric formalism of special relativity.

In this text, *PFR-E&M*, we further specify our understanding of the analogically general aspects of nature related to locomotion[22] by studying two types of fields, i.e. powers of the plana, the electric and magnetic fields and their sources (causes).

The appellations "magnetic" and "electric" come from the two manipulatable sources of magnetic and electric fields: magnetite (lodestone) and amber, the Greek word for the latter is ηλεκτρον, which in Latin is rendered *electrum* and in English is *electron*.

In the 1200's, Peter Peregrinus broke the ground on the study of magnetism, making important discoveries but it was not until Coulomb that a real quantitative study of E&M began. James Clerk Maxwell gave the ultimate empiriometric formulation of classical E&M.

Static electricity arises in ordinary circumstances when normally neutral bodies, i.e. bodies that have the same amount of positive and negative charge, take on a net charge when electrons are moved from one body to another. By simple experiments, one can determine that there are positive and negative charges. Like charges repel and unlike ones attract.

---

[22] This includes rest, for example, in the case of balanced forces.

Electrons can flow through metals such as copper with little resistance. Ben Franklin erroneously thought that electricity was the flow of positive charges, leaving us with the funny convention that current flows opposite to electron flow.

Rubbing two bodies of different types together results in electrons being transferred from one to the other, leaving one body with a net positive charge and the other with a net negative charge. This tribo-electric effect is schematized for many materials in Figure 1-5a.

Electric and magnetic fields are fundamental analogically generic abilities or powers.[23] Those powers are best studied in inanimate substances where their generic nature is most evident. Note, however, when these inanimate substances become parts of living things such as you and I, they become powers of you and I. Thus, the study of E&M is also deeply relevant to all the substances with which we are in immediate sense contact.

Electric fields travel from a source through the plana.

---

[23] Note again this meaning of power should not be confused with the narrower definition in modern physics, i.e., the measure of the time rate of change of energy: dE/dt.

Problems

1. Give examples for two of the categories of properties in which Newtonian physics has provided further specification of our knowledge.

2. In your imagination, put yourself on the surface of Mars. a) Using what you know about Mars, think about what you would see, hear, feel, smell and even taste if you could be sure of what would not harm you to do so. b) Discuss the difficulties in actually being able to directly sense things on Mars in their actual state and how you might overcome them and what limitations your solutions have. c) Imagine further that you have a rabbit's fur and plexiglass as well as a magnet. What might you learn with these rudimentary instruments?

3. Making use of all 10 categories, i.e. substance and the nine categories of properties, as done in *PFR-M*, write a script for a twenty questions game where the "what you are thinking of" is "an electric field."

4. Speculate as to why it took so long for E&M to become a quantitative science even at the simplest level let along one in which measurements are cast into a symbolic and axiomitized system of equations?

5. Why might NASA cancel our trip to Mars using its ***Triboelectrification Rule*** in which they cancel a launch when the launch vehicle would pass through certain types of clouds? *Hint*: Think about *both* the origin and effects of lightning.

6. The pyroelectric effect was noticed in the material tourmaline by Theophrastus in 314BC. When heated it becomes statically charged, able to attract bits of straw and ashes. Hypothesize as to how this happens. Note that pyroelectric materials, which are always crystals, do not have to have contact with a second material for the effect to occur. Simply heating it causes the effect, though it wears out after awhile.

7. Give the two coupled reasons why the empiriometric tend towards the analysis of locomotion? Hint: Consider change and quantity; reviewing Appendix I of *PFR-M* might be helpful here.

8. Pull off scotch tape from a roll and describe what happens? Keeping to the simplest possible explanation, speculate as to why? Go into a dark room pull more tape off the roll and describe what happens? Are these phenomena related? Explain simply? Note that you will not be able to explain this last except by weak unsorted analogy until we study electromagnetic radiation.

9. List some ways that static charge might be generated? How, for example, does the wand shown in Figure 1-4c work?

Chapter II

Static Electric Field and Potential

Part I: Theory

Fundamentals of Charge and Electric Fields

Electric fields are generated by charged bodies. Experiments, such as those discussed in Chapter 1, reveal two types of charge, positive and negative. A positively charged body creates a field in the plana that can cause impetus in a second charged body. In particular, it causes impetus of a type that moves the second body towards the first if the second is negatively charged and away from the first if the second is positively charged. A negatively charged body causes impetus that moves the second body toward it if the second is positive and away from it if the second is negative. Of course, we summarize this all by saying like charges repel and unlike charges attract.

Two other facts about charge that we find experimentally are the *quantization of charge*, i.e. there appears to be a smallest unit of charge, and the *conservation of charge*, i.e. the net charge, |total positive charge| - |total negative charge|, in any *isolated* location never changes. Said another way, conservation of charge means that the net charge in any region cannot change without charge entering or leaving that region. We will further discuss these two facts below as we probe the nature of charge in its relation to electric fields.

Basics of the Electric Field

To probe the nature of the electric field at the simplest level, we will first ignore *local* conservation of charge[1] and consider what would happen in the moments just after a positive and a negative charge are simultaneously created one meter away from each other in the plana. As shown in Figure 2-1, each particle immediately starts causing an electric field in the plana. The positively charged particle activates the power (ability) we call the electric field in the parts of the plana in contact with it. These plana parts then act on the parts next to them causing an electric field in them and so on. When a field is caused in a part of the plana that is in contact with a massive body able to receive the action of the field, i.e. what we have been calling a charged body, e.g. the negatively charged body, something further can happen. Namely, the field can begin causing impetus in the negative charge to move it toward the positive. Thus, the quality of the plana known as the electric field is a power with two distinct acts; it can activate an electric field in neighboring plana parts and/or activate impetus in charged massive bodies that are in contact with it. A similar process occurs from the negative to the positive charge. These physical processes bring four issues related to the nature of charge to the fore.

---

[1] Thus, so far as we know, this could not happen in nature, so that if it did happen, it would have to be a special act of God.

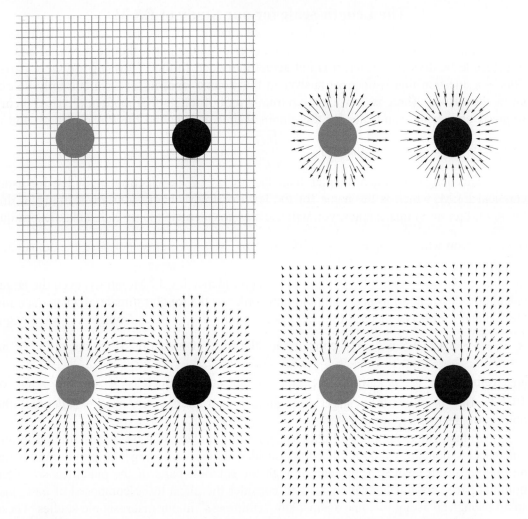

Figure 2-1: a. (*top left*) A positive charge (red) and a negative charge (black) shown at the moment that they are created before they have time to activate even the plana cells immediately in contact with them. For convenience, the plana parts are shown square and very large, but of course, we don't know the shape of the parts and we know they are extremely small. **b-d.** (*top right, bottom left and bottom right*) The charges have now activated the plana cells next to them, and the fields have activated the cells next to those. The three figures show various stages of the propagation. Note how the fields propagating from each particle reinforce each other in the plana cells in the middle along the line connecting them. Note also that the plana cells are not drawn in this series of pictures to avoid too busy of a drawing, and because, as we will see when we discuss the potential later in this chapter, there is, in a certain sense, detail that such *E*-field pictures leave out.

The Length Scale for our Study of E&M

First, our analysis has made us note that plana has parts. Like all physical substances, plana is divided into a certain number of actual parts; though it can, of course, in principle be divided further, it is not actually divided further. How big are these parts? This is a question that we cannot answer in this course, and, in fact, we currently do not know the answer. There is, however, a minimal unit of length that modern physical theory *suggests* as a prime candidate for this minimal unit; it is called the Planck length and is

written (in *cgs* converted to meters): $l_p \sim \sqrt{\dfrac{\hbar G}{c^3}} \sim 1.6 \times 10^{-35}\, m$.

This length is much smaller than the radius of the electron predicted by using classical E&M, which is the name for the level of E&M theory you are learning in this book. Further on in this chapter, you will see how to calculate the so-called classical radius

of an electron which is: $r_0 = \dfrac{e^2}{m_e c^2} \sim 2.8 \times 10^{-15}\, m$. Measurements show that the actual size

of an electron is less than $2.8 \times 10^{-19}\, m$ at 95% confidence level.[2] Moreover, even the larger calculated classical radius is already the same order as the rough estimate of when quantum mechanical effects need to be accounted for, the Compton wavelength

$\lambda_c = \dfrac{h}{m_e c} \sim 2.4 \times 10^{-12}\, m$.[3] By comparison, the size of an atom is on the order of an

angstrom, $1 A = 10^{-10}\, m$, while the size of an atomic nucleus is on the scale of $10^{-4} A = 10^{-14}\, m$. Now, these larger dimensions are, of course, very much smaller than the objects we deal with in ordinary life.

In fact, since our study of classical E&M in this text will be largely limited to this macroscopic domain, we do not need to consider the parts of charges or parts of the plana on this angstrom level scale. Hence, though the smallest parts of the plana (its so-called minimal parts) are much smaller, we will consider the plana to be composed of parts just small enough so as not to come against any "graininess" in our macroscopic studies. These larger "parts" will now chiefly be the only parts we discuss. We average over each of these parts, ignoring smaller scale effects, or said in our language we abstract or "switch off" smaller scale effects as they are not our focus of attention. Furthermore, if we choose our scale small enough relative to the scale of our observation, we can make a further idealization and treat the parts as if they are points and thus define an electric field at each point. Furthermore, since the electric field in a given part, which we idealize to a point, has (1) a certain strength, i.e. can exert a certain magnitude of force on a unit charge (which in turn is the ability to cause a certain change in the intensity of impetus per unit), and that (2) force has a definite direction in space, we can and usually will represent the electric field with a vector field. Again, as we have discussed multiple times, such representations involve analogies and idealizations that cannot exist as such in the physical world, but, as

---

[2] Bourilkov (2000)

[3] It turns out that, quantum adjustments to E&M are not significant down to distances less than $10^{-12}\, m$, Purcell (1965, pg. 3; 1983, pg 2). And, activity in the atom can be described using quantum mechanics for massive particles, but the same Maxwell action for the electromagnetic field.

long as we remember what we have left out and what its limits are, this is a very helpful and indeed needed way to understand the physical aspects we have chosen to analyze.

Discrete Nature of Charge

Second, and related to the length scale just discussed, we note that because of the small size of elementary charges (i.e. the massive bodies with the least charge, electrons and protons[4]), when dealing with macroscopic bodies[5] we are dealing with many such charges. These charges are parts of macroscopic bodies, which tend to be neutral, having some multiple of Avogadro's number (6.02×10^{23}) of positive and negative charges (in an end of chapter problem you will be asked to calculate the number of charges in a drop of water). When you rub electrons off such a neutral macroscopic body, you end up with a deficit of roughly some multiple of 10^{10} electrons.[6] These numbers are so high that we can and will treat the charge on bodies as continuous. This of course follows our tactic of leaving out what is not important for the *general* causal understanding of a given physical situation, while capturing enough detail to still allow a good approximation of the magnitude(s) of the force(s) acting.

In this same vein, one should remember that we have also "switched off" the distinction between substances, groups of substances and parts of substances. Hence, we will ignore whether or not a given electron is part of a substance such as yourself. In that case, for example, we would otherwise say such electrons are virtually present, present by their power because they are powers of you. Indeed, it is essential, i.e., first nature, to all ordinary substances (which are composite) to have virtual charge and to *be able* to have excess charge, but only second nature to *actually* have it. By contrast, it is essential to an elementary particle existing on its own, such as an electron, to actually have charge. However, using our analogically general definition of body, we will say a body has charge, when we might simply mean that there is a group of substances that have a net charge. In short, in this spirit, we can,[7] and often will, treat bodies as if they were nothing more than these charges stacked or held in position in some configuration.

Coulomb's Law and Equality of Passive and Active Charge

Thirdly, we see that there are two distinct properties of charged bodies: the ability to cause an electric field in the plana and the ability to receive the action of the plana's field. Though we use the same word "charge" for both, "charged" bodies have two distinct properties: (1) active charge, which is their ability to cause an electric field in the plana and (2) passive (or receptive) charge, which is their ability to receive the action of an already established field in the plana. Furthermore, since they are receptive to the action of the field to cause impetus in them, they must, of course, have mass.

This discussion should remind you of the similar discussion on gravitational fields in *PFR: Mechanics* (*PFR-M*). Indeed, the similarity goes further, for careful experiment has revealed that active and passive charge, like active and passive mass, are always

---

[4] Quarks, the constituents, of protons and neutrons, have 1/3 and 2/3 of the electric charge of the proton and electron, but are not in principle (as far as we know) able to exist on their own. Quarks are, of course, also parts of other particles, such as the π-meson which consists of an up quark and an anti-down quark.

[5] Here we mean macroscopic bodies *with mass*, for plana is not composed of virtual protons and electrons.

[6] Note that this is an extremely course approximation as the exponent is only accurate to one significant figure.

[7] At one level of analogical abstraction.

proportional to each other no matter what the material. As in the case of gravity, this is a result of the need to satisfy Newton's third law. But we are getting ahead of ourselves. Before we can establish this connection, we need the experimentally verified result that relates the force caused by an active charge of measured strength, q^A, on a passive (receptive) charge of measured strength, q^P, that are in plana separated by a distance r from each other. (We also assume the bodies are much smaller than the separation between them so that we can ignore their size.)[8] For charge *one* acting on charge *two* and vice-versa we find:[9]

2.1
$$F_{1 \to 2} = q_2{}^P E_1 = q_2{}^P \frac{q_1{}^A}{r^2}, \quad F_{2 \to 1} = q_1{}^P E_2 = q_1{}^P \frac{q_2{}^A}{r^2}$$

where: superscript A indicates the measure of the strength of the action of the charge on the plana, i.e. the active charge, and P indicates the measure of the receptivity to the action of the field, i.e., the passive charge. E is a measure of the electric field strength and r is the distance between the two "charges."

This inverse square result was first proposed and measured by Auguste Coulomb (see history box). Fundamentally, such experiments involve looking at the effect of one fixed charge on a second charge that is allowed to move. One then applies just enough "external" force to prevent impetus from being activated in the second charge. That is, one balances the force caused (via the plana) by the first charge with this external force— Coulomb used a torsion spring to supply this external force—thereby making a static situation. In an end of chapter problem, you will be asked to design a dynamical experiment in which you measure the effect of the changing impetus directly; i.e., you measure the acceleration of the second body.

Now, we can prove the proportionality of active and passive charge using Newton's third law between the massive bodies *by* prescinding from the activity in the plana by assuming the speed of propagation of the field in the plana is infinite; note again the mental construct (being of reason) of a completed infinity. In general, as we will see in Chapter 6, if we do not assume infinite speed, Newton's third law (for the massive bodies alone) will *not* be satisfied at all times. By contrast, if we do this, the forces act instantaneously, one particle on the other via the plana, and we can simply write: $F_{1 \to 2} = F_{2 \to 1} \Rightarrow q_2{}^P q_1{}^A = q_1{}^P q_2{}^A$. This can be written:

2.2
$$\frac{q_1{}^A}{q_1{}^P} = \frac{q_2{}^A}{q_2{}^P}$$

Hence, the ratio of the strength of any given charged massive body's ability to cause a field to its receptivity to action by a field is the same for every charged massive body. Now, through Newton's second law, we can absorb the proportionality constant in our definition of force[10] (by the units we pick) or of the charges themselves, and thus take $q^A = q^P \equiv q$

---

[8] We also generally assume, for simplicity, that (a) the active and passive charge is uniformly distributed through each body and (b) both bodies are spherically symmetric.

[9] A much more detailed expression accounting for more of the physical proprieties is given in Appendix II.

[10] Since the units of distance, mass, and time are usually decided by the scale size naturally accessible to our senses (e.g., mks or cgs) and because the units of force are fixed by our selection of these units, we will usually choose to modify the definition of charge rather than force. Of course, this determination of the units of force by the basic units mentioned follows from simple mechanics. In particular, by measuring the speed

for all charges. In this way, we may choose to simply speak of the charge of the particle, without distinguishing between active and passive. Moreover, because of this, we will often follow standard usage and just call a charged massive body a "charge." To further probe the meaning of the equality of passive and active charge see Appendix II. *Exercise*: Think of experiments to test this equality (see end of chapter problem).

Of course, this identification of active and passive charge does not change the reality but is only an analogical abstraction; i.e., it leaves out of consideration certain aspects so that we may focus on others. Again, it in no way eradicates the real physical distinction between action and reception.

With this simplification, we can write Coulomb's law in standard form, where we have used $\vec{F} \equiv \vec{F}_{1\to2} = \vec{F}_{2\to1}$:

2.3 $\qquad \vec{F} = k \dfrac{q_1 q_2}{r^2} \hat{r}$,

where $\hat{r} = \dfrac{\vec{r}_2 - \vec{r}_1}{\left|\vec{r}_2 - \vec{r}_1\right|}$, or more simply, $\hat{r}$ is the unit displacement vector pointing from charge one to charge two. Here we have introduced the proportionality constant, k, to reconcile the chosen system of force units with the chosen system of charge units.

In *cgs*, $k = 1$, with q measured in *esu* (electrostatic units), where $e = 4.803 \times 10^{-10} esu$. In this text, impetus is measured in *phor*, which comes from the Greek word, $\varphi\acute{o}\rho\alpha$, for impetus, while, also using Greek based words, applying standard usage, force is measured in dynes, from the Greek word *dynamis*, meaning force ($1 dyne = 1 phor / s$), and energy is measured in ergs, from the Greek word *ergon*, meaning work ($1 erg = dyne \cdot cm$).

In *SI* units (which we can loosely call *mks*), $k = \dfrac{1}{4\pi\epsilon_0} = 8.988 \times 10^9 \dfrac{Nm^2}{C^2}$,

with q measured in units of coulombs, where the charge on an electron is $e = 1.602 \times 10^{-19} C$. As in *PFR-M*, impetus is measured in buridans, force is measured in newtons and energy is measured joules.

Also, notice our notation automatically incorporates the correct signs to indicate which types of impetus are activated. For example, with two positive charges as shown above, the equation predicts that the force caused by particle one (via the plana) on particle two is in the direction of $\hat{r}$, which is the correct *repulsive* type action between two like charges. By contrast, if particle two is negative the correct *attractive* type force is predicted.

---

and mass of the body experiencing the action of the electric field at two closely separated times, we can calculate the momentum at each time and thus calculate the effect of the force (and thus its strength and directional type) due to the field. For example, we get a body receiving so many (kg m/sec)/sec, or Buridan/sec.

With this understanding in hand, we can give the generic answer, which we will specify later, to the implicit question that launched this chapter: "What is an electric field?"

Charge is the power or ability (quality) of a massive body to cause an electric field in the plana or receive the action of an electric field. Because these two abilities are proportional, we can speak simply of charge, without referring to the real distinction between active and passive charge. There are two types of charge in this generalized sense, which we designate by positive and negative, since the fields generated by the "negatives" cause impetus in a given test charge of the opposite type of that generated by the "positives."

The measure of the charge of a body, irrespective of whether it is a *net* charge (which arises from the effect positives *and* negatives together) or not, is represented by the symbol q.

Electric Field is a power or ability (quality) of the plana caused in it by a body with charge. The relation between the measure of the electric field, $\vec{E}$, the measure of the distance, r, and the measure of charge, q_1, causing it is:

$$\vec{E} = \frac{q_1}{r^2} \hat{r} \; .$$

The relation of the measure of the force caused on a charge of measure q_2 receiving the action of the field of the first charge is written:

$$\vec{F} = q_2 \vec{E}$$

Note two key facts about the force law in equation 2.3. First, the magnitude of the force, F, is proportional to both the magnitude of "charge" of the first body, q_1 and that of the second, q_2. Or more explicitly, as we've already seen above (under the appropriate simplifying assumptions--see below and Appendix II), one can take q_1, to be the magnitude of the active charge of body one and q_2 as the magnitude of the passive charge of body two. And, since the active and passive charge are proportional we can simply talk of the charge of each. The second key fact is that F falls off as $1/r^2$.

The Inverse Square Law

The inverse square law works over 24 orders of magnitude of length scale.[11] The fall off is due to the field's decreased ability to cause field in farther parts of the plana as it moves away from the source causing it in the plana. In particular, the inverse square law corresponds, in Euclidean geometry, to the increase of boundary area as one considers successively more distant radii from a center point. To see this, imagine concentric spheres, such as shown in Figure 2-2a, drawn at increasing radius from the chosen point, say the location of a charge and consider the construct shown in Figure 2-2b. This implies that as the field propagates in the plana away from its source, it loses ability to activate field in neighboring plana cells in proportion to the boundary (an area) over which that power must act. Such weakening is analogous to the way an incompressible fluid flow decreases as it moves from a small source (see Figure 2-3). The rate of fluid flow per unit area across the

---

[11] J.D. Jackson (1975). Indeed, classical E&M, which are expressed in Maxwell's equations, work over a wide scale range, suggesting that there is something deeper than mere scale at issue.

boundary decreases because the same amount of fluid is spread thinner as it moves outward. Again, the field is not a fluid, but, the power of the field to propagate itself is analogically like the parts of a fluid that must thin out as it expands.

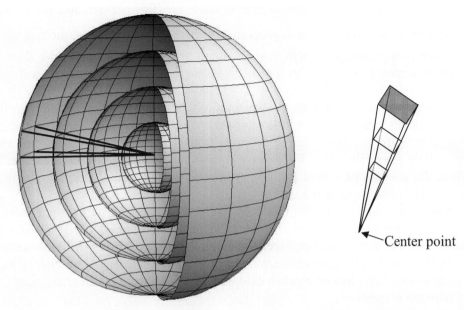

Center point

Figure 2-2: **a.** *(left)* Note how the area of the concentric spherical shells increase as the square of the radius. This is reflected in the formula for the surface area of a sphere, $S_{sphere} = 4\pi r^2$. **b.** *(right)* To understand this, consider an arbitrarily small[12] (red) square at some distance from the center. Consider moving along one of the lines that connects a boundary point of the red square to the center point. At each distance, r, from the center point another square is defined by these lines. As one moves out from the center point along a given line, the area of these squares increases as r^2. To see this, take the angle subtended by the red square as $d\theta$, then the length of the side of the square at distance r is $rd\theta$ and its area is $r^2 d\theta^2$, i.e. it is proportional to r^2.

Figure 2-3: Three frames of a thin spherical shell of incompressible liquid, such as water, expanding under the action of impetus that is gradually reduced as it expands. Only a slice thru the *y-z* plane of the concentric shells is illustrated, revealing the donut-like cross-section. As the water expands, its total volume must stay the same so the thickness, Δr, must decrease as shown; in fact, $\Delta r \propto 1/r^2$. Furthermore, thinking of the shell as having layers, as it expands, the lead layer must move out of the way to make room for the next layer. This means that the volume of water inside the shell, no

---

[12] Or as we say, forming a being of reason, an "infinitesimally small" square.

matter how expanded it is, must cross a sphere of a given radius in the same time, say Δt, as it takes to cross a sphere of any other radius. That is, the full thickness of the shell, Δr, must cross a given $r = \text{constant}$ in time Δt, so that, taking $v_r = \Delta r / \Delta t$, the volume of water crossing per unit area per unit time in the radial direction is proportional to $v_r \propto 1 / r^2$. That is, the rate of water flow crossing through a given small area perpendicular to the flow must decrease according to the inverse square of the radius of the water shell.

We can use this idea with our vector field representation of the electric field to indicate the field's strength in the following way. As illustrated in Figure 2-4, the electric field line starting from a point charge diverges from the field line next to it, and thus the density of field lines falls off as the inverse square by the same arguments used in Figure 2-3. Thus, the strength of a field can be represented by the surface density of field lines, i.e., the number of lines per unit area.

It is interesting to note that this density changes when the particle is in relativistic motion, i.e. moving near the speed of light. It becomes more pronounced perpendicular to the motion and less pronounced in directions along the motion, still correctly describing the electric field strength. This happens because of propagation effects and because of an effect that is only analogically like an electric field that we will include in our analysis in Chapter 6 and Chapter 3 respectively.

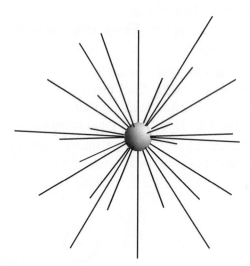

Figure 2-4: Field lines represented by equally spaced (in solid angle) lines emanating from a spherically symmetric charged massive body. Any spherical surface concentric with the charge (with radius larger than the charge) will intersect all of these field lines, as long as each of the lines is continued indefinitely. Hence, the field line density, i.e. the number of lines per unit area (dn / dA), will decrease as the inverse square of the radius of the sphere. Thus, if we make the number of lines proportional to the charge (say, $dn = q d\Omega$, so that $n = 4\pi q$), then, because of Coulomb's law's

inverse square fall off, we can use the field line density in such pictures to represent the strength of the E-field, $\left|\vec{E}\right|$.

Note that two other mechanisms, in addition to the weaken effect described above, could be at play (cf. Appendix II) in the fall-off. First, the plana's receptivity could be affected by the strength of the field trying to activate field in the next piece of plana; a stronger field, might for example, be harder to propagate than a weaker one. Second, the plana's receptivity could also be affected by the field already present in the plana, being, for example, more receptive when field is already activated in a given region. Though the first effect is worthy of further thought, classical E&M ignores this distinction ("switches it off") by making the simplifying assumption that the field fall off is only due to distance weakening.[13]

The second effect, which is fundamentally the effect of the field of one particle on that of another, is more easily accessible. To understand it, consider measuring the effect of a charge acting on a test charge that has an arbitrarily small charge, so that we can always make the field it causes at any given point in the plana as small as we want, thus, in turn, making whatever effect it has on the already existing field of the other particle at a given plana cell also as small as we like. Now, as we increase the charge on this "test" particle, we will see that the field it causes simply adds to the field caused by the first particle. Whatever field is generated by the first particle at a given place in the plana simply adds to the field generated by the second and third etc... There are no interactions between them. The fact that the total field at a given place in the plana can be obtained by simply adding the independent field contribution of each particle is called the superposition principle. It is called also linearity; so, for example, we say E&M is a linear theory.

Conservation of Charge

Lastly, recall that the initial thought experiment that spawned these last three sub-sections of discussion violated charge conservation in order to simplify the subsequent analysis and understanding. We posited that two oppositely charged bodies suddenly appear in two places that are separated by 1 meter. In nature, as we've said, there is no evidence that such a thing can happen; all the evidence verifies net charge conservation.

Mathematically, we write, analogous to the way we write the conservation of net momentum (i.e. $\sum_{i=1}^{N} \vec{p}_i = \text{constant}$):

2.4
$$\sum_{i=1}^{N} q_i = \text{constant},$$

where q_i refers to the magnitude of the charge of the ith body.

Note that here we follow the sign convention that we will use throughout E&M and beyond. *A negative sign is used with magnitudes of charges of the "negative" type.* This sign convention incorporates into our mathematical formalism the fact that charges of opposite types cause impetus of opposite types.

---

[13] Indeed, to avoid this and other issues, we can lump this affect into $\dfrac{q_A}{r^2}$, not distinguishing the various causes that lead to this net result. Again, this does not obliterate the possibility of the effects, but only allows us to postpone consideration of them so we may focus on other more immediate realities.

Thus, the nature of physical interactions are such that if charged particles are created, say by interactions within the plana, they must be created in pairs with opposite charges *at the same place,* i.e. within a region smaller than the macroscopic length scale with which we are concerned. For example, positron-electron pairs can be created during an interaction in a particle accelerator. The creation of a positron-electron pair from the plana excited by external electromagnetic radiation is shown in Figure 2-5. Already existing pairs can annihilate each other and produce electromagnetic radiation ("light"); this happens, for example, after the positron that results from the $\beta+$ (symbol for a positron) decay of an atomic nucleus runs into an electron that is nearly at rest relative to it.

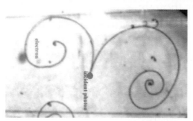

Figure 2-5: Creation of an electron-positron pair from the plana as seen in Fermilab's 15 foot Bubble Chamber (tracks were discovered in the data at the University of Birmingham). [14] The red dot marks the place where an incident photon precipitated the creation of the pair. The green spiraling line on the left is the track of the *electron*, and the purple spiral on the right is the track of the *positron*. You will learn both why each of the particles spiral and why the tracks spiral in opposite senses in the chapter on magnetic fields.

Thus, if we were to modify our starting thought experiment so that it respected conservation of charge while still allowing us to discuss the propagation changes of the electric field, we would have to consider the charges as appearing in one place and then separating. Thus, we would have to analyze the fields of moving charges, and we cannot do this until a later chapter. Hence, we leave this modification of our thought experiment aside until an understanding of moving charges and several other requisite principles have been introduced. For the moment, we leave aside propagation issues and turn to better understand Coulomb's law itself by applying that law to concrete examples.

---

[14] Picture from http://teachers.web.cern.ch/teachers/archiv/HST2002/Bubblech/mbitu/electron-positron.htm

Charles-Augustin de Coulomb (1736-1806 AD) was born in Angoulême, France. Over 500 years after Peter Pilgrim's (Peregrinus) pivotal work, Coulomb is the one that finally breaks the barrier and accomplishes the *first quantitative measurements that probe an essential aspect of E&M*. Namely, he annunciates Coulomb's Law (c. 1785). Coulomb, like Peregrinus, was French, Catholic, and an engineer in the army (though by then, the war with Islam was over). Coulomb was an officer in the army of the French King Louis XVI. There, he applied himself to, among other things, a number of construction problems using variational calculus. Some of his techniques, including a sliding-wedge theory of soil mechanics, are still in use today.

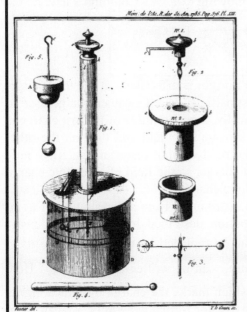

He also did important work in the study of friction as well as wrote: "The Theory of Simple Machines." The torsion balance that he used both for his Coulomb law work and for magnetic work is shown to the left. Coulomb also demonstrated that "non-conductors" pass electricity to some extent.

The great scientist James Clerk Maxwell later found that the reclusive experimenter Henry Cavendish had anticipated Coulomb's law by some years.[1] Cavendish is usually not given credit largely because he never published, nor made any effort to make known these results. Cavendish made use of a torsion balance in determining the force between massive bodies and "weighting the earth," verifying the "Coulomb's law"-like gravitational force law.

[1]The Electrical Researches of the Honorable Henry Cavendish written between 1771 and 1781 by James Clerk Maxwell, Cambridge Univ. Press, 1870

Understanding and Using Coulomb's Law

The Full First Extended Meaning and Form of Coulomb's Law

An alert reader may wonder under what static circumstances Coulomb's law, $\vec{F}_{1\to2} = k\dfrac{q_1 q_2}{r^2}\hat{r}$, equation 2.3, applies. The answer is that it applies, in an analogy to the gravitational case, to charges whose diameters are much smaller than the distance between them. Said another way, Coulomb's law applies to point charges, i.e. massive bodies that have a fixed charge whose size is allowed to approach arbitrarily close to zero. Point particles are indeed mental constructs, since, for them, one *completes the limit* that takes the size of a massive body with fixed charge to zero. Furthermore, since the inverse square relation and the superposition principle both apply to charge as well as to gravity, the reasoning we gave there showing that all mass outside any spherically symmetric mass distribution cause exactly the same field at any point outside the radius of that mass distribution as would a point particle of the same mass at the center of that body. That is, every spherically symmetric charge can be treated as a point charge located at the center of that charge. Coulomb's law thus is most written as applying to two point masses; however, using the superposition principle and assuming a positive test charge, we can write, for a series of point masses labeled by the index *i*, the following *extended* Coulomb's law:

2.5
$$\vec{E} = \sum_{i=1}^{N} \frac{k\, q_i}{r_i^2}\hat{r}_i$$

In this representation the coordinate system origin is effectively at the test charge. Note again the use of the word "test" charge or test particle, which means a charge used to "test" the field at a given region in the plana. Typically, the magnitude of the test charge is made arbitrarily small so that its impact on the surrounding field and thus on the other particles is negligible; in this way one samples the field without disturbing it. Also, note that positive is the conventional choice for the test charge, so the field of a positive charge always acts to give impetus away from such a positive charge.

Now, we can obviously create any charge distribution from a distribution of point particles. That is, if we assume a charge distribution described by the charge density function $\rho(\vec{r}')$, the charge, q_i, in equation 2.5 can be replaced by the charge in a given arbitrarily small volume dV; this charge can, in turn, can be written as $\rho(\vec{r}')dV$. Here the charge label, *i*, is replaced by the vector $\vec{r}'$, which marks the charge by indicating its location. Then by moving the origin off the test charge for formal generality, we can write the discrete equation above in continuous form:[15]

2.6a
$$\vec{E} = k\int_{Charge} \frac{\rho(\vec{r}')}{|\vec{r}-\vec{r}'|^2}\,\overrightarrow{(\vec{r}-\vec{r}')}\,dV$$

$$= k\int_{Charge} \frac{\rho(\vec{r}')}{|\vec{r}-\vec{r}'|^2}\left(\frac{\vec{r}-\vec{r}'}{|\vec{r}-\vec{r}'|}\right)dV$$

---

[15] To get the less general case shown in the picture associated with equation 2.5 , set $\vec{r} = 0$.

2.6b $$\vec{F}_{net} = q_{test}\vec{E}_{net}$$

Where the unprimed variable, $\vec{r}$, is the vector from the origin to the field point, i.e. to the point at which one would like to know the field strength and direction. The primed variable, $\vec{r}'$, indicates the location of a very small part of the charge.

Spherical Symmetry

With this in hand, we could now prove the above statement that a spherically symmetric charge distribution acts as if all of its charge were at a point at the center. However, since this was already done in detail for the analogous case of gravity in *PFR-M*, we will instead prove, in an intuitive way not done in the gravity chapter, that there is no net force inside of a uniformly charged spherical shell (i.e., a shell that is uniformly charged on the surface, not necessarily in the radial direction, which after all is arbitrarily small).

To prove it, first note that, because of the symmetry of the shell, any field point or viewpoint (i.e., a place in the plana at which we seek the field value point) that we pick inside the shell will be surrounded by a cylindrically symmetric charge distribution. To see this, imagine the cylinder of symmetry that has an axis that passes through the field point and the center of the sphere. Next, draw a line through the field point at an angle θ from the symmetry axis, and imagine rotating this line around that axis. This traces out a circle at the places where the moving line intersects the spherical shell as shown in Figure 2-6. Repeat this procedure for a line that makes an angle of $\theta + d\theta$. These two lines form an angle of $d\theta$ that slices two ring-like pieces out of the shell, one on the top and one on the bottom. Now, we would like to determine the field at the field point where we put the test charge (the red dot in Figure 2-6) due to these two ring pieces of the shell. We only need consider the z-component of the field, i.e. the up-down component shown in the figure since the other components are zero. This is so because for each piece on the right "pushing" the charge away toward the left there is an equal charge piece pushing it back to the right from the same radius and thus with the same strength. The ring piece on the top has a charge given by $\sigma\big((2\pi r_1 \sin\theta)(r_1 d\theta)\big)$ and is at a distance of r_1 from the field-point, thus tries to activate downward field of strength given by:

$E_{z-top} = -k\dfrac{\sigma\big(2\pi r_1^2 \sin\theta\, d\theta\big)}{r_1^2}\cos\theta$, where σ is the surface charge density, i.e. charge per

unit area, of the sphere. The bottom piece has a charge given by $\sigma\big((2\pi r_2 \sin\theta)(r_2 d\theta)\big)$ and is at a distance of r_2 from the field-point, thus tries to activate upward field of strength

given by: $E_{z-bottom} = k\dfrac{\sigma\big(2\pi r_2^2 \sin\theta\, d\theta\big)}{r_2^2}\cos\theta$. Thus, the top ring cancels the effect of the

bottom ring and no field operating in the z-direction is created. Now, our construction of the top and bottom rings can be repeated for all angles until we have divided the entire shell into rings that align with the axis of cylindrical symmetry. Since the above argument is true for each angle and thus for each ring that makes up the shell, the shell can create no field inside the shell at the field point. Lastly, since all points interior to the shell can be thought of in exactly the manner just discussed, there can be no field anywhere in the interior of the shell. Mathematically, we say $\vec{E} = 0$ everywhere inside a uniformly charged spherical shell.

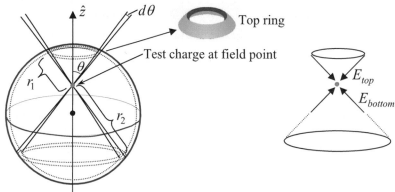

Figure 2-6: a. Inside a uniformly charged spherical shell, no points (parts) of the plana have a field activated in them, i.e., there is no field inside. This occurs because the affect of one part of the shell is canceled by that of another. The cancellation occurs because the nature of the field propagation in the plana agrees with the nature of a key aspect of three dimensional extension (Euclidean space). In particular, for each ring, the field strength fall-off as the inverse square of the distance from the charge is exactly compensated by the surface area increase by the square of that distance.[16] **b.** The affects of the charge of the top and bottom rings cancel in such a way that no field is created at the field point in the plana. E_{top} (and E_{bottom}) represents the field that would be caused by the top (bottom) ring respectively if only the top (bottom) ring were present.

Energy and Coulomb's Extended Law

If we consider charge distributions such as the above spherical shell as composed of pieces that can be modeled as point charges, such as electrons and protons, we might wonder: How much energy does it take to put together such a charge distribution out of point charges? Note that this question assumes that the point charges are initially separated enough that their mutual interaction can be neglected. By assuming the point charges are brought in from infinity,[17] we can calculate how much energy is needed. For example, in making the spherical shell, we can ask how much radially inward kinetic energy, i.e. total rate of transfer of intensity of impetus, must the positive particles initially have in order to just come to rest at the radius of r that defines the shell. Or, still another way of saying it is: what work must be done on the particles so that they only are given arbitrarily small momenta to move them in from infinity to the radius r. These particles that make up the spherical shell are then said to have a certain (electrical) potential energy. That potential energy can be turned into actual (kinetic) energy when the restraining forces that keep the

---

[16] As seen in the body of the text, this compensation is only needed for z-component of the field.

[17] "Bringing in from infinity" is another example of a mental construct. We start by considering the point particles as arbitrarily far away from each other and then complete the infinity by saying they are infinitely far away. This is then a mental construct or being of reason, because, as we showed in Chapter 2 of *PFR-M*, such a completed infinity cannot exist in reality.

shell from exploding are "turned off." We will discuss potential energy in more depth shortly, but for now our task is to calculate the potential energy given Coulomb's law.

Potential energy of two Point Charges

Before calculating the energy of the shell, let's start with the simpler case of calculating the energy required to move two equal-charge point particles in from infinity. Suppose one body is much heavier than the second so that the first can be considered at rest, and define the direction of the vector $\vec{r}$ as pointing from the first to the second particle. Then, the work done by the external force is given by:

$$\int_{\infty}^{r_0} \vec{F} \cdot d\vec{r} = \int_{\infty}^{r_0} \left(-k \frac{q_1 q_2}{r^2} \hat{r} \right) \cdot \hat{r} \, dr = -k \int_{\infty}^{r_0} \frac{q_1 q_2}{r^2} \, dr = k \frac{q_1 q_2}{r_0}$$

Note that, for charges of the same sign, say positive charges, the work done is positive because the test charge has to be *pushed* towards the fixed charge (while for charges of the opposite sign the work is negative because the test charge has to be restrained from moving inward). Said another way, the work is positive because the impetus that is created, though (infinitesimally) small, is in the same direction as the external force (while for charges of opposite sign the work is negative because the impetus that is created is in the direction opposing the external force). In the case of charges of the same sign, the external force allows the (small) impetus to act in that it stops the electric field from deactivating that (small) impetus. Also, the positive work means positive potential energy which means that there is potential or "stored" energy so that impetus and hence kinetic energy will be created when the restraining force is "turned off." Note further that for the charged particle with the large mass, it is natural to take the potential energy of the second "test" particle as zero "at infinity."

Work Required to Make a Spherically Symmetric Charged Shell
(for more advanced students)

To calculate the work required to construct a uniformly charged spherical shell, we imagine a shell with an arbitrarily large radius. We then compress it inward towards its center against the pressure caused by the parts acting on each other via the plana to bring it to its final size. Because of the symmetry of the distribution of charges, only the radial component of the field survives, and we only push against it. Because of this same symmetry and because the shell does have a finite, though small, thickness, it is useful to divide the shell into radial layers, i.e. the spherical shell is itself divided into spherical shells. To calculate the force that is caused by the shell on itself, we need to calculate the field's action on each layer, since it can vary from layer to layer.

Considering the shell fixed in its final state and noting that only the radial direction is non-trivial, we designate the linear distribution of charge in this radial dimension, summed over the entire volume of a layer of radius r, by the function $\lambda(r)$. Keeping in mind that this charge density is defined *only for the shell in its final state*, by this definition, a layer, i.e. a spherical shell within the shell, at radius r' of thickness dr' has a total charge $\lambda(r') dr'$. In the cross-section of the complete shell illustrated below, we can see the radial dimension of the shell, dr, centered on its radius r.

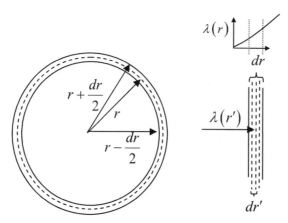

Using this figure to picture the integration over the layers, we can write the net charge of the whole shell as:

2.7
$$q = \int_{r-dr/2}^{r+dr/2} \lambda(r')dr'$$

Where $\lambda(r)$ is the total charge per unit length in the radial direction of a given layer of the spherical shell at the radius r.

Using this fact and by integrating over all the layers of the shell, we can write the effective radial force acting on the shell of charge q, $F_{eff}(r)$ at the radius r as:

$$F_{eff}(r) = \int dF$$

$$= \int Edq$$

2.8
$$= \int_{r-dr/2}^{r+dr/2} \left(\frac{k\int_{r-dr/2}^{r'} \lambda(r'')dr''}{r'^2} \right) \lambda(r')dr'$$

$$= k\lambda(r) \frac{\int_{r-dr/2}^{r} \lambda(r'')dr''}{r^2} dr$$

On the third line, we have written the electric field term in brackets. This field comes *only* from the charge below the given radius, r', since, as we've shown above, there is no field inside of a spherical shell of charge. That electric field term is multiplied by the charge of the given layer to get the force acting on that layer. In the last equality, we make use of the fundamental theorem of calculus, which can be written: $\int_{r-dr/2}^{r+dr/2} f(r')dr' = f(r)dr$, to do the integral over all layers, i.e. over r'.

Next, using the fact that $\lambda(r)dr = q$ and doing the integral over r'' gives:

2.9
$$= kq \frac{\int_{r-dr/2}^{r} \lambda(r'')dr''}{r^2} = kq \frac{\frac{dr}{2}\lambda(r)}{r^2}$$

$$= \frac{k}{2} \frac{q^2}{r^2}$$

The second equality introduces the important factor of two, which reveals that the relevant field is the average field between the inside of the shell where the field is zero and the outside where it obeys Coulomb's law.

Note that this seemingly complex result hinges on the simple fact that we can approximate $\lambda(r)$ by a constant function in the limit of infinitesimally small shell thickness. This, in turn, fundamentally depends on the reasonable assumption that $\lambda(r)$ is an analytic function. All analytic functions posses series expansions; hence, we can write $\lambda(r)$ near r as:

2.10 $$\lambda(r+\Delta r)=\sum_{n=0}^{\infty}\lambda^{(n)}(r)\frac{(\Delta r)^{n}}{n!}=\lambda^{(0)}(r)+\lambda^{(1)}(r)\Delta r+O\left((\Delta r)^{2}\right)$$

Thus, to first order, we only need the first term which is independent of the thickness Δr.

Now, we are ready to calculate the work needed to shrink this shell from infinity to r_0. We get:

2.11 $$W=\frac{1}{2}\int_{\infty}^{r_0}\vec{F}\cdot d\vec{r}=\frac{1}{2}\int_{\infty}^{r_0}\left(-\frac{k}{2}\frac{q^2}{r^2}\right)dr=\frac{k}{2}\frac{q^2}{r_0}$$

Thus, the potential energy of a shell of radius r_0 is one half that of two charges of the same charge at a distance r_0 from each other:

2.12 $$E_{r_0\_shell}=\frac{1}{2}E_{two\_charges\_r_0}$$

After having posited "point" particles for use in Coulomb's law and then using them to create other distributions of charge and even to calculate forces and potential energies of those distributions, you may be wondering what more we can say about these "point" particles. These particles are, after all, some kind of elementary particle, out of which others are made. What for example is their potential energy? What is its structure? Before taking the point particle limit, we obviously need to know the potential energy, also called self energy, of a finite radius particle. The simplest known charge that can be isolated is the electron; the electron seems to be as elementary as one can get. So, we will first consider a spherically symmetric model of the electron and then return to our primal question about the nature of an actual elementary particle, asking the central questions that the point particle construct brings to the fore.

Self Energy of a Classical Electron

If we model the electron as a uniformly charged sphere (which, again, is not a good model since we leave the very domain of classical E&M that such a description invokes), we can calculate its *self energy*, the energy the sphere would release if its charged parts were allowed to fly apart, in the following way. This potential energy stored in the structure is equivalent to the amount of energy it takes to put it together, i.e., to bring all the little bits of charge in from arbitrarily far away to their final position in the small ball.

Instead of bits, we take the electron to have layers like an onion. We take the sphere to have charge Q with a constant charge per unit volume given by ρ. We begin with a sphere of radius r_0, which we divide into equal shells of thickness dr so we can write:

2.13a,b $$Q=\frac{4\pi r_0^3}{3}\rho \text{ and } \int_0^{r_0}\lambda(r)dr=Q$$

Where $\lambda(r) = 4\pi r^2 \rho$ is the charge density per unit length in the radial direction.

We suppose that each of these shells, each of these layers of the onion, is brought in from infinity to build up the sphere. In doing this, because of the symmetry, it will also be convenient to use $\lambda(r)$ to write the total charge *of a given shell defined by its radius, r, when it is part of the completed sphere*; namely, the charge of that shell can be written: $\lambda(r)dr$. Now, because of the spherical symmetry, the work done in bringing any one of these (infinitesimally thin) shells in from infinity is exactly the same as the work done in bringing in a point charge of the same magnitude to the same radius. Hence, all that matters is the total charge on each shell, i.e. the product $\lambda(r)dr$. Again, note well that the function $\lambda(r)$ only gives the radial density of the shells when the sphere is assembled, and, generally, the thickness of the shells may change as one brings it in from infinity, converging to dr when it reaches its final place.

For concreteness, we can imagine that each shell maintains its density, ρ, as it contracts. In this case, to conserve charge, the thickness of the shell must increase as its radius contracts towards its final size in the sphere. (See end of chapter problem for case of constant shell thickness, in which the density must increase as it comes in from infinity.)

Now, we are ready to dive into the details of the calculation of the work. To reinforce how we move from finite thickness shells to infinitesimal ones, we proceed methodically, showing the steps in building up the sphere. First, using the equation for the self energy of a spherical shell, equation 2.11, we note that the work that must be done against the parts of the shell itself to bring the parts in from infinity is second order, since its charge is infinitesimal, say dQ. Hence, the only work that must be done is against the fields caused by the charge already laid down on the sphere. The first shell then, to first order, costs no energy to put down. The second shell, which has charge $\lambda(2\Delta r)\Delta r$, must be pushed in against the charge of the first shell, $\lambda(\Delta r)\Delta r$. The third shell, which has charge, $\lambda(3\Delta r)\Delta r$ must be pushed in against the charge of the first shell *and* the second shell, $(\lambda(\Delta r) + \lambda(2\Delta r))\Delta r$, thus the work required to put down the first three shells, to first order, is:

2.14
$$W_2 = k\frac{(\lambda(\Delta r)\Delta r)(\lambda(2\Delta r)\Delta r)}{\Delta r} + k\frac{((\lambda(\Delta r) + \lambda(2\Delta r))\Delta r)\lambda(3\Delta r)\Delta r}{2\Delta r}$$

This then is easily generalized to n-shells to get:

2.15
$$W_n = k\sum_{n=1}^{N}\frac{\left(\sum_{m=1}^{n}\lambda(m\Delta r)\Delta r\right)\lambda((n+1)\Delta r)\Delta r}{n\Delta r}$$

Note that the term in parenthesis is the charge contained within the radius r, which is the charge one must push against to bring the next shell (which has charge: $\lambda(r+dr)dr \sim \lambda(r)dr$) into its place at radius r. This is so because the charge in the shells outside the radius r causes *no* net electric field inside that radius.

Now, introducing the definition of Δr, r and r': $N\Delta r = r_0$, $n\Delta r = r$, $m\Delta r = r'$, keeping only first order terms, and taking the limits gives:

2.16
$$W = k \int_0^{r_0} \frac{\left(\int_0^r \lambda(r') dr' \right) \lambda(r) dr}{r} = k \int_0^{r_0} \frac{\left(\int_0^r 4\pi r'^2 \rho \, dr' \right) \lambda(r) dr}{r}$$

Now substituting $\lambda(r) = 4\pi r^2 \rho$ and using equation 2.13a to get a result in terms of Q, the charge on the electron, rather than ρ we get:

2.17
$$W = k \left[\int_0^{r_0} \left(\frac{\frac{4}{3} \pi \rho r^3 \left(4\pi r^2 \rho \right)}{r} \right) dr \right] \left(\frac{Q}{\frac{4}{3} \pi \rho r_0^3} \right)^2 = \frac{3Q^2}{r_0^6} k \int_0^{r_0} r^4 dr \text{, thus}$$

2.18
$$W = \frac{3k}{5} \frac{Q^2}{r_0}$$

If we now equate this "self energy" to the rest energy of the electron, which can be written as $\frac{m_e c^2}{e} = 5.11 \times 10^5 \, J/C$, with $Q = e = 1.6 \times 10^{-19} C$, recalling that in *SI* units $k \sim 9 \times 10^9$, we get the approximate classical radius of the electron as:

2.19
$$r_0 = \frac{3}{5} k \frac{e^2}{m_e c^2} \sim 10^{-15} m$$

Self Energy of a Point Particle

Clearly, in light of equation 2.18, we can now see that the completed point particle limit makes no sense for the energy goes to infinity in that limit. This is no surprise, for as we've seen, extension is one of the nine categories of properties of all physical things as we know them directly through the senses. Furthermore, we have only included electric forces in our analysis; all other forces (as well as much else) have been left out, "switched off." In fact, to consistently apply Coulomb's electrical law at all levels, we need a non-electrical force. That is, if Coulomb's law is valid even down to the parts of the simplest particle (say, for the sake of concrete argument it's the electron), there must be some other force to balance the electrical force so that the electron does not fly apart. Of course, such a balancing force is not absolutely necessary for we have not verified Coulomb's law for the parts of such an elementary particle. It may be possible that at that elementary level there are no inter-part forces; i.e. that may simply be the nature of those parts at that level.

Energy and Scalar Potential

You may have noticed that the field caused by electrical charges is a conservative force field and so can be represented as the gradient of a potential. This will have consequences, as we've seen for gravity, on how we represent potential energy. We will discuss this and other related issues after we develop some of the mathematics we need to make full use of Coulomb's law in the extended form that includes the superposition principle.

A Mathematical Aside:
Importing information into our formalism from the first property of all physical things

Purely Mathematical Version of Gauss's Law

Ultimately, we want to incorporate into our formalism the fact that the inverse square fall-off of the electric field is directly related to an essential aspect of Euclidean (generic) extension, i.e. the fact that the surface area, S, increases according to the square of the distance, r, from any point in three dimensional space, i.e., $S \propto r^2$. You know this latter from the formula for the surface area of a sphere: $S = 4\pi r^2$. The resulting incorporation leads to an extremely important law, Gauss's law. As preparation for this, in this "a mathematical aside", we will use our ordinary experience with fluids to more deeply digest the elements of extension that are at play and derive, among other things, the fluids analog of Gauss's law.

To investigate this aspect of quantity (extension), we consider again an incompressible fluid for which we will take water as our iconic case. Such a fluid is helpful because it is an easy way to visualize all the different ways a fixed volume can fill space (three dimensional extension). For example, 1 m^3 of water can fill a cube with 1 m edges, or it may fill a sphere of radius $\sqrt[3]{\dfrac{3}{4\pi}}$ meters. Because water can flow around as well as deform, one can easily visualize the equivalence of these two volumes in the following way. Imagine a pitcher of water filling the cube to capacity, then take the cube and use it to fill the sphere. We cannot of course do this experiment in the kitchen, but the point is we can see in principle that the water of the same volume will be able to just fill both containers, and we can thus, for example, analyze these facts about extension with the vector/calculus formalism we have already established.

Water flow, our iconic fluid flow, can be represented by the so-called "current density", $\vec{J} = \rho \vec{v}$. The current density, $\vec{J}$, is the measure of the rate of flow of water across a surface per unit time per unit area, where ρ is usually taken to be the mass density in (cgs) units of $\dfrac{g}{cm^3}$, i.e. mass/(unit volume),[18] thus giving current density the units of $\dfrac{g}{cm^2 \, s}$. However, our interest in this section is in mathematics, *not in qualitative properties* such as mass, so we will speak here about the volume, rather than the mass, of water; hence, the current density will be in units of $\dfrac{cm^3 of \ water}{cm^2 \, s}$, where the cm^2 on the bottom refers to a unit of area of a surface in the environment thru which the water travels,[19] and ρ will have units of $\dfrac{cm^3 of \ water}{cm^3 of \ environment}$. Because it would get confusing to try to distinguish

---

[18] Later in this course it will be used for charge density.

[19] Note that the speed is in *cm/s* through the environment, so that $\vec{J} = \rho\vec{v}$ has units of $\dfrac{cm^3 of \ water}{cm^2 \, s}$.

in our notation between the volume of the water and the volume of the space in which the water is moving (i.e., the volume of the environment), we will sometimes simply use grams as a place holder for the volume of the water. Remember water is an incompressible fluid and in fact: $1g$ of water[20] is about $1\ cm^3$ of volume. Note also that our uses of moving volumes and changes of shape are simply methods to visualize the various ways space can be divided and put together. So, what we are doing is, indeed, fundamentally an investigation of extension (quantity), though it also, thus, has obvious analogical uses, for example, for expressing mass or charge conservation.

Now, we would like to understand how volumes are preserved as they move across surfaces. The figure below begins this analysis.

It shows water going downward at a uniform speed[21] into a screen tilted at an angle from the horizontal. Note that since the screen is at an angle, some component of the impetus of the water (represented by a green arrow) is moving the water *along* the area of the screen. That component of the water's motion thus cannot contribute to the flow across the area of the screen. In other words, only a fraction of the velocity is effective in bringing water across the screen. Hence, the current density of the water, J, say in grams of water *per unit area* crossing the screen in a second, is less when the screen is tilted. This component goes as the cosine of the tilt angle, θ. Thus, the current density of the water per unit area thru a given portion of the screen is reduced by a factor of $\cos\theta$: $J_{screen} = J\cos\theta$. So, for example, when $\theta = 0$, there is no reduction, and the impetus of water is such that no component is wasted, i.e. all is used to carry water into the screen. By contrast, when θ is close to $90°$, i.e. when the screen is aligned nearly perpendicular to the flow direction, the current density is very *small*, because most of the velocity takes the water along the screen rather than through. Note, by contrast, assuming we have a large enough screen, the area of the screen that the water crosses is now very *large*. Indeed, until we get to the extreme case of $\theta = 90°$, the total grams (or the volume of water in cm^3) per second going through the screen is the same as the flow rate (flux) out of the faucet, since though the current density is decreased by $\cos\theta$, the area of the screen that water crosses is increased by $1/\cos\theta$. The current density is less at anyone place on the screen, but the amount of area is proportionally larger so the total flux across the surface is the same. *Whatever volume of water comes out of the faucet must cross any surface that intercepts the whole column of water* since the total volume of the water is preserved, despite whatever different shape the screen might define for the water.

---

[20] This is approximately true for liquid water at 4°C and atmospheric pressure.
[21] Note that here, for simplicity, we assume an approximately uniform speed of flow, which would not be the case for water under the pull of gravity.

Now, consider what happens when we let the water expand out from some central location, say, for simplicity, from a sphere. Suppose that the water is surrounded by an arbitrarily shaped surface, S, that encloses it so it must go through at least some part of the surface to get to the other side. Clearly then, whatever water does not cross the surface must remain inside. We now proceed to understand better how this happens using what we discovered above and the inverse square law. This process will also aid us in writing this conservation law down formally.

We use $\vec{J}$ for the flow of water (in terms of volume or mass) per unit surface area per unit time, and $\Delta\vec{A} = \hat{n}\Delta A$ for the small patch of area of the S with $\hat{n}$ as the direction of the normal to the patch of area pointing outward (see Figure 2-8). We then can write the amount of water that crosses that small portion of the surface's boundary (which to first order is flat) per unit time as $\vec{J} \cdot \Delta\vec{A}$. Then, using what we learned above, we can see that the local shape of the surface will not matter. In particular, if we imagine locally deforming the surface so that the surface is locally tilted relative it its original state (see Figure 2-7), we will note that, for the deformed surface to remain closed, the area must stretch according to the cosine law discussed above so that the water crossing the original ΔA_1 also crosses the ΔA_2 of the deformed surface giving: $\vec{J} \cdot \Delta\vec{A}_1 = \vec{J} \cdot \Delta\vec{A}_2$. Note that here we have assumed that the scale is chosen small enough so that we can neglect any spreading of the water by taking the surface deformation to remain arbitrarily close to the original surface.

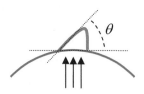

Figure 2-7: The green surface can be deformed *locally*, giving it a "wart" shown in red. Clearly the flux through the green differential area dA_1 must equal the flux through the red differential area dA_2 and, as discussed relative to the water faucet figure above, $\vec{J} \cdot d\vec{A}_1 = \vec{J} \cdot d\vec{A}_2$.

This explains how the total flux remains the same under changes of shape (i.e. a local tilt) of the surface, but what if the surface moves further away? To answer this, consider again our small patch of square area on a given surface, say a sphere. Keeping in mind that we already have dealt with the effect of the surface's orientation, we consider only the effect of pulling this patch of surface away from our baseline surface, the perfect sphere. The simplest possibility that comes to mind is when the water has only impetus perpendicular to the patch as shown in the figure to the right. In that case, clearly no change results if we distort the surface by pulling this patch away from its initial location, say by stretching the parts of the surface near it, without changing the patch's size (or orientation). By contrast, if one considers a source of water that emits uniformly in all directions as shown in the next figure, then the water will expand within some angle[22] through the patch. As discussed earlier,

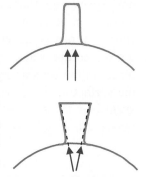

---

[22] In fact, there will be an angle in the direction perpendicular to the one shown as well; these two angles define a two dimensional analog of an angle, called a solid angle, through which the water will flow.

this means, as illustrated in Figure 2-3, the rate of flow per unit area will decrease as the inverse square of the distance from the dispersion point. However, as also discussed earlier and as illustrated in Figure 2-2, the patch of the deformed surface must be larger than the original patch by the square of that distance. Hence, these two effects cancel, so that, as required, the water that crossed the small patch on the first surface also crosses a corresponding patch on the deformed surface. Again, this conservation of water volume (or mass) obtains because of the inverse square fall off of the $\vec{J}$ of the water. Notice that if we abstract from moving volumes by ceasing to consider an incompressible fluid and consider an arbitrary vector field $\vec{J}$, then we can imagine a $\vec{J}$ that does not fall off as $1/r^2$ so that conservation of whatever flux is represented by $\vec{J}$ would no longer hold.

Further argument is needed for the general case in which the source does not emit water uniformly. In such a case, the flow rate will decrease inversely proportionally to the product of the linear dimensions perpendicular to the flow. Why so? To understand, we need to recall that the volume of a right cylinder is equal to the area of its base times its height. We then consider a small part of water that is approximately taken to be cylindrically shaped. Now, no water volume is lost or gained, so if the area of the surface of the outgoing part of the water increases by a certain factor then the thickness of the part must decrease by the same factor to preserve the volume. *As in the case of uniform expansion, since the shell gets thinner, the flow rate per unit area decreases, but since the area is bigger the net flow rate is the same.* Using the following diagram, we can show this effect for a small area by recalling the area of an ellipse is $A = \pi ab$, where a and b are respectively half the long and short dimensions of the ellipse.

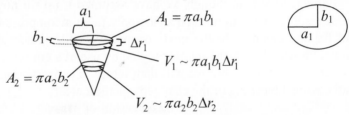

Now, starting with our sphere, we can create any surface we like by locally deforming and/or radially moving each small patch that makes up that surface. Thus, because of the nature of space (extension) revealed through the use of the incompressible fluid water, independent of the shape and size of the surface, if we add all such differential areas over any given surface S, which we write as $\int_S \vec{J} \cdot d\vec{A}$, we know that the volume of water coming out must equal the volume which is being pumped across it from the interior. Of course, we already knew this generally but we did not know the local details of extension which that reality demands. Our analysis in this section has uncovered those realities, and the formalism we have introduced has encapsulated them so as to facilitate calculation and serve as mnemonics for them. Mathematically, i.e. in our formalism, we write "the current" across the surface as:

2.20 $$\int_S \vec{J} \cdot d\vec{A} = \frac{d(\text{volume (e.g., of water)})}{dt}$$

This, of course, implies a source of water somewhere inside the closed surface S. Typically, this source is pushed back to an arbitrarily small region in the center; this, of course, means we have to start with some highly compressed matter and make water from it.

We can also get rid of the time dependence (by, for example, thinking of what happens in one second) and write, in a suggestive notation, $\vec{J}dt = \vec{e}$. Thus, we get the time integrated form of equation 2.20:

2.21
$$\int_S \vec{e} \cdot d\vec{A} = \text{volume (e.g. of water)}$$

> Here, $\vec{e}$ specifies how much volume per unit area and in which direction it is pushed at each point on the surface, while the dot product projects out only the portion that crosses the surface and the integral adds all the contributions.

This is the mathematical version of Gauss's law, but is *not* the standard version. In the section titled "The Fully Extended Coulomb's Law", we will give the electrical analog which is what most people refer to as "Gauss's law." As we have said, this mathematical equation helps to formally convey the fact that the same volume, i.e. three dimensional extension, can be limited or contained by an uncountable number of different shapes. We discover this by having the parts of that volume move, as specified by $\vec{e}$, in different ways across a series of surfaces till it is reshaped and in a different place. For example, we considered above a series of surfaces one larger than the next, starting with one that enclosed the only sources in the system. In terms of our incompressible fluid, the $\vec{e}$ field specifies a reshaping, for example, of some oddly-shaped shell of the fluid as it moves outward from some source. Note that equation 2.21 remains valid even if S is not a closed surface.

Further, note that, though we have assumed it up till now, the $\vec{e}$ and $\vec{J}$ fields do *not*, of themselves, have to be such as to apply to an incompressible fluid for equation 2.21 to hold for any given S. In the next section, we will connect the right hand side of the equation with what leaves the interior region by use of a conservation law. We will initially use conservation of mass, but we will then return to conservation of volume, so as to make the mathematical meaning of the changed equation clear.

Conservation of Mass

At this point, advanced students may have recognized equation 2.20 as closely related to the equation of conservation of mass or charge: $\int_S \rho\vec{v} \cdot d\vec{A} = \dfrac{d(\text{volume})}{dt}$. Indeed, if we use mass (a quality), rather than volume, thereby leaving pure quantity, we can write, taking S to be *only* those surfaces that enclose a region of space:

2.22
$$\oiint_S \vec{J} \cdot d\vec{A} = -\frac{dm_{inside}}{dt},$$

Here we have introduced a minus sign, because we have switched from the change in the quantity (i.e., the measure of the mass) *outside* the closed surface to the change of the quantity *inside*. In particular, conservation of mass means that whatever mass increase occurs outside must correspond to mass decrease inside, and, formally this corresponds to a change of sign. Equation 2.22 can also be written:

2.23
$$\oiint_S \vec{J} \cdot d\vec{A} = -\frac{d}{dt}\left(\int_V \rho dV\right)$$

Note also that this remains valid if we allow the density to change with time and position, so that the mass of the fluid (no longer water which is largely incompressible) can be more or less compactly packed. If we consider a small volume in space, we can then write:

2.24
$$\oiint_{S_{\Delta V}} \vec{J} \cdot d\vec{A} = -\frac{d\rho}{dt}\Delta V$$

Where: $S_{\Delta V}$ is again the surface area of the volume ΔV

We can state the meaning of this relation in the most direct terms by coming back again to incompressible water to allow us to discuss what happens as we move parts of a volume (3-dimension extension) from place to place in space (the environment), i.e. by returning to an analysis within the category of quantity. In particular, we can think of the J-field as describing water flow only and thus as only existing where there is water. We can think of the J-field in a given small volume as describing how one moves the parts of the water in that volume. The magnitude and direction of each of the vectors tells how much (per unit area) and in which direction to move that part of the volume of water to which it refers. If, for example, two parts that are next to each other along a horizontal line both are assigned vectors that point up, but the first has twice the magnitude of the second, this means the first is to be moved twice as far as the second. Sometimes the J-field in a region will demand the parts move such that the water only undergo changes of shape and/or place, while other times one must create (or destroy) water (volume), indicating some source (or sink) of water in a region over which it is passing. In the former case, there is no input of water (volume) needed so the right hand side of equation 2.24 is simply zero. However, in the case in which there is creation (or destruction) of water, we need to define a "ρ," the water volume density, which is given in (cm^3 of water)/(cm^3 of environment). Normally, ρ does not change much. However, if, for example, we were able to destroy water in a given place, then we would leave a hole where the water used to be, a spot in the plana where there would be no water unless other water rushed in to fill the spot. This is an effective change (decrease) in the water volume density, ρ. Then, since ΔV tells how many cm^3 of environment is under consideration, the right hand side of equation 2.24 is $-\frac{d\rho}{dt}\Delta V = -\frac{d(cm^3 \text{of water})}{dt}$. Integrating over time then gives: $-\Delta(cm^3 \text{of water})$, which tells how much water must be created (or destroyed) in the given small region for our description of the meaning of the J-field to make sense.

Thus, again, we see that the flux meaning of J gives equation 2.24 clear mathematical (category of quantity) meaning, but we also see that meaning can be easily extended by analogy to intensities of qualities such as mass.

Gauss's Theorem (The Divergence Theorem)
Basic Notion

The next piece of math results from dividing the volume, V, enclosed in the surface, S, over which one is doing the surface integral $\int_S \vec{J} \cdot d\vec{A}$, into smaller volumes. We are basically looking for a way to express the essential quantitative facts about flux in the limit of arbitrarily small size (this will then also lead to an arbitrarily small size version of the flux theorem of equation 2.20). Through such a mental construct, we can encapsulate the essential aspect of flux, without explicitly introducing a surface.

To begin, consider an arbitrarily shaped volume such as shown in Figure 2-8. If we divide it in two pieces as shown in Figure 2-8a, the surface integral over the top part will include the flux out of the dividing surface S_D, while the surface integral over the bottom part will have a similar term. However, since outward across the bottom is "up", while outward of the top is "down", these two terms will cancel. Since the remaining two terms are integrals that together cover all S, we have $\int_S \vec{J} \cdot d\vec{A} = \sum_{i=1}^{i=2} \int_{S_i} \vec{J} \cdot d\vec{A}$, where S_1 and S_2 are, respectively, the *outer* bounding surfaces of the top and bottom divisions of the volume. The same reasoning applies for all such divisions, so that we can write:

2.25
$$\int_S \vec{J} \cdot d\vec{A} = \sum_{i=1}^{i=N} \int_{S_i} \vec{J} \cdot d\vec{A}$$

Where, N is the number of divisions and S_i is the *outer* surface of ith division of the volume.

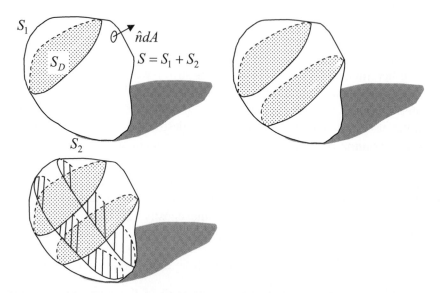

Figure 2-8: a. Top left shows an oddly shaped volume divided into two parts with a typical differential area patch, dA with its defining normal, $\hat{n}$ pointing outward. **b.** Top right and bottom left show further divisions of the same volume. Note that, in all three figures, what flows through each boundary that divides the volume flows out of one region of the volume into another. Hence, each term in the surface integral performed over *any* given region cancels with one in an adjacent region, *except on the boundary of the entire volume*, where there is no adjacent region with which to cancel.

Note that if we divide the volume, V, into N roughly equal parts, as we do in preparation for the limits needed for integration, each division of the volume is approximately $\Delta V = V / N$. Now, if we want to use one of the terms in the integral in equation 2.25, i.e., $\int_{S_i} \vec{J} \cdot d\vec{A}$, in our small limit flux statement, we are in trouble since each will go to zero as $N \rightarrow \infty$, i.e., as $\Delta V \rightarrow 0$. Note, however, when we cut any given division in half, say in a process of taking N to $2N$, the surface integral of each subdivision is about half the original, so there seems to be some proportion to volume. Hence, to avoid going to zero, our infinitesimal flux statement needs to involve: $\lim\limits_{\Delta V \to 0} \dfrac{\int_{S_{\Delta V}} \vec{J} \cdot d\vec{A}}{\Delta V}$, where,

again, $S_{\Delta V}$ is the surface area of the volume ΔV . This does indeed converge in the limit. We call this quantity the divergence of the current density $\vec{J}$ and formally write:

$$2.26 \qquad \operatorname{div} \vec{J} \equiv \lim_{\Delta V \to 0} \frac{\int_{S_{\Delta V}} \vec{J} \cdot d\vec{A}}{\Delta V}$$

This is our sought after differential flux statement, though it's not much use until we have a coordinate form for calculating $\operatorname{div} \vec{J}$. After we develop one such form in the next sub-subsection, we will illustrate the divergence by using it on a particular functional form of $\vec{J}$.

For now, note that the divergence tells us about the current density vector field, $\vec{J}$. It tells us at a given point in space, what amount of mass per unit time per unit volume moves out of (or into) a given small volume of space. Again, the divergence is determined by what the $\vec{J}$ field does in the region. If, for example, the field is pointing towards a given location from all directions, then, clearly, we will have influx into the volume. And, the greater the magnitude of those vectors, the greater the influx per volume will be.

If we just want to consider extension alone, we consider volume flow and let the field represent how much volume, say of water, and in what direction we push it. We multiply the equation 2.26 by dt to get: $\operatorname{div} \vec{e} = \lim\limits_{\Delta V \to 0} \dfrac{\int_{S_{\Delta V}} \vec{e} \cdot d\vec{A}}{\Delta V}$. Loosely, think of this as revealing what volume of water, for the given $\vec{J}$ field, comes into (or out of) the volume around the point in one second. More precisely, the numerator tells how effectively the given field is in moving the water in (out) at the given location. This effectiveness has to do with how the field is directed from all points around the location and how much volume it's trying to push relative to the other points; we will see this more clearly shortly when we analyze the divergence in Cartesian coordinates. The denominator simply scales this effectiveness to some reference volume. This excludes the non-essential information that, once we've attained a small enough scale, the more volume we consider the proportionally more surface area, and thus flux, there will be. Thus, using equation 2.25, we can write:[23]

$$2.27 \qquad \int_S \vec{J} \cdot d\vec{A} = \sum_{i=1}^{i=N} \int_{S_i} \vec{J} \cdot d\vec{A} = \lim_{\Delta V \to 0} \sum_{i=1}^{i=N} \frac{\int_{S_i} \vec{J} \cdot d\vec{A}}{\Delta V} \Delta V = \int_V \operatorname{div} J \, dV$$

$$2.28 \qquad \boxed{\textbf{Gauss's Theorem:} \qquad \oiint_S \vec{J} \cdot d\vec{A} = \int_V \operatorname{div} \vec{J} \, dV}$$

This is the celebrated *Gauss's Theorem*. Note that it is *different* from Gauss's *law*.

---

[23] One can start with Cartesian coordinates and show that the limit in equation 2.27 is independent of the shape of the surface.

Also, by noting that the integral on the left is an integral over one dimension less than the one on the right, we can see by analogy to the fundamental theorem of calculus ($\int \frac{df(x)}{dx} dx = f(x)$) that it makes sense to write the divergence as $\nabla \cdot \vec{J}$. We will see the full meaning of such notation in this next sub-subsection.

Cartesian Coordinates

      To understand in terms of the basic coordinate system that most manifestly brings out the three dimensional nature of space,[24] consider a rectangular box such as shown in Figure 2-9. Suppose fluid is coming in from left at a certain rate per unit area per unit time, i.e. with a certain $\vec{J}$. Look, not at volume flow first, but mass flow. We then see that if mass cannot disappear (and classically it is conserved) and for the moment we ignore the perpendicular faces, whatever goes into the box must either stay in the box or must come out the other face. (Of course, if the fluid stays in the box it will compress thus changing the density ρ at that "point."). As seen in equation 2.24, this is the same as the mass flux into the volume, and thus by equation 2.28, this is another approach to the divergence. By using Figure 2-9 and equation 2.24, we can write the formal mathematical presentation of these fact, as follows.

$$2.29 \quad -\dot{\rho}\, \Delta x \Delta y \Delta z = J_y\left(x+\frac{\Delta x}{2}, y+\Delta y, z+\frac{\Delta z}{2}\right)\Delta x \Delta z - J_y\left(x+\frac{\Delta x}{2}, y, z+\frac{\Delta z}{2}\right)\Delta x \Delta z$$

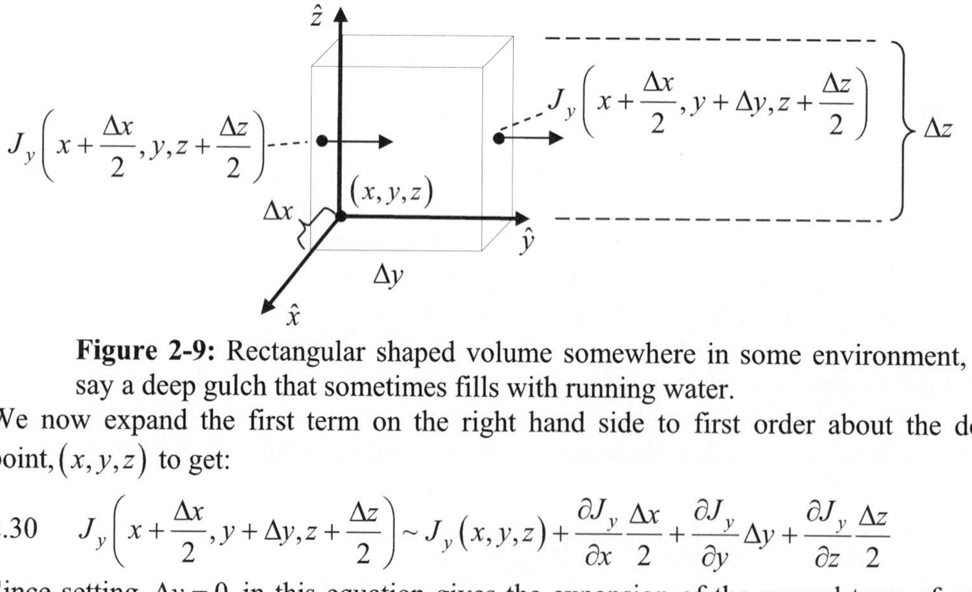

Figure 2-9: Rectangular shaped volume somewhere in some environment, say a deep gulch that sometimes fills with running water.

We now expand the first term on the right hand side to first order about the defining point, (x,y,z) to get:

$$2.30 \quad J_y\left(x+\frac{\Delta x}{2}, y+\Delta y, z+\frac{\Delta z}{2}\right) \sim J_y(x,y,z) + \frac{\partial J_y}{\partial x}\frac{\Delta x}{2} + \frac{\partial J_y}{\partial y}\Delta y + \frac{\partial J_y}{\partial z}\frac{\Delta z}{2}$$

Since setting $\Delta y = 0$ in this equation gives the expansion of the second term of equation 2.29, revealing that the Δy term is the only difference between two expansions, equation 2.29 can be written:

$$2.31 \quad\quad\quad\quad\quad\quad -\dot{\rho}\, \Delta x \Delta y \Delta z = \frac{\partial J_y}{\partial y}\Delta y \Delta x \Delta z$$

---

[24] Remember we abstract Euclidean space from our immediate experience.

Now, we must include the other pairs of faces, i.e. one pair with their normals in the x-direction and the other in the z-direction. Since these pairs are replica's, in the relevant sense, of the other two dimensions, the above analysis gives terms just like the one above, so that we can write:

2.32
$$-\dot{\rho}\,\Delta x \Delta y \Delta z = \left(\frac{\partial J_x}{\partial x} + \frac{\partial J_y}{\partial y} + \frac{\partial J_z}{\partial z}\right)\Delta x \Delta y \Delta z$$

Note that this is our earlier sought after differential form of flux in a more complete form.

Now, using this last equation and equation 2.24 , we can write the divergence as:

2.33
$$\text{div}\vec{J} = \lim_{\Delta V \to 0}\frac{\int_{\Delta V}\vec{J}\cdot d\vec{A}}{\Delta V} = \left(\frac{\partial J_x}{\partial x} + \frac{\partial J_y}{\partial y} + \frac{\partial J_z}{\partial z}\right) = \nabla\cdot\vec{J}$$

Or simply: $\boxed{\text{div}\vec{J} = \nabla\cdot\vec{J}}$ where the definition of the del-operator in

Cartesian coordinates is: $\boxed{\vec{\nabla} \equiv \hat{x}\dfrac{\partial}{\partial x} + \hat{y}\dfrac{\partial}{\partial y} + \hat{z}\dfrac{\partial}{\partial z}}$.

The Conservation Law

Notice also that using, equations 2.32 and 2.33, we have the conservation of mass (or charge as) equation:

2.34
$$\boxed{\textbf{Conservation Equation:}\quad \nabla\cdot\vec{J} = -\dot{\rho}}$$

In terms of volume rather than mass, i.e. mathematically in the fundamental sense, we, by integrating out the time dependence as earlier and changing the sign on the right hand side by considering what leaves rather than what changes inside, we can write $\nabla\cdot\vec{e} = dV$. This equation says that $\vec{e}$ is such that volume dV comes out of the given small region.

We can also write equation 2.34 in integral form as:

2.35
$$-\int_V \dot{\rho}\,dV = \int_V \nabla\cdot\vec{J}\,dV$$

Curl and Stokes' Theorem

The last piece of math we need is the curl of a vector field. We will discuss it briefly here, leaving a more in depth discussion for later chapters when physical reality will call for its use. The curl, written as $\nabla\times\vec{J}$, is analogous to the divergence of a vector field, $\nabla\cdot\vec{J}$; instead of the surface integral of the field over a small volume per unit volume, the curl is the line integral around a small area per unit area. Formally, we write this as:

2.36
$$\text{curl}\vec{J} \equiv \lim_{\Delta A \to 0}\frac{\int_{C_{\Delta A}}\vec{J}\cdot d\vec{l}}{\Delta A}\hat{n} = \vec{\nabla}\times\vec{J}$$

Where, $\hat{n}$ is the normal to the plane of the curve $C_{\Delta A}$ that is obtained by the right hand rule applied to the direction of the line integral.

The form of $\nabla \times \vec{J}$ in Cartesian coordinates was already proven in Chapter 5 of *PFR-M* in the section titled "Mathematics of Conservative Forces;" for convenience the relevant section is reproduced in Appendix III. In that section, we also proved the following theorem, called Stokes theorem:

2.37
$$\oint_{Path} \vec{J} \cdot d\vec{x} = \iint_{Surface\,defined\,by\,Path} \left(\vec{\nabla} \times \vec{J} \right) \cdot \hat{n}\, dA$$

Where

$$\vec{\nabla} \times \vec{J} = \left(\hat{x}\frac{\partial}{\partial x} + \hat{y}\frac{\partial}{\partial y} + \hat{z}\frac{\partial}{\partial y} \right) \times \left(J_x\hat{x} + J_y\hat{y} + J_z\hat{z} \right)$$

$$= \left(\frac{\partial}{\partial y}J_z - \frac{\partial}{\partial z}J_y \right)\hat{x} + \left(\frac{\partial}{\partial z}J_x - \frac{\partial}{\partial x}J_z \right)\hat{y} + \left(\frac{\partial}{\partial x}J_y - \frac{\partial}{\partial y}J_x \right)\hat{z}$$

The surface integral on the right is done over any area bounded by the closed path specified on the left. The circulation is taken to be positive if the integral is positive when carried out in the direction around the path defined by the right hand rule applied to the direction of the normal to the surface, $\hat{n}$.

From the definition given in equation 2.36, it is clear that the curl tells the amount of circulation of the field *near* a given point. To get the circulation far from a point, we need the curl of all the points out to the boundary defined by the line integral that defines the circulation. With understanding of curl at a point, it is easy to draw flow pictures, such as shown in Figure 2-10, that have varying curl and divergence. It is helpful to have a strong visualization of these properties of fields because these mathematical concepts (i.e. concepts having to do with extension) will be used many times and in various analogous ways throughout your study of physics. Study the pictures and try to discover what curl and divergence they have, then try to think of some of your own.

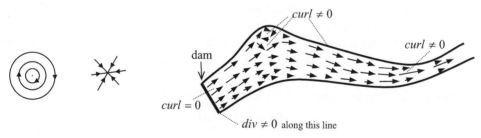

Figure 2-10: a. (*left*) Curl at center of field, no divergence anywhere. **b.** (*middle*) Divergence at center of field no curl anywhere. **c.** (*right*) Divergence and curls for a portion of a 2-dimensional river that begins at a dam where water pours in from a third dimension; the arrows indicate the magnitude and direction of flow at various points.

It is also helpful to note that the notation "$\nabla \times \vec{F}$" borrows from the cross product notation, which we introduced and explained in Chapter 2 of *PFR-M*. As we saw there, that notation involves two vectors, and was incorporated to analogically extend ordinary multiplication to include aspects of two dimensions of space. In particular, the cross product incorporates the "common" or perpendicular region between the two vectors and specifies the orientation of that two dimensional plane in the full 3-space. This new notation involves an operator and a vector field. In Cartesian coordinates, the del operator (∇) is like a 3 dimensional vector, $\nabla = \hat{x}\dfrac{\partial}{\partial x} + \hat{y}\dfrac{\partial}{\partial y} + \hat{z}\dfrac{\partial}{\partial z}$, in that it defines a derivative to be taken along each of the *three directions*. Unlike a vector, to have a magnitude and to even make sense, the del operator requires something on which to operate, namely a field. In the case of the curl, it operates on a vector field and reveals how that field changes perpendicular to itself (cf. Chapter 5)[25], i.e. how it "curls." For example, for $\vec{F} = F_z\hat{z}$, $\nabla \times \left(F_z\hat{z} \right)$ tells nothing about $\dfrac{\partial F_z}{\partial z}\hat{z}$, but instead tells $\dfrac{\partial F_z}{\partial y}\hat{x} - \dfrac{\partial F_z}{\partial x}\hat{y}$.

Generalized Stokes Theorem: Grad, Curl and Div

Thus, we can summarize the results of the curl and the divergence in a so called generalized Stokes theorem $\int_{\Omega} d\omega = \int_{boundary\,of\,\Omega} \omega$, where Ω is an n-dimensional volume.

Generalized Stokes Theorem: $\int_{\Omega} d\omega = \int_{boundary\,of\,\Omega} \omega$		
Dim	**Name**	**Equation**
1	**Fundamental Theorem of Calculus**[26]	$\int_{C=[a,b]} \vec{\nabla}f \cdot d\vec{x} = f(\vec{b}) - f(\vec{a})$
2	**Stokes' Theorem**[27]	$\iint_S \left(\vec{\nabla} \times \vec{J} \right) \cdot \hat{n}\, dA = \oint_C \vec{J} \cdot d\vec{x}$
3	**Gauss's Theorem**	$\iiint_V \vec{\nabla} \cdot \vec{J}\, dV = \oiint_S \vec{J} \cdot \hat{n}\, dA$

Note that in the fundamental theorem of calculus, the right hand side is indeed the boundary of the line, C, since the boundary of a curve (one dimensional) is simply its end points.

---

[25] Among other things, we will see in Chapter 5 that because a change always involves two points and because there are three dimensions, we must look at the field at nearby points *along each of the three dimensions* (or in two dimensions along each of the two) to determine if there is a change in the field direction perpendicular to itself. This is important because, in terms of rectangular coordinates, some component of the field may be zero at one point but not at a nearby point along one of the three dimensions. Thus, if we only considered the first point where the particular component of the field is zero, and *not* the second one where it is not zero, then we would clearly, erroneously, decide that directions perpendicular to that component are not perpendicular to the field.

[26] For 1-dimensional case, we have: $\int_{C=[a,b]} \dfrac{df(x)}{dx} dx = f(b) - f(a)$.

[27] Note that Green's theorem is a two dimensional special case of the standard Stokes' theorem.

The Fully Extended Coulomb's Law: Gauss's Law

With this information, we can now return to our study of the electric field. Remember that a small charged particle causes an electric field in the plana surrounding it in such a way that the strength of that field decreases as the inverse square of the distance from the charge.

Each such charge pumps a certain strength E-field that, in *cgs*, is measured in *esu* of charge per cm^2 into the plana, and the charge must maintain the "static field" in the plana at all times, otherwise it will cease. In this limited sense, we *can* say the field is "propagating" all the time,[28] though, following standard practice, we will usually reserve the word "propagation" for the transmission of *changing* fields. So, *we can think of the charge as a kind of source of electric field strength that is analogous to water flow.* Just as the water flux in volume/cm^2 (cf. discussion before equation 2.20) moves out uniformly in all directions decreasing as the inverse square of the distance as shown in Figure 2-3, so does the strength of the electric field in esu / cm^2. Thus, by the similar arguments used in deriving equation 2.21, we can write the analogous equation for the electric field:

2.38 $$\int_S \vec{E} \cdot d\vec{A} = \text{charge}$$

Now, we need to connect this definition of charge with the definition in Coulomb's law: $\vec{E} = k\dfrac{q}{r^2}\hat{r}$, where we have not specialized to *cgs* units ($k = 1$) yet. Since the electric field of a fixed charge is spherically symmetric (note that we will see later that this is not the case for a moving charge, even a uniformly moving one), the left hand side of equation 2.38 integrates out to: $4\pi k\,q$. Hence, we get **Maxwell's "first" equation** *in integral form*:

2.39

> **Gauss's Law** $\qquad \oiint_S \vec{E} \cdot d\vec{A} = 4\pi k q$
>
> With $k = 1$ in *cgs* and $k = \dfrac{1}{4\pi\varepsilon_0} \sim 9 \times 10^9 \dfrac{Nm^2}{C^2}$, in *mks*

This extension[29] of Coulomb's law is very important and is called Gauss's law. It is an extension because it is more inclusive than Coulomb's law. If Coulomb's law is true Gauss's law must be true by the reasoning just given above; however, Gauss's law does *not* imply Coulomb's law. There are fields that satisfy Gauss's but not Coulomb's law. Consider, for example, the non-uniform field around one of two closely separated charges. Coulomb's law along with superposition can deal with it but not alone. Gauss's law is an extension beyond even the inclusion of superposition, because we will see it prove true even when we take account of another electric-like effect in Chapter 3. To get Coulomb's law from Gauss's law, we will need to also specify $\nabla \times \vec{E} = 0$, which we will discuss later.

---

[28] The gravitation field gives a helpful macroscopic analogy for this field-maintenance type of propagation. Namely, if the sun were to suddenly disappear, it would be 8 minutes before the earth left its orbit around the sun, for the field acting on the earth at the time the sun vanished was sent out 8 minutes beforehand--the 8 minutes it takes to get to the earth.

[29] Note this might be called a generalized versions (as might the intermediate extension discussed in the section before this), but this is only with regard to "extension," i.e., the range of things to which it is applicable, not its "intension," i.e., not with respect to the information it contains. In that more fundamental sense, it is a further specification, not a generalization.

We've seen in our analysis of water that Gauss's law preserves the net flux, and so, by analogy, we could say that as the static electric field thins itself out in a given direction it is less able to propagate itself in that direction. Whereas, if it is somehow directed more strongly in one direction, it can more effectively propagate in that direction. In effect, as we will see in another chapter, this is what happens for a uniformly moving charge; because of propagation effects and a deeply related distinct effect to be discussed later in the context of magnetism, the effective E-field is reinforced along the direction of propagation and is strengthened perpendicular to it. It turns out that even for fields of non-uniformly moving charges and in all free field situations of which we know (at our classical level of abstraction) Gauss's law is obeyed, though Coulomb's law is radically incorrect for such cases.[30] Gauss's law tells us something about the way an active charge causes E-fields (as well as the distinct effect mentioned above which is related to an effect of impetus on charge) and how those fields propagate in the plana.

We should not leave this analysis without recalling, as illustrated in Figure 2-4, that the density of the field lines can be used as indicators of field strength. It is a handy way of representing the inverse square loss of field strength with distance because it so closely links it with the nature of three dimensional extension from which we have argued that the strength fall-off fairly directly derives. Because of this close connection which is *not* limited to uniform "flow," field lines maintain their value for conveying Gauss's laws embodiment of the deeper connection between charge and E-field, including that above mentioned E-like effect related to impetus activated charge and magnetism.

Gauss's law can also be formulated locally in the following way. Applying Gauss's theorem (the divergence theorem), equation 2.28, i.e., $\int_S \vec{J} \cdot d\vec{A} = \int_V \nabla \cdot \vec{J}\, dV$, to equation 2.39 gives:

2.40
$$\int_V \nabla \cdot \vec{E}\, dV = 4\pi k\, q = \int_{V \text{ of charge}} 4\pi k \rho\, dV$$

Where we have also introduced a charged distribution with density, ρ, over some confined volume.

Now, if we equate the volume on the left with that on the right and then make the volume very small, we see the integrands must be equal, giving a local form of Gauss's law, i.e. Gauss's law in differential form, *Maxwell's "first" equation in differential form*:

2.41

> ## Gauss's Law
> ### (Differential Form)
> $$\nabla \cdot \vec{E} = 4\pi k\rho$$

Comparing this with the mass conservation law, equation 2.34, i.e., $\nabla \cdot \vec{J} = -\dot{\rho}$, we see that our fluid analogy can be extended to mass flow, if we note that $\dot{\rho}$ is a kind of source for the mass density flux of water, $\vec{J}$, since the mass leaving the surface is provided by the mass inside of it. Similarly, the charge is a source of electric field as it continual "pumps," i.e., sustains, the electric field across the surface. In the case of mass flow, there is a mass transfer and thus a loss of mass from the region in the process, so we have $-\dot{\rho}$ as

---

[30] Coulomb's inverse square law only applies to static cases (except for the anomalous "semistatic case" when the charge density increases linearly in time).

the source. However, in the case of the charge, there is no charge loss, or even loss of field, the charge is a continual source of field of a certain strength (esu/cm^2). This power to cause a field, which is measured in *esu*, "spreads" out over space like a fluid but without diminishing the actual amount of *esu*'s of the charge. Schematically, the charge continually generates the lines of field shown in Figure 2-4. So, we do not need to specify a loss of charge strength inside the region, but an ability to generate field emanating from that region. Formally then, in the electric field case, the source side of the equation is a quantity proportional to the charge or in the case of a measure of the *ability to cause* an *E*-field *per volume* like given above in equation 2.41, it is the charge density (esu / cm^3).

As adumbrated above, Gauss's Law in differential form, equation 2.41, is the first equation in Maxwell's system of equations in their simplest form (i.e. the differential form). To give some further insight into its meaning and how to use it, we will now proceed to use it in two ways: first, by simple calculation, second by applying it to better understand what the field says about the energy.

Example Usage of Divergence and Curl for Electric Field

Here we apply the Cartesian coordinate version of equation 2.41 to the point charge field given by Coulomb's law:

2.42 $$\vec{E} = k\frac{q}{r^2}\hat{r} = kq\left(\frac{x}{r^3}\hat{x} + \frac{y}{r^3}\hat{y} + \frac{z}{r^3}\hat{z}\right), \text{ where } r = \left(x^2 + y^2 + z^2\right)^{1/2}$$

Assuming a uniform spherical charge distribution of radius r_0, which we will eventually let be arbitrarily small ($r_0 \rightarrow 0$) to reach our point limit, we get for the right hand side of Gauss's law: [31]

2.43 $$\begin{array}{ll} 4\pi\rho k & r \le r_0 \\ 0 & r > r_0 \end{array}$$

We calculate the right hand side outside the particle as follows.

$$\frac{\partial}{\partial x}\left(\frac{x}{r^3}\right) = \frac{1}{r^3} - 3x^2 r^{-5}, \text{ using } \frac{\partial}{\partial x}(r)^n = n\ r^{n-1}\frac{\partial r}{\partial x} = n\ xr^{n-2} \text{ with } \frac{\partial r}{\partial x} = \frac{x}{r}, \text{ and similar}$$

equations follow for y and z. Thus we get: $\vec{\nabla}\cdot\vec{E} \propto \dfrac{3}{r^3} - 3\dfrac{\left(x^2 + y^2 + z^2\right)}{r^5} = 0$. This is, of

course, what we expect because there are no sources of field, no charge in any small volume element outside the lone charge, and so what comes out is only what goes in.

It still remains to find the divergence for a point in the interior of the particle; this is left for an exercise (see end of chapter problem). It is left as an exercise to do this calculation in the natural coordinates of this problem, spherical coordinates.

We can also apply the curl to equation 2.42. Because of the spherical symmetry, x, y and z appear in completely interchangeable ways, thus we only need to show one of the three components is zero. For the exterior of the shell, we get:

2.44 $$\left(\nabla \times \vec{E}\right)_z = \left(\frac{\partial}{\partial x}E_y - \frac{\partial}{\partial y}E_x\right) = kq\left(y\left(\frac{-3x}{r^5}\right) - x\left(\frac{-3y}{r^5}\right)\right) = 0$$

---

[31] For the integral form, the right hand side would be: $4\pi kq'$ where q' is the charge inside the given radius.

Indeed, we can see the curl is zero without calculation because the E-field of a point particle has no circulation anywhere. This is, of course, true for the interior of the particle as well.

Electric Field Energy Density

The energy of putting together a charged spherical shell and a uniformly charged sphere, which we took as a classical model of the electron, has already been discussed and calculated. We will now analyze these results by looking at the fields they produce with the help of Gauss's law.[32]

First, applying Gauss's law in integral form (equation 2.40) to the outside boundary of the spherical shell, we get the following equation for the electric field, which only has a radial component E_r: $E_r 4\pi r^2 = 4\pi k q$. Rewriting this by defining the surface charge density, i.e. the charge per unit area, $\sigma \equiv \dfrac{q}{4\pi r^2}$, we get the radial field:

2.45
$$E_r = 4\pi k \sigma$$

Now, in the sub-subsection for advanced students, "Work Required to make a Spherically Symmetric Charged Shell," it was proved that the net electric field acting on the shell by its own various parts is half this value, so the force on a shell of radius r and surface charge density, σ, is:

2.46
$$F = \left(2\pi k \sigma\right)\left(4\pi r^2 \sigma\right)$$

This means the potential energy stored by the act of decreasing the shell's radius from r to $r + dr$, which is equal to the amount of work required to accomplish the compression, is:

2.47
$$dU = dW = F dr = 8\pi k\, r^2 \sigma^2 dr$$

As we decrease the size of the shell, regions inside where there is no field now become regions outside that do have a field.[33] Hence, the effect of decreasing the size of the shell in the way specified is to cause a new field to be present in the region between r and $r + dr$. This new field then is a manifest sign of the new potential energy, U, created by the compression. We thus rewrite equation 2.47 in terms of this sign, i.e. in terms of E_r rather than σ; using $\sigma = \dfrac{E_r}{4\pi k}$, we get:

2.48
$$dU = 4\pi r^2 \frac{E_r^{\,2}}{8\pi k} dr = \frac{E_r^{\,2}}{8\pi k} dV$$

---

[32] Of course, Coulomb's law could also be used, but our aim here is to both show the meaning of field energy and illustrate the use of Gauss's law.

[33] What happens to the plana during this compression is not relevant to the current discussion of energy density, nor to any of our discussions at the level of analogous abstraction of classical E&M. Remember there are certain considerations we just omit so as not to complicate the analysis and understanding of what we are addressing. However, to give some idea, one can make a couple of hypothesizes. Note that the simplest explanations are the most likely (and the least intrusive on the current reasoning) because one ought to account for what is known by positing *only* what is necessary to explain a given effect. In this spirit, one might say the plana passes right through the shell since in reality the shell is composed of atoms, i.e., discrete charges separated by plana or virtual plana. Another possibility is that the interior plana simply is annihilated or "absorbed" into the interior plana and the exterior plana is created or "stretched," for recall there is no need to suppose that plana is conserved. This last seems much more likely, but again the issue is way outside classical E&M.

Thus, we can treat $\dfrac{E_r^{\ 2}}{8\pi}$ (in *cgs* units, $k=1$) as a potential energy density or simply the field energy density. Thus, we have yet another analogical generalization of energy. Here, we say the field has energy only because this field is associated with the energy of putting the given charge distribution together in the way just discussed. In fact, there are other reasons for taking this as the field energy density, one of which we will see when we study electromagnetic radiation.[34]

Let's check this with calculation of shell energy done previously. We first write the integral form and then apply it to the shell:

2.49
$$U_{shell} = \int\limits_{Entire\ field} \frac{E_r^{\ 2}}{8\pi k}\,dV = \frac{1}{8\pi k}\int_{r_0}^{\infty}\left(k\frac{q}{r^2}\right)^2 4\pi r^2\,dr = \frac{k}{2}\frac{q^2}{r_0}$$

which is exactly the result given in equation 2.11.

Now, the energy of the uniform *spherical* charge of the same radius will clearly have the same potential energy contribution from fields outside of it as does the *shell* charge. However, unlike the shell, the sphere will have contributions to the field density inside. By drawing a "Gaussian surface," i.e. a surface to use in Gauss's law, interior to the sphere as shown below, we can write the following equation for the

field inside the sphere: $E_{r\_inside}\,4\pi r^2 = 4\pi k\left(\dfrac{4}{3}\pi r^3 \rho\right) = 4\pi k\,q\left(\dfrac{r}{r_0}\right)^3$,

which gives:

2.50
$$E_{r\_inside} = \frac{4}{3}\pi r k \rho = \frac{k\,q}{r_0^{\ 3}}r$$

Thus, the total field energy inside the sphere is given by:

2.51
$$U_{inside} = \int\limits_{r<r_0} \frac{E_r^{\ 2}}{8\pi k}\,dV = \frac{1}{8\pi k}\int_0^{r_0}\left(\frac{k\,q}{r_0^{\ 3}}r\right)^2 4\pi r^2\,dr = \frac{k}{10}\frac{q^2}{r_0}$$

Adding this to the outside contribution, which is given by equation 2.49, we get:

2.52
$$U_{total\_sphere} = \frac{3k}{5}\frac{q^2}{r_0}$$

This is the same as result obtained by directly calculating the work required to put the sphere together given in equation 2.18.

Field energy can be further understood as can the nature of the field itself by investigating the electrical potential, which we mentioned earlier results from the conservative nature of the electric field.

---

[34]In more advanced studies, it becomes a component of the so-called stress energy tensor, $T_{\mu\nu}$.

Electric Potential and Energy

The Potential, ϕ

The electric field, because of its very nature, can be written as the gradient of a function we call the potential; formally we write:

2.53
$$\boxed{\vec{E} = -\vec{\nabla}\phi}$$

"The potential" is a scalar field symbolized by $\phi(x, y, z)$. Sometimes for clarity we will refer to it as the potential field. However, often people will simply call it the potential and use the word "field" only for the electric field (or later, the magnetic field). By integrating equation 2.53, we can write the potential as: $\phi = -\int \vec{E} \cdot d\vec{x}$.

The directional nature of the electric field naturally suggests the idea that specifying that E-field requires at least two plana parts,[35] for one needs at least two points to specify a direction.[36] And, two consecutive plana parts specify a direction within the error of the sizes of those parts. Further, remember that it is the nature of any given quality of a body to specify, in some way, the extension of that body, e.g. the disposition of one part with respect to another. Thus, it is natural to think of the quality of the electric field as specifying the underlying extension by in someway unifying at least two parts according to the directional type of impetus it can cause. For a large region, each part of the plana is qualitatively proportioned to the next according to the directional type and strength of the field.

To help in visualizing this in a simple case, Figure 2-11 shows a rectangular-shaped "test" charge (i.e. one that does not significantly affect the surrounding field) in a uniform field. Suppose that the uniform field is created by a very thin, but large-area, slab of uniform negative surface charge, i.e., an "infinite" charged slab that is perpendicular to the page. In the figure, the slab is shown in cross section as a dark line on to the right of the grid. Also, the intensity of the ϕ-field, which is some disposition of the given plana cell, is assigned a number starting from zero at say the plane of origin of the field, the slab surface. These dispositions are characteristic of a uniform electric field. If the potential were constant then there would be no electric field; however, the constant potential may, in principle, cause other effects that we leave for other courses, except to say that quantum mechanics leads many to conclude that there are such effects.

Note that, as we discussed in *PFR-M*, the zero of potential is nearly always picked in accordance with: (1) the given physical situation (within the parameters of concern) and (2) ease of calculation. In actual fact, of course, if the potential is a real quality, its intensity is determined by the full physical situation, not by the convenience of a certain abstractions and/or formalisms.

---

[35] Remember by a "part" we do not here mean a minimal part, but a region of the plana that is large enough so as to allow averaging out all but the effects relevant to classical E&M for the given problem, i.e. the given physical situation under analysis.

[36] The shape of each part also suggests itself, but this then naturally leads to considering the disposition of the parts of the part and this is against our whole intent which was that these parts were the smallest we wish to consider.

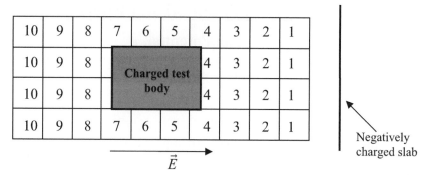

10	9	8	7	6	5	4	3	2	1
10	9	8				4	3	2	1
10	9	8				4	3	2	1
10	9	8	7	6	5	4	3	2	1

Charged test body

$\vec{E}$

Negatively charged slab

Figure 2-11: A test charge in the presence of the quality in the plana called the potential, which we can view as a further specification of the causal agency of the electric field. The large square plana parts are labeled by numbers which represent the potential assigned to each part. The size of the (massive) charged body is not shown to scale relative to the plana parts-- they are actually much smaller than the size of the massive body.

What does Energy say about the Nature of the Potential, ϕ?

Some thought about energy will further elucidate the physical import of the potential. To facilitate this, we need to do a few calculations.

We start by calculating the general formula for the work *the electric field does* in taking a test charge of charge q from a point A to a point B. (Note this energy has the opposite sign to that of the work *an external force does* on the test charge in bringing it from point A to point B with arbitrarily small impetus). We get:

2.54
$$W = \int_A^B q\vec{E} \cdot d\vec{x} = q\int_A^B -\vec{\nabla}\phi \cdot d\vec{x} = q\left(\phi(A) - \phi(B)\right)$$

We now need to learn how to calculate ϕ. We first take the case of the field outside of a *positive* uniformly charged sphere, we have:

2.55a,b
$$-\vec{\nabla}\phi_{sphere} = \frac{kq}{r^2}\hat{r} \quad \text{or} \quad -\frac{\partial}{\partial r}\phi_{sphere} = \frac{kq}{r^2}$$

Integration yields:

2.56
$$-\int_0^{\phi_{sphere}} \frac{\partial}{\partial r}\phi\, dr = -\phi_{sphere} = \int_\infty^r \frac{kq}{r^2}\, dr = \left[\frac{kq}{r}\right]_r^\infty = -\frac{kq}{r}$$

Which means:

2.57
$$\phi_{sphere} = k\frac{q}{r}$$

Where we have taken zero of potential to be zero at infinity for convenience.

We next proceed to the case of special interest, the uniform electric field near an infinite slab of uniform *positive* charge density σ. We take the slab to lie along the y-z plane centered on the origin, and we focus on the negative x-axis so as to facilitate comparison with Figure 2-11. This gives: (*exercise:* check using equation 2.53)

2.58
$$\phi_{slab} = 2\pi k \sigma x, \quad (\text{for } x > 0)$$

In some system of units for any particular σ, this can be written: $\phi_{slab} = -x$. Hence, the energy of the test charge can be written as: $U = -q\,x$.

In terms of Figure 2-11, this means that if you move from a given right cell to the one immediately to the right of it, the potential, i.e. the potential energy per unit charge, drops by unit.

From this knowledge of the energy drop, we of course, can get the field strength, i.e. the increase in strength of impetus per second per unit charge caused in a test charge; mathematically $\vec{E} = -\vec{\nabla}\phi$. In particular, in the above discrete terms, we write: $\phi(n_x l)$, where n_x is the number of the cell or part of the plana in the x-direction for $x > 0$, and l is the length of the cell. The field strength in the x-direction is then written, $E_x(n_x) = -\dfrac{\phi(n_x l + l) - \phi(n_x l)}{l}$. Of course, to reach the continuum limit that we use throughout this text, we must take the limit in the standard way. In particular, in the limit in which these parts are smaller than anything we try to examine, we obtain:

$$E_x(x) = -\lim_{l \to 0} \frac{\phi(x+l) - \phi(x)}{l} \ .$$

Hence, we see that the potential explicitly manifest the nature of the field and its ability, specifying in a natural way, both the field's force and energy, i.e. how that force "gives out" its impetus. Namely, any two plana cells proportioned with ϕ-intensities as discussed specifies both (1) how much energy the field can cause in a test charge, as well as, since the two plana cells together set a length scale, (2) the E-field's strength. By contrast, the value of the E-field at a *single* place (two or more cells) tells us nothing about the energy. Furthermore, if the potential is set according to the physical situation (e.g. the zero potential is the place where there is no further potential for kinetic energy for the given state of the field), then the potential also says something about the *overall* field-- which is, in turn, reflective of the charge that caused it and the plana of which it is a property. For example, two different fields might have the same potential at the same place, (say at $(x,y,z) = (1,0,0)$ for $\phi_{sphere} = \dfrac{1}{r}$ and $\phi_{slab} = x$) but they both imply that there is a unit of kinetic energy per unit of charge available if the charge is somehow brought from the current location to the zero potential location.

For these reasons and others that we will discuss later (especially when we discuss the so-called gauge constraint in the full dynamics of E&M), we will take the potential, ϕ, to represent a real quality of the plana. It should be noted that the things we have said here do not prove the ϕ represents an actual quality of the plana, but do give strong reasons for thinking so. You will find further evidence that this is the case as your study advances into other courses, and you also will see the potential move closer to the center of that study.

At this point, we can complete the answer to the question "What is the electric field?," which was started with equation 2.3 and the definitions below it, by summarizing what we have learned about the potential:

> The ___Potential___, ϕ, is a further specification of the ***electric field***; it reveals that the disposition of each "cell" is really (given our above caveat) a deposition of at least two parts of that cell. The relation between the measure of the electric field and the measure of the potential is:
>
> $$\vec{E} = -\vec{\nabla}\phi$$

Knowing that the electric field is the quality, but the potential is a certain determination of that quality, we can ask why it is that the potential falls off only linearly for the fundamental case of the point charge.[37] The field acts from one region of the plana on another in the manner described earlier, resulting in a certain intensity of the potential at each cell; why that certain intensity? The key to the answer is to note that one can speak analogically, as we often will, of the potential as some kind of independent property. However, this way of speaking should not make us forget that the power (quality) we are focusing on in classical E&M is the power to cause impetus (and hence energy). Since, in a given region, only a *potential* (a ϕ field) that is a specialization of an electric field (only those ϕ fields with non-zero gradient) can cause impetus, clearly only such potentials are relevant in understanding how that power is dispersed through the plana. In other words, it's only to the extent that the power acts on the next plana cell so as to dispose it relative to the current cell to cause impetus--i.e. to the extent it causes an electric field-- that it "uses up" or "disperses"[38] some of its ability to cause more field. As we already said, as the field propagates outward it has a larger surface area over which it must act, hence is proportionally less effective in causing a field. Since the electric field is reflective of its specialization, the potential, the potential propagates in such a way that the field falls off as $\frac{1}{r^2}$; this is the part of the nature of the power of the plana that we call the electric field.

Poisson's and Laplace's Equations: Gauss's Law in terms of ϕ

Having decided that the potential gives the full specification of the field and knowing its utility in physics generally, it behooves us to write our key result, Gauss's Law, the first Maxwell equation, in terms of ϕ. Gauss's law (equation 2.41) gives:

$\nabla \cdot \vec{E} = \nabla \cdot \left(-\vec{\nabla}\phi\right) = -\nabla^2\phi = 4\pi k\rho$. Thus we have:

2.59

> **Poisson's Equation** $\qquad \nabla^2\phi = -4\pi k\rho$

In Cartesian coordinates, the operator above, called the Laplacian, is written:

2.60
$$\nabla^2 = \frac{\partial^2\phi}{\partial x^2} + \frac{\partial^2\phi}{\partial y^2} + \frac{\partial^2\phi}{\partial z^2}$$

It can be shown that the solution for a point charge is exactly that given by equation 2.57. It can also be shown that the solution for any arbitrary distribution of charge, ρ , is exactly what one would expect from adding together, using superposition principle, the

---

[37] Recall the point charge approximation means we are far enough away that the structure of the charge within the particle can be ignored.
[38] Note our use of the fluid analogy again.

contribution to the potential from various differential elements of a given charge distribution using equation 2.57 for each piece. We write this as:

2.61
$$\phi(\vec{r}) = k \int_{Charge} \frac{\rho(\vec{r}')}{|\vec{r} - \vec{r}'|} dV$$

The special case of a charge free region gives:

2.62

Laplaces's Equation $\nabla^2 \phi = 0$

The solutions to the Laplace's equations, such as ϕ, are called *harmonic functions*. A solution to Laplace's equation minimizes the field energy, $U = \int_{all\ space} (\vec{\nabla}\phi)^2 dV$ for a given system of particles and constraints, i.e. "finds" the static placement of particles that lowers their field energy as far as it can be within the given constraints. This is physically clear if one remembers the field is the sign of the kinetic energy (activity) that would be released if the particles that make up the system were allowed to move.

The simplest example is that of two positive small charges of radius r_0. The least energy solution is when the particles are both "at infinity." Figure 2-12 shows that this lower energy is reflected in the lower potential of the separated particles and hence, the lower gradients of the potential, i.e., the electric field. In particular, the two charges brought near the same location generate a potential (and, hence, a field) like that of a charge of $2q$ in all but the small region very close to the charges. Thus, in nearly all of space, the nearby case has twice the field of the isolated charges and thus 4 times the energy of each alone. This agrees quantitatively with the energy of putting the charges together, for if each separated spherical charge has energy E_0, then, using equation 2.52, $U \sim (2q)^2 / r_0$, for the energy of a spherical charge of measure $2q$, we can see that the two charges forced into the same radius, r_0, would have an energy of $4E_0$.

Figure 2-12: a. (*left*) The potential of a positive charge far from a second such charge. **b.** (*right*) Potential of two positive charges very close together.

You should be aware of two other important properties of harmonic functions, such as ϕ when it describes the potential in a charge free region. For both properties, we consider the case of interest for us, the potential, and we leave aside the mental construct of charges at infinity. The two properties are: (1) the uniqueness property: given potentials defined on some surfaces and charge on others, there is only one solution for the potential in the charge-free regions, so if you find one solution to equation 2.62 with requisite boundary conditions, you have the solution.[39] (2) In a charge free region, the average of the potential over the surface of *any* sphere is equal to the potential at the center of that sphere; formally, we write:

2.63
$$\frac{\int_{\substack{surface\,of\,sphere \\ centered\,at\,(x,y,z)}} \phi(x', y', z')\,dS}{\int_{surface\,of\,sphere} dS} = \phi(x, y, z)$$

Charge q is spread uniformly around spherical surface

This can be understood physically using the above figure. Suppose a point charge Q is resting at the origin causing a potential field given by $\phi(r) = \dfrac{kQ}{r}$. One brings in some charge, q, piece by piece and spreads it uniformly over the surface of a sphere of radius R centered at some distance x_0 from the origin. The total potential energy of such a configuration is the average of the potential over the surface of the sphere. This can be seen formally by writing:

2.64
$$U = \int_{Sphere} \phi(\vec{r}) \underbrace{\left(\frac{q}{S}\right)}_{dq} dS = \frac{q\int \phi(\vec{r})\,dS}{\int_{Sphere} dS}$$

Now, the energy is also clearly equal to the work it would take to bring the point charge Q into its position with the sphere already charged. As far as the external field of the sphere is concerned, we can treat the sphere as a point particle of charge q. Hence, the energy, using equations 2.54 and 2.57 in *cgs*, is qQ / x_0, and, upon equating it with U above, we find that the average of the potential over the sphere is, indeed, equal to the potential at its center Q / x_0.

The Static Case of the "Third" Maxwell's Equation

Knowing that the electric field can be written as a gradient of a (reasonable)[40] function makes it clear that the electric field can have no curl. We can see this by noting that the field, i.e. the gradient of the potential, tells the amount and direction of the steepest increase in the potential at the given point. Therefore, in traveling around a small closed curve, one will encounter as much increase in the potential as decrease. That is, there will be as much field along the direction of travel as against it, and so there will be no net circulation. For example, always going perpendicular to the gradient around a local maximum (or minimum) will mean that there is no field in one's direction of travel as indicated by the constancy of the potential, and thus there will be no net circulation. Off the local maximum (or minimum), you will get a certain amount of "circulation" as you travel

---

[39] There are various techniques for solving Laplace's equation boundary value problems, for example, conformal mapping methods that arises out of the study of complex functions, variational methods that minimize the energy, and approximate numerical methods.

[40] This is made mathematically precise in more advanced studies. For now, smoothness is a good intuitive condition.

along the curve in the direction in which the potential is increasing only to get it taken away on the return "downward" part of the curve, since the component of the gradient along your direction in a given distance tells you how much potential you gain or lose. In particular, since, in a closed curve, one comes back to the same point and hence the same potential, there again will be no net circulation. This is seen most simply in the case where the potential is a function of only one dimension. For example, consider the following line integral, $\oint \vec{\nabla}\phi \cdot d\vec{x}$ along a small closed loop in a potential $\phi(x)$ which yields an electric field in the $-x$-direction: (1) moving Δx forward, one will gain $\Delta\phi = \dfrac{\partial\phi}{\partial x}\Delta x$, while (2) there is no change in the potential when one moves over Δy, but (3) on the way back, moving $-\Delta x$, one *loses* $\Delta\phi = \dfrac{\partial\phi}{\partial x}\Delta x$ and the last leg (4) in returning to the original y position by moving $-\Delta y$ does not further change the potential. Hence, the integral, which is proportional to the curl at a given position, is zero.

Thus, we write the static case, i.e. unchanging electric and magnetic field case, of what we will call the *third Maxwell's Equation*:

2.65
$$\boxed{\text{3}^{\text{rd}} \text{ Maxwell's Eqn. (static case)} \qquad \nabla \times \vec{E} = 0}$$

This equation coupled with the "first" Maxwell equation, given in equation 2.41, captures implicitly the key aspects of the static electric field, and, using them along with the various theorem's given in the table at the end of the "Mathematical Aside" section we can solve many problems.

Part II: Applications

The best way to begin to digest the meaning of the potential and its relation to the *E*-field as well as to gain expertise in the use of the new formalism is to begin to apply this knowledge to certain important cases. This will force us to specialize our understanding and build good problem solving habits with the formalism.

Of course, this half of the chapter is only the beginning of this process, the end of chapter problems are also very important in developing both your understanding and your skill in problem solving. And, in later chapters, we will learn more about current, circuits and components, but now we will cover much of the fundamentals of those subjects.

Conductors and Capacitors

Conductors of Electricity

A basic question that arises in treating the electric field is the mobility of charges, for clearly an electric field can only cause impetus in charges that are free to move. There is a whole area of study that addresses the complexities of charge mobility in quantitative detail. However, here, we will start and stay simple.

If you bring a charged object near a given body, the charges within that body will be allowed to move significantly in a given period of time or they will not; they will be highly mobile or they will not. We say that the given body is, respectively, either a good conductor of electricity or it is not. The former are called conductors, while the latter are called insulators (non-conductors). There is a third category worthy of mention, semiconductors, which we will discuss briefly in Chapter 10.

We begin by considering good conductors, such as metals. These carry electrical currents very easily because they have mobile electrons that can be moved under the action of an *E*-field. This was seen in the early electrical experiments and in simple experiments that you can reproduce yourself.

For example, consider the "inductive charging" of an electroscope such as shown in Figure 1.5b. In thinking about what follows, remember that macroscopic objects tend to have no net charge, having an equal number of electrons and protons.[41] If one charges a comb and places it near the metal disk outside the jar, the positively charged comb (via the plana) pulls electrons up the metal rod into the disk, leaving the electroscope leaves with an excess positive charge, which causes the leaves to repel and thus separate. Touching the metal disk with a wire that runs to a large metal object (or any kind of "ground") will result in the negative charge being pulled in through the wire onto the disk. If one then pulls the wire away the result will be to leave the metal disk/rod negatively charged. Also, once the comb is removed, the extra electrons will no longer accumulate near the disk but spread throughout the metal, trying to maximize their mutual distance.

The freedom of electrons to rearrange in this way under their mutual interaction as well as under the action of an external field is a key characteristic of the nature of metals. To understand this behavior in its most primal form, we consider ideal metals or

---

[41] Recall, at our level of analogical abstraction, we leave out whether the electrons and protons are parts of a substance, such as an animal, or individual substances themselves. For example, an electron that was part of a substance would properly be called a virtual electron to indicate this status, but we will sometimes leave out such words to make evident in as simple a way as possible that we are making the abstraction just described.

conductors. We make the idealization that leaves out ("switches off") any forces internal to the metal that might act against the external field (e.g., from the comb) other than the field of the "free" or so-called "conduction band" electrons themselves.[42] In such a material, the electrons will move away from their initial equilibrium positions until the external electric field is *canceled* inside the metal by the internal electric field of the electrons it moves. We also look on time scales that are slow enough that we do not have to pay attention to their intermediate positions, and we do the already discussed averaging over large enough volume to be able to ignore the discrete nature of the charges (most especially, electrons and nuclei) and our plana parts.

So, repeating the crucial point, the electric field inside an (idealized) metal is zero; i.e., the potential, ϕ, is a constant inside of a metal. Figure 2-13 below shows how this is so in the simple case of a slab of metal. Some electrons are free, so the external field activates impetus in them, moving them in the opposite direction of the field toward the surface to the positions shown in the figure. This leaves some atoms on the opposing surface absent the electrons they need for neutrality, resulting in a positive surface charge density there due to these positive ions. Only enough electrons are moved so as to cancel the field within the metal. If more were moved, there would be a net field in the metal, due to the extra free electrons, that would push them back. In other words, a static situation inside the metal is only reached when the inside of the metal has a net charge of zero, and there are just enough surface electrons (and corresponding surface ions on the other surface) to impede the external field from forming inside the metal.

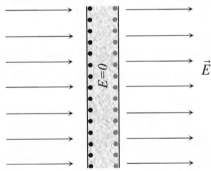

Figure 2-13: A thin slab of metal in a uniform electric field. Note that the external field is exactly canceled inside the metal by the field of the internal "free" electrons (black dots) that move to the left surface, leaving positively charged ions (red dots) on the right surface.

Hence, we have established that there is no electric field inside of an ideal conductor such as approximated by a metal, and this implies such a conductor is at one potential throughout. Using similar reasoning to the above, we can also say that, arbitrarily close to the surface of the metal, there is no field parallel to the surface, for if there were,

---

[42] We have, following our procedure of leaving out all but our focus of interest, greatly idealized the situation in other ways as well. For example, the electrons do not even move in the same way that a macroscopic body does, and there is thermal activity inside the metal so that the atoms are not simply at rest but vibrating about in various ways within their respective locations. Again, having left out or switched off such effects, they will have to be addressed at some point.

the charges would move along the surface, thus violating our static assumption. However, there can be an E-field perpendicular to the surface, because the edge forces can impede the field from activating impetus to move the charges. We can give quantitative statements of both these facts in the following way.

Stokes' Law and Gauss's Law Applied to Conductors

Using Stokes' and Gauss's laws and the two figures below, we can calculate the electric field parallel and perpendicular to the surface of a conductor.

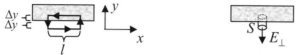

The parallel component can be found using Stokes' law with a line integral such as shown above on the left. From Stokes' law and the fact that E has no curl we get:

$$\iint_{Surface\ defined\ by\ Path} \left(\vec{\nabla} \times \vec{E} \right) \cdot \hat{n}\, dA = 0 = \oint_{Path} \vec{E} \cdot d\vec{x}$$

Taking the E's to be positive quantities in the x or y-direction, and using unprimed symbols for the fields in the left leg of the path and primed symbols for the right leg, we get:

$$\oint_{Path} \vec{E} \cdot d\vec{x} = -E_{inside\|}\, l - \left(E_{inside\perp}\Delta y + E_{outside\perp}\Delta y \right) + E_{outside\|}\, l + \left(E_{outside\perp}'\Delta y + E_{inside\perp}'\Delta y \right)$$

Noting that the inside fields are zero and taking the width of the path to be arbitrarily small, i.e., taking $\Delta y \to 0$, gives $E_{outside\|} = 0$, as expected.

We can calculate the perpendicular component using the Gaussian surface, called "Gaussian pill box", S, shown in the right figure above. In particular, applying $\int_S \vec{E} \cdot d\vec{A} = E_\perp A = 4\pi k\, q$ and using the surface charge density $\sigma = q / A$, we can write the field on the surface of a conductor as:

2.66
$$E_\perp = 4\pi k\sigma$$

You should be aware that this equation does not mean that the surface charge density inside the "pill box" causes the field at that point. It is, in fact, a net effect of *all* the charge, nearby and far away. Part of the power of Gauss's law is its ability, through taking advantage of various symmetries of a system of charges, to incorporate the effect of charges through out that system.

Inside a Metal Box

We next consider an arbitrarily-shaped solid chunk of metal with a net positive charge on it. Drawing a Gaussian surface just *inside* the surface of the chunk, and noting that there is no flux through this surface, we see that Gauss's law verifies our earlier finding that there is no excess charge inside of a metal. All of the excess charge in the chunk resides on the *outer* surface. Hence, if we take out the middle and only leave the thin layer of charge fixed in place on the outer surface, obviously no change in the field will result. Of course, the charge will not, in general, be uniformly distributed, but will depend on the shape of the shell. Inside the shell, however, the field will be zero no matter what the shape.

Benjamin Franklin (1706-1790 AD) was (likely) the first to discover the fact that the charge on a conductor (he used a "silver pint cann") resides on its outer surface and that there is no field inside of a conductor. Franklin was unable to explain this effect, and

Joseph Priestly (1733-1804 AD), apparently aware of the fact that the inverse square law is responsible for the gravitation field being zero inside of a uniform density spherical shell, says: "May we not infer from this that the attraction of electricity is subject to the same laws with that of gravitation and is therefore according to the squares of the distances; since it is easily demonstrated that were the earth in the form of a shell, a body in the inside of it would not be attracted to one side more than another?" [43] Of course, this is somewhat misleading because the reason that the gravitational field inside of a spherical shell is zero is not only because of the inverse square law, but also because of the symmetry. Furthermore, more must be at play in the case of an electric field inside of a conductor, otherwise the field would not be zero inside of Franklin's "cann" or any other non-spherical shell. Indeed, the mobility of the charges, the existence of *both* negative and positive charges, along with the inverse square law (and superposition) are required in general. Michael Faraday (1791-1867AD, see history box in Chapter 4), verified that there was no field inside of a conducting shell by actually going to "live in" a metal covered box that he charged to a high potential above ground. He testifies he did all kinds of electrical experiments, verifying the absence of the field inside despite the lightning bolts that flew from the outside. Because of his work, conductors that are used to shield static electric fields are now often called "Faraday cages."

Method of Images

Because metals (and conductors generally) have a constant potential (equipotential) throughout, thin shells of metal can be put along a given equipotential surface generated by any given charge distribution without changing the physical situation. Indeed, note that putting a metal on an equipotential surface ensures that, as required, the field lines will be perpendicular to it. Perhaps the simplest example of an equipotential surface can be found in the plane lying midway between a positive and a negative point charge. One can prove this plane is an equipotential by simply recalling the potential for the positive and negative point charges, $\phi_+ = \dfrac{q}{r}$ and $\phi_- = -\dfrac{q}{r}$, and noting that any point on the plane is the same distance from the negative charge as it is from the positive (see figure below). Hence, the point charges attempt to cause potentials of equal and opposite type and so "cancel each other out," making the potential along the plane zero everywhere.

In fact, one can use the nature of this field, along with the uniqueness of the previously mentioned boundary-value solution to Laplace's equation to solve the problem of a single positive charge next to an infinite conducting plane for the region to the left of the plane. Namely, given *a solution* to Laplace's equation for the latter physical situation, such as the potential for the two point particles in the given region, that solution must be *the solution*. Physically, the point particles generate a certain potential in the plana, and that potential has equipotential surfaces, including the mid-plane between the particles, that

---

[43] J. Priestly, *The History and Present State of Electricity,* 3rd ed. London, 1775 (from *pdf* on Google books). Note the spelling in this quote has been altered to conform with modern usage.

can either have a metal surface along it or not. If it does not and such a metal surface is inserted, it will not change the potentials, because, by definition of an equipotential surface, the field only has a component acting perpendicular to the metal. If the metal is inserted along the mid-plane, the charges on the left (right) surface arrange in such a way to create, on the left (right), exactly the potential that was caused by the point particle on the right (left). It does this to exactly compensate internally for the perpendicular field caused by the two particles. In this way, the metal, in a certain sense, divides the system into two separate pieces. In particular, removal of the negative point charge will not affect the field on the left side, but only on the right. The left surface will still have the same surface charge density, since it still must cancel the field parallel to its surface due to the positive point charge on the left side. However, the right side no longer needs to accommodate itself in that way. Indeed, a solution to the right hand side is $\phi = $ constant, therefore this is *the* solution, and there is no field on the right hand side.

This method of solution is called the method of images because a charge near an infinite plane looks as if it has an image reflected in the metal of equal but opposite charge. This method can be helpful, saving much work in some limited types of situations, but it is of most value to us for the insight it gives into the nature of conductors, charges and the fields they cause.

Capacitors

Lastly, we can consider a particular useful combination of metals, which is called a capacitor. Capacitors probably get their name from their ability to store charge. In its simplest form, a capacitor consists of two large thin slabs of metal in proximity to one another. Figure 2-14 shows such a capacitor with a certain potential difference across it. The potential difference is often applied by a battery, which we will discuss in the next section. Potential difference (and the potential itself, if it is referenced to the zero potential) is measured (in *SI* units) in *volts*; one volt is *1 J/C*. In fact, because the charge on the electron ($e \sim 1.6 \times 10^{-19} C$), not the Coulomb, is the fundamental unit of charge, we have a special unit called an electric volt, *eV*; $1eV = 1.6 \times 10^{-19} J$, which is the amount of energy that an electron gets when it is allowed to be accelerated through a 1 volt potential difference. The potential difference is established by the battery at its terminals. If some of the electrons that the battery forced to its negative terminal are transferred to its positive terminal thus attempting to change its potential, a chemical reaction will occur to supply more electrons. Clearly, when a bare wire is connected to a bare terminal, the wire is then forced to the same potential as the terminals because such a connection effectively makes the wire and the terminal one conductor. Similarly, when each wire is then connected to a plate (see Figure 2-14), the potential difference of the battery is imposed on the plates. This is done by pumping electrons onto the bottom plate, which forces electrons into the positive terminal until the battery potential is reestablished.

Using the Gaussian surface shown in Figure 2-14, we can calculate the strength of the field activated in the plana by the charges, $\pm q$, put on the plates as: $E = \dfrac{4\pi kq}{A}$, where A is the area of the plate and q is its charge. The potential difference, $\Delta\phi$, also called the voltage, is: $V = \dfrac{4\pi kq}{A}d$, where d is the distance between the plates.

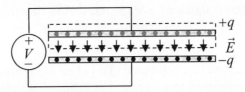

Figure 2-14: Parallel plate capacitor being held at constant potential difference, $\Delta\phi = V$. The cross section of a Gaussian surface to be used in calculating E is shown with dotted lines.

This equation reflects the relationship between the potential difference enforced by the battery and the charge needed to accomplish it. This relationship is determined, given the nature of the field, metal and charge already discussed, by the geometry and size of the two plates. We symbolically write the general relation between charge and voltage as:

2.67 $Q = CV$, where C is called the capacitance and is measured (in SI) in Farads,

which is 1 coulombs/volt or 1 C^2/J (in *cgs* it is measured in centimeters, which nicely reflects of its geometric nature).

So, the parallel plate capacitor has a capacitance of $C = \dfrac{1}{4\pi k}\dfrac{A}{d}$. For fixed d, the greater the area, A, the greater the capacitance because more charge is needed to maintain the same field which is, in turn, specified by the potential difference applied by the battery. In other words, the E-field must cause V units of energy in a unit of charge accelerated from one plate to the other. On the other hand, if the distance, d, between the plates is reduced, in order to cause this same amount of activity, there must be a stronger electric field in the region between the plates, because there is less distance over which the field can operate. Now, according to equation 2.67, to make a stronger field, given the same plate geometry, one needs more charge; hence the capacitance is also larger.

To get an idea of how much capacitance a parallel plate configuration has, consider making a capacitor with $A \sim 1cm^2$, $d \sim 1cm$, with vacuum or air between the plates;[44] this gives:

2.68 $C = \varepsilon_0 \dfrac{A}{d} = 10^{-12} F \equiv 1pf$,   where   we   have   used: $k = \dfrac{1}{4\pi\varepsilon_0}$

$\varepsilon_0 = 8.85 \times 10^{-12} F / m$

This geometry is simple, but, of course, more complex ones are possible. For example, we could simply have two spheres, nested or next to each other. Furthermore, because the earth is close by and in fact often there is all kinds of conductors of various degrees around us one can treat those as a second conductor "at infinity" and any single conducting object on your desk can become a capacitor with respect to this second conductor at infinity, the earth, which is also simply called "ground."

The Capacitor applied to Static Discharge

We can apply our capacitor analysis in a simple way to the case of the static discharge that you feel when touching some metal appliance after scuffling across the carpet on a dry cold day. In particular, we would like to know the amount of charge that is

---

[44] In Chapter 8, we will see that we can increase the capacitance by putting certain materials, called dielectrics, inside the capacitor.

transferred in one of those shocking events. Air breaks down and becomes a conductor, i.e. electrons are broken free from atoms in the air and become mobile, at about $E_{break} \sim 3 \times 10^6$ V/m. This event is basically the discharge of the capacitor formed between your body and, for example, an appliance. To get an approximate answer, we will take the appliance as solidly connected to a "ground at infinity" all around you. We also make the gross but simplifying approximation that your body can be approximated as a sphere of radius $.5m$.

Next, we calculate the capacitance of two nested spherical shells, expecting to take the radius of the outer shell to infinity at the end. Assuming that we have already charged the capacitor, so that the inner shell has charge $+q$ and the outer one $-q$, we know that, inside the shells, the potential caused by the inner shell is: $\phi = \dfrac{kq}{r}$, whereas the potential from the outer shell contributes nothing. This means the potential difference between the two shells is: $V \equiv \Delta\phi = kq\left(\dfrac{1}{r_i} - \dfrac{1}{r_0}\right)$, where r_i and r_0 are respectively the radius of the inner and outer shells. The definition of capacitance gives:

2.69 $$ C = \frac{1}{k\left(\dfrac{1}{r_i} - \dfrac{1}{r_0}\right)} \sim \frac{r_i}{k} \sim 50\,pf, \quad \text{for } r_0 \gg r_i. $$

(Note, this is equivalent to about $50cm$ in cgs units.)

Taking the arc distance to be $d \sim 1cm$, and noting that the field has to be at least the breakdown level for the arc to happen we get:

2.70 $$ E \sim \frac{V}{d} > E_b \sim 3 \times 10^6 V/m, $$

which implies $V > E_m d$

so that:

2.71 $$ Q = CV > CE_b d \sim 10^{-6} C = 1\mu C $$

We could do similar calculation using a uniformly charged line density, and we reserve this for the end of chapter problems. Instead we move on to look at the most well known usage of electricity, electronics. All electronics begins with an understanding of circuits.

Basic Circuit Theory

Batteries

The capacitor brought us to consider the battery, which is an essential ingredient of most modern circuits. Now, you may have wondered why, in our circuit shown in Figure 2-14, we applied a potential difference rather than a certain charge to the plates. The reason is a potential difference is naturally caused by certain chemical reactions. This, in turn, reveals the central importance of energy, in this case potential energy, which specifies the amount of ability a body has to cause activity. Just giving the field, as we discussed, does not specify the activity, the energy. As mentioned in the capacitor sub-subsection the standard way a potential difference or voltage is maintained is by a battery; the first of which was Alessandro Volta's (1745-1827AD) "Voltaic Pile." Batteries undergo chemical reactions that force (approximately) a certain potential difference, even when a current is flowing. Said another way, they have the power to cause a field that can cause a certain amount of activity (rate of transfer of intensity of impetus per unit charge). Figure 2-15 shows lemons being used as a battery.[45] In outline, a lemon battery works as follows.

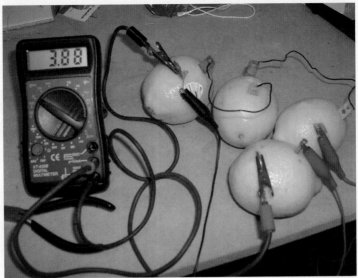

Figure 2-15: 4 Lemons, each punctured with one copper and one zinc electrode produces a total potential difference of 3.88 volts. The lemons can light the green LED. The lemon/electrode arrangements are a type of battery in which chemical reactions support a potential difference. You can make this battery using a zinc plated nail and a penny as done in the bottom right lemon.

First, the fundamental principle underlying any chemical battery derives from the nature of certain atoms to give electrons to others. When an atom gives electrons to another, chemists call it oxidation; this name arises, for example, because in the oxidation of iron, oxygen takes two electrons from iron (giving them to oxygen) to form rust, for instance $Fe^{+3}_2O^{-2}_3$ (or notably, even black lodestone, Fe_3O_4 also written $FeOFe_2O_3$). Reduction refers to the opposite process of an atom receiving electrons. Oxidation-reduction or so called redox reactions thus occur in pairs. In the lemon battery above, the nature of zinc (Zn) is such that, in contact with the solution, a couple of its outer parts, the electrons, are peeled off according to the reaction $Zn(s) \rightarrow Zn^{2+}(aq) + 2e^-$ while the nature of copper (Cu) is to attract a couple of extra parts, two electrons, according to the reaction, $Cu^{2+}(aq) + 2e^- \rightarrow Cu(s)$. In short, copper "wants" electrons and zinc wants to give them away. Forces, springing form the nature of the acidic juices of the lemon along with the zinc and copper, act in such a way as to cause this transfer. As we've said, this "force," called an *electromotive force* and often symbolized by $\mathcal{E}$, acts in such a way to cause a certain energy (activity). Remember that the potential is a specification of how the force of

---

[45] Technically, a battery is composed of multiple cells (or sometimes a single cell) and a cell (wet or dry) is a single electrode pair along with its electrolyte.

a field acts. *The force "tells" the momentum per unit time whereas the potential "tells" how much activity (energy) the force causes.* Thus, the electromotive force is not just a force but that further specification of the force that shows how much activity it causes per unit charge. For our lemon battery, the electromotive force or potential difference it forces is about 1 volt,[46] which is roughly what is seen in Figure 2-15 above. Processes analogous to the above are found in nature, e.g. living cells. In fact, as we pointed out in Chapter 1, the first mentions of electricity appear in reference to animals.

Current and Resistors

Now that we understand something of how batteries push charges, we should discuss more what happens when they are successfully pushed. At the most general level, when they are successfully given impetus, they move causing flow of charge, or current. We have already discussed current density and current in reference to the flow of water, for charge it is similar except now it is not the flow of the body itself, but of the strength or intensity of a property of the body, its charge. This is analogous to the case of energy which is the rate of transfer of the intensity of the impetus, except in that case: we were neither interested in (1) how the impetus was spread over a volume, nor (2) the direction of activity but only the total activity. Furthermore, in the energy case there is (3) a dependence of the impetus on speed (so we used a differential to write $dK = \vec{v} \cdot d\vec{p}$) and there is no such dependence needs to be applied for charge, which appears the same no matter how fast it moves.

To switch gears from water to charge, we will slow down some here to explain the formula previously introduced for the current density, $\vec{J} = \rho \vec{v}$, where the current density $\vec{J}$ is the direction and rate of flow of charge per unit area in some region and ρ is the charge per unit volume in that region. Eventually, we will return to our continuum idealization, by averaging over large enough regions that the effects of the "clumping" of charge are small, but let's start by considering the basic charges that we know are flowing. Consider a distribution of N electrons of charge e flowing at velocity $\vec{v}$. Further, assume this distribution can be represented by an approximate number density, n, i.e. number per unit volume, in a given region. Now, suppose that we insert a non-interacting mesh (analogous to the one pictured in the water flow at the beginning of the "Mathematical Aside" section) at some angle θ to the flow. We can then calculate the rate of charge per second, i.e. the current, that crosses through this mesh as:

$$I = \frac{\text{amount of charge in volume that will pass through in time } \Delta t}{\Delta t}$$

2.72
$$= \frac{\text{charge/volume} \cdot (\text{volume})}{\Delta t}$$

$$= \frac{ne(\vec{v}\Delta t \cdot \vec{A})}{\Delta t} = ne\vec{v} \cdot \vec{A}$$

We then define the current density as:

---

[46] In fact, the reaction above is calculated by so called standard potential electrode potentials given at 25C. For the Zn reaction that number is .76Volts, while for Cu, it is .34 Volts, giving a total of 1.1Volts. Note that these reactions do depend on temperature. Cf., for example, http://hyperphysics.phy-astr.gsu.edu/HBASE/tables/electpot.html#c1

2.73 $\vec{J} = ne\vec{v} = \rho\vec{v}$, where ρ is the charge per unit volume.

Hence, we can easily generalize equation 2.72 using differential areas to get:

2.74 $I = \int \vec{J} \cdot d\vec{A}$

The current I is measured in ampères or coulombs/sec (or in cgs in esu/sec). You may have noticed that the coulomb and now the ampère have been left undefined; to understand their definitions, we will need the material of the next chapter, magnetic fields, so we will wait till then. We have also glossed over an important point above. Technically, since we are dealing with electron flow, if we stuck to standard notation, we should have introduced a minus sign in our equations above. Franklin, as we mentioned, incorrectly believed positive charges were mobile, and as a result left us with a funny sign convention. Namely, the currents we typically talk about, i.e. the flow of electrons, are written as negatives. Again, this is a notational matter, but notation is best chosen to illuminate the physics, and this can do the opposite if one is not careful. In particular, the notational choice means that when you draw an arrow indicating the direction of current flow in an electrical wire, the actual flow of electrons will be in the opposite direction! For notational accuracy, you can think of the flows we discussed above as flows of positrons rather than electrons, (or, for the equations, you can think of the minus sign as inside the e).

Because charge is conserved, current is conserved and equation 2.34, the conservation equation, applies to electrical currents, as long as we keep our convention of treating positive charges with a "+" and negative charges with a "−", (remember they are called such because they cause, via the plana, opposite types of impetus). This means, for any circuit, the total current coming into any node must be zero. For example, pictorially and symbolically, we have:

$$I_1 + I_2 + I_3 + I_4 = 0 \qquad\qquad I = I_1 + I_2 + I_3$$

This is called *Kirchhoff's current law*, which we can write in the following way:

2.75 $\sum_i I_i = 0$, for every node (junction point) in a circuit

We can write a parallel rule for the potential difference across various elements in a circuit, using the fact that $\nabla \times \vec{E} = 0$ everywhere for the *static* case (and it can be generalized to the case of sinusoidal voltages under the so-called lumped element approximation) such as shown pictorially and symbolically below. V_C, V_R, and V_D represent the voltages across three different types of components, a capacitor, a resistor (which we discuss below), and a diode (which we will discuss in Chapter 10).

$$V = V_C + V_D + V_R$$
$$or$$
with consistent definition of signs:
$$V_1 + V_2 + V_3 + V_4 = 0$$

This fact that the sum of the voltages around any given loop in such a circuit is zero, is called *Kirchhoff's voltage law* and is written:

2.76 $\qquad \sum_i V_i = 0$, around every closed loop in a circuit.

Now, connecting a thick piece of metal, e.g. a large gauge copper wire, across a battery, will of course result in current flow, but it will not last very long and can blow out your battery, because, of course, metals conduct so well that they overtax the capacity of the battery. There are, however, other materials that conduct much less; these materials are used to make circuit components called *resistors*, because they resist the flow of current. Resistors obey "Ohm's Law." Indeed, there are many materials that obey this important, but not fundamental, "law" at some level. In such materials, the current produced is proportional to the voltage applied. Namely, we have:

2.77 $\qquad$ $\boxed{\text{Ohm's Law } V = IR}$

Where (in SI) V is in volts, I in coulombs per second and R is in ohms, written using the Greek letter for "o," omega, Ω.

At first thought, you may think that an electron flows in such materials like a ball falls under the action of gravity, i.e. smoothly accelerating as the field causes a stronger and stronger impetus in the electron. This is not the case. Neither do the electrons move under the action of an unchanging impetus. Instead, they are accelerated by the field in the material until they "collide" with an atom or the like in the material. This creates an effective or average speed (we leave the filling out of this model for end of chapter problems) of electron travel, which we can then use as our choice for the speed in our equation $\vec{J} = \rho \vec{v}$. Said another way, the average speed is proportional to the field. A more principled, though less practical, form of Ohm's law that brings this out expresses the current density in terms of the field that causes it:

2.78 $\qquad$ Ohm's Law $\vec{J} = \sigma \vec{E}$ $\qquad$ (version II)

Here σ is called the conductivity (this symbol is the same as that used for a surface charge density but they are, of course, fundamentally different quantities, being measures of *qualitatively* different aspects of a body) and is measured in $1/(\Omega m)$, inverse ohm-meters (or siemens/m) or more often in inverse ohm-centimeters. Conductivities of a few substances are given in the table below.

Material	Conductivity[47] $1/(\Omega m)$	Material	Conductivity $1/(\Omega m)$
Silver (best metal conductor)	6.30×10^7	Sea Water[48]	4.8
Copper (used in wires)	5.69×10^7	Drinking Water	.0005 to .05
Gold	4.52×10^7	Deionizied Water	5.5×10^{-6}
Aluminum	3.5×10^7	Air	$3\,to\,8 \times 10^{-13}$

Note that Ohm's law will cease to be valid if (1) the field gets so high that the maximum kinetic energy given by the field before colliding is about equal to the average thermal kinetic energy of the parts, or, if it gets even higher, it can cause an avalanche of carriers to be released and cause a catastrophic break down such as occurs in electrostatic discharge in air, or (2) if the field is applied over a time much shorter than the average collision time, which is very small, much less than 100 picoseconds.

Circuits

Now, we are ready to analyze some circuits. Reserving more complex analysis and circuits for later, we here analyze two very simple circuits only to bring out the fundamental principles and set the basic rules in place. We start with two resistors wired in series as shown below in Figure 2-16a.

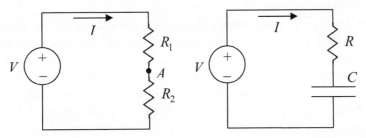

Figure 2-16: a. *(left)* Voltage source driving a current through two resistors in series. **b.** *(right)* Voltage source charging (an initially uncharged) capacitor through a resistor.

Using Kirchhoff's voltage and current laws, respectively, and Ohm's law we have:

2.79 $V = V_1 + V_2$, $I = \dfrac{V_1}{R_1} = \dfrac{V_2}{R_2}$

Giving: $I = V / (R_1 + R_2)$

Thus, *the total resistance of resistors in series is the sum of their resistances.* Also, the voltage at the test point A, is:

2.80 $V_A = V_2 = \dfrac{V}{R_1 + R_2} R_2$.

Lastly, we analyze an uncharged capacitor charging through a resistor as shown in Figure 2-16b. Again using Kirchhoff's voltage and current laws and Ohm's law, we have:

2.81 $V = V_R + V_C$, $I = \dfrac{V_R}{R}$

---

[47] See http://en.wikipedia.org/wiki/Electrical_conductivity.

[48] Corresponds to an average salinity of 35 g/kg at 20° C.

Since the capacitor starts with no charge on it, the voltage across it, V_C, obviously, also starts at zero. Thus, the voltage across the resistor, V_R, will be equal to V, and the initial current will be: $I = V / R$. However, after a time Δt, the capacitor will have a charge $\Delta Q = I \Delta t$, and thus, using $\Delta V_C = \Delta Q / C$, so the *resistor* voltage will change by: $\Delta V_R = -(I \Delta t) / C$. This *change* in resistor voltage will be the same for every increment of time; thus, taking limits, we have:

2.82
$$\frac{dV_R}{dt} = \frac{I}{C} = -\frac{V_R}{RC},$$

This has the solution:

2.83
$$V_R = V\, e^{-\frac{t}{RC}} \quad \text{(Where we used the initial condition } V_R = V \text{)}$$

Hence, the voltage across the capacitor as a function of time is:

2.84
$$V_C = V\left(1 - e^{-\frac{t}{RC}}\right).$$

Hence, the voltage on the capacitor asymptotically approaches the supply voltage with a time scale of $\tau = RC$. Here it is helpful to remember that, with R in $k\Omega$ and C in pF, the time is in μs, i.e. microseconds, millionths of a second.

With these last touches of basic circuit theory, we have in place the fundamental understanding of charge and the power in the plana it causes, the static electric field, including the potential which, in our view, further specifies the nature of the field, as well as important tools for understanding many particular systems in some quantitative detail.

Of course, we are not done with the electric field, because we have only considered *static* electric fields in this chapter; we will return to it and the potential in a later chapter. We first need to introduce a second power (quality) that can be caused in the plana, a power that is deeply complementary to the electric field, the magnetic field. Indeed, without the magnetic field, special relativity, as we will see in Chapter 7, could not be true. We now turn to the magnetism.

Summary

Part I

Incorporating charge into our diagram categorizing substances, we have:

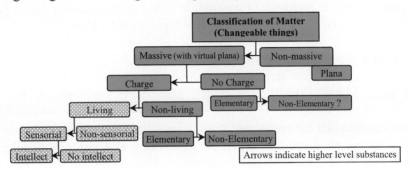

As with gravity, there are two co-related types of ***charge***, active charge and passive charge, that a massive substance can have. Active charge is a quality that causes *a power (ability) (also in the category of quality) of the plana called the **electric field** that can cause impetus* in bodies that are receptive to that activity, i.e. have the quality known as passive charge.

Static electricity is manifested by the action of charged bodies (or parts of bodies) on each other via the plana as described above. Sparks from a door knob, for example, result when an excess of charged particles, electrons, on one surface flow to the other under the action of electric field caused by the excess.

Electric fields propagate at the speed of light. Even static electric fields are maintained by their sources (charges) at this speed by constantly causing an *E*-field in the parts of the free plana in immediate contact with them. Those parts of the plana then maintain the field in the parts in contact with them and so on. Nothing new happens unless the field reaches a body that is receptive to its action, i.e. has passive charge, at which point the field activates impetus in that body.

Thus, if the charges causing the field disappear, it will take a finite amount of time for the field to disappear at a given field point, because it is still being maintained by activity started before the charge disappeared.

As in *PFR-M*, because of the fundamental nature of the powers we are discussing, we generally, by an analogical abstraction, ignore the distinction between individual substances, groups of substances, and parts of substances, using the word "body" to refer to any member of this group.

Proportionality of active and passive charge

The strength of the active charge is proportional to that of passive charge for *every* charged body that we know of. Also, because of this proportionality, we conclude that active (which always precedes passive in principle) charge in massive bodies essentially includes the idea of a proportional passive charge. The reason for this proportionality (final cause) is that, without this essential relation, Newton's third law would no longer be true, even for the infinite speed of propagation idealization. This, in turn, means that we would lose *both* conservation of momentum and energy, with the attending losses of intelligibility in nature mentioned in *PFR-M*.

Because of the proportionality, we can, and do take, $q \equiv q_{active} = q_{passive}$. Thus, it is standard to refer simply to a charge of magnitude q, not distinguishing active and passive charge. Further, a charged massive body is often simply called a charge.

A *point charge* is a massive body of charge q is defined as follows. A given body is said to be point-like if its typical scale size, D, is much smaller than the distance, r, at which we measure its field. Mathematically, we write: $D \ll r$. If we consider the charge as *arbitrarily* small in size compared to a fixed distance r, then by a mental construct (being of reason), we can consider the completed limit of an infinitely small particle with the same charge q. This means we think of the particle as existing at a point, with infinite charge density at that point; hence, the term *point charge*.

Coulomb's law

The relation between the measure of the *electric field,* $\vec{E}$, at a given field point which is a measured distance, r, away and the measure of charge, q_1, causing it is: $\vec{E} = \dfrac{q_1}{r^2} \hat{r}$.

The relation of the measure of the force acting on a charge of measure q_2 receiving the action of the field of the first charge is written: $\vec{F} = q_2 \vec{E}$. These together give **Coulomb's**

Law: $\vec{F} = \dfrac{q_1 q_2}{r^2} \hat{r}$, $\hat{r} = \dfrac{\vec{r_2} - \vec{r_1}}{\left| \vec{r_2} - \vec{r_1} \right|}$

where, from the proportionality of active and passive charge, $\vec{F}$ is the force caused, via the plana, by q_1 on q_2 or vice-versa. Point charges are assumed for Coulomb's law.

The inverse square law above corresponds to the way the surface area of a sphere, centered on a given point charge, increases as its radius increases. Coulomb's law indicates that the point charge's ability to cause field strength diminishes in proportion to the surface area over which it must act; its ability is, in a way, getting thinned out.

Because of this correspondence, in three dimensions, we can use the density of field lines to represent the strength of the field.

There are two *types* of charge, each having the same amount of ability to cause an E-field as to receive the action of an E-field as described above. These are called "positive" and "negative", because they cause oppositely directed types of impetus in a given test charge. We summarize the behavior of charges as: "Unlike charges attract and like charges repel."

A *test charge* is a charged body that we place in the field to test what the magnitude and direction of the field is at a given point. The test charge needs to be small enough in size to be able to sample the field strength as closely in space as required, which is why a point charge is usually chosen. The strength of the test charge, q, also needs to be small enough so as not to significantly move the charges in the source, so as not to significantly alter (i.e. within the required level of approximation) the very field that one is trying to sample. Further, if multiple test charges are being used simultaneously, the test charges must not be strong enough to significantly affect the other field points being sampled.

Charge is discrete

The magnitude or strength of the charge of a massive body is always discrete. That is, the smallest known substances, such as isolated electrons and protons, always come in a certain magnitude, never more or less. The only exception known is quarks, which do not exist stably apart and which cannot be handled by classical E&M; they do however have a charge that is multiples of 1/3 of that of an electron. Furthermore, all known composite substances are observed to have a charge strength whose measure is an integer multiple of that on an electron.

Charge, i.e. essentially related active *and* passive charge, is first nature to all ordinary *massive* bodies, for example, those which we can see, hear and touch directly. Charge is first nature in an even more primary way to electrons, protons and other elementary particles. Charge is an essential property of ordinary massive bodies because, of necessity, electrons and protons are virtually present in such bodies; they would not be the bodies they are without them. Excess or net charge, however, is *second nature* to ordinary massive bodies, for most (perhaps all) such bodies can have net charge without losing their proper nature. By contrast, it is *essential*, first nature, to a charged elementary particle existing on its own, such as an electron, to actually have charge.

Furthermore, using our analogically general definition of body, we will say a body has charge, when we might simply mean that there is a group of substances that have a net charge. In this spirit, we can, and usually will, treat bodies as if they were nothing more than these charges stacked or held in position in some configuration.

Length scale of E&M

Classical E&M is valid over a wide range of lengths scales from the very small to the very large. To avoid complexities outside of the domain of E&M, we make the idealization to always consider large enough regions that the minimal parts of the plana and elementary charges can be ignored. Quantitatively, this means there are limits to the accuracy of our results, but the errors this averaging introduces are very small and will be neglected here.

Charge is conserved

Net charge cannot be created or destroyed. This means that if, at a given place, negative charge is created, then an equal strength positive charge must also simultaneously be created at the same place (within the length error specified above). We write this conservation law as: $\sum_i q_i = \text{constant}$. Without charge conservation, in a given region, positive charge, say 50 positrons, could appear and instantly potential energy would be created and we would lose conservation of energy.

Principle of Superposition
Electric fields caused by one charge add (or cancel) with those caused by others. In terms of their measures; mathematically, we can simply write: $\vec{E}_{net} = \sum_i \vec{E}_i$, where each $\vec{E}_i$ is the field that would be generated at a given point (arbitrarily small place) in the plana by the ith charge if it were alone.

　　Physically, this means that the actions caused on one part of the plana by multiple parts in contact with it are in some sense not affected by each other. One could imagine, for example, that two parts of the plana, each attempting to activate a field E_0 at a given part between them, might hamper each other's attempt, rather than "work completely together." Any hampering of one by the other would lead to a field less than $2E_0$ being generated in the given part, rather than the full $2E_0$ which we would actually observe. This principle is called *superposition* or *linearity* because one can simply (linearly) add the fields that each charge would cause separately to determine the field that actually results when they are all present.

First Extended Coulomb's law:
Using superposition, we write the discrete form of Coulomb's law:

$$\vec{E} = \sum_{i=1}^{N} \frac{k\,q_i}{r_i^2}\,\hat{r}_i$$

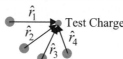

And, the continuous form:

$$\vec{E} = k\int_{Charge} \frac{\rho(\vec{r}')}{\left|\vec{r}-\vec{r}'\right|^2}\,\widehat{(\vec{r}-\vec{r}')}\,dV$$

$$= k\int_{Charge} \frac{\rho(\vec{r}')}{\left|\vec{r}-\vec{r}'\right|^2}\left(\frac{\vec{r}-\vec{r}'}{\left|\vec{r}-\vec{r}'\right|}\right)dV$$

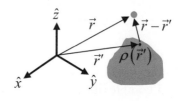

$$\vec{F}_{net} = q_{test}\vec{E}_{net}$$

Here, $\rho(\vec{r})$ is the charge density, the charge per unit volume at a given location marked by the vector $\vec{r}$. In *cgs*, $k=1$, with q measured in *esu* (electrostatic units), where $e = 4.803\times10^{-10}\,esu$. In this text, impetus is measured in *phor*, which comes from the Greek word, *φόρα*, for impetus; force is measured in dynes ($1\,dyne = 1\,phor\,/\,s$), and energy is measured in ergs ($1\,erg = dyne\cdot cm$). In *SI* units, $k = \dfrac{1}{4\pi\epsilon_0} = 8.988\times10^9\,\dfrac{Nm^2}{C^2}$, with q measured in units of coulombs, where the charge on an electron is $e = 1.602\times10^{-19}\,C$. As in *PFR-M*, impetus is measured in buridans, force is measured in newtons and energy is measured in joules.

Results from the Extended Coulomb's law
As with a gravitational field, because of the inverse square law, the *E*-field inside of a uniformly charged spherical shell is zero. Also, like gravity, the field generated outside of a

uniformly charged sphere is the same as would result if, instead of the sphere, there were a point charge located at the former center of the sphere. The work done to bring together two point charges is: $k \dfrac{q_1 q_2}{r_0}$. The energy to put together a uniformly charged spherical shell of charge q and radius r_0, i.e. its self energy, is $\dfrac{k}{2} \dfrac{q^2}{r_0}$.

The current density at a given point, $\vec{J}$, represents the direction and magnitude of a current flow per unit area in a given time at that point. The current could be, for example, the number of bodies or the measure of the strength (or intensity) of a property crossing through a given surface per unit area of that surface in a given time. For a fluid, $\vec{J}$ could, for example, tell us how many grams of water (or how many units of volume, cm^3 since water is (nearly) incompressible) cross a given surface per cm^2 in one second. We can write: $\vec{J} = \rho \vec{v}$, where ρ is the mass density of the fluid and $\vec{v}$ is its velocity at a given point. We can write the current per unit area through a surface with a normal $\hat{n}$ as: $\hat{n} \cdot \vec{J}$.

Mathematical form of Gauss's law

Letting $\vec{J}$ represent the *volume* current density, we can write the following: $\int_S \vec{J} \cdot d\vec{A} = \dfrac{d(\text{volume (e.g., of water)})}{dt}$. Since we are interested in quantity only, we integrate out the time to get the *mathematical form of Gauss's law*:

$$\int_S \vec{e} \cdot d\vec{A} = \text{volume (e.g. of water)} ,$$

Where $\vec{e} = \vec{J} \Delta t$ specifies at each location on S, how much volume per unit area one is pushing around and in what direction. The dot product projects out only that part that is pushed across the surface and the integral adds up all the contributions.

This formula simply gives the total volume that will get pushed through the surface, S, if one applies the set of instructions for pushing volume around given by the $\vec{e}$-field. When coupled with the idea of moving a volume, i.e. the extension of an incompressible fluid like water, from place to place, it can be used to study how volumes can be pushed through different surfaces, resulting in rearrangement of the original volume. This is, of course, a very simple aspect of quantity. However, there are infinitely many details involved in the infinite number of ways this can happen and all are captured in this formula. Note that in the purely mathematical viewpoint, the focus is on the different ways space can be put together that is revealed by the rearrangement of parts of a volume specified by $\vec{e}$, not on the movement itself.

Of particular interest is the case in which S encloses a region that has a source of water within it, but none outside of it. Then, assuming the flow has been going on long enough, it is clear that the amount of water exiting any surface, S', that encloses S, will equal that which leaves S. Integrating over time, we say that the volume moved across S is the same as that moved across any S', so that we can write the equation as: $\int_{S'} \vec{e} \cdot d\vec{A} = $ constant for all S' enclosing or coinciding with S.

Most particularly, if I push a certain volume of water *uniformly* away from a dispersion point, the amount of volume that crosses each unit of area of a spherical surface centered on that point will fall off as the inverse square of the radius. This is because the given amount of water always takes up the same volume, so that, for example, a shell of water must thin out as it moves outward; that is, in general, it must take on different shapes to accommodate the way it moves. In general, the water flow (current) per unit area falls off as the inverse of the magnitude of the surface area through which the water flows.

The *divergence* is defined as follows:

$$\nabla \cdot \vec{J} \equiv \text{div}\, \vec{J} \equiv \lim_{\Delta V \to 0} \frac{\int_{S_{\Delta V}} \vec{J} \cdot d\vec{A}}{\Delta V}$$

Mathematically, it tells us, given $\vec{J}$, the specification of the volume movements at each location, i.e. how much volume comes out of (or goes into) a given arbitrarily small region of the environment *per unit volume of that small region*. If, for example, there is a net outward flux of water, i.e. the creation of new volume that "uses up space" not used before in the environment, say by displacing air, then there must be creation of material, in this case water, in the small region. In such a case, we say there is a *source* at this location. If there is net inward flux, we say there is a *sink*. Like the mathematical form of Gauss's law, the divergence is a way formally handling the different ways that volume can be rearranged. In Cartesian coordinates it is expressed as: $\nabla \cdot \vec{J} = \left(\dfrac{\partial J_x}{\partial x} + \dfrac{\partial J_y}{\partial y} + \dfrac{\partial J_z}{\partial z} \right)$, where

we can think of the del operator as: $\vec{\nabla} \equiv \hat{x}\dfrac{\partial}{\partial x} + \hat{y}\dfrac{\partial}{\partial y} + \hat{z}\dfrac{\partial}{\partial z}$.

Gauss's Theorem
It can be shown, and indeed it is intuitively clear, that adding up all divergences in a given region will give the net flux out of that region. This is written as:

$$\oiint_S \vec{J} \cdot d\vec{A} = \int_V \text{div}\vec{J}\; dV$$

This formula deals only with closed regions. Mathematically, it connects the sources and sinks of volume (e.g., the extension of some water) within a finite region with the flux at the surface, thus allowing us to work with the infinitely many ways that volume can be made to fill space.

Conservation Law (integral and differential forms)
Using a conservation of mass, thereby leaving pure quantity, we can write: $\oiint_S \vec{J} \cdot d\vec{A} = -\dfrac{dm_{inside}}{dt}$, where we use the first form of the mathematical form of Gauss's law given above and the fact that the flux of mass out of the region enclosed by S, must result in a decrease of mass within that region. The differential form of this law is:

$\nabla \cdot \vec{J} = -\dot{\rho}$, where ρ is the mass density. Both these equations are formal ways of expressing the details of mass conservation.

Maxwell's First Equation *Gauss's law (Differential and Integral form)*

Since the electric field weakens in the same inverse square law way as the incompressible fluid described by the *mathematical form of Gauss's law*, we can take the propagation of the electric field strength to be analogous to the movement of an (irrotational) incompressible fluid. In other words, the same quantitative facts about volume movement that the mathematical form of Gauss's law incorporates can be analogically applied to the strength and direction of the electric field. Hence, we can write Gauss's law in *integral* form:

$$\oiint_S \vec{E} \cdot d\vec{A} = 4\pi k q \qquad \text{where } q \text{ is the charge inside the region enclosed by } S$$

This is the relation between the measure of the *net* strength of the "static" electric field propagating through the surface S and the measure of the strength of the source inside the region enclosed by S that is responsible for sustaining the propagation of that field. If there is no charge inside there will be no *net* field strength propagating through the surface. Other charges, like other water sources, can change the flow patterns, but cannot destroy what emerges from the charges inside S. Each charge is a continual source of field of a certain strength (esu/cm^2). This ability (power) to cause a field, which is actually part of the power of the field itself and which is measured in *esu*, "spreads out" over space like a fluid but without diminishing the actual amount of *esu*'s of the charge. This is unlike the case of a conserved quantity which can only appear in one place by leaving another.

In *differential* form, i.e. for each point in space, we can write using, Gauss's Theorem:

$$\nabla \cdot \vec{E} = 4\pi k \rho, \text{ where in } cgs, \ k = 1; \text{ in } SI, \ k = \frac{1}{4\pi\epsilon_0} = 8.988 \times 10^9 \, Nm^2 \, / \, C^2$$

This equation gives the relation between the measure of the *net* electric field and the measure of the strength of the source (per unit volume) in an arbitrarily small region, idealized to a point.

Gauss's law extends Coulomb's law beyond the incorporation of the superposition principle, which we will understand after we incorporate the A-field, which has an E-like effect.

To recover Coulombs law from Gauss's law, even for a single point particle, we need $\nabla \times \vec{E} = 0$.

The *curl* is defined as follows:

$$\text{curl } \vec{J} \equiv \lim_{\Delta A \to 0} \frac{\int_{C_{\Delta A}} \vec{J} \cdot d\vec{l}}{\Delta A} \hat{n} = \vec{\nabla} \times \vec{J}$$

Where, $\hat{n}$ is the normal to the plane of the curve $C_{\Delta A}$ that is obtained by the right hand rule applied to the direction of the line integral.

It tell us, given $\vec{J}$, the degree to which water (volume) flow in a region "curls," i.e., the degree to which the water flow in a given direction at a given point changes as one moves to a nearby point perpendicular to that direction. Water rotating around a point is the most intuitive type of "curl" that one can have at a point. However, currents circulating around a point are not the only flows that have curls, as can be seen in the case of a straight stream that moves parallel to its banks, but more swiftly in the middle than near the banks.

We will give a more complete understanding of the mathematical meaning of the curl in Chapters 3 and 5, where it is needed. We will also see in the next chapter that one can judge whether there is a curl at a given point in the flow by putting a tiny symmetrical cross, called a curl meter, in the flow. If it turns, then there is a curl at that point in the flow.

Like the divergence, the *curl* tells us something about the nature of the *J*-field that we are dealing with. Later, we will see that under certain "reasonable conditions," specifying the curl of a field and its divergence at every point is equivalent to specifying that field. In Cartesian coordinates, it is expressed as:

$$\vec{\nabla} \times \vec{J} = \left(\frac{\partial}{\partial y} J_z - \frac{\partial}{\partial z} J_y \right) \hat{x} + \left(\frac{\partial}{\partial z} J_x - \frac{\partial}{\partial x} J_z \right) \hat{y} + \left(\frac{\partial}{\partial x} J_y - \frac{\partial}{\partial y} J_x \right) \hat{z}$$

Stokes Theorem
The following is the analogical equivalent of Gauss's theorem for two dimensions. Namely, it can be shown (see Appendix III), and it is intuitively clear, that adding up all curls in a given slice of space will give the net tendency to have motion around that slice. Formally, we write:

$$\oint_{Path} \vec{J} \cdot d\vec{x} = \iint_{Surface\,defined\,by\,Path} \left(\vec{\nabla} \times \vec{J} \right) \cdot \hat{n} \, dA$$

Energy density of static electric field
The electric field of a charged body is a sign of the total self energy, i.e. the energy that would be released if the parts of that body were allowed to fly apart under the action of their electric fields. The relation between the measure of the energy per unit volume, U, and the measure of the electric field is: $U = \int_{particle} \frac{\vec{E} \cdot \vec{E}}{8\pi k} dV$. To get the total field energy one can think of adding up the differential contributions of the field energy: $dU = \frac{\vec{E} \cdot \vec{E}}{8\pi k} dV$. In this limited sense, one can think of the electric field as having an energy density at each point.

The *gradient* of a function called $f(x, y, z)$, $\vec{\nabla} f$, gives the steepest rate of assent at the given point.

Electric potential
The *potential*, ϕ, is a further specification of the *electric field*; it reveals that the disposition of each "cell" is really (given the caveat mentioned in the text) a deposition of at least two parts of that cell. The relation between the measure of the strength and direction of the electric field and the measure of the intensity (strength) of the potential is: $\vec{E} = -\vec{\nabla} \phi$.

The *E*-field tells something of the ability of a charged body to act via the plana, while the potential further specifies that ability (power). In particular, at a given point, the *E*-field tells how much momentum per second the charge body, via that field, causes in the test particle per unit charge of that test particle. Whereas, the potential tells how much total energy (activity) that the charged body will cause, via the field, in the test particle if it is allowed to act freely for an arbitrarily long distance. Loosely, in understanding an ability or power to act, we want to know both its strength and its total output, i.e., in a way "how long it will last;"[49] the electric field provides the first and the potential provides the second.

Gauss's law, in terms of the potential, is *Poisson's equation*: $\nabla^2\phi = -4\pi k\rho$, which has the solution $\phi(\vec{r}) = k\int_{Charge}\dfrac{\rho(\vec{r}')}{\left|\vec{r}-\vec{r}'\right|}dV$. In Cartesian coordinates, *the Laplacian*, is written:

$$\nabla^2 = \frac{\partial^2\phi}{\partial x^2} + \frac{\partial^2\phi}{\partial y^2} + \frac{\partial^2\phi}{\partial z^2}.$$

In a region with no charge, Poisson's equation reduces to *Laplaces's equation*: $\nabla^2\phi = 0$. Solutions to this equation, called harmonic functions, have three important properties.[50]
(1) the *uniqueness property*: given potentials defined on some surfaces and charge on others, there is only one solution for the potential in the charge-free regions, so if you find one solution with the requisite boundary conditions, you have *the* solution, (2) in a charge free region, the average of the potential over the surface of *any* sphere is equal to the potential at the center of that sphere, (3) A solution to Laplace's equation minimizes the field energy, $U = \int_{all\,space}\left(\vec{\nabla}\phi\right)^2 dV$ for a given system of particles and constraints, i.e. "finds" the static placement of particles that lowers their field energy as far as it can be within the given constraints.

Maxwell's Third Equation

Because a static electric field can be written as the gradient of the potential, it's curl is always zero, i.e., $\nabla\times\vec{E} = 0$. Using Stokes Theorem, we get the integral form: $\int_C \vec{E}\cdot d\vec{s} = 0$, where *C* is the boundary of a given surface.

Part II

Good conductors such as metals have free electrons that can move under the influence of an electric field. Non-conductors, such as plastic or wood, do not.

Ideal Conductors
An electric field cannot be maintained in an ideal conductor, which a metal approximates, because the mobile electrons, called *conduction electrons,* rearrange so as to precisely

---

[49] Your strength doesn't last or runs out because you get tired, whereas a charge never gets tired, but its influence gets weaker as it successfully moves the test charge away and, in this sense, gradually "runs out."
[50] For both these, we leave aside the mental construct of charges at infinity.

cancel whatever field one attempts to apply. That is, for the static case, $\vec{E} = 0$ in an ideal conductor, which means the potential is constant, $\phi = \text{constant}$, throughout the body of a conductor. Because of this, all the excess charge, free electrons and the ions they leave behind, must reside on the outer surface of the ideal conductor, where they are only stopped from moving further by the forces that hold the metal together.

The electric field is perpendicular to the surface of an ideal conductor, because fields parallel to it are canceled by motion of conduction electrons on the surface.

Inside of a metal box, a Faraday cage, there is no static electric field.

The method of images takes advantage of the properties of an ideal conductor. In particular, a positive point charge brought near a plane-like slab of conductor, creates a field as if there were a negative point charge symmetrically on the other side of the plane, as if the plane were a mirror. This is because two point charges of opposite sign make a constant potential surface in the plane midway between them, and a plane-like conductor near a point charge must accomplish the same effect, which it does by the rearmament of its surface charges so as to mimic the presence of the second "image" charge.

Through internal chemical reactions, batteries supply *electromagnetic force*, EMF, that can separate charge.

Electrical current, *I*, is defined as the amount of charge that passes through a given surface in a unit of time, $I = \dfrac{dQ}{dt}$. Given $\vec{J} = \rho\vec{v}$, where ρ is the charge per unit volume, $I = \int \vec{J} \cdot d\vec{A}$. Current is *defined* as positive in the direction of the flow of positive charges.

A *capacitor* is an electronic component that stores charge. The relationship between its *capacitance*, *C*, and the charge, *Q*, dumped on it and the resulting voltage, *V*, is given by: $Q = CV$. Capacitance is measured in farads (F) in the SI system. The capacitance of the parallel plate capacitor, in SI, is: $C = \epsilon_0 \dfrac{A}{d}, \epsilon_0 = 8.85 \times 10^{-12} F / m$

Kirchhoff's current and voltage laws
Because, current is conserved:
$\sum_i I_i = 0$, for every node (junction point) in a circuit

Because, statically, the potential ϕ specifies the electric field:
$\sum_i V_i = 0$, around every closed loop in a circuit.

A *resistor* is an electronic component that resists the flow of electrons. The relationship between its *resistance*, *R*, the applied voltage, *V*, and the resulting current, *I*, is called *Ohms law* and is given by: $V = IR$. In more fundamental terms, we write: $\vec{J} = \sigma\vec{E}$, where σ is the conductivity in inverse ohm-meters.

Helpful Hints

Keep in mind that charge can be thought of as distributed in three ways, according to the three dimensions of space. We can have a certain amount of charge per unit length, i.e. a linear charge density, symbolized by λ, charge per unit area, which is called a surface charge density and symbolized by σ, and charge per unit volume, symbolized by ρ.

There is generally some ambiguity in the use of symbols that is only resolved by the context of the physical situation. For example, λ is also used for a length, such as a wavelength, σ is also used for conductivity in the more fundamental form of Ohm's law (equation 2.78), and ρ is also used for the radius in cylindrical coordinate systems.

It is helpful to remember the following standard potentials for positive and negative point (or spherical) particles: $\phi_{+} = \dfrac{q}{r}$ and $\phi_{-} = -\dfrac{q}{r}$. The signs in these potentials can, in turn, be recalled using the rule: "negative for negative, positive for positive."

The key intuitive reason for the validity of Gauss's theorem and Stokes theorem can be remembered, respectively, by the following figures, showing arbitrarily small regions into which a volume and an area are divided. The first figure shows that what leaves one volume enters the adjoining volume, except of course on the outer boundary where there is no adjoining region. The second shows, analogously, that the contribution to the curl at the boundary of a region is canceled by the contribution of the loop around the region that shares that boundary, except on the outer boundary where there is no adjoining region.

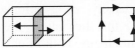

Problems

1. Calculate how many negative charges and how many positive charges are in a drop of water.

2. Consider two positive charges, A and B, that are near each other. Explain why the field created by A on *itself* does not come into our calculation of the net effect on A.

3. Make an analogy between mass flow and the propagation of the power of the field by using the units of gram for the first and *esu* for the second.

4. How does a stud detector work?

5. Assuming two equal mass, equal sized, $1cm$ diameter aluminum balls are hanging from thread such that they are just in contact. After someone charges the balls equally to an unknown level, q, they separate settling in to make an angle of about 45 degrees with the vertical. Calculate q, in terms of the angle, θ, and the length, l, of the thread and the ball's mass, m. Take: $\rho_{aluminum} \sim 2.7 g / cm^3$ and $l \sim 1m$ give an estimate of the charge in coulombs and in number of electrons.

6. (a) Calculate the surface charge density, σ, as a function of the distance along the planar surface of a large metal plate that lies along the surface of constant potential between a positive and a negative point charge, both of charge q. Do this for both sides of the plate. (b) Can you do the same for two *positive* charges of charge q? What about for charges of $10\,q$ and $-q$? Explain.

7. Using the standard zero at infinity definition for the potential of a point particle, give an equation describing the shape of the zero potential surface between a point charge of strength $10q$ and $-q$. Describe the cross section of this shape in a plane that passes through the two charges.

8. Consider a volume, V, that, as discussed in the subsection "Purely Mathematical Version of Gauss's Law", emits a certain amount of X. a) Write an equation for this effect and explain its meaning. b) For both the case of conserved X and non-conserved X, explain what can be said about the amount of X inside the volume, V. c) Do the same for the case in which the volume has a *net* amount of X going in.

9. Consider shells like discussed in calculating the self energy of the classical electron. a) Write down the relationship between the total charge q, the radial charge density $\lambda(r)$, and the shell thickness, dr, for the completed sphere. b) Write down the relationship between $\lambda(r)$ and dr on the one hand and the radial charge density, $\bar{\lambda}(r')$, as a function of the radius of a *given* spherical shell at various radii, r', as it is being brought in from infinity and its thickness dr' on the other. c) Explain the case in which we bring the shells in from infinity with a constant thickness dr. What happens to $\lambda(r)$? ρ? Give quantitative

as well as qualitative answers. d) Do the same for the case of ρ remaining constant as one brings each shell in from infinity.

10. Related to the above problem, consider a constant mass density fluid ($\rho = $ constant) such as water. a) Show that the speed, v, hence the current density, J, of a uniformly expanding spherical shell of such fluid decreases as the inverse square of its radius. b) Show how the fall-off of the shell's thickness, Δr, reveals the inverse square law behavior of J.

11. a) Describe the mathematical meaning (in terms of 3D volumes) of this relation:

$$\oiint_S \vec{v} \cdot d\vec{A} = \frac{d(\text{volume of water})}{\rho \, dt} = \frac{d(\text{volume of environment})}{dt}$$, where ρ is the ratio of the

volume of water to the volume of the environment for a given amount of water (say 1 gram of water, which makes the ratio unity). b) Contrast this with the meaning of the relation:

$$\oiint_S \rho \vec{v} \cdot d\vec{A} = \frac{dX}{dt}$$, where we first take ρ to be volume density of an incompressible fluid

as above, then as a mass density.

12. Given the field of the Earth, as ~100 V/m (which is what is measured over level ground) calculate the charge on the Earth, assuming it is only due to an approximately uniformly distributed surface charge.

13. Using the identity $\oiint_S \vec{\nabla}\phi \cdot d\vec{A} = \oint_C \phi \, dl$, and what you know about the appropriate

Maxwell equation obtain the equation for the potential of a point particle of charge q.

14. As we have mentioned in the text, water flow provides a reasonable analogy to the propagation of the static electric field. To better understand the curl of a field that specifies a current flow, consider the problem of eddies and backflow. Draw a typical Eddie current with its typical backflow around a log or rock in the river. In simple terms, explain how it arises.

15. a) Does a current density field given by $\vec{J}$ that circles around a certain point and falls off as $1/r$ have a curl anywhere? If so where? b) Suppose it does not fall off as $1/r$; for example, suppose the magnitude does not change as a function of radius or angle? c) Treating the field as a force field, rather than a flow field, calculate its effect on an arbitrarily small curl meter for the general case of a field that makes a circular pattern centered at the origin. Draw and discuss the particular cases of $\vec{F} = \frac{k}{r}\hat{\theta}$ and $\vec{F} = F_0 \hat{\theta}$,

where $k, F_0 = $ constant .

16. Think of an experiment to determine the inverse square distance law of Coulomb's law using the impetus that the field activated plana causes in a test particle, instead of a balancing force. Comment on the difficulties of each experiment. Which is easier and why?

17. Calculate the classical radius of a proton, a top and bottom quark? Discuss your results and the limits of classical E&M.

18. How much charge would Mars need to have in order to exactly counter act the force of gravity for a small bit of paper charged with $1\mu C$ of charge, given the density of paper as $80g/m^2$?

19. Explain how a large enough electric field, in for example a parallel plate capacitor, could "blow out" a single birthday candle. How would you go about estimating how large a field would be needed?

20. a) An interesting region of the *Paschen curve* for CO_2, the major constituent of the Martian atmosphere, is shown in black below.[51] Roughly, the x-axis is the product of the pressure and the distance between two parallel plates, while the y-axis shows the potential difference at which the gas breaks down under the action of the electric field. In terms of this Paschen curve, explain why electrical breakdown at a given voltage can sometimes occur easier for contacts that are more separated. Also, qualitatively explain the physical origin of the effect. b) The red curve is the result with known important Martian trace gases (N_2, Ar, O_2, CO) included. In the regime represented by the graph below, what is the optimum spark distance given the Martian pressure is approximately 7 mm of Hg. c) In that same regime, what pressure of Martian atmosphere should we put equipment under to best avoid sparks on Mars power stations?

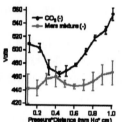

21. a) Describe how to measure the ratio of the charge to the mass of a small particle by suspending them against gravity by an electric field. Also, give the formula relating the ratio to simple parameters of the experiment. This experiment was first done by Millikan by charging oil drops. b) Assuming a singly charged drop with a mass which is found to be $5\times10^{-16}kg$ while the stopping voltage is 50 volts for a 1 *cm* parallel plate configuration, estimate the charge of the electron assuming the drop has only one electron of net charge. How did Millikan get the mass?

22. a) Taking the electric field strength that air breaks down at as $3\times10^6 V/m$, estimate the amount of charge transferred during a single bolt of lightning. Hint: approximate the ground and clouds as two parallel plates. b) For the case of a lightning bolt during discharge, i.e. while its current is flowing, estimate how far out the air would be ionized from the charge of the lightning bolt itself. To get your estimate, assume, in a gross idealization, that the lighting follows a straight line of length $\sim 5km$ and the charge is uniformly distributed.

---

[51] http://empl.ksc.nasa.gov/CurrentResearch/Breakdown/Breakdown.htm

23. Calculate the field around and in an infinite area slab of width w and uniform charge density ρ.

24. If one considers a point particle idealization at the origin, then we have $\nabla \cdot \vec{E} = 0$ every where except at the origin. By a mental construct, we conceive the point at the origin as having a finite strength charge, which means the charged density is infinite there. In our mathematical formalism, we write this infinite charge density at the origin as $\delta(x,y,z)$, so have: $\nabla \cdot \vec{E} = \delta(x,y,z)$. a) Derive Coulomb's law from this equation. b) Take the divergence of it using the standard Cartesian form for the $\vec{\nabla} \cdot$. Why is the answer identically zero and what does it say about Coulomb's law?

25. We often assume for simplicity of exposition that electrical current consists of a flow of electrons, with a fixed strength impetus, i.e. constant momentum, and hence speed, proportional to the applied voltage. For concreteness, further assume a long cylindrical device where the voltage is applied across the ends and that this voltage causes a uniform E-field throughout its length. a) Give the simplest physical explanation that is evoked by this picture of a constant unchanging impetus under the action of a constant E-field. b) A more accurate view notes that "free" electrons actually accelerate for a certain time. Given the mass m_e and charge e of an electron, the time, t_i, that the ith electron accelerates and assuming N electrons, write down the equation for the *average* momentum of a free electron, $\langle p_e \rangle$. c) Assuming after each "smash" the impetus given by the field is, on the average, lost, give an expression for the speed, $\bar{v}_e$, averaged over times long compared to the average collision time. Assume that each electron can, *on the average*, accelerate for τ seconds before its *thermal* velocity, not its velocity acquired from impetus given by the field (since, for fields below a certain strength $v_t \ll v_{field}$), "smashes" it into an atom. d) Write down the current density and show this is a proof of Ohm's law.

26. Calculate the approximate capacitance of a small metal sphere of radius a above an infinite metal plane. Check your result by comparing it with the capacitance of that sphere incased in a conductive concentric sphere of arbitrarily large radius, which models "a ground plane at infinity."

27. The device below on the left is an *electrostatic* motor invented by Ben Franklin; on the right is a homemade version constructed from two spice jars, aluminum foil, stick glue, 2 nails, a sharpened pencil, and the bottom of a plastic cup. Explain how the one on the left works using the following outline. The cylindrical jars on the left and right of the figure are called Leyden jars. These jars each have metal (e.g. aluminum foil) lining the inside and outside of the jar, with the contact to the inside surface being brought out by a rod through the rubber stop on the top. The spokes of the rotor are terminated with metal caps. a) Taking the jars to be about $10cm$ high, very roughly calculate the capacitance of the Leyden jar (assume here that the glass "dielectric" effectively multiplies the vacuum capacitance by a factor of 5) and the capacitance of each spoke cap to ground and show that $C_{Leyden} \gg C_{spoke}$. b) Draw a schematic representing the circuit through which charge

flows from the spherical tip of the rod emerging from the jar to the ground. c) Assuming that the Leyden jar is charged to 10kV, estimate the charge, q, that will arc from the jar to the cap. d) Noting that one jar is charged with the inside positive and the other with it negative, explain qualitatively how the device works. e) Treating the spoke cap/jar tip as a capacitor, assuming all the voltage of the jar appears across it and that the gap between them is about .5 cm, estimate the total torque on the rotor. f) Taking the mass of the whole rotor assembly to be about 5 grams, and assuming the rotor, when spinning at the typical speed of the motor, ω_0, stops after one rotation due to friction, very roughly approximate the rotational speed of the motor. For simplicity, assume the rotor is a uniform density disk.

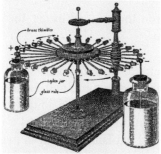 

28. Suppose a particle existed that had twice the passive charge as active; how would this show up in an experiment?

29. Discuss the diagram below which attempts to incorporate the weak and strong interactions into our schema.

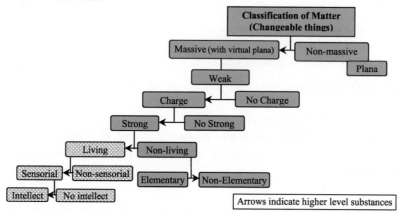

30. Electrostatic discharge can and does damage electronic components. Grounded conductive mats with wrist bands are often used to protect components. Discuss how such mats could be helpful and how to use them. Name at least one problem that such mats might introduce.

31. In thinking about average current flow, it is important to decide on a time scale, since the flow can appear much different on different scales. To see this effect, consider the following simple example. Two men take the same Dallas to Houston trip. The first travels

the 240 miles nonstop at 40mph. The second has a problem with his car that allows him to travel at 80mph but only for only 12 minutes, then he must cool his car off for 12 minutes. He cannot leave his car during cool-off, so this period should be counted as part of the travel time. For each of the two travelers, plot the average speed versus time for each of the following time scales: a) 6 hours b) 10 minutes c) 1 minute. Explain the differences and their importance in understanding the flow. What is the advantage of instantaneous definition of speed first conceived by the medievals[52] and put into symbolic mathematical form by Newton and Leibniz by introducing the derivative?

32. Water can be bent under the influence of a charged comb (see Figure 8-1a). Given that the water is composed of dipole parts (molecules), in qualitative terms what causes the water to bend?

33. Calculate the energy required to bring together three point charges in an equilateral triangle configuration with side length d.

34. Explain why a uniformly charged sphere of total charge $2q$ will cause twice the field at any point but four times the field energy relative to a sphere with charge q of the same radius.

35. Explain how the discrete nature of charge, i.e. that there is a *smallest* strength of charge from which stronger charges are made, is related to the properties of ordinary objects, especially their neutrality.

36. Explain how this water dripping device generates high voltage.

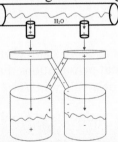

37. Draw a monopole moment, a dipole moment, and a quadrupole moment of point charges. Explain why the simplest type of radiation comes from a varying electric dipole moment, not from an electric monopole moment. In contrast to E&M, in terms of this same type of argument, why is the lowest order radiation from quadrupole sources in gravitation. Hint: think about conservation of linear and angular momentum. Assume a gravito-magnetic field results from impetus activated charge, and argue from conservation of angular momentum.

38. Discuss what the world might be like if negative charges did not repel each other exactly as much as positive charges did.

52 See "Oxford Calculators" box in Chapter 2 of *PFR: Mechanics*.

39. Calculate the field inside of a uniformly charged sphere.

40. Calculate the curl and gradient of field inside and outside of a uniformly charged sphere in Cartesian coordinates.

41. We have not proved that the curl of a vector field, as defined in equation 2.36, is indeed a vector, though it is and can be proven to be one. When one says a vector has magnitude and direction, one must be careful because though this makes good intuitive sense, this intuitive grasp can, if one is not careful, easily be mistranslated into the formalism. If for example, one were to attempt to define a vector as $\hat{n} \cdot \vec{V} = |\vec{W}|$, where $\hat{n}$ is any unit vector and $\vec{W}$ is another vector. Explain the problem with such a formal definition.

42. a) Using Coulomb's law extended by the superposition principle, show how to derive the formula for the work required to assemble a charge distribution in terms of the potential, $\phi(\vec{r})$, and $\rho(\vec{r})$ that specifies the distribution: $U = \dfrac{1}{2}\int \rho\phi dV$. b) Verify the same result starting with the field energy density equation expression for the self energy:
$U = \int_{Everywhere} \dfrac{\vec{E} \cdot \vec{E}}{8\pi} dV$; prove and make use of the vector identity:
$\vec{\nabla} \cdot (\phi\vec{V}) = \vec{V} \cdot \nabla\phi + \phi\nabla \cdot \vec{V}$.

43. Establish the laws: a) for summing resistors in parallel for two resistors, b) for N resistors in parallel and series c) do the same for capacitors.

44. a) Show it is impossible to have a *stable* static charge distribution, i.e. one that holds a charge in a given position under slight perturbations. For example, a ball at the top of a hill is not stable but one at the bottom of a valley is. *Hint:* The extension of the condition for stability in one dimension, $\dfrac{\partial^2 E}{\partial x^2} > 0$ to three is $\nabla^2 E > 0$. b) Given a positively charge point test particle of charge q located at the halfway point between two *fixed positive* point particles of unit charge separated by a distance d. Give the equation for the potential energy as a function of displacement from the center point along the line connecting the particles. Is the test particle's position stable to perturbations along that line?

45. We did not prove that $\vec{E} \cdot \vec{E} / 8\pi$ is the electric field energy density for any distribution of charge. Prove it by use of the superposition principle and calculating the field energy of two point particles of elementary charge q in the following way. a) In spherical coordinates, write down the field, $\vec{E}_1$, for the first particle at the origin and the field, $\vec{E}_2$ for the second particle lying on the z-axis at distance a from the origin. b) Calculate $\int_0^{2\pi} \int_0^{\infty} \int_0^{2\pi} \dfrac{\vec{E} \cdot \vec{E}}{8\pi} r^2 d\phi dr \sin\theta d\theta$ and show, neglecting the self energy of the point charges, that it is the same as the direct calculation of the self energy. Which terms correspond to

the self energy of the point particles, and how can we justify ignoring them? c) Complete the proof for the case of a general distribution of N such charges.

46. Look up a Wimshurst machine, and a Van de Graff generator; draw a diagram and describe how each works. How might you approach giving quantitative analysis of these devices?

47. Knowing electric fields can be helpful in deciding whether lightning might be a threat to a launch vehicle. The worries are much more intense for vehicles that have human lives at stake, such as the manned mission to Mars. The Electric field mill can measure this field. Look it up and describe how it works. See, for example, July 1999 issue of Scientific American.[53] NASA has thirty one ground-based field mills, such as pictured to the right, around Kennedy Space Center (KSC) and Cape Canaveral Air Station.

48. Vacuum tubes were the first active electronic components. A diode allows current to flow one way but not the other; it's a kind of electronic one way valve. The vacuum tube diode, such as diagramed at left,[54] consists of a plate called the *cathode*, which is heated to emit electrons that are accelerated to the higher voltage plate, called the *anode*. The potential does not fall off linearly as it does in a capacitor because of the built up of electrons called the space charge effect. When the current is limited by this effect, the voltage from cathode to anode can be approximated, neglecting the cylindrical geometry and

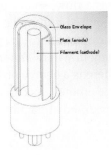

treating it as two parallel plates separated by distance d. $V(y) = V_{anode} \left(\dfrac{y}{d} \right)^{4/3}$

a) Calculate the E-field as a function of the distance, y, from the cathode. b) Calculate the divergence of $\vec{E}$ and determine the charge density as a function of y. c) Recalling that the current density at the anode is given by $J = \rho v_y$, where v_y is the final velocity of the electron, determine the relationship between the voltage applied and the current produced.

49. Play one round of the following two-player game dealing with 2-D flows and divergences. A given region of space is divided into 25 squares (see game board below). Like the figure to the left, each square is drawn using solid lines for the left and bottom sides and dotted lines for the top and right sides. The solid lines represent the flow region *within* the square above and to the right of those lines. The dotted sides distinguish the given square from the squares on top or to the right of it. The small empty circle in the middle of each solid line indicates the possibility of a source or sink point along the given flow direction. The filled circle at each grid line intersection marks where the vector current density is to be specified. In this two player game, each player will alternate turns trying to fill in the flows throughout the grid. Each turn consists in

---

[53] http://www.scientificamerican.com/article.cfm?id=detecting-the-earths-elec
[54] Drawing by Ojibberish under *Creative Commons Attribution-Share Alike 2.5 Generic.*

marking a flow by putting an arrow extending from any one of the solid circles with a number by it; (think of this as specifying a component of the current density, $\vec{J}$). The number cannot be greater than 9, unless it is required in a manner that we will see, by already specified flows. In addition, each player is allowed, during any moves of his choosing, to specify the magnitude of two (and only two) sources (or sinks) per game. This is done by putting a positive number for a source or a negative number for a sink near one of the empty circles. When putting down a flow arrow if the source/sink on the given line is not specified it is taken to be zero.

The points on the left and right edges of the grid, i.e. along what might be called the banks of the river, have zero flow both horizontally and vertically. The bottom and top are specified as 1 unit coming in and one unit leaving. The object of the game is to be the one that actually completes the last flow line for the entire grid.

How to Play: The first player begins by putting in required flows in the second box to the right of the bottom left corner and then picks one of his own choosing. Once that square is complete, players move on to any box touching that square. In general, once a given box has a flow, source or sink, you can move to any box touching it and so on. You can fill out required moves (i.e. moves required by flows and sources already defined) before your turn, not after, except in your last move of the game. If, on a given turn, you can see the way to win *in that turn*, you are allowed five moves to do so, not counting forced moves. If there is a source or a sink on a line, any flow that goes through that line must lose or gain the amount specified by the number on the empty circle. Again, if a source or sink is not specified at the time of your turn, it must be taken to be zero.

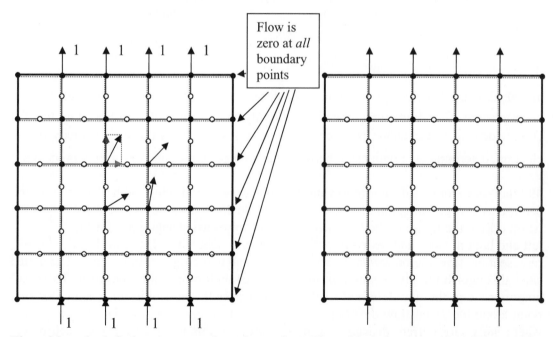

The grid on the left shows vectors for a few points. The grid on the right should be used for your game.

50. Consider again the two point oppositely charged particles separated by metal plate discussed in the text. a) Using the method of images calculate the surface charge density, σ, as a function of the distance away from the line connecting the two charges. b) Show that if one draws a Gaussian hemispherical surface around the positive surface charge on the plane and the negative point charge (q), as the radius of that surface approaches infinity, the net flux through that surface goes to zero so that the total charge on the planar surface is q. *Hint*: show that $\int \sigma dA = q$.

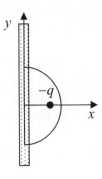

51. Use uniqueness of the boundary valued solution to Laplace's equation to prove $\vec{E} = 0$ inside of a metal enclosure.

52. After walking across a carpet in a dry room with rubber soled shoes on, you experience a shock, hear a "pop" and see a spark. In terms of charge and force, explain why this happens. Explain the arcing in terms of capacitance and the potential, ϕ.

53. a) Using the distances specified in the drawing, find the capacitance of two spheroidal shell's that are resting at the equipotential surfaces of two oppositely charged particles. b) By taking a limit, give a first order approximation for the capacitance between two spherical shells that are far apart. Express the results in both *cgs* and *SI* units. (*Hint*: Note if the charges are sufficiently separated the equipotential surfaces will be approximately spherical.)

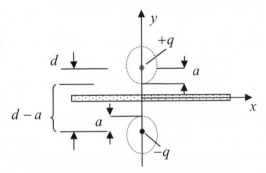

54. Calculate the electric field for an infinite line charge of uniform linear density, λ.

55. Radiation shielding is critical for interplanetary trips as astronauts must travel in the solar wind, composed of high energy electrons and protons. NASA engineers are investigating a system of spheres made of a flexible material and coated with a thin conductive (e.g. gold) coating. The spheres would be charged to high electrostatic voltages; those around the outside would be positive, while those in the middle would be negative. Their design for a moon base would include a planar conductive net to make image charges, thereby increasing the effectiveness of the shield. a) Describe the general effectiveness and limits of such a device. b) How is this method like the standard method of using thick layers of material such as concrete or lead to block radiation? c) Assume a very simple design with two 1 *m* radius balloons. Each is charged to a given voltage before being brought in proximity to the other. One balloon is negatively charged to -100 *MV* and is placed just above the astronaut, and the other is positively charged to 50 *MV* and is placed about 10 *m* above the first. If a 20 *MeV* proton and a 20 *MeV* electron came exactly

along the axis defined by the centers of these two balloons, what would happen to them on their way to the astronaut? Explain using the plot of the potential as a function of distance.

56. In a time much before the advent of multimeters, the brilliant experimenter Henry Cavendish used shocks to the elbow to determine the relative resistance of rain water to "saturated sea water." How could this be done? *Hint*: Null experiments, i.e. ones that only give a result when a signal is present, are often the best.

57. Discuss the geometric ideas of expansion, shear and twist of fluid flows.

58. Quantitatively explain the flow of water in terms of mass (which is equivalent to volume) per unit time of a spherically symmetric shell of water that expands uniformly in all directions. In particular, show how the speed of the water must decrease as the water expands.

59. Imagine a world in which water and all other fluids are compressible. Imagine, for example, that water was easily compressed by gravity so that water that filled a cake pan to the top, would only measure an inch when poured into a tall narrow pitcher. Discuss what it would be like to discover the quantitative ideas of Gauss's law (mathematical version) in this world without incompressible fluids?

60. Using the following the steps, show the analogy between mass flow and the electric field. a) Consider a $1g$ clump of water that moves isotropically across a spherical surface of radius $1/4\pi$; calculate the "current" density, $\vec{J}$ in units of kg/cm^2 for this case, then show the surface integral to calculate I.
b) Consider $1esu$ in some sense moving across a spherical surface of same radius, is the calculation any different? Explain. c) In the analogy as given so far, the *esu*, i.e. the charge, would leave the enclosed region; what distinction is needed to preserve the analogy? Explain in detail its significance.

61. A key element in the proof of the divergence theorem (illustrated in Figure 2-8) is that cutting a volume in half halves the total flux integral. Prove this for the case of a cube for surfaces that cut the cube in half perpendicular to the faces of the cube.

62. Rewrite the continuous limit definition of the electric field in the x-direction, $E_x(x) = -\lim_{l \to 0} \dfrac{\phi(x+l) - \phi(x)}{l}$, as a discrete limit.

63. Given that the internal resistance of a AA battery is .3 ohms, what is the maximum current it can produce?

64. In *cgs* units, we can easily introduce the "fine-structure constant," which is an indicator of the strength of the electromagnetic interaction that is important in particle physics. a) Calculate the fine structure constant, α, by finding the ratio between the amount of work required to bring two electrons within a distance d of each other and the energy of a photon

with a wavelength of $2\pi d$. b) Show that the same number results when one takes the square of the ratio of the charge on the electron, e, to the Planck charge, which is the unit of charge made from the fundamental constants, c, G, h.

Chapter III

Static Magnetic Field and Vector Potential

Introduction

St. Augustine (354-430AD), who is credited with first clearly reporting the distinction between electricity and magnetism, was "thunderstruck" the first time he saw a magnet pick up a ring that in turn picked up another ring and so on up to a fourth ring and when he heard from someone he trusted who had firsthand experience with a magnet that moved a bit of iron through a silver plate, without any noticeable effect on the plate.[1] If static electricity with its invisible pull can be startling, magnetism can be even more so. We are usually not much amazed by such things, because they are now commonplace. However, if we reawaken our minds to ask why, how and even exactly what happens in more fundamental and clear terms, then they are once again a wonder.

Magnets are different from charged bodies in obvious ways. Magnets always, as far as we know now, have poles, one north, one south. Like poles repel, unlike attract, which on the surface is similar to charge, but does not reach very deep. These simple facts will be explained at a deep simple level by the end of this chapter. Furthermore, though understanding the detail of the fascinating phenomena discussed by St. Augustine will have to wait till we reach the chapter on dia-, para-, and ferro- magnetism, we will discuss the fundamental physical principle of magnetism which is behind those effects.

At that fundamental level, we need to ask, as we did for electricity: What is magnetism? Magnetism, at the level of classical E&M, is related to moving charge, i.e., charge with impetus. It is the ability to cause a quality, called the magnetic field, in plana. The magnetic field, like the electric field, is activated from one part of the plana to the next and must be maintained in that way. We will see that the definition of the magnetic field is:

> The ***Magnetic Field*** is a power (quality) activated in the plana by a charged massive body that has impetus which can cause impetus in a receptive body, i.e. a charged body which has impetus of a certain type. Hence, like the electric field, there is an active power, the impetus-activated active charge, and a receptive power, an impetus-activated receptive charge. This power is most specified by the A-field, which reveals details of the nature of the disposition of the plana known as the B-field, as well magnetic properties that do not relate directly to B.

Note again we call a massive body that has the properties of active and passive electrical charge "a charge," or a charged body, because charge, at our level of abstraction, is that body's most specifying property.[2] The details of what type of impetus of what strength and what type of charge of what strength will cause how much impetus are, of course, necessary for a complete definition and will be specified shortly.

In fact, we can distinguish two types of magnetism: (I) that caused by currents and (II) that caused by quantum mechanical "spin." However, because spin may be seen, even

---

[1] "The Intellectual Rise in Electricity: A History" by Park Benjamin, quoted from pg 87, from "City of God," book 21, chap. 4.

[2] Confer discussion of virtual charge in section titled "Discrete Nature of Charge" in Chapter 2.

by its suggestive name,[3] to be some kind of analogous motion, we treat both as charge with impetus and indeed at our level of abstraction, i.e. classical *E&M*, that is all we can do from a principled point of view. Hence, from this point on we will generally leave out of consideration those complexities that properly belong to case II and consider magnetism as simply caused by currents, i.e. impetus activated charge. We will start with currents properly so-called and then make our way back to permanent magnets.

The Force due to the Magnetic Field

Ampère's Force Law

The work of André Ampère (1775-1836AD, see history box) on the force between wires with currents flowing in them is the starting point of a truly fundamental understanding of the nature of magnetism and its relation to electricity, for though they are distinct, there is a deep unity between them. A picture of an apparatus that you can make to see the repulsion of two currents is shown in Figure 3-1.

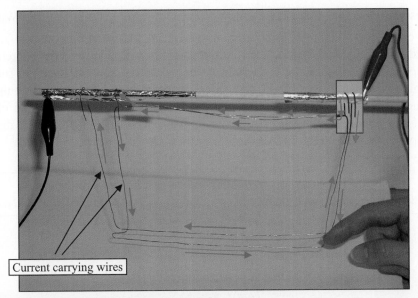

Current carrying wires

Figure 3-1: Ampère type experiment with two parallel current carrying wires at bottom. Green (red) arrow shows the current path in the front (back) loop. The wire loop (made of a thin coated wire) is constructed so that, when voltage is applied, the charge carriers in the wire will have impeti of opposite types, thus causing, via the plana, the wires to repel each other and act against gravity. The current is feed by the black and the red wires, which clip on to the aluminum foil to make contact with the bent and stripped wire that holds the rectangular-shaped loop.

---

[3] One should not carry the use of the word "spin," for example, for the quantum mechanical spin of an electron, too far, because it is after all only analogous, probably in a very loose sense, to ordinary spin, not identical with it. Indeed, it may ultimately not be motion properly, but only cause effects as if it were spinning.

An Ampère type experiment finds the following law for the force per unit length (using c for the speed of light):

3.1 $\qquad \dfrac{F}{l} = k_A \dfrac{2I_1 I_2}{r}$, where $k_A = \begin{cases} \dfrac{1}{c^2} & \text{in cgs} \\[4mm] \dfrac{\mu_0}{4\pi} = 10^{-7} \dfrac{N}{A^2} & \text{in mks} \end{cases}$ $\begin{array}{l} k_A = k_{mag} k_{force}, \text{ where} \\ k_{mag} \text{ and } k_{force} \text{ are defined} \\ \text{in equations 3.2 and 3.3.} \end{array}$

To understand this behavior, think of the wire's current as due to a uniform arbitrarily-thin line of positive charge (with linear charge density λ) moving under the action of impetus. And, to model the *neutrality* of the wire, think of the line of charge as surrounded by a stationary and arbitrarily-thin sheath of negative charge (charge density $-\lambda$) as shown below. Wire W1 (shown below) has impetus-activated positive charge and produces a change in the plana that propagates outward towards wire W2. We call this property of the plana the magnetic field, or, for precision that we will find helpful later, the B-field. The B-field generated by W1 near W2 is labeled $\vec{B}_{1\to2}$ in the figure below; its direction is shown and will not be discussed now, but will be shortly. Wire W2, also has a current. The B-field it generates near W1, $\vec{B}_{2\to1}$, is shown below as well. We find that if W1 and W2 have the same type of impetus, then there is an attractive force, as indicated by equation 3.1, pulling W2 toward W1 (and vice-versa). For example, a line of positive charge activated by impetus along the x-direction causes a field in the plana that acts on a second such line of charge, which is also activated by the same type of impetus, in such a way that that second moving line charge is given impetus toward the first. However, if the second charge has impetus of the opposite type, i.e. along the negative x-axis, that charge is disposed to receive impetus away from the line of motion of the first charge.

We can summarize this as follows:

> *Like impeti attract, while like unlike impeti repel.*
> (If the charges are of opposite type, then unlike impeti attract and like impeti repel.)

$$I_1 = \lambda_1 v_1 \quad \vec{B}_{2\to1} \odot$$
$$\text{W1}$$
$$\vec{B}_{1\to2} \otimes$$
$$\text{W2}$$
$$I_2 = \lambda_2 v_2$$

This accounts for the direction but not the magnitude of the force. If we increase the charge in wire W1 by a factor of α, so we go from $\lambda \to \alpha\lambda$, we will see that the magnitude of the force also increases by α. Furthermore, if we increase the speed of the line charge W1 by a factor (say $v \to \alpha v$), we will see the force also increases by that same factor. Lastly, we will note the force goes as the inverse of the distance from the wire. Hence, we can write the field strength as:[4]

---

[4] Note here we see a hint of the origin of the factor of "2" in Ampère's force law. Namely, the magnetic field has a factor of "2" in it. This, in turn, arises when we treat the current, as we will later, as composed of infinitesimal pieces. That formula for $\vec{B}$ as a general sum of all the infinitesimal current pieces, i.e. an integral over all the current sources (see equation 3.17), will *not* have a "2" in it, but a factor of "2" will arise from the integration if we specify to the straight line current source.

3.2 $$B_1 = k_{mag} \frac{2\lambda_1 v_1}{r} = k_{mag} \frac{2I_1}{r} \quad \text{where:} \quad k_{mag} = \begin{cases} \dfrac{1}{c} & \text{in cgs} \\ \dfrac{\mu_0}{4\pi} = 10^{-7} \dfrac{N}{A^2} & \text{in mks} \end{cases}$$

We can see that the magnetic field is dependent on the effectiveness of the action of the impetus, i.e. how much the impetus is actually able to result in speed of the body; that is, it takes into account not just the strength of the impetus but also how effective the body receives that strength, i.e. the mass. This makes sense as the more effective the impetus is the more it can cause its effect in the plana as well. This shows that the mass is again at the core of the nature of massive bodies.

To completely determine the force caused by this B-field, we have to do similar experiments varying the speed and charge of the second wire, the wire that receives that action of the field, the wire upon which the force is exerted. Such experiments would reveal that the receptivity to the action of the field is also dependent on the effectiveness of the action of the impetus and the amount of charge. Both of these make sense because impetus-activated charge is the source of the receptivity of a body to magnetic field. The linearity is just an indication that having a certain amount of receptivity does not affect further receptivity. Hence, we have the force per unit length (cf. equation 3.1) on the second wire is:

3.3 $$\frac{F_{1\to2}}{l} = k_f B_1 \lambda_2 v_2 = k_f B_1 I_2 = k_f k_{mag} \frac{2I_1 I_2}{r} \quad (\text{Thus: } k_A = k_f k_{mag})$$

$$\text{where,} \quad k_f = \begin{cases} \dfrac{1}{c} & \text{in cgs} \\ 1 & \text{in mks} \end{cases} \quad \text{and } k_{mag} \text{ is given in equation 3.2}$$

Note that by symmetry, i.e. the fact that there is no difference between the physical setups when we reverse the roles of the two wires, this expression is also valid for the force of the second on the first.

We could also learn by experiment that the B-field, like the E-field, is linear. That is, it obeys the superposition principle, so that one can obtain the field caused by a given point in the plana by N sources, by simply summing the contributions caused by each source alone.[5]

---

[5] Initially, we would take wire W2 as a "test" current, i.e. as an arbitrarily small current so that W2's effect on the B-field produced by the source, W1, (and on W1 itself) is arbitrarily small. As with "test charges" to probe an electric field, we use test currents to probe or "test" a B-field. In this way, we test the B-field produced by W1 with only an arbitrarily small influence due to W2.

André-Marie Ampère was born in 1775AD in Lyons, France (d. 1836). Being born just before the American Revolution and reaching manhood during the French Revolution, he lived in radically changing times, what Dickens called "the best of times" and "the worst of times." Ampère's father took time off to teach him, including Latin. Ampère used his Latin to digest the work of the great mathematicians Euler and Bernoulli.

At eighteen years old, he identified three key points in his life: his First Communion, a decisive moment in his Catholic life, the reading of Antoine Leonard Thomas's "Eulogy of Descartes", which helped ignite his interest in science and mathematics, and the taking of the Bastille, which signaled the beginning of the French Revolution, which he apparently thought expressed his ideals of liberty and the unity of man. However, he must have questioned this when his beloved father, along with multitudes of other innocents, was killed by that very revolution in 1793 and when he was so positively moved by his beloved future wife's refusal to be married by priests who did not denounce the revolution. His wife died 10 years later, moving him to write: "O Lord, God of Mercy, unite me in Heaven with those whom you have permitted me to love on earth."

Ampère's work alternated between metaphysics, physics, and mathematics, including publication of *The Mathematical Theory of Games*.

In 1820, he learned of H. C. Ørsted's discovery that a magnetic needle is affected by an electrical current. One week later, Ampère presented a paper to the Academy that contained a much more complete account of this and related phenomena. On the same day, Ampère also demonstrated before the Academy that parallel wires carrying currents attract or repel each other, i.e. "Ampère's law." Ampère's law was the starting point in the deep understanding of magnetism. Ampère's seminal work in the field of electromagnetism culminated in the publication in 1830 of the *Memoir on the Mathematical Theory of Electrodynamic Phenomena, Uniquely Deduced from Experience.* It included a rigorous mathematical derivation of the electrodynamic force law and was substantiated by four experiments. J. C. Maxwell will call him the "Newton of Electricity." In tones already set by Newton, Galileo and their medieval predecessors, Ampère intimates at his ultimate motivation for doing science:

> *We can see only the works of the Creator but through them we rise to a knowledge of the Creator Himself. Just as the real movements of the stars are hidden by their apparent movements, and yet it is by observation of the one that we determine the other: so God is in some sort hidden by His works, and yet it is through them that we discern Him and catch a hint of the Divine attributes.[1]*

Ampère retained his Catholic faith throughout his life and would take refuge in the reading of the Bible and the Fathers of the Church.

In 1881, the Paris Conference of Electricians named the practical unit of electric current in his honor.

[1]*Essai sur la Philosophie des Sciences II*, Paris 1843, 24 f.

The Direction of $\vec{B}$ and its Point Particle Force Law

Now, to understand more generally the force that a given B-field causes, suppose we have a single small positively charged test body moving at speed v. We will soon see a more complex pattern of behavior than for electric fields. Indeed, we should expect it to be qualitatively different, because magnetic field unlike the electric field depends on the impetus for the way it *acts* and the way its action is received, and impetus is by nature directional.

Because of the symmetry of the line charge and its impetus, it cannot matter where around (or along) the wire the test particle is placed. The electrical case had the lines pointing radially outward (or inward) from the wire. Using this same pattern for the new impetus caused field (the B-field) might mean, given the results of Ampère's law, its effects would be closely similar to that of the electric field, thus not providing the complementarity that we might expect from a second type of field. There are actually three possibilities that respect the cylindrical symmetry of the line: the radial direction just mentioned, $\hat{\rho}$, the angular direction, $\hat{\theta}$, and the direction along the line, $\hat{z}$. The second of these, we will see, is intimately related to the third through the A-field, which we will take as a further (analogical) specification of the B-field, so we choose the second. Hence, we draw the B-field pointing angularly around the wire as shown in Figure 3-2. We will now see that this drawing actually does represent the B-field, provided we know what we mean by it. That meaning will be discussed as we investigate the actual behavior.

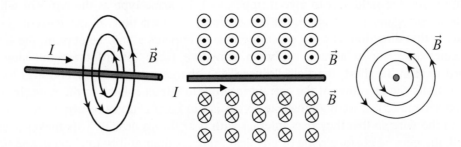

Figure 3-2: Three different views of a current carrying wire with the B-field it creates around itself. The left view shows that the circular field pattern goes in a right hand rule fashion illustrated in Figure 3-3. The middle image shows the cross section through the length of the wire; here we see that above the wire, the field comes out of the page, while below the wire, the field pokes into the page. The right image shows the current coming straight at you. The field, of course, is the same all along the length of the wire, but in the first figure, we have only shown one plane of lines to keep the drawing simple. The closer the field lines are to each other the stronger the field.

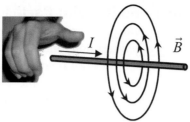

Figure 3-3: A *right hand rule* can be used to determine the field created around a current. Hold your right hand with the thumb extended and pointing in the direction of the current, then your fingers will naturally curl in the direction of the *B*-field. We will see in a later section that this relation can be written $\nabla \times \vec{B} = \dfrac{4\pi \vec{J}}{c}$ (in *cgs* units).

Six distinct directional cases arise and are shown in Figure 3-4. Of course, there are only three if we leave out the lesser distinction between motion along a given direction versus motion against that direction (formally, of course, we account for this real but lesser distinction by the simple use of a plus sign for a motion along and a minus for against).[6] The three cases of the particle moving *along* each of the axes are shown in Figure 3-4a. First, the charge moves with impetus along the direction of W1's current (*x*-direction); hence, it is attracted as shown. Second, the charge moves away from the wire (*y*-direction), and the field acts to give it impetus of the same type as the wire for which the field is a surrogate, i.e. it applies a force along the direction of the current, giving it impetus in that direction. Having given the charge impetus along the current, the field will then, since like impeti attract, apply an attractive force toward the wire, slowing its outward motion. In short, the field "senses" the charge moving away and acts to give it impetus along the direction of the current in order to attract it back to the wire. In both the first and second cases, we note that the forces act to keep charges nearby or to bring them closer to the wire so that they may interact with it or things near it. This makes sense, for, after all, the generic characteristic of physical things is their ability to change, and nearness and contact facilitate interaction that leads to change. Furthermore, keeping multiple things nearby also facilitates interaction between them. There is no effect for the third case for which the particle follows the field.

The second three cases, motion in the direction against each of the three axes (i.e. $-\hat{x}, -\hat{y}, -\hat{z}$), are shown in Figure 3-4 b. First, the particle with impetus of type opposite the source current is repelled in what we will see shortly is also part of an attempt to keep the particle from leaving the vicinity of the wire. In the second figure, the particle already has impetus towards the wire and thus "threatens" to move past it; thus, the surrogate for the wire, the field, acts to give the particle the impetus of the opposite kind as the source

---

[6] As you may have already realized, keeping track of minus signs can be the difference between a calculation done correctly and one done radically wrong. They can appear trivial but they represent an irreducible aspect of reality; for example, they are the difference between fields from two sources acting together or completely inhibiting ("canceling") one another. In the original movie (1961) "The Absent-Minded Professor," the professor blows up his lab and later realizes it was caused by a minus sign error. This movie classic can remind you to watch for such mistakes; checking the reasonableness of your answer can often go a long way in this regard.

current, i.e., leftward impeti in this case. Unlike impeti repel, so this action will tend to cause the particle to be pushed away. The last case, like the last in the group above, is not affected by the field at all. The field at a given point *only* has the ability to affect impeti with a direction in the plane defined by that point and the line defined by the direction of the momentum of the source charge.

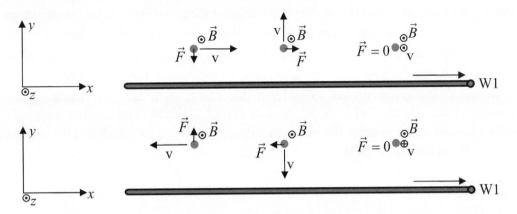

Figure 3-4: a. (top) From left to right is shown the force exerted by *B*-field on positively charged particle moving, respectively, in the direction of the x, y and z axis. **b.** (bottom) Illustration of forces for particle moving in the *negative* x, y and z direction.

In thinking through the above, you may have noticed that the behavior we have described will tend to try to force a moving particle into a circular motion around the magnetic field.[7] In the simple case of uniform *B*-field with a test particle with no component of its velocity along the direction of $\vec{B}$, this fact is most clear, as the particle will move in simple circular motion with radius $r = \dfrac{mv}{qB}$ (*exercise*: prove this using the force law below). This fact along with the previous mention of the *B*-field defining a plane recalls the mathematical fact that rotation defines a plane of rotation and an axis in space. In particular, in this case, the $\vec{B}$ is the axis of rotation around which the particle moves and also defines the plane in which the velocity vector, $\vec{v}$, changes under the action of the force, $\vec{F}$, caused by the field. The force centripetally supplies the change in impetus needed to keep the velocity vector bending around in a circular manner. We further recall the cross product notation that we introduced into our formalism to capture these facts about extension (space), the first property of all physical things. In particular, we can write the force caused by a *B*-field of given magnitude and direction on a point particle of charge q and velocity $\vec{v}$ as:

3.4	**Magnetic Force Law:** (For point particle)	$\vec{F} = k_f q\,\vec{v} \times \vec{B}$ where $k_f = \begin{cases} \dfrac{1}{c} & \text{in cgs} \\ 1 & \text{in mks} \end{cases}$

[7] The positive (negative) test particles will tend to follow a left (right) hand rule circle around the field lines.

This shorthand encapsulates all six results discussed above, yielding the proper direction and magnitude for the forces. To better understand the meaning of the cross product in this new context, consider the following.

Using the figure below and recalling that the B-field is always the normal to the plane formed by the field point in the plana and the line along the source's velocity,[8] the equation shows that the force the B-field causes remains in that plane, thus keeping all *new* activity in that plane as found above. The cross product picks out only vectors in this plane because the normal, $\hat{B}$, of any given plane crossed into any vector, $\hat{v}$, produces a vector *normal* to a new plane defined by $\hat{B}$ and $\hat{v}$,[9] and, hence, normal to $\hat{B}$ and thus *in* the plane defined by $\hat{B}$. As for the magnitude, the cross product yields only the commonality, i.e. the "area between" the given vectors ($\vec{v}$ and $\vec{B}$ in this case), so it correctly excludes any component of the velocity that is along the direction of the field. Said another way, only those components of the impetus that are in the plane of activity, for which $\hat{B}$ is the normal, can receive the action of the field.

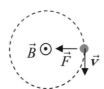

Figure 3-5: Positively charged particle shown moving in (left handed) circular pattern around the B-field in a region of approximately uniform $\vec{B}$. The velocity can of course also have components of velocity out of this plane of rotation, but they are not affected by the field at this point in the plana.

So, what happens when a positively charged particle enters a region of field $\vec{B}$? If the field is constant or varying slowly enough, a positive (negative) charge will follow the field line spiraling in a left-handed (right-handed) manner as shown in Figure 3-5. The result for a field that does not vary significantly over the radius of the spiral or over the time the particle travels in one period is shown in Figure 3-6a below. In general, the result can be quite complicated and can be quite important, for example, in plasma physics.

With this understanding in hand, we can easily answer the question evoked by the electron-positron pair creation image discussed in the "Conservation of Charge" subsection in Chapter 1. Namely, why do the positron and electron spiral as they move? The experiment is done with a magnetic field present for just the purpose of making them spiral so they can be distinguished. Electrons spiral in the opposite direction of the positrons.

This phenomenon of charges following fields is responsible for the northern lights, also called generally Aurora Borealis, shown in Figure 3-6b (and also depicted on the cover). Charges follow the Earth's field lines to the Earth's polar regions where they strike the atmosphere to cause fabulous colors and patterns of light visible from the ground.

---

[8] This can be written $\hat{J} \times \hat{r}$, where $\hat{r}$ is the vector from the source point to the field point. We will see this used directly in the Biot-Savart law discussed later in this chapter.

[9] In terms of the vector notion used in footnote 8, we have $\hat{F} = \hat{v} \times \left(\hat{J} \times \hat{r} \right)$

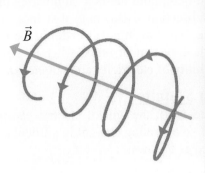

Figure 3-6: a. *(left)* A positively charged point particle will tend to follow the magnetic field lines, spiraling around them in left handed manner as they go. The magnetic field, unlike the electric field, does not change the energy of the particle, but rather gives activity it took away in one dimension to another. **b.** *(right)* The northern lights, shown here above Bear Lake, Alaska, are caused by charged particles from the solar wind being funneled down the earth's field lines in a way similar to shown in the left figure. The light is emitted when these charged particles collide with atoms that make up the atmosphere.

In summary, we see that a line of impetus activated charge, i.e. a current, causes a *B*-field that acts to keep other moving charges near it. It acts perpendicular to the motion of the charges so as to maintain the energy of the particles while generally striving to guide it along the field lines, thus keeping it nearby and often causing it to interact with the bodies related to source, as in the case of the Aurora above. This contrasts with the *E*-field which does somewhat direct the particle, but also either gives it energy (activity) or takes away and stores its energy. In these static field cases, the electric field then seems to take "a store and give back role," whereas the magnetic field takes a "shepherding" role. These roles seem complementary like the roles of attraction and repulsion, which are generically needed, respectively, to bring particles together so that they may interact and form complex substances that stay together, and to keep all substances from ending up all together in one place, not having any space for their proper activities. For example, without some repulsive force to balance the attractive force, things, if they started to interact by colliding, would generally pass right through each other, thus destroying each other.

Lastly, note that in this section, we have analyzed currents that have no net charge (e.g., wires). This is, of course, the general case for macroscopic bodies; however, we also needed this assumption because the magnetic field is generally much weaker than the electric field. We will discuss this further contrast between the two fields in the next section.

Units and Some Primary Definitions

We can now finally address the question of the definition of the units of charge and through this better understand the formal definitions of the forces caused by the magnetic as well as the electric field.

For those who don't require the details given here, note the box summaries in the subsections below and read the "Force Laws..." subsection. Also, note that in *SI* units electric *charge* is defined through the magnetic force, whereas in *cgs* it is defined directly through the electric force itself. *SI* is a more practical system of units, whereas *cgs* has theoretical advantages, revealing more of the formal unity of electricity and magnetism.

Fields and Charge in *SI* units

In *SI* (which is, again, a kind of standardized *mks*), the unit of electric *charge* is defined by the force caused by a static *magnetic field* on a moving charge. In particular, the unit of charge, the coulomb, abbreviated, *C*, is defined as follows:

$$1\ coulomb = \begin{cases} \text{the amount of charge that must flow in } one\ second \\ \text{in each of two arbitrarily (infinitely) long parallel wires } one\ meter \text{ apart} \\ \text{to cause a force per unit length of } 2\times10^{-7}\ N\ /\ m \text{ on the other.} \end{cases}$$

This definition of charge fixes the constant k_A in Ampère force law equation 3.1 giving:

3.5

$$F_B\ /\ l = k_A \frac{2I_1 I_2}{r},\ \text{where } B = k_A \frac{2I_1}{r}\ \text{ with } k_A = \frac{\mu_0}{4\pi} = 10^{-7}\ N\ /\ A^2$$

For comparison purposes, we also write the electrical force law:

3.6

$$F_E = k \frac{q_1 q_2}{r^2},\ \text{ where } E = k \frac{q_1}{r^2}\ \text{ with } k = \frac{1}{4\pi\varepsilon_0}$$

Where *r* is the distance between the sources (either a charge *q* or a current, *I*) and *l* is the length of the current segment receiving the action of the *B*-field.

Recall that in *SI* (*mks*): *F* in newtons, *I* in ampères (coulombs per second), *r* and *l* in meters.

These imply that *B* is measured in $\frac{N}{Am}$ and *E* is measured in $\frac{N}{C}$.

To clarify the meaning of the constants and the definitions, we will now examine the above equations in some detail. Taking the second current in Ampère's force law as measuring the receptivity to the action of the field due to the first current, we can then, as shown in equation 3.3, write: $F\ /\ l = k_f B_1 I_2 = k_f \left(k_{mag} \frac{2I_1}{r} \right) I_2$. Thus, we see that, without

k_f, the units we pick for the strength of the action of the field, *B*, and the receptivity, *I*, will determine the units with which *F* is measured. If we want to have freedom to use other units, we need the constant, k_f, to account for the unit change. In the case of *mks (SI)*, we let the chosen unit of force, newtons, determine the unit for the currents through our choice of units for *B*; hence, we don't need k_f, i.e., we take $k_f = 1$. Now, *B* is determined by our chosen unit of current[10] and distance as well as our choice of k_{mag}. In particular,

$B_1 = k_{mag} \frac{2I_1}{r}$. Like the constant k_f for the force, the constant k_{mag} provides for any needed

---

[10] Recall the choice of unit for passive and active charge is the same, because of the universal proportionality of active and passive charge. In the same spirit, active and passive currents are also given in the same units.

transition from the chosen unit of current and distance to a desired unit for B. Without k_{mag}, the units for the field is A/m, which will not yield the correct units for force, so we take $k_{mag} = k_A = 10^{-7} N / A^2$, so that B is measured in $\dfrac{N}{Am}$. Note that if we had let $k_{mag} = k_A = 1 N / A'^2$, our new unit of current, A' would represent the amount of current that needs to flow in each wire to cause one newton of force. This means that the new unit would represent $\sqrt{10^7}$ times more charge flowing in one second than the old. Said another way, $\sqrt{10^{-7}} A'$ must flow *in each wire* to cause $2 \times 10^{-7} N$ of force so that the old unit would be $\sqrt{10^{-7}}$ times smaller than the new.

Explaining the meaning of the constants in electric field for this case of *SI* units is left as an exercise for the student.

Fields and Charge in *cgs* units

In *cgs*, the unit of charge is defined by the force caused by a static *electric field* on a stationary charge.

$$1\,esu = \begin{cases} \text{the amount of charge that} \\ \text{each of two arbitrarily small (infinitesmal) bodies separated by 1 } cm \\ \text{must have to cause a 1 } dyne \text{ of force} \end{cases}$$

Thus, this definition of charge defines the constant, k, in Coulomb's law:

3.7

$$F_E = \frac{q_1 q_2}{r^2}, \text{ with } E = \frac{q_1}{r^2}, \text{ where } k = 1, \text{ so we do not even show it.}$$

For comparison purposes we also write the magnetic force law as well:

3.8

$$F_B / l = k_A \frac{2 I_1 I_2}{r}, \text{ with } k_A = \frac{1}{c^2}, \text{ with } B = \frac{2 I_2}{cr},$$

(Note that in *cgs*: F in dynes ($g\,cm / s^2$), I in electrostatic units per second (*esu/sec*), r in *cm*).

These imply that E and B are *both* measured in $\dfrac{esu}{cm^2} = \dfrac{dyne}{esu} = \dfrac{\sqrt{dyne}}{cm}$, where

the last equality follows from the first. This is a real advantage of *cgs* units, for it highlights their similarity, namely they are both fundamentally powers of the plana to cause impetus. Because of special relativity, we will see that there is a profound link between E and B and, due to that link, the fully formal view of E&M will only be manifest when their units are the same.

To understand these definitions, we follow a line of argument parallel to that given in the *SI* subsection above.

First, we consider the *electric field* since, in *cgs* units, it defines the unit of charge. Taking the second charge in Coulomb's law as measuring the receptivity to the action of the field due to the first, we can then write: $F = \bar{k}_f E_1 q_2 = \bar{k}_f \left(k_{elec} \dfrac{q_1}{r} \right) q_2$. Without $\bar{k}_f$, i.e.

$\bar{k}_f = 1$, the units we pick for the strength of the action of the field, E, and the receptivity, q_2, will then determine the units with which F is measured. And, as before, if we want to have freedom to use other units, we need a constant, $\bar{k}_f$, to account for the unit change. In the case of *cgs* for the *electric* force, we do want that freedom. Indeed, we want *both* the charge, q, and the field, E, to be *directly* determined by the choice of units for the force and the distance. With this requirement, two 1 *esu* charges separated by 1 *cm* will cause 1 *dyne* of force, and each will cause a field of $1 esu / cm^2$. Hence, $\bar{k}_f = 1$ and $k_{elec} = 1$; that is, these constants neither have dimensions nor change the scale.

The *cgs* magnetic force law, by contrast, makes use of both its constants. Taking the second current in Ampère's force law as measuring the receptivity to the action of the field due to the first, we can then, as shown in equation 3.3, write: $F / l = k_f B_1 I_2 = k_f \left(k_{mag} \dfrac{2 I_1}{r} \right) I_2$. Again, we see that, without k_f, the units we pick for the strength of the action of the field, B, and the receptivity, I, will then determine the units with which F is measured. As before, if we want to have freedom to use other units, we need the constant k_f to be available to account for the unit change. In the case of *cgs*, we have already defined the charge, and hence the current, before we come to study the magnetic field. So, we cannot expect F in the equation to be in the *cgs* unit of dynes without use of the constants. Now, we would like the units of the magnetic field to be the same as the electric field, esu/cm^2. Noticing that, in this case, the ratio between $F_B = l \cdot k_f I_2 B_1$ and $F_E = q_2 E_1$, namely F_B / F_E, not considering the k_f, has units of speed, we see that k_f must have units of speed (cm/s) in order for the ratio to be dimensionless. Now, anticipating the essential role of the speed of light, c, in E&M, we choose

$$k_f = \frac{1}{c} \sim \frac{1}{3 \times 10^{10} cm / s}.$$

Now, B is determined by our chosen unit of current and distance as well as our choice of k_{mag}, via $B_1 = k_{mag} \dfrac{2 I_1}{r}$. In particular, we cannot get B in the *cgs* units of esu / cm^2, with I in esu / s and r in cm without giving k_{mag} the proper dimensions to make this transition. Note here again we need units of inverse cm / s, so we take $k_{mag} = \dfrac{1}{c}$. This

gives: $F_B / l = \dfrac{1}{c} I_2 B_1 = \dfrac{1}{c} I_1 \left(\dfrac{1}{c} \dfrac{2 I_2}{r} \right) = \dfrac{1}{c^2} \dfrac{2 I_1 I_2}{r}$.

Force Laws for a Point Particle in *SI* and *cgs*

We can summarize the force laws for *SI* and *cgs* with the force law equation already mentioned.

3.9   SI: $\vec{F} = q \vec{v} \times \vec{B}$

3.10   cgs: $\vec{F} = q \dfrac{\vec{v}}{c} \times \vec{B}$

In the *SI* case, it is evident that the magnetic field and charge are defined so that the force equation is a simple function of the key variables. However, in the *cgs* case, we see that a factor of c is introduced so that the magnetic field will have the same units as the electric field, (i.e., the units of $B \propto F_B / q$ are the same as $E \propto F_E / q$).

We can also now write the total force law incorporating both electric and magnetic fields. This law is called the *Lorentz force law* and is written:

3.11 $\qquad$ *SI*: $\qquad \vec{F} = q\vec{E} + q\vec{v} \times \vec{B}$

3.12 $\qquad$ *cgs*: $\qquad \vec{F} = q\vec{E} + q\dfrac{\vec{v}}{c} \times \vec{B}$

Because of the advantages of *cgs* already mentioned, we will now use it to address the issue of the relative strength of electricity and magnetism. To make this comparison, we use Ampère' law for the *magnetic* force due to a line current acting on a segment of a second line current and the equation for the *electric* force due to a line charge acting on a second charge. This latter is obtained by considering the electric field, $E = \dfrac{2\lambda_1}{r}$ of a line charge of linear charge density λ_1 acting on a segment of line charge of length l and linear charge density λ_2 thus giving a total charge $\lambda_2 l$, so that $F_E = \dfrac{2\lambda_1\lambda_2 l}{r}$. Hence, the ratio then is:

3.13
$$\frac{F_B}{F_E} = \frac{\dfrac{1}{c^2}\dfrac{2I_1 I_2}{r}l}{\dfrac{2\lambda_1}{r}\lambda_2 l} = \frac{1}{c^2}\frac{\lambda_1 v_1 \lambda_2 v_2}{\lambda_1 \lambda_2 l} = \frac{v_1}{c}\frac{v_2}{c}$$

In everyday experience, massive bodies that we see move much slower than the speed of light. Indeed, many massive bodies move at speeds much less than the speed of light, and thus the magnetic field is generally much weaker than the electric force. Furthermore, as we hinted at earlier, it is because most macroscopic bodies tend to be neutral that we are able to see the effects of the magnetic field. If they were heavily charged, the electrical effects would swamp the static magnetic effects.

Helmholtz Decomposition Theorem

So far we have only discussed the magnetic field of an infinitely long straight wire. We would like to determine B for more general current sources. To help in this process, we have available a mathematical theorem, Helmholtz Theorem, that says that any "reasonable"[11] vector field, $\vec{B}$ can be decomposed as follows:

3.14 $\qquad\qquad \vec{B} = \nabla \times \vec{A} + \vec{\nabla}\phi_m$

Where $\vec{A}$ is a vector field and ϕ_m is a scalar field.

In words, we will see this means that any "reasonable" field can be decomposed into a sum of its curling part and diverging part.

Now you can (and should do) show that dotting each side of equation 3.14 with the operator ∇ gives:

---

[11] The field needs to be twice differentiable and fall off faster than 1/r at infinity.

3.15 $\nabla \cdot \vec{B} = \nabla^2 \phi_m$

Note the absence of $\vec{A}$.

One can also show that crossing each side of equation 3.14 with the operator ∇ gives:[12]

3.16 $\nabla \times \vec{B} = \nabla \times \left(\nabla \times \vec{A} \right) = \vec{\nabla} \left(\nabla \cdot \vec{A} \right) - \nabla^2 \vec{A}$

Note the absence of ϕ_m.

You might guess from the way these two operations pick out the two fields, ϕ_m and $\vec{A}$, that the curl and divergence can specify a field. In fact, it can be shown that specifying the curl and the divergence of a vector field (that both fall off faster than $1/r^2$) is also sufficient to completely specify that vector field as long as that field is known to go to zero as one approaches infinity.[13] Intuitively, we can see why the divergence and curl are sufficient to completely specify a field by looking at fluid flow, first curl-less flow then divergence-less flow.

Analysis of Curl-less Fields

Consider, for example, a river that consists only of water and remember that the river flow can be described by what each small region of it does. For any small region within the river, the flow of water (according to volume or mass) into the region will either equal the flow out or it will not; i.e. the divergence of the current density of the water, say by mass, is either zero or not. If the divergence is not zero this means there must be a source (or sink) of water within the region. For instance, imagine a point source of water inside the river. If there is no curl in the region, so that we have $curl = 0$, $div \neq 0$, and the river is stagnant (which we will see means no curl *anywhere*), this can only happen in one way. That is, in our case of a *source* of water, the water must move *uniformly*, i.e. the same in all directions, outward into the river, falling off as the inverse square of the distance from the source (analogically like the electric field is activated in the surrounding plana by a positive point charge). Or, in the case of a sink (i.e., a "point" object that leaches water without changing its volume), the water must move *uniformly* onto the leaching object. A two dimensional analog of a uniform "sink" is, of course, the draining of *non-rotating* water down the drain of a kitchen sink, illustrated in the figure below on the left.

If the river is flowing, such sinks behave the same except that the uniform outward or inward motion is at each point augmented by the velocity of the overall river flow, which must be in a fixed direction to maintain $curl = 0$ in the region. This tilts the flow pattern as shown below on the right for the two dimensional case. However, note that in order to add such a uniform flow to the field pattern, one necessarily introduces curls at the edge of our river, violating our curl-less field assumption. Thus, the right hand figure is ultimately *not* a curl-less field, and our river must be completely "stagnant." In this way, we begin to see that any curl-less field can be constructed through the choice of placement

---

[12] Note the interesting vector calculus identity given by the second equality in this equation. The first level E&M student should not worry about trying to prove it, but only note it.

[13] Thus, to specify the field, F, one needs to specify $\nabla \cdot \vec{F}$ and $\nabla \times \vec{F}$ everywhere, not just locally. For example, for a point charge at the origin, we specify $\nabla \cdot \vec{E} = 0$ and $\nabla \times \vec{E} = 0$ everywhere except at the origin where the divergence is singular, but the curl remains zero.

of sources and sinks, along with boundary conditions consistent with the curl-less assumption.[14]

Curl and an Analysis of Divergence-less Fields

Before discussing the $curl \neq 0$, $div = 0$ case, we note that the curl can be heuristically conceived as the result of the measure of the tendency to rotate a "curl meter," which is basically a rod with small crossed blades on the end, such as shown in Figure 3-7, that can be inserted into the flow (or into any field).

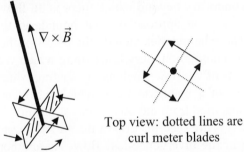

Top view: dotted lines are
curl meter blades

Figure 3-7: Typical "curl meter;" side view on left, top view on right. This meter is imagined to be arbitrarily small so that it can be placed into an arbitrarily small place in the flow. The four faces then are influenced by the differential flow in the two directions of the plane perpendicular to the rod. The greater the tendency of the rod to rotate under this influence, the greater is the curl of the field at that point.

Now, suppose we do not allow sources or sinks, so that whatever comes into a region, and only what comes in, must come out. In such a case, we can just consider plain water flow. Recall that, as we discussed in Chapter 2, an incompressible fluid, like water, cannot lose any volume. Whatever volume of water goes into a region must come out, which, means that the divergence of the velocity vector for the current flow (which is

---

[14]As an example of the latter, we can make the field in the right figure curl-less by, for instance, bounding it with a circle centered on the point source and adding appropriate sources along the boundary to keep from introducing curls on that boundary. Thus, we move away from the original case of a straight flowing river with a *single* point source in the middle to get curl-less flow.

proportional to the current density $\vec{J}$)[15] is zero. We also, for simplicity, focus on two dimensional flows. Now, consider the special case of two small volumes of water that have been moving parallel to each other for some period of time. If they are moving at *different* speeds then, the curl will be non-zero; this case is shown in the left figure below for various spots in the river. On the other hand, if the two volumes of water are moving parallel and have the *same* speed, they have no curl. If the curl remains zero in the next region that the water enters, then the two paths must remain parallel.

Now consider what happens in the slightly more complex case illustrated in the middle figure below. In particular, if the left volume of water continues forward while a portion of the right one moves to the left and joins the forward flow of the left volume, then (excluding influence from other regions for the moment) the left flow will increase in magnitude by the amount of flow that moved over, while the right will decrease by that same amount. In order to conserve the volume of water, which is coming in at a certain rate from the left and right streams, the speed of the left portion must now be faster and the right portion slower (this happens in air flow around a wing). This will mean there will be a curl in this region, and, assuming the flow steadied out again after this, there will be a permanent curl established as seen at the top of the middle picture below. We would specify the opposite type turn of the right portion of the flow by an opposite sign curl. In this way, one can begin to see how specifying the curl at each point specifies the direction and magnitude the water flows. Of course, to get a flow velocity to start with, we must remember that there is a boundary beyond which there is no flow so that one must specify a flow that falls to zero as one approaches that boundary; this, in turn, translates to conditions on the curl. (These boundaries include the banks of the river, and, in the case of a non-circulating flow, the chosen boundary lines along which we define the flow to avoid reaching the diverging parts of the flow, such as the source, which we have excluded from the river). Or, in the case of a magnetic field, which weakens as it proceeds from the source, the curl must be specified from the boundary "at infinity."

Keep in mind that it is the specification of the curl at *every* point that determines the direction and magnitude of the divergence-less flow. As another example, in an initially straight flow such as shown in the left figure below, if the curl is a fixed non-zero magnitude and direction (e.g., out of the page in the top portion) along the initial flow direction, then the direction and magnitude of the flow (i.e., the current density) remains the same along that direction. Note that a fixed curl everywhere would mean that perpendicular to that direction the flow steadily increases, so that to keep from developing arbitrarily large flows the curl must change at some point.

Two dimensional flows can become quickly complex as both components become involved as in the middle case. One can create a simple flow that involves both components by bending a uniformly flowing body of water; for example, if a uniformly flowing stream at some point turns in a gradual curve, keeping the current per unit area, $|\vec{J}|$, constant, then a curl *will* be introduced. However, if the flow is forced in such a way that $|\vec{J}|$ falls off as the inverse radius from the center of curvature, i.e. $|\vec{J}| \propto 1/r$ so the

---

[15] In particular, $\dfrac{1}{\rho}\vec{J} = \vec{v}$, where $\vec{v}$ is the velocity of the water.

water's speed goes as: $|v| \propto 1/r$, then it can be shown (see section titled "Curl of B...") that there will be *no curl in* the river. Indeed, even if the river curves into a complete circle, there will still be no curl anywhere, excluding the side boundaries of the river. The curl can be zero everywhere even though the water circulates because the curl is a *local* feature of the flow, and we have excluded the center.[16] Of course, such a case is artificial for several reasons, including that neither the bank of the river, nor the water are frictionless. Thus, the water's speed will gradually drop to zero parallel to the bank, and even if one initially establishes a water speed that falls off as the inverse of the radius ($|v| \propto 1/r$), the friction of one part of the water on another will work against it, tending to create curls throughout the flow.

The most complete curl is the one in which the water in the small region makes a complete circuit around a "stagnation point," as shown in the right figure below. I say it is the most complete for the obvious intuitive reason that the water is actually circling (and it includes its center). This shows up in the fact that, in this case, there are contributions to the curl from the difference between adjacent paths in *both* directions that can cause curl around the axis out of the page. In particular, in the drawing, we see that the curl components arise from (1) the horizontal change of the vertical flows and (2) the vertical change of the horizontal flows corresponding, respectively, to the two terms in the $(\nabla \times J)_z$, namely $\partial_x J_y - \partial_y J_x$.

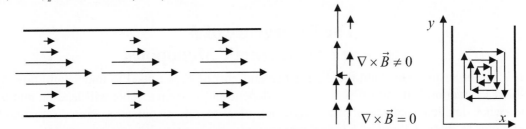

More complex fields such as the circulating ones, which explicitly involve both components (in two dimensions), can be built in the way described earlier, i.e. using a given specification of the curl at each point. With the above thoughts in mind, one may now wonder how such a construction can work since any given value of the curl can be accomplished in an infinite number of ways given that we are allowed to use both terms in the curl: $\partial_x J_y - \partial_y J_x$. In particular, one can use only one of the two terms to get the given value or any of the infinite combinations that give the specified value. The reason the construction can be done has already been seen, though implicitly. Namely, given that whatever water comes in must come out, as we build outward from the predefined boundaries, using our given plan for the flow, which is in the form of a map which gives the curl at every point (i.e. the vector field $\nabla \times \vec{J}$), the small flow regions that we have already put together constrain the possible combinations, leaving only one flow option, i.e. one possible J at the point, that can realize the specified curl there. (Of course, our plan,

---

[16] Note that we have excluded the boundaries, which become this point when they are shrunk to the middle. Also, even if we included the center, the inside edges of such a circular "river" cannot be allowed, given our $1/r$ fall off, to shrink to a point without requiring an arbitrarily large velocity.

i.e., our choice of $\nabla \times \vec{J}$, must be consistent with the given boundary conditions, for not every flow will be possible with any given boundary conditions).

Summary and What's Next

We have now seen that there are locally only two things a flow can do, curl or diverge, and thus any flow can be built out of knowledge of how much of each it does at each point. We now have real insight into how specifying the sources and sinks and their magnitudes, i.e. the divergences, along with the curls specifies the entire flow. Using this insight and the mathematical formalism that expresses those realities, we now proceed to understand the origin of $\vec{B}$ by investigating its divergence and curl. In the case of the electric field, we saw that the divergence described, via Gauss's Law, the relationship between the static $\vec{E}$ activated in the plana and its source; for $\vec{B}$, we will see that the curl will take on this role. After studying the divergence and curl of $\vec{B}$, we will move to study the A-field, and we will see that A will finally give us the deepest insight into the nature of B. We will then lastly attack a series of examples in which we investigate B with certain simple sources. These will give you further access to the specific ways impetus activated-charge causes B-field and thus more insight into the nature of A and B-field. The examples will also provide practice using the mathematical formalism that, if done thoughtfully, will further develop your understanding of the physics and the formalism that expresses that understanding.

The Divergence of *B*:
There are No Magnetic Monopoles

In order to understand the divergence and curl of B, we need to understand how a B-field's direction and fall-off with distance depend on its source. This will lead us into a somewhat detailed analysis of the divergence of B to rule it out as a source equation, leaving us to study the *curl* of B in the next section.

If a student wishes to skip the detail of this section, he should note that the magnetic field has no divergence, as expressed in symbolic form in Maxwell's second equation, equation 3.19, and also that its divergence-less nature is related to its being caused by impetus-activated charge. Such a student also should take note of the "Biot-Savart" form of the B-field for a current density, $\vec{J}$, given in equation 3.17 (and 3.18). Those staying with the section will now begin to understand these things by attacking the question of the B-field's dependence on distance and direction from its source.

If the static magnetic field is the result of impetus-activated charge, we expect that the influence of the field will be felt in the plane of the line of the impetus and the line of the field's outward propagation from the source. In other words, the impetus defines one line along which it is natural to expect influence, and the propagation direction outward defines a second line along which we might expect influence; the two together form a plane. Thus, for example in the simplest case of a long straight wire, we expect the action of the field to be in the series of planes that pass through the wire as shown here:

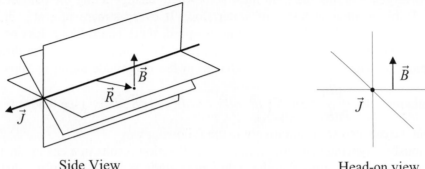

Side View   Head-on view

As we have already explained, such planes can be picked out by their normals. The direction of the B-field does that. We can write the direction of these normals, i.e., the unit vectors for the B-field, as: $\hat{B} = \hat{J} \times \hat{R}$, where $\vec{R}$ is the vector from the source point to the field point and where $\vec{J}$ is the current density of the source ($\vec{J} = \rho \vec{v}$). As for the distance dependence of the B-field, it makes sense that the B-field should fall-off as some function of the distance, R, from the source to the field point; in particular, we expect it to get weaker as it gets more distant from its source.

In the case of an infinitely long wire, our analysis implies that the B-field always points along the angular direction around the wire as shown in Figure 3-2 and Figure 3-3. From earlier discussions of the wire, we know that this prediction of the field's direction and its fall-off is correct.

Indeed, the direction and magnitude dependence for steady sources magnetic fields has been so successfully verified that it is enshrined in Biot-Savart's law, which we derive formally later:[17]

$$3.17 \qquad \boxed{\vec{B} = \frac{1}{c} \int \frac{\vec{J} \times \hat{R}}{R^2} dV \qquad (SI: \vec{B} = \frac{\mu_0}{4\pi} \int \frac{\vec{J} \times \hat{R}}{R^2} dV)}$$

We can write this in differential form if we are careful:

$$3.18 \qquad \boxed{d\vec{B} = \frac{1}{c} \frac{\vec{J} \times \hat{R}}{R^2} dV}$$

We need to be careful because, in the case we are considering, there is no build up of charge, so that every current loop has to be closed. In other words, it makes no sense to consider a single element of current by itself, for it must go somewhere. Indeed, the whole derivation of the Biot-Savart law depends on this fact.

The important point is to notice that each differential element points in the direction we predicted and falls off, like the static electric field, as $1/R^2$, and thus also meets our fall off prediction.

Now, because the differential field element, $d\vec{B}$, has the given direction and magnitude, we can show that its divergence is zero. In particular, a necessary condition for

---

[17] We will see later that, in the non-stationary case, this needs to be extended to:

$\vec{B} = \frac{1}{c} \int \frac{[\vec{J}] \times \hat{R}}{R^2} + \frac{[\partial \vec{J} / \partial t] \times \hat{R}}{cR} dV$. This can be misleading because it might make one think the

changing $\vec{J}$, i.e. changing impetus, causes B-field.

a non-zero divergence is that the field must somewhere change along the path of its own direction.[18] Think about it in terms of water flow. If every stream of water in a river continued in a certain path without changing its speed, there could be no loss or gain of water, since every bit of water will be accounted for. We can verify that the differential current element produces a field with no change in the magnitude along its direction by

direct calculation of $\dfrac{\partial \left| d\vec{B} \right|}{\partial d\hat{B}} \equiv d\hat{B} \cdot \vec{\nabla} \left(d\vec{B} \cdot d\vec{B} \right)^{1/2} \propto d\vec{B} \cdot \vec{\nabla} \left(d\vec{B} \cdot d\vec{B} \right)$, (see end of chapter

problem), but we can also see it intuitively in the following way.

Obviously, requiring that motion in a given direction results in a change in a field's magnitude is the same as saying that the field's magnitude is a function of the distance in that direction. Thus, for motion in the direction of $d\vec{B}$, which is by definition perpendicular to $\hat{R}$, to result in a change of its magnitude, that magnitude, $\left| d\vec{B} \right|$, must depend on the perpendicular to $\hat{R}$ direction (which is an angular direction), not simply on $R = \left| \vec{R} \right|$, which will not change as one moves in the direction of $d\hat{B}$. By contrast, the magnitude of a field whose direction is along $\hat{R}$, such as the electric field of a point particle, will clearly change along that field's direction ($\hat{R}$), if its magnitude is a function of R. Again, for $d\vec{B}$ to change in the direction of its field, its magnitude would have to have a dependence on the angular direction, such as shown in Figure 3-8. Since there are no such functional dependences in the magnitude of the given field (equation 3.18), there can be no change in magnitude in the direction of the field. Hence, there can be no divergence of the B-field produced by a differential current element. Furthermore, since any B-field is a sum of such differential fields (cf. equation 3.17), which each have no divergence, the net field must have no divergence. Hence, we always have $\nabla \cdot \vec{B} = 0$.

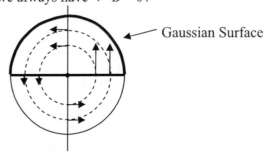

Gaussian Surface

Figure 3-8: A cylindrically symmetric field whose magnitude depends on the angle around the center would have a divergence. Consider a Gaussian surface with a semicircular cross section and a flat bottom. No field flows out the ends. However, note that flow up into right bottom side of the Gaussian surface is greater than that out of left bottom side, so there is a net flux into the volume.

---

[18] In the form of a logical proposition one can say: $\nabla \cdot \vec{B} \neq 0 \Rightarrow \vec{B}$ changes along its own direction. Said another way, a field that nowhere changes along its own direction is a sufficient (though not necessary) condition for a divergence-less field. This is the logical contra-positive of the first proposition.

Note well that we have said a differential current element produces a field that does not change magnitude along its own direction. However, the constancy of the B-field along its own direction is a sufficient not a necessary condition for a divergence-less B-field. Thus, although the zero divergence of the B-field is preserved when one sums the contributions of the many differential elements of an extended source, the constancy of the B-field along its own direction is not necessarily preserved.[19] The reason is that the summed field is: $\vec{B} = \frac{1}{c} \int \frac{\vec{J} \times \hat{R}}{R^2} dV$; hence, the square of the magnitude at a given point

is: $\vec{B} \cdot \vec{B} \propto \int \frac{\vec{J} \times \hat{R}}{R^2} dV \cdot \int \frac{\vec{J} \times \hat{R}}{R^2} dV$, which is not the same as $\int \left(\frac{\vec{J} \times \hat{R}}{R^2} \cdot \frac{\vec{J} \times \hat{R}}{R^2} \right) dV$.

Of course, one can mathematically show, as you will be asked to do in an end of the chapter problem, that the divergence of $\vec{B} = \frac{1}{c} \int \frac{\vec{J} \times \hat{R}}{R^2} dV$ is zero. One sees that the zero result arises because of the two key points just discussed: (1) the direction $\vec{J} \times \hat{R}$ and (2) the magnitude solely depends on a function of R. It is also informative to see how it works with a particular Gaussian surface, e.g. triangular wedge, as is discussed in another end of chapter problem.

Of course, this section can be summarized by writing what we will call *"the second" Maxwell equation*:

3.19

> 2nd Maxwell's Eqn. $\qquad \nabla \cdot \vec{B} = 0$

One can immediately recognize its similarity to the "first" Maxwell equation: $\nabla \cdot \vec{E} = 4\pi\rho$. The major difference is that while charges cause static electric fields, only *impetus-activated* charges cause magnetic fields. Or, said another way, there are no magnetic charges. There is no simple property of a body that, without impetus, can cause a magnetic field in the plana. This means there are no sources of the field direction, so that the field must circle back on itself. This, in turn, means that there are "poles:" a place where the B-field leaves the source (called, roughly, the north pole) and a place where it returns to the source (the south pole). Hence, in this language, equation 3.19, one says there are no magnetic monopoles, no charge density of monopoles $\rho_{magnetic}$. There have been many attempts to find such monopoles. One famous experiment by Cabrera[20] for a while looked like it had detected a lone monopole. However, it like all others before and since, have yet to yield convincing evidence of magnetic monopoles.

Before we leave this subject, it is worth touching an interesting point of theory. It can be shown[21] that one can hypothetically introduce magnetic monopoles in the following way. We introduce them into the complete system of Maxwell's equations[22] and simultaneously define a new E-field, say E', to be a linear combination of the original E

---

[19] The magnetic field pattern known as a "dipole field," which is approximated, for example, by the far field of a standard bar magnet, is an example proving this point.

[20] *First Results from a Superconductive Detector for Moving Magnetic Monopoles*, Phys. Rev. Lett. Vol. 48, No. 20, 1378, May, 1982

[21] See end of chapter problem for Chapter 5, "The Last Quasi-Static Effect"

[22] The complete Maxwell's equations are beyond the equations that we discuss in this and the earlier chapters, since they allow for non-static fields and arbitrary motion of charges.

and *B,* i.e., $\vec{E}' = \vec{E}\cos\theta + \vec{B}\sin\theta$. We also introduce such a definition for $\vec{B}' = -\vec{E}\sin\theta + \vec{B}\cos\theta$. The new system of equations would describe a world with monopoles and would be experimentally indistinguishable from the standard Maxwell's equations. However, to achieve it, we would have to posit new properties for no real reason. In particular, we would need to introduce magnetic charge in addition to electric charge and impetus-activated magnetic charge as well as impetus-activated electric charge. Furthermore, the electric (magnetic) field would be produced in two ways (possibly implying they are distinct types of fields?), by electric (magnetic) charge and impetus activated magnetic (electric) charge. Because it does multiply entities with no reason, the existence of such a transformation remains a curiosity of Maxwell's theory, though an interesting and possibly one pregnant with insight for new physics.

Given that the divergence of the static *B*-field is zero (we will see later that it is zero even when the field is changing), the curl of *B* must specify *B*.

The Curl of *B*:
Ampère's Circuital (Path) Law: $\nabla \times \vec{B} = \dfrac{4\pi}{c}\vec{J}$

We begin to understand the curl of *B* by returning to Ampere's force law. First, we notice that according to *B*-field fall-off involved in that law (i.e., $B = \dfrac{2I}{cr}$, given in equation 3.2), the *B*-field falls off as $1/r$. This behavior is exactly the same as that of the *E*-field of a uniformly charged line, $E = \dfrac{2\lambda}{r}$. Now, recall that we decided that the electric field strength decreases from a point charge as the inverse square because of the increasing area over which it must activate field. The line charge decreases in the way it does because of the reinforcement provided by the extension of the charge source from a point to a line. We can reasonably ascribe the same reason to the fall-off of the *B*-field for a line of current, especially given the $1/r^2$ fall-off seen in the differential Biot-Savart formula (equation 3.18).

However, there is an important difference; the static *B*-field, as we have discussed in the previous section, always points *perpendicular* to the direction in which it spreads out. Now, assuming the field strength decays with distance because the field must act on a larger areas as it moves out, we would naturally want to express this using a divergence as: $\nabla \cdot (\text{propagation vector}) = source\ strength$. However, we clearly cannot connect the strength of the source causing the *B*-field to the direction and strength of the *B*-field in this way, even leaving out its divergence-less nature, because the direction of the field's propagation is not directly represented by the direction of the field. Instead, we need a feature of the flow that captures the component of the field that is perpendicular, rather than parallel, to the vector from the source to the field. The curl can naturally do this for us.

In particular, there is a preferred direction around which we will be interested in the curl, namely the direction of the impetus that moves the charges that form the current, which is the direction of the current density, $\vec{J}$. And, as just mentioned, for a steady current in a long straight wire, the angularly directed *B*-field that it generates should fall-off as $1/r$ and be proportional to the current. How then do we incorporate $\vec{J}$ generally using the curl?

Again, our source equation for the electric field was the extended Coulomb's law, Gauss's law: $\nabla \cdot \vec{E} = 4\pi\rho$, whereas we will now take the source equation for the magnetic field as $curl\ \vec{B} = \text{something}$.[23] What then is the something? Here's where $\vec{J}$ comes in, for we've seen that the flow of current causes B-field, making $\vec{J}$ a preferred direction and implying a proportionality between the magnitude of the current and B. So, we take the current density, $\vec{J}$, at a given point as a representative of the source of circulation, which is *all* of the differential current elements of the source, and write the following equation, which is the *static* form[24] of the "last" or ***"fourth" Maxwell equation*** in *differential form*:

3.20

$$\boxed{4^{\text{th}} \text{ Maxwell's Eqn (static case)} \qquad \nabla \times \vec{B} = \frac{4\pi\vec{J}}{c}}$$

> Where the constants are given in *cgs* and are chosen for convenience in certain situations, but can be chosen differently in other units.

This then implies the circular B-field pattern around the direction $\hat{J}$ of an isolated differential current element. Thus, for example, a current loop will cause a B-field around each small element of current. Note the implications of this equation are somewhat more complex than might appear at first blush. Most importantly, we implicitly assume, even when we are only focusing on one small element of the current, that the current indeed travels in a complete loop, so that the B (and thus the curl of B) at any one point is due to the effect of *all* parts of closed circuit of current. The curl of B at a given point is then, under the assumption of closed circuit and infinite speed of propagation of the B-field, represented by the current density at that point but not caused by it alone. This point will be developed in Chapter 5. For now, we simply note that this equation, in the proper context, expresses the relationship between the source and the B-field because of the special nature of the B-field, especially its being generated in "curling-only" fashion by $\vec{J}$.

B-field around a Straight Wire

And, we can now easily see that equation 3.20 gives the required results for a long straight current[25] by taking the surface integral of the $\nabla \times \vec{B}$ over the circular region perpendicular to the wire (shown below to the right). We use Stokes' theorem, $\oiint_S \nabla \times \vec{B} dA = \oint_C \vec{B} \cdot d\vec{s}$, to get ***Maxwell's fourth equation*** in *the integral form*:

3.21

$$\boxed{\oint_C \vec{B} \cdot d\vec{s} = \frac{4\pi}{c} \oiint_S \vec{J} \cdot d\vec{A}}$$

---

[23] Later we will also find that there is an analogical source equation for E, for changing fields:

$$curl\ \vec{E} = -\frac{1}{c}\frac{\partial \vec{B}}{\partial t}$$

[24] This ceases to be valid when $\vec{E}$ changes with time. We are missing a term on the right hand side that will be discussed in Chapter 5.

[25] We can take the circuit of the long wire to be completed at "infinity," i.e., sufficiently far away that its effects can be neglected at the level of accuracy we choose.

Then, we write:

$$\oint_C \vec{B} \cdot d\vec{s} = B(2\pi r)$$

3.22

$$= \frac{4\pi}{c} \oiint_S \vec{J} \cdot d\vec{A} = \frac{4\pi I}{c}$$

So, we get: $B = \dfrac{2I}{cr}$, exactly the result obtained earlier from Ampère's *force* law. Note in particular, that the fall-off is the same as we would expect based on area arguments because the length of the line around the circumference of a cylinder is $2\pi r$ and the circumference times the height gives the surface area of the cylinder. Hence, the curl of B in the region can also describe this fall-off. We will see in another chapter that, when we allow for quickly changing currents, the curl of B will not be given by equation 3.20 because account must be taken of finite propagation speed of the B-field. And, that equation needs modification to account for diverging currents. For now, we continue to ignore both.

The "curling" field created by the current moves outward from the long straight wire, where it is more properly called a curling field because the B-field has no curl outside the wire. This is clear from equation 3.20, since there is no current, i.e., no J, off of the axis, which is the wire. The $1/r$ fall-off ensures this. We can see this in the following way.

A curl of zero at a point means that any arbitrarily small line integral around that point is zero. If the curl is zero in a region, this is also true for finite size curves in the region. To show this, we start with curves in the plane perpendicular to the long wire. We use the following sequence of figures that show the B-field circulating in a counterclockwise sense because I is coming out of the plane of the page.

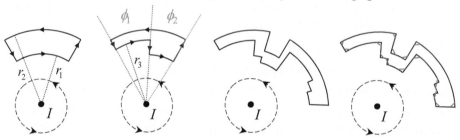

The first path (shown with solid lines) on the left, starting with the bottom part of the path, yields the following line integral: $-B(r_1)\phi_0 r_1 + 0 + B(r_2)\phi_0 r_2 + 0 =$ $\phi_0\left(-\dfrac{2I}{cr_1}r_1 + \dfrac{2I}{cr_2}r_2\right) = 0$, where we take ϕ_0 as the angle subtended by the path. Notice that the cancellation due to the $1/r$ fall-off happens without having to take the limit to a small path since the curl is zero in any region that does not enclose the wire. The second diagram yields: $\phi_1 r_3\left(-\dfrac{2I}{cr_3}\right) + 0 + \phi_2 r_1\left(-\dfrac{2I}{cr_1}\right) + 0 + (\phi_1 + \phi_2)r_2\left(\dfrac{2I}{cr_2}\right) = 0$, where ϕ_1 and ϕ_2 are, respectively, the angles subtending the left and right part of the path. Hence, it is easily seen that any pattern that makes radial steps in any complicated pattern, such as shown in the third drawing, will give zero. The last diagram shows the third diagram with the corners cut-off by (red) lines. To get the path integral of the new curve, we only have to subtract off the path integrals of the triangular shaped regions that are created by introducing the lines. Now, we can get arbitrarily close to any curve using an arbitrarily large number of radial type steps with lines to cut off the corners. And, thus, if the triangular path integrals created are each separately zero in the limit of small triangles, we will have shown that any arbitrarily shaped loop will have no curl as long as it does *not* encircle the current.

So, we now show the triangular loop integral is zero using the figure below:

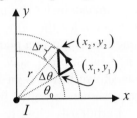

We first write the ends of the hypotenuse of the triangle in terms of polar coordinates:

$$(x_1, y_1) = (r\cos\theta_0, r\sin\theta_0), \quad (x_2, y_2) = \left((r + \Delta r)\cos(\theta_0 + \Delta\theta), (r + \Delta r)\sin(\theta_0 + \Delta\theta)\right)$$

This gives the difference ($P2 - P1$) as:

$$(\Delta x, \Delta y) = \left((r + \Delta r)\cos(\theta_0 + \Delta\theta) - r\cos\theta_0, (r + \Delta r)\sin(\theta_0 + \Delta\theta) - r\sin\theta_0\right)$$

$$= \left((r + \Delta r)(\cos\theta_0 - \sin\theta_0\Delta\theta) - r\cos\theta_0, (r + \Delta r)(\sin\theta_0 + \cos\theta_0\Delta\theta) - r\sin\theta_0\right)$$

$$= \left(\Delta r\cos\theta_0 - r\sin\theta_0\Delta\theta, \Delta r\sin\theta_0 + r\cos\theta_0\Delta\theta\right)$$

Where, because $\Delta\theta$ and Δr are arbitrarily small, to first order, $\sin\Delta\theta = 0$ and $\Delta\theta\Delta r = 0$

So that, for the hypotenuse of the triangle, we have, given our convention for positive direction along the path: $d\vec{\ell} = -\left(\Delta x\,\hat{x} + \Delta y\,\hat{y}\right)$. Using this along with Ampère form of the magnetic field, $\vec{B} = \dfrac{B_0}{r}\hat{\theta}$ with $B_0 = 2I/c$, gives, for the line integral along the hypotenuse of the triangle:

$$\int_1 \vec{B}d\vec{\ell} = -\frac{B_0}{r}\left[\hat{\theta}\cdot\hat{x}(\Delta x) + \hat{\theta}\cdot\hat{y}(\Delta y)\right]$$

$$= -\frac{B_0}{r}\left[-\sin\theta_0\Delta x + \cos\theta_0\Delta y\right]$$

$$= -\frac{B_0}{r}\left[-\Delta r\sin\theta_0\cos\theta_0 + r\sin^2\theta_0\Delta\theta + \Delta r\sin\theta_0\cos\theta_0 + r\cos^2\theta_0\Delta\theta\right]$$

$$= -B_0\Delta\theta$$

Where to get the second line we used: $\hat{\theta} = -\sin\theta\hat{x} + \cos\theta\hat{y}$.

The next leg of the triangular path is the radial leg, and it clearly gives no contribution. For the last leg, the line integral over the arc that is subtended the angle $\Delta\theta$ shown in the figure above, we get:

$$\int_3 Bd\vec{\ell} = \frac{B_0}{r}r\Delta\theta = B_0\Delta\theta$$

Thus, the two non-zero terms cancel, and the total path integral is zero.

Thus, we have shown that, for the given $1/r$ law for the magnetic field, the path integral around any arbitrarily shaped curve within the plane that does not enclose the current the curl is zero. This is easily extended outside the plane since there is no B-field in that direction.

We can also show that our $1/r$ law will be enough to make the line integral around an arbitrarily shaped loop that encircles a current yield an answer proportional to the current and independent of r. Of course, we know this will be the case because of the way we have defined the curl to be zero everywhere except at the origin, but important insight into the content and use of the formalism can be had by understanding how it works out.

Take an arbitrary curve that encircles the wire, such as shown in the left figure below. For *any* such curve (even one that is not in a plane), we can draw a circle centered on the wire touching the curve at its nearest point, such as shown. We can then break the funny shaped loop and the circle as shown in the right hand figure below. We know that the line integral around the new path, shown in dotted lines, will be zero since it does not encircle the wire. Indeed, we can get arbitrarily close to closing the gap and it will still be zero. However, since the integral around the *circle* does give $4\pi I/c$, this must be canceled by the line integral around the funny shaped curve: $\int_{Circle} + \int_{Funny} = 0$. Hence, the integral around the funny curve in the indicated direction is $-4\pi I/c$. Hence, the path integral around any closed curve that encircles the wire will yield $4\pi I/c$, which is the curl of the field at the origin, the only non-zero curl inside the closed curve.

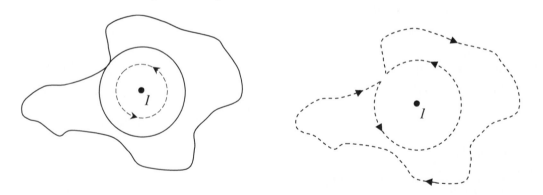

B-field around Bent Wires

Up till now, we have been considering the B-field around a long straight wire, which does not appear to be a closed loop. For it to be so, we must imagine a return wire connecting its right end to its left end arbitrarily far from the region in which we probe the field of the straight wire. We say the wire is closed "at infinity."

We have yet to consider the B-field in the more general case of an *arbitrarily-shaped current* loop. This means we have to bend our long straight wire. Now, since we have posited that the cause of the B-field is only impetus-activated charge, we do not expect such bends to qualitatively change the situation *other than* by (1) how such bends force the impetus to be different at different points along the wire and (2) how it changes the distance the field must propagate to the given field point. By contrast, one might think that a *changing* impetus of given type and magnitude could cause a different field than an unchanging impetus of that same type and magnitude; i.e., empiriometrically, one might expect $\dot{\vec{J}}$ to contribute as well as $\vec{J}$. However, only the impetus of the given charge at the given moment,[26] not whether or not that impetus is changing (which it must be in a bent wire), should affect the field, except in so far as such a change, for finite propagation speeds, can result in different affects arriving at different times. For the steady state case that we are considering, every point of the plana is acted on in the same way at every moment, so we do not have to consider such propagation effects. Moreover, we also take, in this chapter, the propagation speed to be infinite. Hence, for such cases, we do not expect there to be any *fundamental* difference between a curved wire and a straight wire. Again, the changing impetus doesn't cause any effect, it's the impetus that does.

However, we might still expect that a change in equation 3.20 is required to account for the changed directional type of the impetus of the various elements of the circuit and of the changed distance of those elements from the field point that accompanies the change in shape. Still, as we have seen, this equation captures two key aspects of the B-field, in the sourceless/sinkless static[27] case including (1) its circulation-only nature and (2) the need to respect the direction $\hat{J}$ as the local axis of that circulation. Furthermore, the equation captures the behavior of the straight line B-field, which, in one sense, plays, for B-field, the role of primary part that the point charge does for the E-field (because B is caused by impetus-active charge rather than simply charge). Because it captures the straight line behavior, it also implicitly captures a third key aspect of the B-field, the special $1/R^2$ fall-off around a differential current element. Indeed, if we go one step further and consider a differential current element (which, for the souceless/sinkless case under consideration, is a being of reason), we see the equation goes very deep. In particular, it can be proven that the key aspects of the B-field manifested in the Biot-Savart law are caught by the formula (equation 3.20). Thus, we expect that the formula that originally encapsulated all these realities will be robust enough to hold for any shape closed current path ($\nabla \cdot \vec{J} = 0$) or even more generally any closed circuit of current.

Now, it can be shown that this is the case if the way that the currents generate the B-field is accurately described by Biot-Savart (equation 3.18), and if the B-field obeys the superposition principle.[28] In short, the Biot-Savart integral given in equation 3.17 is the ("static"[29]) solution of equation 3.20.

In fact, equation 3.20 is experimentally borne out, and, we have the remarkable fact, that our equation defining the strength of the static field $\nabla \times \vec{B} = \dfrac{4\pi \vec{J}}{c}$ holds for any

---

[26] Note well that this is a shorthand way of talking, for, of course, it is finally only the impetus as it is actually able to act, as manifested by how effective it is in causing locomotion, that matters. Remember action and reception are co-relatives; one cannot exist without the other. Thus, for impetus one needs to account for both the activity (impetus) and the way it is received, i.e. the bodies receptivity or resistance to the action of the impetus (mass); empiriometrically, $v = p/m$.

[27] There are a few non-static exceptions for which the equation works, which we do not discuss here.

[28] Recall that the superposition principle says that the quality known as the B-field generated in a given part of the plana from different source elements is as if each of these elements acted independently, so that the field caused there has the same strength and directional type as the vectorial sum of fields that would be caused by each of the parts of the source acting independently.

[29] More exactly, it is the solution in the infinite speed of propagation approximation when $\dot{\vec{J}} = 0$

shaped wire loop, and thus we get the complete statement of Ampère's circuital (closed path) law: *For a closed circuit source ($\nabla \cdot \vec{J} = 0$ everywhere), the line integral around any arbitrarily shaped closed curve is equal to the total current that pierces the surface bounded by that curve; in our formalism, or as we say, mathematically, it is expressed by:*

3.23
$$\oint_C \vec{B} \cdot d\vec{s} = \frac{4\pi}{c} \oiint_S \vec{J} \cdot d\vec{A} = \frac{4\pi I}{c} \qquad \left(\hat{n} \cdot \circlearrowright \right)$$

Where, as already defined (see Chapter 2), the circulation sense is taken to be positive when directed in the right handed sense along the normal vector to the area such as shown above.

Of course, it should be clear that if the integral is carried out N times around the path, one will get $N I$ rather than I.

Lastly, we could formally show that the curl of the field given by an arbitrary field given by the Biot-Savart law (equation 3.17) yields required result. Instead, we will leave it

as a problem to show: $\nabla \times \int d\vec{B} = \nabla \times \oiiint_V \left(\frac{\vec{J} \times \hat{R}}{R^2} \right) dV' = \frac{4\pi I}{c}$.

The Magnetic Potential, the *A*-field

More about the relation between Divergence and Curl

Before we introduce the key concept of this section, we need to recall and understand the fact that the divergence of a curl of a field is always zero; namely, $\nabla \cdot \left(\nabla \times \vec{A} \right) = 0$. One can get a firm intuitive feel of why this must be the case in the following way. To keep the principles in the forefront, we consider the very simple case of a vector field, call it $\vec{A}$, pointing solely in the z-direction in a given region of space, such as shown by green arrows in the figure below. Furthermore, first assume that the *A*-field is the same in all four corners of the base of a cubed shaped region. Note that, at this point, the field clearly has zero curl because its z-component (the only component) does not change as we move in any perpendicular direction. Now, imagine shrinking one of the vectors at one of the corners so that the field is the same at only three points such as shown in the middle figure below.

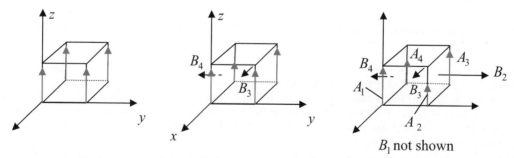

B_1 not shown

Using the figure, one can easily see that shrinking the left back arrow results in a curl, which we will call $\vec{B}$. Actually, it causes *two* curls, one coming *in* the back side, B_3 and one coming *out* the left side, B_4. The crucial point is that shrinking one always causes both curls, and they always will be of equal magnitude with one coming in one side and another going out the other. Thus, we have created an A-field with a curl and from that another field, namely its curl, i.e. $\vec{B} = \nabla \times \vec{A}$, and that field has zero divergence. In other words, we see that the divergence of the curl must be zero. We can easily make the generality of this approach more manifest by making the A-field different at each of the four corners, as shown in the right portion of the above figure. Then, calculating the divergence, assuming the cubes have sides of unit length and recalling that the normal to a part of a closed surface is positive when it points out of the enclosed region, to get:

$$B_1 = A_2 - A_1, B_2 = A_3 - A_2, \; B_3 = -\left(A_3 - A_4\right) B_4 = -\left(A_4 - A_1\right)$$

flux $= \sum B_i = A_2 - A_1 + A_3 - A_2 - \left(A_3 - A_4\right) - \left(A_4 - A_1\right) = 0$. Hence, we have the basic understanding of why the divergence of the curl of a field is always zero.

The Definition of the A-field

We summarize the results of the previous two sections succinctly using our formalism:

$$\nabla \cdot \vec{B} = 0 \quad \text{and} \quad \nabla \times \vec{B} = \frac{4\pi \vec{J}}{c}$$

By Helmholtz's theorem, this specifies the B-field throughout the plana. The latter describes the relation of the field to its source. The former tells us that the B-field can simply be defined by one vector field, $\vec{A}$, defined such that:

3.24 $\qquad\qquad \vec{B} = \nabla \times \vec{A}$

This ensures, as just discussed, that $\nabla \cdot \vec{B} = 0$. By comparison, the static E-field, with:

$$\nabla \cdot \vec{E} = 4\pi\rho \text{ and } \nabla \times \vec{E} = 0$$

required only one scalar field, ϕ, because $\vec{E} = -\nabla\phi$ automatically satisfies the second.

Substituting $\vec{B} = \nabla \times \vec{A}$ into the equation for the curl of B gives (using a vector identity that you are not expected to memorize but only be aware of):

3.25 $\qquad \nabla \times \left(\nabla \times \vec{A}\right) = \vec{\nabla}\left(\nabla \cdot \vec{A}\right) - \nabla^2 \vec{A} = \frac{4\pi \vec{J}}{c}$

We will later discuss the so-called Lorenz Gauge which posits the following relationship between $\vec{A}$ and ϕ: $\nabla \cdot \vec{A} = -\frac{1}{c}\frac{\partial \phi}{\partial t}$ (in *cgs* units). We also will have to wait to learn why we have chosen it. We will see the logic of choosing it in the context of propagation, which we, at this point, have "switched off." In our current case, ϕ does not change, so the Lorenz condition becomes $\nabla \cdot \vec{A} = 0$ and thus equation 3.25 becomes:

3.26 $\qquad\qquad \nabla^2 \vec{A} = -\frac{4\pi \vec{J}}{c}$

This is actually three equations. In particular, it is three instances of Poisson's equation, one for each component of $\vec{A}$. Using the solution to the scalar Poisson equation given in Chapter 2, it is clear we can write the solution to this vector Poisson equation as:[30]

---

[30] Note that, in a later chapter we will see that, in the Lorenz gauge, the A-field and the ϕ-field each propagate outward at the speed of light; for now, as we have said, we neglect this and make the infinite speed approximation.

3.27 $$\vec{A}\left(\vec{r}\right)=\frac{1}{c}\int\frac{\vec{J}\left(\vec{r}'\right)}{\left|\vec{r}-\vec{r}'\right|}dV\,,$$

where $\vec{r}$ is the vector to the field point, i.e. the vector pointing from the coordinate origin to the point in the plana at which one wants to calculate the field, and $\vec{r}'$ is the vector to the source point, i.e. the vector pointing to the point in the source of which one is currently attempting to determine the influence.

Thus for the differential case, we can write:

3.28 $$d\vec{A}=\frac{1}{c}\frac{I\,d\vec{l}}{R}$$

Where $d\vec{l}$ is a differential element of length of the source current, and, as usual, R is the distance between the source point and the point in the plana where one wants to calculate the field.

We can now readily show that the curl of this equation gives a differential form Biot-Savart law. We need the following vector identity, which we leave as an exercise for you to verify: $\nabla\times\left(\psi\vec{V}\right)=\vec{\nabla}\psi\times\vec{V}+\psi\,\nabla\times\vec{V}$. Since $d\vec{l}$ does not depend on the field point $\left(x,y,z\right)$, its derivative with respect to these variables (as represented by the del operator, ∇) is zero, and only the first term will contribute. Thus, we get (in *cgs*):

3.29

$$d\vec{B}=\nabla\times d\vec{A}=\frac{1}{c}\nabla\times\frac{I\,d\vec{l}}{R}=-\frac{I\,d\vec{l}}{c}\times\nabla\left(\frac{1}{R}\right)$$

$$d\vec{B}=\frac{I}{c}\frac{d\vec{l}\times\hat{R}}{R^2}$$

Where to get the last line, we use: $\nabla R=\hat{R}$, (show this as an exercise).

We can also show that the divergence of $\vec{A}$ is zero, as required (see end of chapter problem).

Now, the A-field is to the B-field as the ϕ-field is to the E-field. That is, we can and will take the A-field as the full specification of the B-field. Of course, as with the ϕ-field and the E-field, if the B-field is given everywhere, one can deduce the A-field. However, often a B-field is given only in a certain region without specifying its source. In a given region, different sources can approximate the same B-field, but, the A-field will be different. Thus, the use of the B-field can be a way of abstracting from the particularities of a source. In any case, the A-field makes explicit what is, at best, only implicit in the B-field. The best way to see the insight that the A-field gives into the behavior and nature of the B-field is to analyze some specific examples. We now turn to calculating the A and B-field for various sources.

A-field of a Long Straight Wire

We start with our well worn infinite wire because it gives the simplest but, at the deepest level, a fundamentally complete picture for the steady state case. It has impetus of a single directional type, say rightward, moving charge and causing an A-field with a curl, and hence a B-field. The B-field, we already know, is given by Ampère's law and shown in Figure 3-2. The A-field is shown in Figure 3-9 where it is represented using arrows as well as a green shading which we will explain shortly; note that brighter green means a stronger A-field. The wire causes an A-field parallel to the wire that falls-off as $\ln R$, which we obtain as follows. Using the orange path shown in Figure 3-9 and equation 3.23, we get:

$$\int_S \vec{B}\, dS = \int_S \nabla \times \vec{A}\, dS = \int_C \vec{A} \cdot d\vec{l}$$

$$\frac{2I}{c\, r}\Delta r \Delta z = -A_z\left(r + \Delta r\right)\Delta z + A_z\left(r\right)\Delta z,$$

$$\frac{dA_z}{dr} = -\frac{2I}{c\, r}$$

Integration yields:

3.30 $$A_z\left(r\right) = A' - \frac{2I}{c}\ln r$$

The constant, A', will be determined, in this case, by the zero of the A-field. Of course, if the A-field is real, the constant must be ultimately determined by physical reality, not by our choice. However, when seeking to determine $\vec{B}$, which is not effected by the constant (like the case of the ϕ-field for $\vec{E}$), we will often decide the constant based on what is convenient for the focus of the given problem.

In this case, we seek a physical explanation. To determine the constant in the most sensible way, we need to deal with the inadequacy of the "infinite" line current. If we had approached the calculation in a more direct way using equation 3.28, we would have run head-on into this issue.

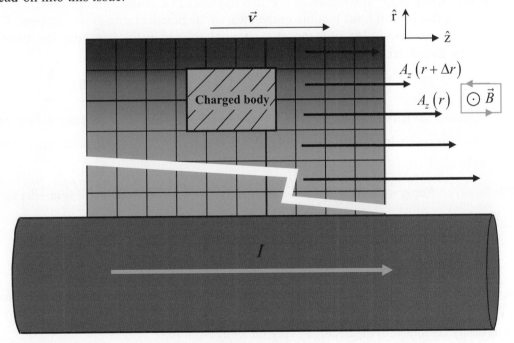

Figure 3-9: Close up of a copper wire (brown cylinder) carrying a current I. The green arrow indicates the rightward impetus that charges in the copper wire must have to cause the current. The green squares, which decrease in intensity with distance from the wire, indicate the A-field type and intensity.

In particular, integrating $d\vec{A} = \frac{1}{c}\frac{I\,d\vec{l}}{R}$ along the z-axis to get the total $\vec{A}$ yields infinity. This unphysical result stems from the fact that an actual current must circle back at some point in order to feed the charge around the loop so as not to have any charge accumulation.[31] To complete the loop and to preserve the symmetry, we close the loop with a cylindrical conductor, making a kind of coaxial wire with hemispherical caps that carry the current to and from the center wire and outer cylindrical shell, as shown in the figure below.

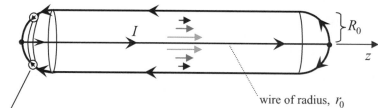

Current source across gap in hemispherical cap

We take the radius of the inner cylinder (i.e. the wire), r_0, to be very small, the radius of the outer cylindrical shell, R_0, to be very large and the caps are taken to be very far away, leaving lots of room around the interior wire, so that it is indeed an *isolated*, long wire. In principle, one could now do the integral of $d\vec{A}$, however, the actual calculational work is very much easier if we stick to our first approach, i.e. using Stokes theorem. Note that weighing this type of issue is some of the meat of the practical side of doing physics.

Using our knowledge of Ampere's circuital (closed path) law near the middle region of the circuit, we get the magnetic field inside the wire as:

3.31   *For* $r < r_0$: $B_\phi = \frac{2Ir}{cr_0^2}$.   Giving: $A_z(r) = A_z(0) - \frac{Ir^2}{cr_0^2}$

Outside the wire ($r_0 < r < R_0$), where equation 3.30 holds, if we take the A-field as zero at the outer cylindrical shell, i.e., $A_z(R_0) = 0$, we get, neglecting the thickness of the outer shell:

3.32   *For* $r_0 < r < R_0$: $B_\phi = \frac{2I}{rc}$ and $A_z(r) = \frac{2I}{c}\left(\ln R_0 - \ln r\right) = \frac{2I}{c}\left(\ln\frac{R_0}{r}\right)$

Lastly, in the remaining domain of r, we have:

   For $r > R_0$: $\vec{B} = 0$, $A_z(r) = 0$.

We can see now the physical reason for our choice of $A_z(R_0) = 0$. In particular, since the A-field must be constant outside the circuit (i.e. for $r > R_0$), any number but zero for this constant would mean that the impetus-activated charge would maintain an A-field just as easily right next to the circuit as very far away from it!

---

[31] Of course, it is also related to the non-physical nature of actually existing infinities. But, we have seen, as for the case of the volume of Gabriel's horn discussed in the problem section of Chapter 2 of *PFR-M*, that some such integrals which extend out to infinity are finite, because some infinite series do mathematically converge.

We can now graph the field for the entire interior region of the coaxial cable circuit. In Figure 3-14, we show the result when one includes the actual thickness of the outer conductor. Such a calculation is discussed in an end of chapter problem.

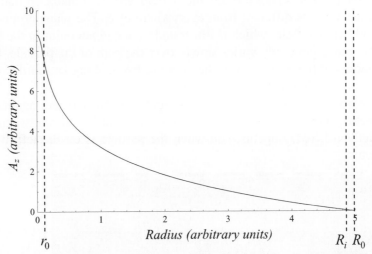

Figure 3-10: A_z versus the radial distance from the center of the coaxial conductor in the left-right middle of the coax cable. r_0 is the radius of the inner wire, and R_0 and R_i are the outer and inner radius of the outer cylindrical conductor. In the text, we take, $R_i = R_0$. The field is zero everywhere outside the coax ($r > R_0$).

Now, how does the power in the plana specified by the A-field influence impetus-activated charges? It does so in two fundamental ways. The second type of influence pertains only to changing A-field, and though it is properly a magnetic influence, it is not the same as a B-field; it properly belongs in the next chapter, so we will leave it until then. The first type of influence can be understood as follows and is described by the six cases that we associated with the B-field in Figure 3-4.

Neither of the two types of impetus that are orthogonal to the plane of the A-field, and thus *along* (or against) the B-field, at a given point are influenced by the A-field. For example, in Figure 3-11 , on the left side of the wire, we see that the A-field has a curl along the z-direction and thus B points along the x-direction; so, for this case, positive particles with impetus along $\pm\hat{x}$ are not affected at the point shown.

Figure 3-11: The A-field around a current carrying wire in an orientation rotated 90 degrees to that of Figure 3-9. The black arrow represents current flowing in a long wire in the y-direction. As in Figure 3-9, the orange lines indicate the line integral used for calculating the A-field.

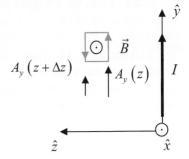

We use Figure 3-12 to discuss the remaining four situations in which the A-field does actually have an effect. The figure shows a positively charged particle in circular orbit around the B-field generated by a wire. The figure shows, using a dashed rectangle, a zoomed view of the particle's path through the A-field activated plana, so that the distance scale inside the rectangle is different from everywhere else. The plana around the wire is activated by a rightward A-field, which is illustrated by the green color in the figure. In the magnified region, the B-field only varies slowly over the path of the particle. Note that we exaggerate the radius of the orbit to keep the drawing from taking an inordinate amount of space (*exercise*: why would it take a lot of space).

When the particle is on the top, it has rightward (which we, again, indicate by green) type impetus, when it is to the right, it has downward (yellow), when it is on the bottom, it has leftward (red) impetus, and when the particle is on the left, it has upward (blue) type impetus.

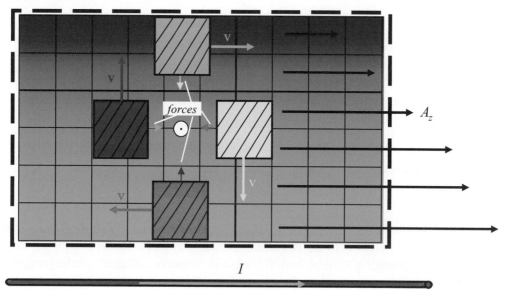

Figure 3-12: A view through a rectangular shaped "magnifier" of a region of the plana above a current carrying wire. A positively charged particle is shown rotating around a point in the plana at four different locations. Note that the forces caused by the A-field activated plana on the charge is always inward (centripetally), which maintains the circular motion of the particle around the magnetic field vector shown in the middle. Note also that the charge takes up multiple plana parts. Each force arrow is shown in the color of the impetus that it causes.

Staring with the charge on top where it has rightward impetus, we see that it is attracted toward the wire, which corresponds with our prescription that *like impeti attract*. Indeed, with the A-field, we can see why at a deeper level. The positive charge in the wire activated by rightward impetus affects the plana by causing a power in the plana that can pull like-impetus towards it. The current in the wire makes a sort of surrogate of itself in the plana. This is why we use the same color for the field as for the impetus of the charges that cause it. The current causes a decreasing strength disposition in the plana which is quantified by $\vec{A}$ and shown in the diagram by decreasing bright green. This sets a path back to the plana measured by the gradient of that field. The steeper that gradient is, the more clearly the path back to the wire is set, and the stronger is the B-field. That is, as we have seen earlier, the A-field pulls like-impeti (positive) particles towards the wire with a strength proportional to this gradient. In our case, the "green" current causes "green" plana which, in turn, attracts the positive particles with "green" impetus along the path of increasing brightness. As we have said, this is part of the "plan" of the particles in the wire to satisfy its "aim" of attempting to keep the "test" particle close without altering its energy.

When the particle is on the right and has downard (yellow) impetus, it is "threatening" to overshoot the wire and thus the plana, acting as the current's surrogate, acts to give it impetus against the motion of the current, which will then cause the plana to exert a force that will decrease the impetus (and hence speed) of the particle. In this way, it changes its direction without changing its energy, since it always acts perpendicular to the particle's motion. Notice that the yellow charge (like the others) covers more than one plana part so that the disposition of the plana parts can and do determine how the plana will act. In particular, the A-field specifies that disposition and, when the impetus is of a type that is perpendicular to the direction of A but *along* the direction of increasing $\left|\vec{A}\right|$, the plana acts to give it impetus of a type against the direction of A. If, on the other hand, the impetus is of a type that is perpendicular to the direction of A but *against* the direction of increasing $\left|\vec{A}\right|$, the plana acts to give it impetus of the *same type* as the direction of A. This later is, of course, what happens to the particle when it is on the left and is blue so moving upward.

When the particle is at the bottom, it has the type of impetus opposite to the current, and thus the plana acts to push it away--perpendicular to the direction of A and along the direction of decreasing $\left|\vec{A}\right|$. *Unlike impeti repel.*

> The A-field of any system can be drawn very simply by noting: (1) A points in the direction of the source (or opposite to it for negative source charges), (2) A decreases in magnitude from the source according to a $1/R$ law for each small element of the current, and (3) A is the linear sum of all the contributions from each piece of impetus-activated charge.

The direction of the B-field can easily be deduced from the A-field by noting the direction that the A-field would push a curl meter. Or, if you would like, you can think of encasing each pair of A-field vectors in a hot tub and think of the A-field like jets of the tub. Then, using the figure here, imagine which way the water would swirl. Applying the right hand rule to that swirl gives you the direction of the B-field.

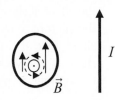

Before we move on to our next example field, we note that the long straight wire can be used to illustrate how the B-field can be used to leave out (analogically abstract) specifications that are obviously necessary in reality, but not relevant for a given problem. In particular, a local B-field of a given magnitude and direction, such as shown in the Figure 3-13, can be created by either a vertical *or* a horizontal wire.

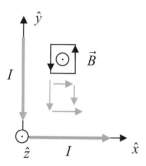

Figure 3-13: Two long wires oriented perpendicular to each other. Note that each wire generates a *B*-field of the same direction at the place indicated, but each does it using *A*-fields of different directional types. In particular, as discussed in the text of the article, each provides one term of the curl of *A* which together yield the total *B*. Yellow indicates downward impetus and green indicates rightward impetus.

Field of a Current Sheet

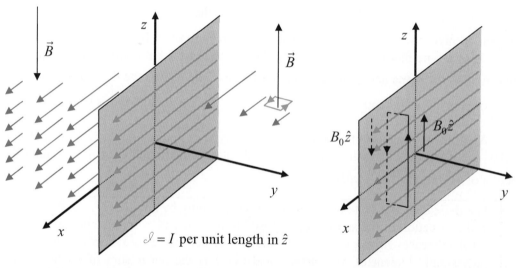

Figure 3-14: a. An *infinite* sheet of current in the *x-z* plane produces an *A*-field with curls and thus also produces a *B*-field. The *A*-field generated by the sheet is shown in some detail on the left side, but, to avoid complicating the picture, only a couple of *A* vectors are shown on the right side of the sheet. The *B*-field on the left everywhere points downward, while, on the right, it everywhere points upward. In each region, the *B*-field is constant throughout. **b.** *(right)* Rectangular path for determining B_0.

Now, consider an infinite sheet of current such as shown in Figure 3-14b. We assume, as usual, that it is flowing in a neutral conductor so we can ignore the electric fields. We calculate the *B*-field by recalling that current produces a curling *B* "at" its own location. Hence, we investigate the line integral around a loop that encircles a typical region of the infinite sheet such as that shown in Figure 3-14. The line integral gives, using equation 3.23: $2B_0 l = \dfrac{4\pi I}{c}$. The left hand side obtains because the top and bottom leg of the integral are both zero since they have no field along them, while the other two legs both have the full strength of B, B_0, pointing along them, thus each giving $B_0 l$. Hence, solving for B_0, we have:

3.33 $$\vec{B}_{\pm} = \pm \frac{2\pi}{c}\frac{I}{l}\hat{z} = \pm \frac{2\pi \vartheta}{c}\hat{z}\,,$$

where the "+" refers to the positive side of the y-axis and the "−" to the negative side, and ϑ is the current per unit of length perpendicular to the flow.

We can intuitively construct the A-field as it is drawn in the figure. The direction is trivial; it is along the current. In any given plane parallel to the x-z plane, the magnitude of A must be the same, because each point in such a plane experiences the activity of identical source elements at identical distances. Lastly, for physical sources (as opposed to the often useful mental constructs which can have infinite extent), the A-field will get weaker as one moves away from its source.

Since, we now know B, we can calculate the A-field using the same method we used for the long wire, i.e., $\int_S \vec{B}\, dS = \int_S \nabla \times \vec{A}\, dS = \int_C \vec{A}\cdot d\vec{l}$. Using the path drawn in orange in Figure 3-14a, we get:

$A_x(x)\Delta x - A_x(x + \Delta x)\Delta x = B_0 \Delta x \Delta y$, which implies $\dfrac{dA_x}{dy} = -B_0$.

This, in turn, gives:

3.34 $$A_x(y) = -B_0 |y| + \text{constant}\,, \text{ where } B_0 = \frac{2\pi \vartheta}{c}$$

The last thing to note is that whatever "constant" we pick, the A-field appears to go negative; because there is no source of negative A-field, this must be an artifact of our chosen configuration. Indeed, we should realize by now that the current in the sheet must circle back at some point so that this solution can only be (approximately) valid within a certain distance from the sheet.

Field of a Circular Loop

Our next example is the circular loop of current lying in the x-y plane shown in the figure below. Such a loop can be created using super-conducting material.

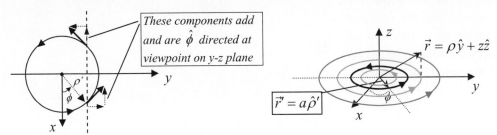

We need the formula for $\vec{A}$ given in equation 3.28 and the figures to the left and right; in the figures, we introduce ρ, which is the distance from the z-axis to a given point. We also note that we can, without loss of generality, choose the field point in the y-z plane (i.e., $x = 0$), since any field point with the same ρ and z but different ϕ would have exactly the same magnitude field and be in the angular direction ($\hat{\phi}$) at that new point because of the cylindrical symmetry of the source. We also need the following unit vectors at the source point located at each angle ϕ': $\hat{\rho}' = \cos\phi'\hat{x} + \sin\phi'\hat{y}$ and $\hat{\phi}' = -\sin\phi'\hat{x} + \cos\phi'\hat{y}$, where we use primed coordinates for source points and unprimed for field points. Then, using equation 3.28, we get:

3.35 $\vec{A}(\rho,\phi,z) = \dfrac{1}{c}2\displaystyle\int_{\pi/2}^{3\pi/2} \dfrac{I\left(a\,d\phi'\left(\hat{\phi}'\cdot(-\hat{x})\right)\hat{\phi}\right)}{|\vec{r}-\vec{r}'|}$

$$(\text{where } \vec{r} = (0,\rho,z), \quad \vec{r}' = (a\cos\phi', a\sin\phi', 0))$$

$$= \dfrac{2I}{c}\hat{\phi}\int_{\pi/2}^{3\pi/2} \dfrac{(a\sin\phi'\,d\phi')}{\sqrt{(a\cos\phi')^2 + (\rho - a\sin\phi')^2 + z^2}}$$

Note that we have here integrated from the y-axis around to the minus y-axis, in order to get the $\hat{\phi}$ (which is $-\hat{x}$ for our choice of field point) contribution from each element of half of the circle that contributes to the total. The other half gives another term just like it in the $-x$-axis direction and cancels all the y-axis directed components caused by the first half, since they are equidistant away from the field point but exactly oppositely directed.

To solve the integral in general we need the solution to, so-called, complete elliptical integrals of the first and second kind. Such "special functions" are a topic unto themselves and are worth some attention if you are interested. Here, we will only state the general solution, which is:

3.36

$$\vec{A}(\rho,\phi,z) = \hat{\phi}\,\dfrac{2I}{c}\,\dfrac{1}{\rho\sqrt{(a-\rho)^2 + z^2}} \times$$

$$\left((a^2 + \rho^2 + z^2)EllipticK[-\dfrac{4a\rho}{(a-\rho)^2 + z^2}] - ((a-\rho)^2 + z^2)EllipticE[-\dfrac{4a\rho}{(a-\rho)^2 + z^2}]\right)$$

In the simpler special case which looks only at the A-field in the x-y plane, we take $z = 0$, and we can plot the result as a function of cylindrical radius ρ to get (in arbitrary units in which we take $a = 1$):

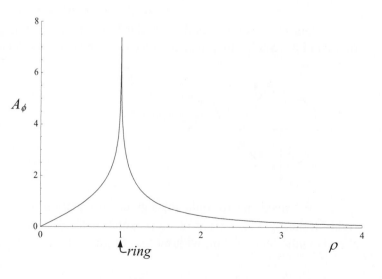

The B-field is of course determined by the curl of the A-field. It turns out we can calculate, as is often possible by taking simplifying cases, the B-field along the z-axis, i.e., $\rho \to 0$. In cylindrical coordinates with only an azimuthal component (i.e., $\hat{\phi}$), we have (which you can look up in appendix VI):

$$\vec{B} = \nabla \times \vec{A} = \frac{1}{\rho}\frac{\partial\left(\rho A_\phi\right)}{\partial \rho}\hat{z} = \left(\frac{A_\phi}{\rho} + \frac{\partial\left(A_\phi\right)}{\partial \rho}\right)\hat{z}$$

3.37

$$= \left(\frac{A_\phi\left(\rho = 0\right) + \rho\left.\frac{\partial A_\phi}{\partial \rho}\right|_{\rho=0}}{\rho} + \left.\frac{\partial A_\phi}{\partial \rho}\right|_{\rho=0}\right)\hat{z}.$$

In the last equality, we used the fact that we are going to take the limit of small ρ, and we expanded the first term in a Taylor's expansion. To zeroth order, the integral in equation 3.35 is zero, so we are left with:

3.38
$$\vec{B} = 2\left.\frac{\partial A_\phi}{\partial \rho}\right|_{\rho=0}\hat{z}$$

Taking the derivative inside the integral, we then get the z-component of B along the z-axis:

3.39
$$B_z = \left.\frac{\partial A_\phi}{\partial \rho}\right|_{\rho=0} = \left.\frac{4I}{c}\int_{\pi/2}^{3\pi/2}\frac{-a\sin\phi'\left(\rho - a\sin\phi'\right)d\phi'}{\left(\left(a\cos\phi'\right)^2 + \left(\rho - a\sin\phi'\right)^2 + z^2\right)^{3/2}}\right|_{\rho=0}$$

$$= \frac{4I}{c}\int_{\pi/2}^{3\pi/2}\frac{a^2\sin^2\phi'd\phi'}{\left(a^2 + z^2\right)^{3/2}}$$

So,

3.40
$$B_z = \frac{2I}{c}\frac{\pi a^2}{\left(a^2 + z^2\right)^{3/2}}$$

Field of a Long Solenoid

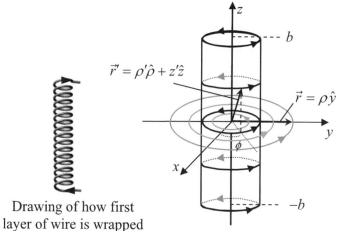

A solenoid wound with   Drawing of how first
thin insulated wire   layer of wire is wrapped

Next, we consider the case of a cylinder that has a small gauge wire wrapped spiraled around it till it is covered from top to bottom, which is typical of solenoids such as shown in the above left picture. In this case, we have a current flowing round and round the cylinder spiraling from the bottom to the top. We can approximate this behavior by a conducting cylindrical shell with a uniformly distributed current moving in the $\hat{\phi}$ direction, with a linear density in the z-direction given by $\vartheta = I \,/\, length$. Such a current distribution is shown in the right figure above. The field point is chosen in the x-y plane. The field point vector, $\vec{r}$, is chosen along the y-direction, but any other direction will give exactly the same result because of the symmetry of the source. Said another way, $\vec{A}$ should not depend on the angle, ϕ, around the cylinder. Also, because of that symmetry, one cannot expect variation along the z-axis as long as we make the mental construct (being of reason) of an infinite length cylinder. Taking the half length of the cylinder, b, to be very large, we then get, extending equation 3.35 for a single loop to integrate all the loops from $-b$ to $+b$:

$$3.41 \quad \vec{A}(\rho,z) = \frac{2}{c}\hat{\phi}\int_{-b}^{b}\int_{\pi/2}^{3\pi/2} \frac{\left(a\sin\phi' d\phi'\right)\vartheta\, dz'}{\sqrt{\left(a\cos\phi'\right)^{2} + \left(\rho - a\sin\phi'\right)^{2} + z'^{2}}}$$

In this integral, we basically sum the contribution to the field at a given point in the plana over the infinity of circular loops that make up the current distribution. Hence, we can use the same reasoning as above to get the integrand, except the current is now determined by the linear density as $dI = \vartheta dz'$.

We do want b to be very large, but, to avoid infinite answers, we do not want to complete the limit. Hence, we do the integral over z (by trig substitution or looking it up in a table) to get:

$$3.42 \qquad A_{\phi} = \frac{2a\vartheta}{c}\int_{\pi/2}^{3\pi/2} 2\,\text{arcsinh}\left(\frac{b}{\sqrt{a^{2}+\rho^{2}-2a\rho\sin\phi'}}\right)\sin\phi' d\phi'$$

And then expand the integrand in $1/b$ to get:

$$\left(\ln 4 + 2\ln b - \ln\left(a^{2} + \rho^{2} - 2a\rho\sin\phi'\right)\right)\sin\phi'$$

Inserting this back into the integral gives:

$$A_\phi = \frac{2a\vartheta}{c} \int_{\pi/2}^{3\pi/2} \left(\ln 4 + 2\ln b - \ln\left(a^2 + \rho^2 - 2a\rho\sin\phi'\right) \right) \sin\phi' d\phi'$$

$$= -\frac{2a\vartheta}{c} \int_{\pi/2}^{3\pi/2} \ln\left(a^2 + \rho^2 - 2a\rho\sin\phi'\right) \sin\phi' d\phi'$$

Now, even this last integral is not easy to solve, but one can use a program, such as *Mathematica*, if one is careful to recast the form to respect the definitions within that program. In particular, we are careful to keep ϕ within the domain $0 < \phi < \pi$ and thus using $\phi \to \phi - \pi/2$, we get:

$$A_\phi = \frac{2a\vartheta}{c} \int_0^\pi \ln\left(a^2 + \rho^2 + 2a\rho\cos\phi'\right) \cos\phi' d\phi'$$

Integration using *Mathematica* gives:

3.43
$$A_\phi = \begin{cases} r < a & \dfrac{2\pi\rho\vartheta}{c} \\[2mm] r > a & \dfrac{2\pi a^2\vartheta}{c\rho} \end{cases}$$

From this we can easily calculate B using the formula for the curl in cylindrical coordinates given in the previous section (see equation 3.37).

3.44
$$B_z = \begin{cases} r < a & \dfrac{4\pi\vartheta}{c} \\[2mm] r > a & 0 \end{cases}$$

Now, we could, in the following way, also calculate the A-field and the B-field for the cylinder without using the formal solution for A.

First, we determine the direction of $\vec{B}$ as follows. Since the direction of $\vec{A}$ is determined by the direction of the current, which is in the ϕ-direction, so is $\vec{A}$. This means A's curl must be due either to changes of A_ϕ in the z or in the ρ direction. Since A cannot vary in the z-direction because of the symmetry of the source, $\vec{B}$, the curl of A, must be in the z-direction.

Next, the magnitude of $\vec{B}$ *outside* the cylinder must be constant, for otherwise the integral around the path shown in orange in the figure on the left below would depend on where one closed the outside part of loop and thus would not be proportional to the current piercing the surface bounded by the path. Furthermore, the B-field outside cannot be a constant other than zero, for this would mean that the elements of the source were causing field at arbitrarily large distances without any diminishing of their ability to do so which we, in turn, know is not true.

Next, knowing that the B-field is zero outside the cylinder, we can simply calculate B inside by taking the integral of B around the closed path shown in orange in the left figure below. We get: $B_z\, l = \dfrac{4\pi I}{c}$ which gives for inside the cylinder ($r < a$);

3.45 $$B_z = \frac{4\pi\vartheta}{c}$$

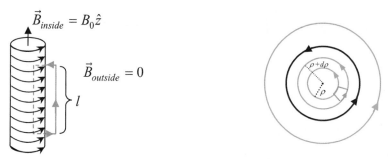

From B, we can now calculate A using the path integral shown in orange in the right figure above. We get, within the cylinder:

$$A_\phi(\rho+d\rho)\cdot(\rho+d\rho)\cdot d\phi - A_\phi(\rho)\cdot\rho d\phi = B_z d\rho\,\rho d\phi.$$

Next, being very careful to note that we have second order terms on each side so we cannot simply divide through by ρ, we write:

$$B_z d\rho\,\rho = A_\phi(\rho+d\rho)\rho - A_\phi(\rho)\rho + A_\phi(\rho+d\rho)d\rho = A_\phi(\rho+d\rho)\rho - A_\phi(\rho)\rho + A_\phi(\rho)d\rho$$

$$= \rho dA_\phi + A_\phi d\rho$$

Thus, recalling the product rule for derivatives, we can write: $\dfrac{d(\rho A_\phi)}{d\rho} = \rho B_z$, which

yields the same A_ϕ for $r < a$ as given in equation 3.43. Determining A_ϕ for $r > a$ by this method is left as an exercise.

Conclusion

In looking over our examples, we have seen that the formal solution for A given in equations 3.27 and 3.29 is analogical to the formal solution for ϕ given in the previous

chapter, (for example, we have: $\vec{dA} = \dfrac{I\,\vec{dl}}{c|\vec{r}-\vec{r}'|}$, $d\phi = \dfrac{dq}{|\vec{r}-\vec{r}''|}$), and gives great insight, but

we have also seen that it is often easier to use other methods to calculate A. In particular,

we found calculating B using symmetry considerations coupled with $\nabla\times\vec{B} = \dfrac{4\pi\vec{J}}{c}$ and

Stokes' theorem often produced straightforward answers. We also found that the analogous equation for A, $\nabla\times\vec{A} = \vec{B}$, gives a powerful way of determining $\vec{A}$; indeed, using the analogy, one might want to think of B as a source of A as a helpful mnemonic to connect the two useful equations. The ability to determine what method of solution will most easily yield the answer and the sought after physical insight in a given physical situation is a skill that every good physicist must develop.

With an understanding of that specification of the B-field that is the A-field, we have completed our study of static fields and are now ready to broach the topic of non-static fields.

Summary

The **Magnetic Field** is a power (quality) activated in the plana by a charged massive body that has impetus which can cause impetus in a receptive body, i.e., a charged body which has impetus of a certain type. Hence, like the electric field, along with the active power, the impetus-activated active charge, there is a receptive power, an impetus-activated passive (receptive) charge. This power is most specified by the A-field, which reveals details of the nature of the disposition of the plana known as the B-field, as well as magnetic properties that do not relate directly to B. The B-field, like the E-field, obeys the superposition principle.

Quantum mechanics will have more to tell us, but here we focus on classical electricity and magnetism, so among other things, quantum mechanical spin will often be treated classically as if it were simply spinning charge. Keep in mind, as we have said many times, these kinds of analogical abstractions have an important place in physics if we are careful about their limits.

In *cgs* units, which are the more fundamental units, the unit of charge, *esu*, is defined directly through the electric force as given by Coulomb's law, while in the more practical *SI* (a kind of extended *mks*) units, unit of charge, the coulomb, is defined through the magnetic (B-field) force given by Ampere's law.

Ampère's force law is:

$$\frac{F}{l} = \frac{1}{c^2}\frac{2I_1 I_2}{r} \; (cgs), \qquad \frac{F}{l} = \frac{\mu_0}{4\pi}\frac{2I_1 I_2}{r} \; (SI), \text{ where } \quad \frac{\mu_0}{4\pi} = 10^{-7}\frac{N}{A^2}$$

It expresses the measure of the force per unit length between two long wires of length, l, one carrying a current of measure I_1 and the other I_2 and separated by a distance r. The fundamentals of Ampère's law are most easily grasped by taking the simplest model of current flow, which is constant velocity movement of a line of charge.

We summarize the effect of the static B-field as: *Like impeti attract, unlike impeti repel.* (If one impetus-activated line charge is positive and the other negative, then unlike impeti attract and like impeti repel.)

The strength of the B-field generated by the impetus-activated charged bodies in the first wire is proportional to I_1, and can be written:

$$B_1 = \frac{1}{c}\frac{2I_1}{r} \; (cgs) \qquad\qquad B_1 = \frac{\mu_0}{4\pi}\frac{2I_1}{r} \; (SI)$$

The more effective the impetus is, i.e. the greater speed the given strength impetus is able to cause, the stronger the field that the impetus-activated charged body generates in the plana. This can be seen by noting that the current can be written: $I_1 = \lambda_1 v_1$, where λ_1 is the charge per unit length and v_1 is the speed of that charge. Note that, I_1, which gives a measure of the capacity of the wire to cause B-field, is proportional to the speed and the charge. Charged bodies can, with no other powers, produces an electric field. The more active charge that a body has, the stronger the electric field it can generate. Following standard practice, we simply refer to charge, ignoring the distinction between active and passive charge; thus, we can simply say: the more charge, the stronger E-field it causes. However, only impetus can change the nature of the charged body in such a way that it can cause a B-field. Thus, it makes sense that a more effective impetus (which includes both its strength and how that strength is received in the body) will generate a stronger B-field.

In terms of the B-field, the force per unit length exerted by the first wire on the second is:
$$\frac{F_{1\to2}}{l} = \frac{1}{c^2} I_2 B_1$$ where $I_2 = \lambda_2 v_2$. The receptivity of a body to the action of the plana-activated B-field is thus proportional to the amount of passive (receptive) charge, which is proportional to λ_2, and also proportional to the effectiveness of the impetus in causing that receptivity as measured by the effectiveness of the impetus in causing motion, which is, in turn, measured by the speed, v_2. The latter makes sense because, like the generation of B-field, the receptivity to the B-field results from the change that the impetus causes in a body with receptive charge; so we expect a stronger and/or more effective impetus to cause more receptivity, resulting in a stronger magnetic force.

The direction of the impetus of a charged body defines a direction around which the B-field "circulates." This is iconized by the field of the long straight wire shown here. The B-field direction, in turn, defines, at a given point, the normal of the plane in which all of the B-field's actions occur. The effects of

the B-field are summarized in the force law: $\vec{F} = q\dfrac{\vec{v}}{c} \times \vec{B}$ (cgs) $\vec{F} = q\vec{v} \times \vec{B}$ (SI).

There are six different effects of a positive impetus-activated charge on a positive test charge, one corresponding to each of the six basic types of impetus that the test charge can have. The net effect, for a constant B-field, is to cause left-handed circular motion shown around the B-field. The circulation is right handed for a negative test charge. Taking a positive test charge positioned above a rightward source current, the force exerted by the field will be: toward the current, for a rightward impetus test charge; away for leftward; rightward for upward; leftward for downward impetus; and no force for impetus into or out of the page (cf. A-field section below for diagram).

The B-field tends to confine moving charges by spiraling them around its lines which close around the source. It does this while preserving their kinetic energy, thus generally keeping them near the source with their original ability (energy) to interact with other particles that are trapped near the source. This happens in the "Northern lights" as charged particles follow the lines of the Earth's field to hit the northern polar region of the Earth's atmosphere. This contrasts with the E-field which, for example, takes kinetic energy away from an incoming particle and has the potential to give it all back in the other direction.

Thus, the E and B-fields appear to have complementary roles, the B-field taking a more "shepherding" role and the E-field a "store and give back" role. Both play fundamental parts in matter's defining feature which is its ability to change by interaction. Both are important in the processes that form composite substances; for example, the magnetic field in formation of stars that form heavy elements, and the electric field in the processes in the star that form those elements.

The total electromagnetic force law, i.e. including both magnetic and electrical effects, is:

$$\vec{F} = q\vec{E} + q\frac{\vec{v}}{c} \times \vec{B} \quad (cgs) \qquad \vec{F} = q\vec{E} + q\vec{v} \times \vec{B} \quad (SI)$$

This law will continue to be valid for all of classical E&M even after we, in later chapters, incorporate changing sources and analogically extend the meaning of $\vec{E}$.

Because ordinary objects travel much slower than the speed of light, the effects of B-fields are, in ordinary experience, much weaker than electrical effects. Indeed, if it were not for the tendency toward electrical neutrality, electrical effects would swamp the magnetic ones we do see in ordinary experience.

Helmholtz Decomposition

Any "reasonable"[32] vector field, $\vec{B}$, can be decomposed into a part with only a curl and a part with only a divergence, as follows: $\vec{B} = \nabla \times \vec{A} + \vec{\nabla}\phi_m$.

Also, one can specify a field, $\vec{B}$, by everywhere specifying $\nabla \cdot \vec{E}$ and $\nabla \times \vec{E}$, which both fall off faster than $1/r^2$. By imagining simple fluid flows, one can develop intuition about the basic quantitative realities encapsulated in this theorem.

A curl-less everywhere flow that has a single source (where $\nabla \cdot \vec{J} \neq 0$), must for example flow uniformly outward from the source as shown here. Thus, intuitively it seems, and the theorem proves, that any curl-less flow can be made out of such sources/sinks.

This leaves only curling flows. Curls can be tested by using the curl meter shown on the right. The curl of a current distribution is, in a loose sense to be made precise in Chapter 5, the degree to which the current changes perpendicular to itself. More accurately, in two dimensions, it has two contributions: (1) the horizontal change of the vertical flows and (2) the vertical change of the horizontal flows corresponding, respectively, to the two terms in $(\nabla \times J)_z$, namely $\partial_x J_y - \partial_y J_x$.

A divergence-less everywhere flow, means that, for example, under ordinary circumstances, whatever water comes in must come out. This restriction, in turn, means that as we build out from predefined boundaries, the predefined curl at each new location defines a new piece of the flow, allowing us to build the entire flow in this way. For example, consider filling out the 2-D flow shown in the figure on the right. Starting with a boundary defined to have no flow on the left and a flow of one unit coming in, then the predefined curl for the shaded region requires

---

[32] The field needs to be twice differentiable and fall off faster than $1/r$ at infinity.

that the missing segment of the flow be as shown by the red line. To better build your understanding of this and the curl generally see the game in the problem section.

Equation for the B-field as a function of its source
Given that the *B*-field picks out the normal of the plane of its action, we can, for the case of a line of current, write: $\hat{B} = \hat{J} \times \hat{R}$, where $\hat{J}$ and $\hat{R}$ are the unit vectors for $\vec{R}$, the vector from the source to the field point and $\vec{J}$ the current density vector. We expect further that the strength of the *B*-field will weaken as it propagates further from the source; that is, *B* will fall-off with the distance from the source, *R*. Noting that the electric field of a line of charge and the magnetic field of a line of current *both* fall-off as $1/R$, it is reasonable to suppose that, like the *E*-field of a differential charge, the *B*-field for a differential element of current falls of as $1/R^2$.

    The field derived from this line of thinking leads to the **Biot-Savart** formula for differential element of current: $d\vec{B} = \dfrac{1}{c} \dfrac{\vec{J} \times \hat{R}}{R^2} dV$ (*cgs*), which is only valid for closed loops

of current (i.e., when charge does not build up) and for currents that do not change. In integral form it becomes: $\vec{B} = \dfrac{1}{c} \displaystyle\int \dfrac{\vec{J} \times \hat{R}}{R^2} dV$ (*cgs*), which in *SI* units is:

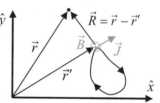

$\vec{B} = \dfrac{\mu_0}{4\pi} \displaystyle\int \dfrac{\vec{J} \times \hat{R}}{R^2} dV$. These give results that correspond to experiment.

Maxwell's Second Equation
The field given in the differential form of the Biot-Savart does not change along its own direction, which means it must have zero divergence. Because divergence-less fields can only combine to create divergence-less fields, all *B*-fields have no divergence, which is *Maxwell's Second Equation in* <u>*differential form*</u>:
$$\nabla \cdot \vec{B} = 0.$$
    This means there are no sources/sinks of the *B*-field, which ultimately derives from the fact that only *impetus-activated* charge causes *B*-field; i.e., there is no simple property of a body that can cause a magnetic field in the plana. This means there are no sources of the field direction, so that the field must circle back on itself. This, in turn, means that there are "poles:" a place where the *B*-field leaves the source (called, roughly, the north pole) and a place where it returns to the source (the south pole). Hence, in this language, Maxwell's second equation says that there are no magnetic monopoles, as is most obvious by comparing to Maxwell's first equation: $\nabla \cdot \vec{E} = 4\pi\rho$.

Maxwell's Second Equation in <u>*integral form*</u>, which is obtained using the divergence theorem, is:
$$\oiint_S \vec{B} \cdot d\vec{A} = 0$$
This means whatever *B*-field "fluxes" into a closed region *S* must flux back out.

Maxwell's Fourth Equation *(with no charge build up)*

Having specified the divergence of *B*, we need to specify its curl to complete its specification. Furthermore, we have yet to give a source equation for *B*, i.e., a differential equation connecting *B* to its source. Since it is natural to think of the *B*-field as curling around a differential line-element, i.e. around the line defined by the impetus that activates a differential piece of charge causing the field, it is natural to look to the curl of *B* for our source equation.

Assuming there is no build up of charge so that the current is conserved: $\nabla \cdot \vec{J} = 0$, we can specify the curl of $\vec{B}$ everywhere simply by requiring it to be proportional to its source $\vec{J}$. The equation with the experimentally verified constants is:

Maxwell's fourth equation in <u>differential form</u>

$$\nabla \times \vec{B} = \frac{4\pi \vec{J}}{c}$$

This means that $\vec{B}$ will have a curl only in those places where there is a source. In this equation, $\vec{J}$ specifies the direction and magnitude of the curl due to *all* currents in the source. Thus, the equation does not *directly* represent how the different parts of the source cause the field. The Biot-Savart formula, which is the static solution to this equation, does not either. The non-causal nature of this equation arises from our setting the curl and divergence everywhere, which is an intrinsically non-local operation, implying immediate effect of distant parts on one another. We avoid concerning ourselves with this in the static realm discussed in this chapter by assuming an infinite speed of propagation.

Notice that by defining the *B*-field by equating its curl to $\vec{J}$ *alone*, we are also implicitly determining the source; namely, we are requiring it to be divergence-less at every point. This is *part* of the way that our mathematical formalism accounts for contribution to the *B*-field at a given source point from $\vec{J}$'s at the other source points.

The conserved current requirement can also be stated as the requirement that all current loops close upon themselves.

Maxwell's fourth equation in <u>integral form</u>, which is obtained from the differential form using Stokes Theorem, is:

$$\oint_C \vec{B} \cdot d\vec{s} = \frac{4\pi}{c} \oiint_S \vec{J} \cdot d\vec{A} \qquad \left(\hat{n} \cdot \circlearrowright \right)$$

(The circulation sense is taken to be positive when directed in the right handed sense along the normal vector to the area such as shown above.)

Maxwell's fourth equation applies to any shaped closed circuit. Bent loops work as well as straight because *B* is caused by impetus, not *changing* impetus, activating charge.

B-field around a straight wire

Because impetus, which moves a body in a straight line, is essential for a charge to cause a *B*-field, the *B*-field of a straight wire (with its circuit completed at infinity) is an important, even iconic, case of the generation and nature of the *B*-field.

The $1/R$ fall-off of the infinite straight wire guarantees that there is no curl in any region outside the wire. This can be shown, using the integral form of Maxwell's 4th equation, by evaluating any line integral along a closed path outside the wire that is made up of lines perpendicular to the wire and circular arcs with the wire as their center. The line integral for *any* loop that does *not* encircle the wire can be shown to be zero by showing, using the $1/R$ fall-off, that the triangles that are created by rounding the corners of the aforementioned loops are zero in the limit of arbitrarily small triangles.

By picking a *circular* path centered on the wire and sufficiently large to touch a single point of a second *arbitrarily-shaped* closed path that also encircles the wire, and by knowing the line integral around the circle is: $4\pi I / c$, we can show that the line integral around the arbitrary-shaped loop is also $4\pi I / c$. This also remains true when the wire is allowed to bend into any shape.

For any closed circuit source, the line integral around any arbitrarily shaped closed curve is equal to the total current that pierces the surface bounded by that curve:

$$\oint_C \vec{B} \cdot d\vec{s} = \frac{4\pi}{c} \oiint_S \vec{J} \cdot d\vec{A} = \frac{4\pi I}{c} \qquad \left(\hat{n} \cdot \right)$$

(The circulation sense is taken to be positive when directed in the right handed sense along the normal vector to the area such as shown above.)

The divergence of a curl of a field is zero, i.e. $\nabla \cdot \left(\nabla \times \vec{A} \right) = 0$. One can get a sense of why this is so in the following way. Start with a region that has no curl, such as a cube with four equal field vectors defined on its base. Then, as shown in the figure, changing one of the four (shown in green) causes a curl in two of the faces. Note further 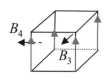 that it creates a curl, B_4, on the right side of the cube, which corresponds to outward flow that is just balanced by the inward flow created by the curl, B_3, on the back face, so that there is no net flux of the curl of the B-field, i.e. there is no divergence of the curl.

A-field: the specification of the B-field

Because $\vec{B}$ is divergence-less, it can be more completely represented by a vector, $\vec{A}$, called the *vector potential*, defined such that:

$$\boxed{\vec{B} = \nabla \times \vec{A}}$$

With it, one no longer needs to separately state $\nabla \cdot \vec{B} = 0$. It is the analog of the scalar potential, the ϕ-field, which, for the static case: $\vec{E} = -\vec{\nabla}\phi$.

Introducing the Lorenz gauge in the static case: $\nabla \cdot \vec{A} = 0$, and using Maxwell's 4th equation, it can be shown that: $\nabla^2 \vec{A} = -\frac{4\pi \vec{J}}{c}$, the solution of which is:

$$\vec{A}(\vec{r}) = \frac{1}{c} \int_{Source} \frac{\vec{J}(\vec{r}')}{R} dV$$

The differential form of this equation is: $d\vec{A} = \frac{1}{c} \frac{I\, d\vec{l}}{R}$, where $d\vec{l}$ is a differential element of length of the source current. The electric potential, ϕ, has an analogous relation to its source, e.g., we have: $d\phi = \frac{dq}{|\vec{r} - \vec{r}'|}$.

Physically, we explain the magnetic effect more completely in the following way. Impetus-activated charge causes a new quality of the plana, which we call the *A*-field. The vector $\vec{A}$ gives the measure of its strength and direction. This quality is caused in the parts of the plana that are next to the impetus activated charge, and it then propagates from one part of the plana to the next. The *A*-field caused in a given region bears a kind of signature of the type of impetus that caused it. For example, a rightward impetus-activated charge will cause a rightward directed *A*-field that will act to draw other rightward impetus-activated charges towards the line along which it moves.

The *A*-field of any source can be drawn very simply by noting: (1) *A* points in the direction of the source (or opposite to it for negative source charges), (2) *A* decreases in magnitude from the source according to a $1/R$ law for each small element of the current, and (3) *A* is the linear sum of all the contributions from each piece of impetus-activated charge.

Infinite straight current

The *A*-field of our iconic long straight wire lying on the *z*-axis is simply calculated by using Stokes theorem and the definition of *A*: $A_z(r) = \text{constant} - \dfrac{2I}{cr}\ln r$. If we take *A* to be a real feature of the plana, as we do, the constant is, of course, determined by physical reality, but, as with the electric potential, we usually choose it according to the convenience of the problem at hand because, we are usually seeking *B* for which the constant is irrelevant. The effect of the *A*-field from a straight wire on various types of impeti is shown in Figure 3-12, and reproduced in miniature here.

Fields of two other current sources

Field of a current sheet in the *x-z* plane with current flowing along the *x*-direction, $A_x(y) = -B_0|y| + \text{constant}$, $\vec{B}_\pm = \pm\dfrac{2\pi\mathcal{I}}{c}\hat{z}$

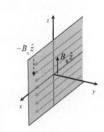

(where $\pm$ refers to the *B*-field on the $\pm y$ side of the sheet, $B_0 = 2\pi\mathcal{I}/c$, and $\mathcal{I} = I$ per unit length in $\hat{z}$).

Field of a long solenoid centered on the *z*-axis.

$$A_\phi = \begin{cases} r < a & \dfrac{2\pi\rho\mathcal{I}}{c} \\[2ex] r > a & \dfrac{2\pi a^2\mathcal{I}}{c\rho} \end{cases} \quad \text{and} \quad B_z = \begin{cases} r < a & \dfrac{4\pi\mathcal{I}}{c} \\[2ex] r > a & 0 \end{cases}$$

Helpful Hints

A way to capture part of the important distinction in the nature of the E and B-fields, while also providing a mnemonic for the source equations for E and B, $\nabla \cdot \vec{E} = 4\pi\rho$ and $\nabla \times \vec{B} = \dfrac{4\pi\vec{J}}{c}$, is the following. Loosely said, the undirected or *"point" nature of charge* results in the E-field's divergence being the relevant aspect of field generation, while the directed or *line nature of impetus-activated charge* results in the curl being the relevant aspect of B-field generation. Going further, we could also add that the specification of the B-field, the A-field, brings in another cylindrically symmetric structure, field lines parallel to the line defined by the impetus.

A good sketch can help your understanding and facilitate calculation, so I recommend that you learn how to do basic sketching, including learning how to do perspective for cubes, circles, spheres and the like, as part of your physics training.

It is helpful in drawing and interpreting pictures of B-fields to think of the arrow indicating the direction of the field as the direction which a "north pole" would be pushed.

In doing calculations always be careful to distinguish between field points expressed in unprimed coordinates and source points in primed coordinates. For example, in the Biot-Savart formula for the field of a differential element of current:

$$d\vec{B} = \frac{\vec{J}\left(\vec{r}'\right) \times \vec{R}}{R^3} dV' = \frac{\vec{J}\left(\vec{r}'\right) \times \left(\vec{r} - \vec{r}'\right)}{\left(\vec{r} - \vec{r}'\right)^2} dV'$$, $\vec{J}$ is *only* a function of source coordinates, but

R is a function of both as indicated. Thus, for example, when we integrate over all such differential elements to get the total field at a given field point, we integrate over $\vec{r}'$, not $\vec{r}$, which remains fixed during the integration.

In defining a constant (whether it be an integration constant, a constant for consolidating a group of constants into one, or any other constant), it is helpful to pick a symbol that matches the units involved, as this allows you to more easily check your results at various stages of a calculation using dimensional analysis. It also helps keep the physical meaning more on the surface.

> Note: In this text, when not otherwise noted units are in *cgs*.

Calculational Techniques:

In calculating the B-field from the A-field, we need:

$$\vec{B} = \nabla \times \vec{A} : B_x = \frac{\partial A_z}{\partial y} - \frac{\partial A_y}{\partial z} \quad B_y = \frac{\partial A_x}{\partial z} - \frac{\partial A_z}{\partial x} \quad B_z = \frac{\partial A_y}{\partial x} - \frac{\partial A_x}{\partial y}$$

To remember this, the following can be helpful.

Mnemonic for Cross product

Using: $\partial_x \equiv \dfrac{\partial}{\partial x}$, $\partial_y \equiv \dfrac{\partial}{\partial y}$, $\partial_z \equiv \dfrac{\partial}{\partial z}$, We have: $B_i = \partial_j A_k - reverse = \partial_j A_k - \partial_k A_j$

In detail, we write the x-component: $B_x = \partial_y A_z - reverse = \partial_y A_z - \partial_z A_y$. We remember the order of the first three indices by putting x on the "B," y on the "∂", and z on the "A," i.e. in order: x, y then z. Then, we simply subtract the "reverse," i.e. the first term on the right hand side with the indices swapped. To get the y-component, we rotate the first three indices of the equation, moving y and z to the left, pushing x off the left side to appear around the right side to get: y, z, x. From this, we write:

$B_y = \partial_z A_x - reverse = \partial_z A_x - \partial_x A_z$. Rotating the indices one last time yields: z, x, y.

From which we get: $B_z = \partial_x A_y - reverse = \partial_x A_y - \partial_y A_x$

Summary of the rotation of the indices that is key to the mnemonic.

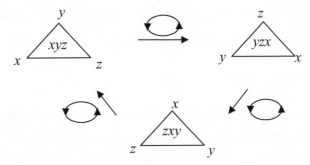

(Note that one can rotate in either direction without affecting the result.)

Integrals of vector functions, such as the Biot-Savart integral, are best done in Cartesian coordinates because curvilinear basis vectors are, in general, a function of coordinates.

See *next page* for a calculational technique for advanced students.

For Advanced Students

The cross product can be complicated to use in calculations. Some shorthand notations that are often very helpful in such calculations are the following.

1) *Repeated index notation for sums.* The summation of components notation where the sum over the components of a vector (even if it is the generalization of a vector that has more than one index, in which case it is called a tensor). For example, the dot product of $\vec{A} \cdot \vec{B}$ is normally written out as $\sum_{i=1}^{3} A_i B_i$; in repeated index notation, we simply write: $A_i B_i$, where the summation is considered implied by the repeated index. If they were not to be summed, we would write: $A_i B_j$

2) *The epsilon "tensor" notation* that can be used for the cross product. ϵ_{ijk} is defined as follows: $\epsilon_{ijk} = \begin{cases} 1 & even \\ -1 & odd \\ 0 & i = j \, or \, j = k \, or \, i = k \end{cases}$, where the two top conditions refer to whether ijk is an even or odd permutation. An even permutation is an ordering of the indices (in this case the numbers represented by i, j and k) that can be put in numeric order by an even number of swaps of adjacent indices; an odd number of swaps of an even permutation would yield reverse numeric order. An odd permutation is one that that can only be put in numeric order by an odd number of swaps of adjacent indices. So, $\epsilon_{123} = \epsilon_{231} = \epsilon_{311} = 1$, whereas $\epsilon_{213} = \epsilon_{132} = \epsilon_{321} = -1$. However, if any two indices are the same, then the result is zero. So, for example, $\epsilon_{112} = 0$. In this notation, using the repeated index convention, we can write the kth component of the cross product as: $\left(\vec{A} \times \vec{B} \right)_k = \epsilon_{ijk} A_i B_j$.

3) *Delta function notation.* Delta is defined as follows: $\delta_{ij} = \begin{cases} 1 & i = j \\ 0 & i \neq j \end{cases}$. So, for example, $\delta_{12} = 0$ and $\delta_{11} = 1$. Thus, we can write the dot product above as $\vec{A} \cdot \vec{B} = \delta_{ij} A_i B_j$. Using the above notations, we write the following very helpful rule:

$\epsilon_{jlm} \epsilon_{jki} = \delta_{kl} \delta_{im} - \delta_{km} \delta_{il}$

Problems

1. Sketch all the different types of combinations of source and test charge impetus and the different resulting fields. Explain how the *B*-field part of the Lorentz force law describes each.

2. In an Ampere type experiment of two parallel currents, explain the effect of a wire's field on itself?

3. Explain roughly whether or not the *B*-field created around a *differential* current element and given by the Biot-Savart law has zero curl in empty space (as it should to obey the differential form of Maxwell's fourth equation) by answering the following. Given the form of the law and the symmetry around the current element, what are the two possible terms that can contribute to the curl of its *B*-field? Under what conditions will the curl be zero? Is it zero? *Hint*: it will be helpful to look up the form of the curl in cylindrical coordinates, but no difficult calculations are involved.

4. Draw the following vector fields and calculate their curls. a) $\vec{A} = -y\hat{x} + x\hat{y}$
b) $\vec{A} = x\hat{x} + y\hat{y}$ c) $\vec{A} = x\hat{x} - y\hat{y}$

5. Play one round of the following two player game dealing with 2-D curls of divergence-less flow patterns. A given region of space is divided into 25 squares (see game board below). The board is divided into three territories. The top two rows belong to the first player, the bottom two to the second player, and the middle row is no-mans land. The object of the game is to create the largest amount of curling in the opposing player's territory. This is defined to be the sum of the absolute value of the curls of each cell in the given territory.

The filled circle at each grid line intersection marks where the vector current density is to be specified. The players take alternate turns trying to fill in the flows throughout the grid. A flow is marked by drawing an arrow extending along one of the grid lines from any one of the solid circles and writing a number not greater than 9 next to it. When a certain number of flows have been marked, certain flows will be forced, and these can, of course, be greater than 9. At the beginning of a turn (and only then), a player can mark any moved forced by his opponent before taking his turn. If a player, during his turn, finds a move that is inconsistent with the rest of the board, he can erase the offending move and substitute whatever value he wishes and then take his own turn. When the last flow is marked, the tally is made and the winner decided. As a test that the game has been accomplished without inconsistency, add up the total curl, this time including the correct signs, and verify that it is zero (why must it be so?).

It is recommended that you photocopy the grid below, play the game and then submit the completed grid with the rest of your problem set.

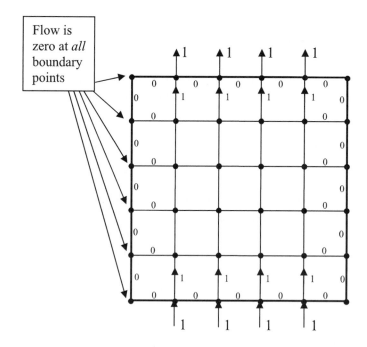

Flow is zero at *all* boundary points

6. Show that one can approximate the flow pattern shown in the figure on the right by using point sources only, which means, ignoring the boundaries, that the flow will have no curls.

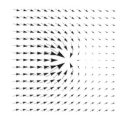

7. It has been found that the mineral magnetite, Fe_3O_4, also called lodestone, that we have mentioned in the text provide the ancient peoples with their first encounter with magnetism. The mineral is naturally occurring but is sometimes found magnetized other times not.   Can you think of what natural phenomena, that still occurs today, might have the power to cause such an effect? *Hint*: when certain materials are very hot and are allowed to cool down in the presence of strong magnetic fields, what amounts to tiny little magnets inside the material can be lined up and cause the material to become permanently magnetic. A similar effect can happen without heat.

8. Given a current, I, flowing in a long straight wire lying on the z-axis, and a test particle of charge q and mass m. a) Using cylindrical coordinates, write down the solution for equations of motion for the test particle for the general case and the case in which there is no initial angular velocity, i.e. $\dot{\phi}(0) = 0$. b) Show that these equations represent a partial

solution: $v_z(t) = \bar{v}\ln\dfrac{\rho}{\rho_0} + v_{z0}$, $\ddot{\rho} = -\bar{v}^2 \dfrac{1}{\rho}\left(\ln\dfrac{\rho}{\rho_0}\right) - \dfrac{\bar{v}\,v_{z0}}{\rho}$ when the initial velocity is:

$\vec{v}_0 = v_{z0}\hat{z} + v_{\rho0}\hat{\rho}$, and where $\bar{v} \equiv \dfrac{2I\,q}{c^2}\dfrac{1}{m}$. c) Taking $\bar{v} \equiv 1$ and $\vec{v}_0(0) = v_{\rho0}\hat{\rho}$ and $z(0) = 0$

and $\rho_0 \equiv \rho(0) = 1$, roughly sketch the trajectory of the particle on a ρ versus z plot, by iteratively figuring out each part of the curve using the proper form of the above equations. Comment on the trajectory.

9. Crystals of magnetite have been found in some animals, including honeybees, termites, salmon, sea turtles, birds such as pigeons and even man. Magnetite is believed to be involved in a "magnetic sense" which seems to be strong enough, for example, in pigeons to help them navigate in flight. Perhaps animals, such as pigeons, can truly sense the magnetism of a body like we taste the flavors of food. a) Given your experience with magnets, would you expect a magneto sensitive organ to have to be on the outside of the body like eyes, ears and noses? *Hint*: why are our organs on the outside? b) Assuming that a pigeon can sense magnetic field analogous to the way we taste flavors and given that the Earth's field is $\sim .5G$, approximately how much stronger (or weaker) would the following "taste" to the bird, if it were approximately 2 *cm* from its magneto-sense organ: i) An ordinary wire running inside the walls of your house lighting a 120W bulb ii) Near three wires carrying their full capacity of 20A each. Answer this question for both the case of a linear response and $\log_{10}$ response. Our ears for example respond in a logarithmic way, so that we can distinguish volume levels for both very loud and very soft sounds.

10. Make a drawing illustrating the cause of the curl for each of the six terms in the cross product. For each term, use a two dimensional drawing with the direction of the curl defining the plane of the page. For example, to illustrate the term $\dfrac{\partial J_y}{\partial x}\hat{z}$ draw:

11. Using Ampère's force law, calculate the force per unit length exerted on an infinitely long current of magnitude I_w located at $x = a, y = 0$ by a sheet of current in the y-z plane with current per unit cm, $\mathscr{I}$ (see figure). Assume both currents flow in the z-direction. Check your result by using the Lorentz force law and the field for a uniform current sheet.

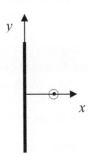

12. a) Give a method for determining the strength of an ordinary disk-shaped magnet which has a radius $a = 2cm$? *Hint*: use the idea presupposed in the next part of this question. b) Calculate the field near the magnet's surface when you know that the magnet just begins to compete with the Earth's field at about 17 *cm*. Assume that the magnetic effects can be accurately determined by taking the field to be generated by a current traversing the edge of the disk.

13. You can make a simple electric motor such as shown here out of two safety pins, a magnet, a battery holder (two AAA), a few feet of thin coated wire, and a wooden base. It is put together as follows. The north end of the magnet points up. The coil is made by winding the wire around an AA battery about 8 times, which makes its diameter very roughly 2 *cm*, the approximate width of the magnet. To hold the coil together, the shorter loose end is wrapped before being brought straight out. The other loose end is wrapped around that same side of the coil, then fed across

the diameter of the coil to the far side where it is also wrapped and then brought straight out, making the configuration shown. These two coil "contact" wires, which now come diametrically out of the coil, are fed through the eye holes in the base of the safety pins. In the working motor, each wire will spin inside its respective eye hole. The head of the safety pins are bent and screwed into the wood base. The key to the motor's operation is that the contact wires are only stripped on one side, so that current only flows through the coil when it is in one of its vertical orientations. Only in that position do the bare sides of the wires contact the safety pins and complete the circuit. a) Explain, in terms of the A-field, how the motor works, including the direction in which it will turn. b) Assuming the strength of the magnetic field in the region of the coil is approximately uniform at 300 gauss, approximating the coil as a square with 2 *cm* sides and assuming the maximum "flash" current two AAA's can give is 4A, give a rough estimate of the torque that the motor can generate. Express the result in *gf-cm*, where *gf* means the equivalent force caused by gravity on a gram mass. c) A torque wrench for tightening bolts could easily measure 20Nm. How does this compare to your result?

14. Calculate the E-field of an infinite line-charge, and the B-field of an infinite line of current. Explain their similarities. In particular, explain why the functional dependence on R, the distance from a differential source element to a given field point, is the same in both integrands. Include in your answer why the Biot-Savart law has the cross product dependence, without which the two would not have the same R dependence.

15. Explain the attraction and repulsion of two long cylindrical magnets in terms of currents and A-fields. *Hint*: take the magnets to be simple objects with a north pole at one end and a south pole at the other and make an analogy with an electromagnet.

16. In terms of the fundamental definition of the cross product given in Chapter 2 of *PFR-M*, explain its use in the Lorentz force: $\vec{F} = q\dfrac{\vec{v}}{c} \times \vec{B}$. In particular, since we cannot say that $\vec{F}$ directly represents the axes of rotation established by rotating $\vec{v}$ into $\vec{B}$ (since neither $\vec{v}$ nor $\vec{B}$ are arrows, the first (quantitative) meaning of vector), what do we say? In your explanation, also discuss why a test particle moving along a field line is acted on differently than one moving in a direction perpendicular to it.

17. Rewrite the magnetic portion of the Lorentz force law in a form that gives the force per unit length caused by a differential current source.

18. By convention, the *magnetic* north pole is defined as the pole of a magnetic compass that points toward the *geographic* north pole. Explain the ramifications of this and sketch a diagram of the field lines around the earth based on your conclusions.

19. Calculate the radius and frequency of rotation of a particle of mass m traveling at speed $\vec{v}$ perpendicular to $\hat{B}$ in a uniform B-field of strength B.

20. Show, by explicitly calculating the divergence of $\vec{B}$, that, as mentioned in the text, the divergence of the differential Biot-Savart law is zero for two key reasons: (1) the direction: $\vec{J} \times \hat{R}$ and (2) the magnitude solely depends on a function of R. *Hint*: use the vector identities: $\vec{\nabla} \cdot (\psi \vec{A}) = \vec{\nabla}\psi \cdot \vec{A} + \psi\vec{\nabla} \cdot \vec{A}$ and $\nabla \times (\phi \vec{A}) = \nabla\phi \times \vec{A} + \phi\nabla \times \vec{A}$ or the advanced techniques given in the "Calculational Techniques" section.

21. Consider the triangular shaped solid region shown here and a B-field generated by a differential element current at the origin with current density $\vec{J}$. Using the Biot-Savart law description of the B-field, show, by calculation, that the net flux flowing out of the triangular-shaped region is zero.

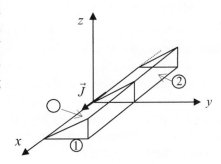

22. Consider a simple universe in which universal repulsion exists between all bodies but no attraction. Given the existence of impetus, explain how things (assuming things could exist at all) could still interact. What problem can you see with having *only* such an (analogically) generic repulsive force?

23. Explain, in the fashion done in the section "Units and Some Primary Definitions," the import of the constants used in the expression for the electric field for the case of *SI* units.

24. Consider the forces, $\vec{F}_B$ and $\vec{F}_E$, that act on a "point" particle (with charge q_T moving at speed v_T in the z-direction) due to the B and E-fields activated in the plana by a second charge, q_S, at the origin that has velocity $\vec{v}_S = v_S\hat{z}$. Use the Biot-Savart law and the Coulomb law to calculate the fields and show that $F_B / F_E = \beta_1\beta_2 \ll 1$ for non-relativistic (ordinary) circumstances.

25. Express an *esu* explicitly using the units of grams, centimeters and seconds.

26. Using either standard vector identities or the advanced techniques given in the "Calculational Techniques" section, rewrite the B-field part of the Lorentz force law by starting with $\vec{F} = q\vec{v} \times \left(\nabla \times \vec{A}\right)$ and eliminating the *curl*. Tell the meaning of each term in the new formula.

27. Using the techniques given in "Calculational Techniques" under "For Advanced Students" prove 3.25, i.e. $\nabla \times (\nabla \times \vec{A}) = \nabla(\nabla \cdot \vec{A}) - \nabla^2 \vec{A}$

28. The statement was made in the text that, in a divergence-less flow, if the curl is constant along a straight direction of the river (such as shown below), then the magnitude and direction of the current density remains the same along that direction. Could we belie this statement and keep the curl constant by introducing changes in the vertical flow to make up for a change in magnitude of the initially horizontal flow?

29. Explain in words why it is somewhat natural, knowing the nature of $\vec{B}$ as specifying the normal to the plane of its activity, to represent $\vec{B}$ in terms of the vector potential in the standard way: $\vec{B} = \nabla \times \vec{A}$.

30. Consider a rectangular flow that has the directional nature shown but whose flow speeds are *not* accurately indicated by the arrows. Instead, we take each place in the flow to have the same speed except at the corners were there is an instantaneous shift in the velocity. Comment on the curl at various places in the flow pattern.

31. Give the formal proof mentioned in the text that the rate of change of the magnitude of the B-field of a differential element of current defined by $d\vec{B} = \dfrac{1}{c}\dfrac{\vec{J} \times \hat{R}}{R^2} dV$ is zero, i.e.

show: $\dfrac{\partial \left| d\vec{B} \right|}{\partial d\vec{B}} \propto d\vec{B} \cdot \vec{\nabla}\left(d\vec{B} \cdot d\vec{B}\right) = 0$. This proof should be done using the advanced techniques mentioned in the "Calculational Techniques" section.

32. Using the vector identity: $\nabla \times (a \times b) = a(\nabla \cdot b) - b(\nabla \cdot a) + (b \cdot \nabla)a - (a \cdot \nabla)b$, or the advanced techniques given in the "Calculational Techniques" section, investigate whether or not $\nabla \times d\vec{B} = \nabla \times \left(\dfrac{\vec{J} \times \hat{R}}{R^2}\right) = \dfrac{4\pi \vec{J}}{c}$ in the following way. a) Taking the concrete case of a differential current element at the origin, show that the curl of the differential Biot-Savart law does *not* satisfy: $\nabla \times d\vec{B} = \dfrac{4\pi \vec{J}}{c}$ at any given point for which it is valid. Start by showing that the Biot-Savart is not valid at the origin. b) Show that, *after integrating over*

all space, the curl of the differential Biot-Savart law does yield the proper curl. c) Explain how your result reveals the restriction on $\vec{J}$ that necessarily arises when one defines curl of $\vec{B}$ to be $\vec{J}$ at a given point, as done in Maxwell's fourth equation.

33. Show $\nabla R = \hat{R}$

34. Extend the reach of the analysis done in the text for the infinitely long wire by also giving the outer cylinder of the coaxial setup a finite thickness. a) Calculate the A-field inside this shell, which has inner radius R_i and, outer radius R_0. b) Use your result to calculate $A_z(0)$ and compare it to the result that you can get from using the equations already given in the text.

35. One of the most serious problems involved in traveling to Mars is interplanetary radiation coming from the solar wind. Such radiation can kill an astronaut or greatly increase his risk of getting cancer. The Earth's magnetic field (as well as its atmosphere) protects us from them, creating the Van Allen radiation belts around the Earth. Once on Mars, shelters can be built to retreat to during the solar storms. What can be done on the way? A shield made of plasma contained by a magnetic field, which is, in turn, generated by the spacecraft, has been suggested by scientists in England's Rutherford-Appleton labs. Comment on this possibility? How does it work? Compare and contrast it with the way the Earth's magnetosphere protects us.

36. Explain why the current density (in $g/cm^2/s$), $\vec{J}$, of water exiting uniformly from a long thin pipe will fall off as the inverse of the distance from the pipe. Contrast it with the case of a point source of water and make the appropriate connections to the fall-off of the B-field and the E-field.

37. Using $\nabla \cdot \left(\psi \vec{J} \right) = \nabla \psi \cdot \vec{J} + \psi \nabla \cdot \vec{J}$, $\int_V \nabla \psi dV = \int_S \psi \hat{n} dS$, (which follows from the divergence theorem), where $\hat{n}$ is the outward pointing normal to the surface S bounding the region V, and index summation notation introduced in the "Calculational Techniques" section, show that the divergence of the solution of vector Poisson equation given in the text, $\vec{A}(\vec{r}) = \frac{1}{c} \int \frac{\vec{J}(\vec{r}')}{R} dV$, gives the static form of the Lorenz gauge, i.e.: $\nabla \cdot \vec{A} = 0$. We assume a conserved $\vec{J}$ and, of course, the static situation required to get the Poisson equation.

38. Henry Rowland (1848-1901) did a first of its kind experiment in which he measured the magnetic field from a rotating charged disk. He measured the magnetic field using two hanging magnetized needles (acting like compasses) on strings that also had mirrors attached to them so that he could accurately monitor their twist. a) Calculate the B-field that one would measure along the axis of a disk of radius a with a uniform surface charged density, σ, rotating at a speed, ω_0. b) With $10^{-6} C$ charge spread uniformly over the surface of a disk with $a = 10cm$, $\omega_0 = 380/s$, how does the B-field at the surface compare to the .5 G field of the earth?

39. Calculate the A-field of an infinite straight wire by direct integration using $\vec{A} = \frac{1}{c} \int \frac{I d\vec{l}}{R}$ on a finite length wire then taking the limit of an infinitely long wire. Compare your result to that given in the text.

40. Using your understanding of the A and B-fields, draw those fields for two parallel sheets of opposing currents. Calculate the A and B-fields.

41. Current flowing *inside* of, for example, a semiconductor in the presence of a magnetic field induces a voltage. This effect was discovered by Edwin Hall (1855-1938), a student of Henry Rowland, and so is called the "Hall effect." Show how the sign of the hall voltage can be used to determine the sign of the current carriers. Using drawings to illustrate the point.

42. Flow down a drain is, in some ways, approximately like the inward spiral of a tetherball as it winds up around the pole due to the impetus already given it. Show that if we model each piece of water of the spiral flow down a drain in this way, the curl of the flow is zero everywhere except at the drain.

43. Excluding the boundaries, consider the following problems concerning two dimensional flows that have everywhere $\nabla \cdot \vec{J} = 0$, $\nabla \times \vec{J} = k$ (where k is a constant). a) Given $J_y \neq 0$ but $J_x = 0$ everywhere and, show that $J_y = kx + b$, where b is an undetermined constant. b) Given that $J_y = cJ_x$ (where c is a constant), find the equation for J_x.

44. The cathode ray tube (CRT) used to be the way everyone saw their favorite TV show, and they are still preferred in other applications. The CRT provides the changing image on a TV screen in the following way. Electrons are boiled off of a piece of metal (called a cathode) inside of an evacuated tube and then accelerated by a strong electric field. The electron beam is steered by electromagnets to a point on the screen. That point of the screen, which is coated with phosphor, lights up under the impact of the electron beam. The electron beam is swept, by the electromagnet, across the entire screen, filling out the image at such a speed that the human eye cannot sense it under ordinary circumstances.

Take the configuration shown below, which uses the same principles as the CRT. Assume the screen and other dimensions are as shown in the figure. Approximate the deflection field, i.e. the *B*-field that is generated in the region of the emission of the electrons, by the field of an infinite plate in the plane perpendicular to the page in the position shown. Assume the field abruptly ends at .1 *m* away from the origin. The origin is also the point of the emission and beginning of the acceleration of the electrons.

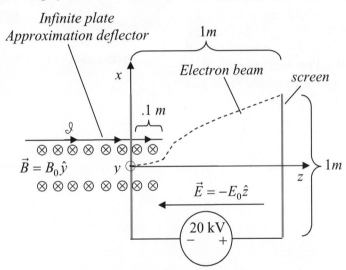

By calculation, give a rough estimate, in *SI* units, of the current per unit length, $\mathcal{I}$, needed in the deflection plate to deflect the beam to the top edge of the screen as shown. Assume the electrons are accelerated by a uniform field generated by parallel plates (shown in red) separated by the 1 *m* between the emission point and the screen with 20kV applied across them. Assume that the arguments of the trig functions obtained in the solution can be taken to be much less than unity and discuss the validity of this assumption and the purpose of rough calculations in general.

45. Show, by direct calculation that a straight flowing river with a point source in it such as shown in the section titled "Analysis of Curl-less Fields" and reproduced here has, *ignoring* the boundaries, no divergence except at the origin defined by the location of that source. Why is the curl not zero at the boundaries?

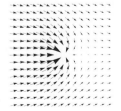

46. The magnetic field experiment on the *Mars Global Surveyor* gave the following results, which show: "Orthographic projections of the three components of the magnetic field (B_r, B_θ, B_ϕ) at a nominal 400 *km* mapping orbit altitude, viewed from 30° S and 180° E longitude (after *Connerney et al., 2001*)." Because there is no substantial planetary *B*-field, the field shown is due to crustal materials, i.e. rocky materials in the outer most layer of the planet, and thus varies widely. Given this data: a) Would an ordinary compass work on Mars? If not, why not and what do you think we have pictured on the cover of this book? That is, how might it detect *B*-field? b) How are the astronauts on the cover using it to navigate? In particular, demonstrate how it would be done by determining, from the data to the right, what the field direction and strength is at a given point and what you would do with it.

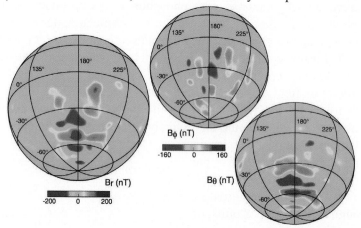

Chapter IV

Magnetic Induction

Introduction

MIT professors have embarked on a program to give power without wires to your cell phone, laptop and all your other devices. They claim they can do this over some distance without wires and without radiation! In fact, they have successfully lit a 60 watt bulb over 2 meters away at 40 percent efficiency; this means the device is not yet ready for the market, for it uses 150 watts to light a 60 watt bulb. Still, how did they even do that? Magnetic induction along with a little help from resonant tuning. In this chapter, we will learn the two fundamental types of induction. Magnetic induction refers to current caused by non-radiative magnetic fields. The three phenomena associated with induction are electric current caused as:

    (1) a conductor moves through an unchanging magnetic field,

    (2) the magnetic field from a *moving current carrying conductor* sweeps over a stationary conductor,

    (3) a changing magnetic field from a *changing current* acts on a stationary conductor. The last two have essentially the same cause. The first we have already encountered in a different context.

The Three Phenomena of Induction and their two Causes

The "infinite" length wire current gives a simple example of the first effect. Recall that the A-field generated by such a long straight wire is as shown in the figure below. Notice, as we saw in the last chapter, that because the A-field has a curl, as is easily seen below, there is a B-field around the wire.

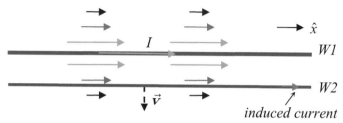

When a second conductor (W2) is moved through the A-field of the first as shown in the figure above, the field-activated plana acts on free positive charges in the second wire giving them the same type of impetus as the charges flowing in the first wire. Of course, actual wires have no free positive charges, but only free electrons, so the field actually causes the electrons to have *opposite* type impetus and thus to move to the left, i.e. in the $-\hat{x}$ direction. However, recall current is defined by the direction of flow of positive charges, or as opposite to the direction of negative charges; hence, the current flow, after all, is as if positive charges were given impetus along the direction of the current in the first wire. In short, the rightward current flow in the first wire causes rightward current in the wire that is moving away from it. We see that the first wire transfers energy to the second. The first is a "transmitter," the second a "receiver." This first transmission/reception, what we call induction, phenomenon is a simple application of what we learned in the last chapter. Indeed, using the B-field force law, we can calculate the magnitude of this effect, though we will wait to do such a calculation until we introduce a closed loop circuit in the next section.

However, the second and third phenomena cannot be explained in this way; they are not directly related to the B-field itself. In the second case, the first wire (the current carrying wire) is moving while the second wire, the receiver, is stationary as shown below.

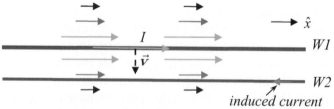

Since W1, the source of the field, is moving, the field at any fixed place in the plana is changing. As we have said, the field actually moves from one part of the plana to another at the speed of light, but we will, for a while, still continue to neglect this finite speed and take the speed to be infinite, which we can functionally do as long as we don't ask for effects that happen on the scale of time it takes for light to travel the distances in question. With this simplification and recalling that the A-field goes as $\ln R$, and assuming W1 moves at a constant speed v, we see that the field at a given point is proportional to $\ln vt + \text{constant}$. Now, we are ready to introduce the new physics. The plana does not "like" the field to change. It resists the decreasing or increasing field by trying to move positive charges so as to, respectively, recreate lost *A*-field or diminish newly generated *A*-field.[1] Thus, when the *A*-field is increasing, as it would be at the location of W2, the plana acts to move positive charges to cause an *A* of the opposite type that already exists to offset the increase. That is, it creates leftward pointing *A* by giving charges in the second wire leftward impetus. Hence, a moving wire with rightward current will cause leftward current in a stationary wire that it approaches.

The third case, i.e. when neither wire moves but a change of the current in the first wire causes a change in the *A*-field at the second wire and thus, as in the second case, the plana acts to oppose this change. In particular, a wire with increasing current causes an increasing *A*-field (as does a wire with constant current moving toward a second wire), which, in turn, causes current to flow in the second wire of the opposite type of the first wire.

These last two cases give a deeper insight into the nature of *A*. We learned in the last chapter that the disposition of the plana called the *B*-field is specified in both terms of its strength and directional type by the curl of *A*. We say the *B*-field is specified "quantitatively" in this way, since it is given in terms of equations that can be used to predict, from certain measurements (which are comparisons that result in (analogical) numbers), the (approximate) measured acceleration, speed and location (which are also numbers). To the *B*-field, we have now added the plana's ability to resist changes in *A*, but we have not yet "quantified" this property in the same way. Experiments verify that the force caused is proportional to how fast the *A*-field changes. The faster the change, the stronger the force is. Furthermore, though the resistance to changes in the *A*-field is clearly a *magnetic*, not an electric phenomena, in the key sense that *A* itself is caused by impetus-activated charge, not charge alone, its behavior is like that of an electric field. Namely, it acts on charges without any impetus and it acts in the direction of the *A*-field not perpendicular to it. Because of this, it is most conveniently incorporated in our formalism as part of $\vec{E}$. Thus, we write in general:

4.1
$$\boxed{\vec{E} = -\vec{\nabla}\phi - \frac{1}{c}\frac{\partial \vec{A}}{\partial t}}$$

Clearly, the first term is a true electric effect and the second term is the inductive effect. To focus on the effect of interest, we also write.

4.2
$$\vec{E}_{ind} = -\frac{1}{c}\frac{\partial \vec{A}}{\partial t}$$

This relation is more profound than one might expect even at this point, and because of that profundity, we now will now see the most important reason why we "paint" the *A*-field activated plana the same color as the impetus that caused it. The reason is most evident in *mks*, where equation 4.2 is written $\vec{E}_{ind} = -\frac{\partial \vec{A}}{\partial t}$. Thus, a point particle of charge *q*, experiences a force according to: $\vec{F} = -q\frac{\partial \vec{A}}{\partial t}$. This means: $\Delta\vec{p} = -q\Delta\vec{A}$. And, if, for example, the particle starts at rest and the *A*-field is some finite value, $\vec{A}$ goes to zero by

---

[1] We will shortly see what *specific* tendency the plana has to act in a direction opposing the change in the *A*-field state. In particular, we will see it *does not* generally activate impetus in the charges in such a way as to completely cancel the change.

the end of the change, i.e., $\vec{A}_f = 0$ and $\vec{p}_i(0) = 0$, we get: $\vec{p}_f = q\vec{A}$. In words, the momentum of the A-field appears to go into the momentum of the particle. The A-field is thus a kind of potential momentum. The field (i.e. the field activated plana) has the potential to cause a certain amount of momentum per unit charge, which is given by the magnitude of A. Hence, the directional type of A is the same as that of the impetus that generated it, clearly giving strong reason to represent both of these with the same color. We thus have a kind of conservation law of a generalized momentum: $\vec{p}_{generalized} = \vec{p} + q\vec{A}$.

Equation 4.1 thus has to do with two different types of potential. The first, as we saw in the "Static Electric Field" chapter, Chapter 2, is the potential energy per unit charge or simply "the potential," ϕ, the second is the potential momentum per unit charge, $\vec{A}$. In special relativity, you will see these two make what is called a four vector. (Indeed, you will see there that, because of the effects of these fields, the first two inductive phenomena, though they have different causes, can be *treated as* due simply to *relative* motion.)

To understand the potential momentum more concretely, consider the following example illustrated in **Figure 4-1**. Here, we consider again the long straight wire with the current moving in the x-direction, which has the A-field previously given of the form: $\vec{A} = A_x(R)\hat{x}$. We further suppose, as shown in **Figure 4-1**a, that we have placed a positively charged body with a hole in it, say a charged bead, which can slide frictionlessly along a thin stiff rod that lies parallel to the wire.

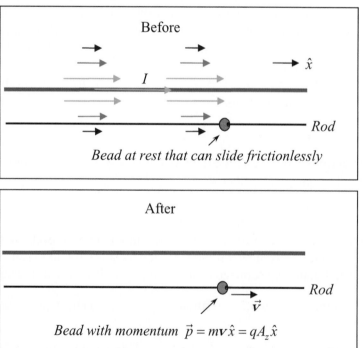

Figure 4-1: a. (*top*) A current carrying wire shown next to a long thin rod with a positively charged bead that can slide frictionlessly along the rod. b. (*bottom*) After the current in the wire has been switched off, there is no A-

field, but the bead moves with constant speed v and momentum $\vec{p} = mv\hat{x} = qA_x\hat{x}$.

If we now turn off the current slowly enough to not cause significant electromagnetic radiation (of which light and radio waves are types), by the time the current is off, the bead will have gained a momentum equal to the potential momentum it had before the current was changed. Mathematically, we have: $\vec{p} = -q(0 - A_x)\hat{x} = qA_z\hat{x}$ so that the bead's velocity is: $\vec{v} = \vec{p}/m$ where q is the charge on the bead and m is the mass of the bead. That is, the potential momentum of the particle in the field, $q\vec{A}$, becomes actual momentum, $\vec{p}$, of the bead.

Before leaving this example, note that the rod serves to keep the bead from moving in any direction other than along the direction of the wire. Without the rod, as soon as the bead gained some impetus along the direction of the wire, the bead would be acted on by the B-field aspect of A giving it impetus perpendicular to the wire. Thus, as you may surmise, the two agencies working together can, in general, cause quite complex effects.

Induction to a Loop

We next need to consider induction from a complete circuit to a complete circuit, i.e. not just the single wire which is only part of a circuit. The setup we will use for this analysis is shown in Figure 4-2. We use a solenoid for the source of the A-field for multiple reasons. It is a commonly used electronic component, and it is easy to make. Furthermore, we have already calculated the field for a solenoid in Chapter 3 (though here we would not want to take the limit as the length of the solenoid gets arbitrarily large). Now, it can be shown, using, for example, *Mathematica* or math tables, that the field of a finite length solenoid in the x-y plane falls-off like the magnetic dipole field if one is far enough away. Indeed, in an end of chapter problem you will be asked to show that, in the far field limit, the field of a small loop of wire (a thin solenoid) reduces to that of a dipole:

4.3
$$\vec{A} = \frac{\vec{m} \times \hat{r}}{r^2}$$

4.4
$$\vec{B} = \frac{m\left(3\left(\hat{m}\cdot\hat{r}\right)\hat{r} - \hat{m}\right)}{r^3}$$

$$= -\frac{m\left(3\cos\theta\,\hat{r} - \hat{z}\right)}{r^3}$$

Where: $\vec{m} = \dfrac{I\,\vec{a}}{c} = m\,\hat{z}$ and $\vec{a}$ is the area of the loop and we take the dipole to be along the z-direction as shown in Figure 4-2. θ is the angle measured from the z-axis and $\vec{r}$ is the vector to the field point from the center of the source. (Note: our use of a lowercase "a" for area, which we sometimes do to avoid confusion with the A-field)

Because there are no magnetic monopoles, the ideal magnetic dipole is fundamental to the B-field the way that the point charge is fundamental to the electric field. Finally, then using the solenoid as a source also allows us to introduce this fundamental entity. If you would like, you can imagine the source as a small loop rather than a full solenoid, since either way we will think of it as an approximation to a dipole. And, since (to simplify our thinking) we will analyze the effect of the field in the x-y plane, we specialize the dipole formula in equation 4.4 to that plane, i.e. $\theta = \pi / 2$, and get: $\vec{B} = \dfrac{m}{r^3}\hat{z}$.

First Case Induction: a *B*-field effect

Starting with the first induction phenomenon, a moving "receiver", we can calculate the voltage generated in a small rectangular loop of length Δy and width Δx using Figure 4-2. (Note that if we take the loop to be arbitrarily small, we can simplify the situation. For example, the *B*-field becomes arbitrarily close to uniform so that, in the case shown in the figure, we can take $B = \dfrac{m}{r^3}$ over the entire loop, which will, in turn, simplify our calculation of the magnitude of the effect.)

Since $\vec{B}$ is in the z-direction and the loop's velocity is in the y-direction, using $F = q\dfrac{\vec{v}}{c} \times \vec{B}$ (*cgs*), we note that the forces on the (positive) charges in the loop are in the x-direction. This means those forces are perpendicular to the two sides of the loop that are parallel to the y-axis. However, it also means that the forces on the charges in the other two sides (i.e., the ones along the x-axis) act along the direction of those wires. In particular, the forces on the sides farthest and nearest from the source, respectively, are:

4.5a,b $\qquad F_x\left(y + \Delta y\right) = q\dfrac{v}{c}B_z\left(y + \Delta y\right) \qquad\qquad F_x\left(y\right) = q\dfrac{v}{c}B_z\left(y\right)$

We then get the electromotive force (EMF), i.e. the voltage, potential energy per unit charge, imposed *along the wire* of the loop by the plana activated field by taking the path integral and dividing by the charge:

4.6 $\qquad \mathcal{E} = \dfrac{1}{q}\displaystyle\int \vec{F} \cdot d\vec{s} = \dfrac{v}{c}\left(B_z\left(y\right) - B_z\left(y + \Delta y\right)\right)\Delta x$

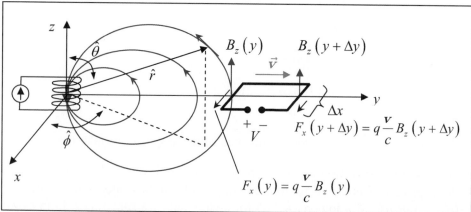

Figure 4-2: Dipole-like field causes electromotive force in the sides of an impetus-activated rectangular loop. When the loop is moving and oriented as shown, the field activated plana only exerts a force parallel to the x-axis, i.e. along the two sides parallel to the x-axis.

Second Case Induction (*the new A-field effect*)
and Faraday's Law

Since the second and third induction phenomena are fundamentally the same, being each the result of the plana resisting the changing of its *A*-field, we will only consider the changing current case. This will immediately allow us to see how a transformer works. Transformers are so important in modern life that you will find one in nearly every electronic device you own, and that usefulness hinges on a key piece of physics, the plana's resistance to changes in *A*, which is interesting in and of itself and which we will now use.

The figure below shows the *A*-field at each of the relevant legs of the loop at a particular moment in time. At the next moment, we suppose the current is decreasing so that the *A*-field (and thus *B*) will be getting weaker.

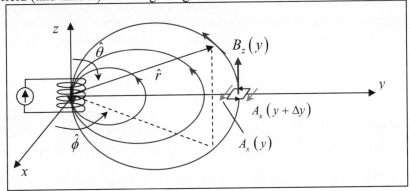

Thus, noting that the net induced fields at the two relevant legs are given by:

$$E_{induced}\left(y + \Delta y\right) = -\frac{1}{c}\frac{\partial A_x\left(y + \Delta y\right)}{\partial t} \quad , \qquad E_{induced}\left(y\right) = -\frac{1}{c}\frac{\partial A_x\left(y\right)}{\partial t}$$

we can calculate the induced EMF to be:

4.7 $\qquad \mathcal{E} = \frac{1}{q}\int_C \vec{F} \cdot d\vec{s} = \int_C \vec{E}_{induced} \cdot d\vec{s}$

$$= -\frac{1}{c}\int_C \frac{\partial \vec{A}}{\partial t} \cdot d\vec{s} = -\frac{1}{c}\int_C \frac{\partial\left(\nabla \times \vec{A}\right)}{\partial t} \cdot d\vec{a} = -\frac{1}{c}\int_C \frac{\partial \vec{B}}{\partial t} \cdot d\vec{a}$$

Note, in the last two equalities, we have used Stokes Theorem and $\vec{B} = \nabla \times \vec{A}$.

Finally, if we take the flux to be defined by: $\Phi = \int_S \vec{B} \cdot d\vec{a}$, we can write *Faraday's Law of Induction*, named after Michael Faraday (1791-1867AD) (see history box).

4.8 $\qquad$ | **Faraday's Law** $\qquad \mathcal{E} = -\frac{1}{c}\frac{d\Phi}{dt}$ |

The minus sign in Faraday's law (equation 4.8) has its own name. It is called *Lenz's law*, and it simply refers to the fact that the plana "tries" to induce a voltage to oppose an increase in flux. In the transformer, in which one side, called a primary, induces voltage into a second side, a secondary, this effect is physically due to the plana resisting the change in the A-field. We will show shortly that the moving secondary also "obeys" Faraday's law. In that case, the negative sign of Lenz's law is physically due to the B-field (indicated by the curl of A) "attempting" to keep the charges close while preserving their energy and not reinforcing the central B-field, but acting to reduce that field to avoid the possibility of positive feedback which is inherently unstable. Indeed, Lenz's law expresses the generally resistive, negative feedback, character of the two inductive agencies: (i) the plana's resistance to changes in A and (ii) the related effect of an impetus-activated conductor being acted on by an unchanging A-field activated plana in which the conductor appears to also try to resist the B-field flux changes it experiences. This last agency acts in the moving secondary case.

To see how the equation for the electromagnetic force (EMF) caused by the moving secondary (i.e., equation 4.6 in the last section) reduces to Faraday's law, we first note that the flux of the B-field through the loop, using the definitions in the above figures, can be written at one moment as:

4.9a $$\Phi(t) = \int \vec{B} \cdot d\vec{a} = \int_{y}^{y+\Delta y} B_z(y)\, \Delta x\, dy$$

At the next moment, the flux is:

4.9b $$\Phi(t+\Delta t) = \int_{y+v\Delta t}^{y+\Delta y+v\Delta t} B_z(y)\, \Delta x\, dy$$

Thus, using the figure above which shows the domains of integration, the difference can be written:

$$\Delta\Phi = \Delta x \left(\int_{y+\Delta y}^{y+\Delta y+v\Delta t} B_z(y)\, dy - \int_{y}^{y+v\Delta t} B_z(y)\, dy \right)$$

$$= \Delta x \left(B_z(y+\Delta y) - B_z(y) \right) v\Delta t$$

This simply says that the difference in flux results from the change in position $\Delta y = v\Delta t$ in the front and back because the flux in the middle is effectively unchanged during the motion.

Hence, in the limit, we get:

4.10 $$-\frac{1}{c}\frac{d\Phi}{dt} = \Delta x \left(B_z(y) - B_z(y+\Delta y) \right) \frac{v}{c}.$$

Thus, we have, by comparison to equation 4.6, Faraday's Law : $\mathcal{E} = -\dfrac{1}{c}\dfrac{d\Phi}{dt}$.

Michael Faraday was born 1791 (d. 1867) in South London. Faraday was responsible for many important discoveries, probably the most important being the law of induction, the fact that magnetic fields can cause currents. Faraday is also responsible for discovering: the magneto-optical effect, benzene, two chlorides of carbon, diamagnetism, laws of electrolysis as well as important contributions to: the electric motor, the generator and the transformer. He invented an early form of the Bunsen burner and also, as we have already discussed, did important electrostatic work. Faraday's work with lines of force changed the path of E&M, opening the way for the idea of fields in space (plana).

Joseph Henry also independently discovered induction, though Faraday published first. Thus, we have "Faraday's law" of induction (Aug. 1831) as well as the unit of inductance, which is called a "henry."

The respect Faraday garnered even 100 years later, can be seen in the fact that Albert Einstein kept a photograph of Faraday on his study wall alongside pictures of Isaac Newton and James Clerk Maxwell. Pieter Zeeman spends several paragraphs on him in his Nobel Prize acceptance speech in 1903.

Faraday was a sincere and committed believer in Christianity; when asked what occupation he might have in the next life, he responded. "I shall be with Christ and that is enough." He said, in the tradition of Galileo and Newton:
"The book of nature which we have to read is written by the finger of God."

Faraday was interested in education and probably out of the deep connection he saw between his Christian belief in the intelligibility of nature and science's effort to find that intelligibility he started a series of Christmas lectures, giving a total of 19, including his famous *Chemical History of the Candle*. Following his lead, Christmas lectures for young people are still given every year at the Royal [Scientific] Institution. His mindset was profoundly that of an experimental physicist as can be seen in his statement "There is no more open door by which you can enter into the study of natural philosophy than by considering the physical phenomena of a candle." Faraday, revealing a more theoretical side, also said:

The exertions in physical science of late years have been directed to ascertain not merely the natural powers, but the manner in which they are linked together, the universality of each in its action, and their probable unity in one...I cannot doubt that a glorious discovery in natural knowledge, and of the wisdom of power of God in creation, is awaiting our age, and that we may not only hope to see it, but even be honoured [sic] to help in obtaining the victory over present ignorance and future knowledge.

We can now easily calculate the actual voltage induced by the dipole field in the *x-y* plane ($\vec{B} = \dfrac{m}{r^3}$) into the small loop of *area a* as follows:

$$\mathcal{E} = -\frac{1}{c}\frac{d\int_a \vec{B}\cdot d\vec{S}}{dt} \sim -\frac{\dot{m}a}{cr^3}, \text{ recalling that } m = \frac{I\,a}{c}, \text{ we have}$$

4.11
$$\mathcal{E} \sim -\frac{\dot{I}\,a^2}{c^2 r^3}$$

Hence, we can increase our "reception" by increasing the rate of change of the current or by increasing the area, but the best way is by decreasing the distance because this gives a cubic increase in the induced EMF. Yes, our use of the word reception does indicate an analogy between TV and radio reception and the induction phenomena; however, radiation, such as is involved in TV and radio, as we will see later in the text, is not merely induction.

The Complete "Third" Maxwell's Equation and Faraday's Law

Faraday's Law deals with integrated regions and thus is useful for large as well as small regions. We can also write a differential form in the following way. Changing the left hand side into an area integral using Stokes' Theorem, and using differential areas for both sides we get:

$$\int \vec{E}_{induced} \cdot d\vec{s} = \int_{dS} \nabla \times \vec{E}_{induced} \cdot d\vec{S} = \nabla \times \vec{E}_{induced} \cdot d\vec{S}$$

$$= -\frac{1}{c}\frac{d\int_{dS} \vec{B} \cdot d\vec{S}}{dt} = -\frac{1}{c}\frac{d\left(\vec{B} \cdot d\vec{S}\right)}{dt}$$

In the third equality, we used $\vec{E}_{induced} = -\frac{1}{c}\frac{\partial \vec{A}}{\partial t}$ and assumed a fixed surface, dS, to pull the time derivative out of the integral. Hence, Faraday's law can be written: $\nabla \times \vec{E}_{induced} \cdot d\vec{S} = -\frac{1}{c}\frac{d\vec{B}}{dt} \cdot d\vec{S}$. Now, this must be true for any area $d\vec{S}$, so that we can write the differential form of Faraday's law, which is also Maxwell's third equation, as:

4.12 $$\boxed{\textbf{Maxwell's Third Eqn.} \quad \nabla \times \vec{E} = -\frac{1}{c}\frac{\partial \vec{B}}{\partial t}}$$

Now, using the definition of E given in equation 4.1 ($\vec{E} = -\vec{\nabla}\phi - \frac{1}{c}\frac{\partial \vec{A}}{\partial t}$) along with $\vec{B} = \nabla \times \vec{A}$, we see that equation 4.12 is automatically true. However, such a substitution makes clear that the formalism of the third Maxwell equation coupled with the companion definitions of E and B only incorporate one of the two physical agencies that can be responsible for the general effect known as induction. It incorporates the resistance of the plana to changes in A, which we will often call the "$\dot{A}$ effect," pronounced "'Adot' effect." However, it does not include the ability of B-field activated plana to cause a force in a moving conductor (i.e., on impetus activated charge). This effect is of course included in the magnetic portion of Lorentz's force law: $\vec{F} = q\frac{\vec{v}}{c} \times \vec{B}$. As broached earlier, if one is careful in the application of Faraday's flux law (equation 4.8), the causal distinction between the "$\dot{A}$ effect" and the "$\vec{v} \times \vec{B}$ effect" can often be *formally* avoided. Of course, as we tirelessly point out, physical reality is not dependent on our formalism, let alone on the convenience of the formalism, and thus cannot be changed by it. However, the formalism can make it easier to solve problems by unifying analogically alike effects (though in this case unlike causes), which can facilitate both calculation and formal development of the theory.

In fact, as you might expect, once you see that the effects are alike, more than this is involved. Special relativity, through the physical principle of integrity, (see *PFR-M* Chapter 10) which posits that all substances behave such that their activity at rest is preserved in uniform motion, requires this similarity of outcome. In particular, in empiriometric special relativity, the three phenomena of induction need be viewed as intertwined. For example, given a changing current acting on a loop, in a frame in which the loop is moving the "$\vec{v} \times \vec{B}$ effect" will appear to occur along with the "$\dot{A}$ effect." Said another way, given the resistive nature of the plana, which causes "the $\dot{A}$ effect", special relativity will require the magnetic Lorentz force law. More generally, given the electric field (which includes both $\dot{A}$ and ϕ based types), special relativity will require the existence of the magnetic field. Faraday's law captures and leans on special relativity, and, thus, the complete power and import of the law will only be seen when we introduce special relativistic considerations later.

Faraday's Disk: A Problem for Faraday's Law?

To see the care that must be taken in applying Faraday's law, we will now examine one interesting case. When in doubt about how to apply Faraday's law, it is advisable to simply calculate using the physical effect that operates in the given case—though you should also keep in mind that sometimes calculational short cuts *can* be had by moving to the "correct" frame and solving a simpler, but operationally equivalent problem.

Faraday's Disk

We take the case of the Faraday disk, which is shown schematically in Figure 4-3a. Figure 4-3b shows an early version of the experiment. Note that it does not have a uniform *B*-field over the entire disk; we will assume the conceptually simpler experiment envisioned in Figure 4-3a, which does.

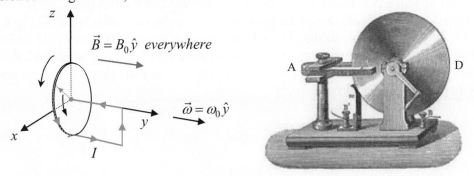

Figure 4-3: a. *(left)* Schematic of "Faraday's" disk. The conductive disk rotates in a uniform magnetic field that is along the axis of rotation. The radial leg of the current travels outward and spins along with the disk. **b.** *(right)* Picture of an early version of Faraday's disk, which is also called a "homopolar generator." "A" is a horseshoe magnet supplying the *B*-field across the copper disk (labeled "D"). The left contact point is connected to the curved piece of metal that pushes up against the disk. The right contact is connected to the center of the disk. Apparently the magnet was recessed more than the crank so that the disk could be spun.

The experiment consists of a conductive disk, e.g., a copper or aluminum disk, of radius b, that is rotating around an axis that is parallel to a uniform B-field. We hook wires to the center of the disk and to the rim as shown. We first calculate in the most straightforward but least general manner by noting the agency at work is the B-field's ability to cause impetus in impetus-activated charge, the "$\vec{v} \times \vec{B}$ effect." In particular, any given element of the disk at a given time has impetus along the angular direction $\hat{\phi}$. Thus, the field applies a radial force to the charges at every point on the disk. This means there will be current paths all over the disk, but we are only interested in the total electromagnetic force, which can be found by integrating along any path from the center to the rim. Noting that the force per unit charge at a given radius, r, is $\vec{F} / q = \dfrac{\vec{v}}{c} \times \vec{B} = \dfrac{1}{c} \omega_0 r \hat{\phi} \times B_0 \hat{y} = \dfrac{\omega_0 r}{c} B_0 \hat{r}$, we get (since there are no forces acting in any of the other legs) the total EMF around the loop shown in orange in Figure 4-3a:

4.13
$$\mathcal{E} = \frac{1}{q} \int \vec{F} \cdot d\vec{s} = \int_0^b \frac{\omega_0 r}{c} B_0 dr = \frac{\omega_0 b^2}{2c} B_0$$

where b is the radius of the disk

This is the voltage induced by the motion. In other words, we have a generator, what is called a homopolar generator. Turning the disk, say with a crank, causes a voltage, i.e. an ability to give a certain amount of energy per unit charge, to appear across the terminals.

Now, you may have already wondered how Faraday's flux law ($\mathcal{E} = -\dfrac{1}{c}\dfrac{d\Phi}{dt}$) can be applied. In particular, you may note that the flux through the disk is constant in time, and ask: Why is there an EMF? Here is where the care needs to be taken. The orange loop in Figure 4-3a bounds a surface composed of two regions, a rectangular region and a sector shaped region which is bounded by the orange radial line and the rim to the point where it meets the orange wire brush. Since the plane of the rectangular region is parallel to the B-field, the only contribution to the flux into the surface is from the sector. If the radial boundary of the sector (shown as an orange arrow) were kept fixed, we would still have no changing flux. However, the charges in the conductor are, in fact, moving, and we need to follow them (we will see why later).

The flux at a given instant through the sector shaped region, and thus the whole region, is B_0 times the area of the arc of angle ϕ: $\Phi = B_0 \pi b^2 \left(\dfrac{\phi}{2\pi} \right) = \dfrac{1}{2} B_0 \phi b^2$. Since the arm moves at angular speed ω_0, the flux change is:

4.14
$$\left| \dot{\Phi} \right| = \frac{1}{2} B_0 \omega_0 b^2$$

Thus, comparing with equation 4.13, we see that Faraday's law gives the correct EMF. However, we have yet to answer how this all works out; in particular in this case: Why do we follow a moving arm? And, what do we do in other cases? And, are there any cases where Faraday's law will not work?

Faraday's Law and the "$\vec{v} \times \vec{B}$ effect"

The answer can be found by looking in detail at how Faraday's law comes about in the case of the plana acting on impetus activated charge, i.e. what we have called the "$\vec{v} \times \vec{B}$ effect." The force due to a B-field acting on a differential length of a circuit, $\vec{dl}$, carrying a current, I is:

4.15
$$dF = \frac{1}{c} I\vec{dl} \times \vec{B}$$

In particular, consider now the case of a closed loop of wire moving at speed v through a uniform B-field. Focus on an arbitrarily small part of a conductor such as shown as a small dotted rectangle on the left in the figure below. In time dt, the conductor moves a length $\vec{dl}$ tracing out part of a current path, as shown on the right in the figure below. Assuming this bit of conductor has a net mobile charge of magnitude dq within it, in a time dt, an amount of charge dq is moved thru a plane perpendicular to $\vec{v}$ so that $I = \dfrac{dq}{dt}$. Since the charge dq has impetus, clearly the B-field-activated plana exerts a force, $\vec{F}$, on that charge. We visualize as shown here:

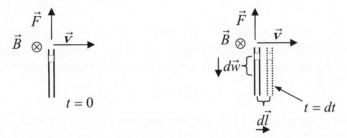

Recall now that the EMF is the work per unit charge that is done in moving each bit of free charge inside the conductor so as to cause an electric field force on that charge that just cancels the force caused by "the $\vec{v} \times \vec{B}$ effect" on it. In particular, moving the charge dq an increment $d\vec{w}$ along a certain path along the wire, we can write the EMF, $d\mathcal{E}$, along that path at a given point as:

4.16
$$d\mathcal{E} = \frac{d\vec{F}}{dq} \cdot \vec{dr} = \frac{1}{c\,dq}\left(I\vec{dl} \times \vec{B}\right) \cdot d\vec{w}$$

Since the right hand side of the last equality in relations 4.16 can be rearranged without changing the absolute magnitude and using $I = dq/dt$, we can write:

4.17
$$d\mathcal{E} = \frac{1}{c\,dt}\vec{B} \cdot \left(\vec{dl} \times d\vec{w}\right)$$

It is important to keep in mind that $\vec{dl}$ is the direction and length along which the initial impetus moves the differential piece of the conductor, while $d\vec{w}$ belongs to the calculation of the potential. In particular, note, for future use, that integrating only over dw will give the total EMF, while integrating over $\vec{dl}$ implicitly implies an integration over time (since $\vec{dl}$ properly belongs associated with dt as $\vec{dl}/dt$, the velocity of the conductor) to get $\int_0^t \mathcal{E}dt$, (which is, using Faraday's law once we prove it, the total flux as a function of time). Now, since Faraday's law connects EMF , $\mathcal{E}$, to the flux, Φ, we need to connect the right hand side of equation 4.17 to the flux.

We note that the term in parenthesis on the right side is the area through which the length $d\vec{w}$ of the conductor is moved in time dt. This then is a differential area through which we need to ask what the flux of B-field is. Mathematically, we write:

4.18 $\qquad d\vec{S} = d\vec{w} \times d\vec{l}$

$d\vec{S}$ is the surface area covered in time dt. In order to get a finite flux change, we need to integrate over both the $d\vec{w}$ and the $d\vec{l}$ direction. If we do this for our simple case, taking $d\vec{l} = dx\,\hat{x}$ and $d\vec{w} = dy\,\hat{y}$, we get the flux as:

4.19 $\qquad \Phi = \int_0^{x(t)} \int_{-b/2}^{b/2} \vec{B} \cdot \hat{z}\, dy\, dx$

Here take the wire to extend from $-b/2$ to plus $b/2$ in the y-direction. However, in Faraday's law we need the differential flux, $d\Phi$. Thus, effectively "undoing" the x-axis integration by taking the differential, we get the change in flux in a small time, dt is:

4.20 $\qquad d\Phi = \int_{-b/2}^{b/2} \vec{B} \cdot \hat{z}\, dy\, dx = \int_{-b/2}^{b/2} \vec{B} \cdot \left(d\vec{w} \times d\vec{l} \right) = \int_{-b/2}^{b/2} \vec{B} \cdot d\vec{S}$

Note, in the second equality, we reinsert the original element of area, $d\vec{S}$ from equation 4.18.

In other words, if we only do the integral over $d\vec{w}$ in $\vec{B} \cdot d\vec{S}$, leaving the $d\vec{l}$ in differential form, we have the differential change in flux in a small time, dt. Physically, integrating over w alone in equation 4.17, gives the differential flux that the entire length of the wire transverses in the time dt, which, in retrospect, is obvious by inspection of the right hand side of that equation.

Hence, doing the integral over w in equation 4.17 we have:

4.21 $\qquad \mathcal{E} = -\frac{1}{c\,dt} \int_{\text{over } w \text{ only}} \vec{B} \cdot d\vec{S} = -\frac{1}{c}\frac{d\Phi}{dt}$

Our work in developing Φ in equation 4.19 now pays off as the last equality gives, upon substituting the flux integral over a general surface of interest, S, (which, since the conductor is moving, is a function of time) for Φ :

4.22 $\qquad \mathcal{E} = -\frac{1}{c} \frac{d\iint_{S(i.e.,\text{over both } w \text{ and } l)} \vec{B} \cdot d\vec{S}}{dt} = -\frac{1}{c}\frac{d\Phi}{dt}$

S is the surface bounded by a real current carrying path, which is defined by its starting point and its tangent at each point (i.e., its magnitude and direction at each point) given by $d\vec{w}$. Faraday's law is valid in fixed field conditions, i.e., *not* changing in time (which is what this subsection is about since we already covered changing fields in the section "Induction to a Loop" under "Second Case Induction…"), as long as we follow a path where actual charge carriers move.[2] The force of a uniform unchanging B-field acting in a simply-connected conductor is conservative, in the sense of path independent, as long as we confine ourselves (and the system allows us to do so) to paths that are perpendicular to the field. We would for example get the same result for Faraday's disk using Faraday's law on a non-radial path (that is instantaneously at rest in the metal, i.e. follows the metal).

We need to be careful in applying Faraday's law but it will work (in the magneto-static approximation) if done correctly. Note well that the problem of the application of Faraday's law is so subtle that even the great physicist Richard Feynman thought that Faraday's law could not be applied to Faraday's disk, the very problem we solved above (see his Feynman lectures on physics). Furthermore, he *thought* he had invented another situation (involving rocking conductors touching at a changing point) that did not come under the "governance" of Faraday's law. You can see in an end of chapter problem that this is not the case and how to properly apply it using the above principles. Of course, since it does matter how Faraday's law is applied and since that application is sometimes tricky, if there is doubt about how to apply it, one should, as we have already said, return to the physical first principles.

Ideal Transformer

Since transformers are such an important application of induction, we will now do a simple analysis of one. We consider the case of two windings wrapped on the same cylinder, such as shown in **Figure 4-1**. Such a cylinder is usually made of a ferrite-like material which, as we will see in Chapter 9, helps confine the magnetic field to within the cylinder to facilitate induction. The two sets of windings are wrapped very close together and in the same direction, the first with N_p turns every l units of length and the second with N_s turns every l units of length. As shown in the figure, we drive the first winding, called the primary, with a sinusoidal voltage and monitor the secondary winding with a voltmeter. The secondary also has a large resister across it so that a small current will flow in it.

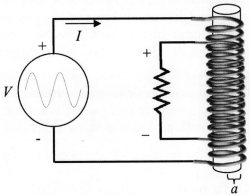

Figure 4-4: A sinusoidal voltage drives a primary coil, which is a wire wrapped N_p times around a core along with a secondary coil, a second wire wrapped N_s times around the core. Each coil is of length l. The secondary is shown hooked to a resistive load. The primary and secondary coil spiral in the same way around the core.

---

[2] Though when the result is path dependent, it does not always yield a result that is easy to interpret or use.

We have already calculated the magnetic field in such a long solenoid of radius a.

It is approximately: $\vec{B} = \dfrac{4\pi N_p I}{cl}\hat{z}$, and $\vec{A} = \dfrac{2\pi a N_p I}{cl}$ where we do not know the current, I

yet (remember we are applying a voltage); however, it will be of the form $I(t) = I_0 \sin \omega t$.
The increasing current during the first part of the cycle will cause an increasing A-field,
which will cause the charges in the secondary wire to move in the opposite way as those in
the primary, causing a voltage of the sign shown in the figure at the secondary output.
Thus, the potential energy per unit charge, i.e. the EMF is:

4.23 $\mathcal{E} \equiv V_s = \displaystyle\int_0^{N_p \cdot 2\pi} \dfrac{1}{c}\dfrac{\partial \vec{A}}{\partial t} \cdot r\,d\phi = \dfrac{2\pi a N_p \dot{I}}{c^2 l} 2\pi N_s a = \dfrac{4\pi^2 a^2 N_p N_s \dot{I}}{c^2 l}$.

Hence, this is the voltage that appears across the secondary, V_s, and hence is applied to the
resister. We would next like to compare this secondary voltage to the primary voltage, V_p.
To do this, we now need the formula connecting the primary voltage to the primary
current.

  If we assume the infinite solenoid field and that the primary and secondary are
wound arbitrarily close to each other, we know that the same "$\dot{A}$ force" acts on the
primary as we just said acted on the secondary. Hence, following the above procedure with
the appropriate substitutions, we can write:

4.24 $V_p = \dfrac{4\pi^2 a^2 N_p{}^2 \dot{I}}{c^2 l}$

Hence, the ratio between the secondary and the primary voltages is:

4.25 $\dfrac{V_s}{V_p} = \dfrac{N_s}{N_p}$

This is the formula for the relationship between the voltages for an ideal transformer.

  Maintaining our ideal conditions, this ratio can be seen even more quickly by using
Faraday's law. If the primary had the same number of turns as the secondary, the flux of B
through each, and hence the change of flux, would be the same; thus by Faraday's law:
$V_s = V_p$. Furthermore, note generally that the flux of the B-field through the core of the
cylinder, around which the windings are made, under our ideal conditions at a given
moment, is independent of what cross section we take. Hence, the total flux of B through a
given *coil* is only different from another *coil* on the same core when it has a different
number of turns. In particular, each new turn means another unit of flux of B through that

coil. So, the ratio of the flux in the secondary to the primary is the ratio of their turns, $\dfrac{N_s}{N_p}$.

Now, Faraday's law, $|\mathcal{E}| = \dfrac{1}{c}\dfrac{d\Phi}{dt}$, says the EMF is proportional to the rate of change of the

flux, so the ratio of the EMF of the secondary to that of the primary equals the ratio of their
turns. And, the EMF in the primary, by definition, must equal the voltage applied to the
primary, while the EMF in the secondary is the voltage applied to the resistor at the
secondary output. Hence, the ratio of the voltage in our secondary to the primary is also the
ratio of the turns as given in equation 4.25.

Transformers are often used to step down the 115v AC (which means alternating current; it is an approximately sinuoisidal varying voltage[3]) voltage of your wall outlets in your home.

Note well that the sign of an induced voltage is very easy to determine physically, though the formalism can sometimes be confusing. The current induced from one wire into a second flows in the opposite way to inhibit changes in the *A-field*. This means that the voltage applied to the first will be the same relative sign as that which appears on the second as shown here.

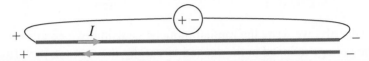

Applying a sine wave voltage to one wire in a configuration such as shown above (twisting the wires to increase their coupling) gives a sine wave output on the second. The input and output voltages from an actual experiment (that you can do with an oscilloscope , a signal generator and a couple resistors with frequency of about 2MHz and about 1 foot of wire) are shown here. The scale is arbitrary and in fact the output voltage is, of course, much smaller than the input.

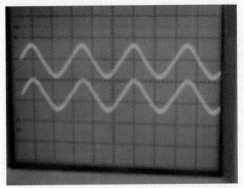

Mutual and Self Inductance and More on Transformers
Mutual Inductance

Two effects are at work in a real transformer: self inductance and mutual inductance. Our emphasis above was on the mutual inductance, which refers to the induction by the primary onto the secondary and vice versa-- though above we focused on the former half. Obviously, we are not limited to our special example above. We can talk about mutual induction between *any* two circuits, which we indicated schematically below by two complicated closed loops of wires.

---

[3] 115V is the rms value. The peak value is of $115\sqrt{2} \sim 162$ volt peak

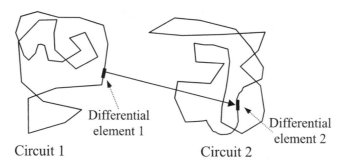

Differential
element 1

Differential
element 2

Circuit 1   Circuit 2

Mathematically, we write the mutual inductance using two equations.

First, the voltage induced in the second circuit by the first can be written:

4.26a $V_{21} = M_{21}\dot{I}_1$, were V_{21} is the voltage induced in "2" by "1" and

Second, the voltage induced in the first circuit by the second can be written:

4.26b $V_{12} = M_{12}\dot{I}_2$, were V_{12} is the voltage induced in "1" by "2"

Physically, current, i.e. impetus-activated charges, flowing in the primary (circuit 1) causes A-field in the plana. When that current changes, the A-field changes and the plana attempts to resist that change in the region of the secondary (circuit 2) by causing a current in the secondary. This can obviously also happen when the role of the primary and secondary are reversed.

In fact, in an end of the chapter problem, using the above equations 4.26a and b, you will be asked to *formally* prove that the ability of the primary to induce voltage in the secondary is exactly the same as the reverse for any two systems of closed conductors. That is, you will show mathematically that $M_{12} = M_{21}$. Hence, the appellation "mutual" inductance is appropriate. This reciprocity can be understood generically by the following analysis.

Two issues matter in the "transmission" from a differential element of one circuit to a differential element of a second (1) the distance between "transmitter" and "receiver" and (2) the relative direction between the two elements. If distance were the only factor, reciprocity must be true, because the distance over which the A-field must propagate from the given differential element to a second is the same whether the A-field is propagated from the first to the second or the reverse. The directional type of the first differential element determines what type of A-field is caused. However, since any effect depends on both the actor and the receiver (recall action and reception are correlatives), we can only say what the effect will be when we look at the receiver. In our current case in which we consider induction by the magnetic effect called by us the "$\dot{A}$ effect," the first does determine the type of A-field but the direction of the second determines how much of that action can move charge. Only the component that is along the direction of the first element can cause current. In other words, it is the angle between the two that matters. Since the angle between the two is the same even if we swap transmitter with receiver, one can expect no difference in effect if we do such a role swap. Hence, the reciprocity obtains; the level of inductance from the first to the second is the same as the reverse.

Self Inductance

Mutual inductance can only occur when there is a second component or system to be considered. As we glimpsed in our analysis of the transformer, if we have only one component, we still have inductance. This is called self inductance, and it happens because an A-field caused by one part of a loop travels to a different part of the loop and causes a force within that part of the wire. Again, we saw this in the transformer, and we found that applying a voltage across the primary (now neglecting the secondary) would force a certain changing current so that the "resistance-less" wire of the primary winding could maintain the voltage difference applied. As already stated generally, this happened as the current caused by the voltage in one part of the cylinder acts on other parts. The result is given in equation 4.24. We can write this in general as:

4.27 $\qquad V = L\dfrac{dI}{dt}$ $\qquad$ Two ports

Where we have dropped the minus sign as a matter of convention.

A two "port" component that has this relationship between the voltage and current is called an inductor. Inductors are crucial in many applications, including our FM radio transmitter project.

Comparing equation 4.27 with equation 4.24 gives the inductance of the cylinder (of radius a, length l, and which is wrapped $N = N_p$ times) as:

4.28 $\qquad L = \dfrac{4\pi^2 a^2 N^2}{c^2 l}$ $\ (cgs)$

Of course, this assumes that there is no ferrite inside the cylinder but only air. With something other than air, L would be increased by a factor μ, called the relative permeability, which we explain later.

If we convert this to *mks* by multiplying by $c^2\dfrac{\mu_0}{4\pi}$, we get:[4]

4.29 $\qquad L = \mu_0\dfrac{\pi a^2 N^2}{l}$

Consider the case of the inductor an air core inductor with 5 turns that is 1 *cm* long and 1 *cm* in radius, which leaves the bounds of our ideal case, but should still give a very rough estimate. Using equation, 4.29 gives:

$\qquad L \sim 1\mu H$

A More Complex Transformer Model

With our new knowledge of self inductance and our further discussion of mutual inductance, we can incorporate more detail into our understanding of the transformer. Transformers can be represented schematically in the following way. The simplest schematic is shown in Figure 4-5a. It shows the inductance of the primary winding and secondary winding, L_p and L_s as well as the mutual inductance, L_m. The dot convention on

---

[4] To find this conversion factor, we first note the appropriate equation is 4.23 with $N_s \rightarrow N_p$. Two things need to be changed. The effective "electric field" in *mks* is $\dfrac{\partial \vec{A}}{\partial t}$, which means we need to multiply by c to convert the given equation to *mks*. Having converted the force per charge, we now need to convert the formula we calculated for A by multiplying by $c\dfrac{\mu_0}{4\pi}$. We obtain this by fixing the calculated definition of B, which is related to A in the same way in *mks* and *cgs* ($\vec{B} = \nabla \times \vec{A}$). Namely, multiplying by c gets rid of the *cgs* definition of B, and the remaining factor provides the constant proper to *mks*.

the diagram indicates that if a voltage is applied to the left coil with positive at the dot, then a voltage will appear on the opposite pair with the positive side being at the dot belonging to that pair. The particular schematic shows the dots both on top implying that the coils are wound in the same direction. The equivalent circuit shown in Figure 4-5b below shows how these inductances function within a circuit. If all the flux stays within the solenoid (an infinitely long solenoid with gaplessly and uniformly wrapped coils), we can show $L_m = \sqrt{L_p L_s}$, (*exercise*). The flux staying within the solenoid is just a way of saying that the primary and secondary wires are as tightly coupled as possible given the solenoidal (air core) configuration.

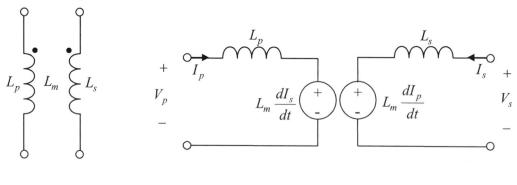

$$V_p = L_p \frac{dI_p}{dt} + L_m \frac{dI_s}{dt}$$

Dot convention = current flowing in one dot means it will flow out of the other.

Figure 4-5: a. (*left*) schematic symbol for non-ideal transformer **b.** (*right*) equivalent circuit for non-ideal transformer. The equation on the bottom left gives the primary voltage as a function of the rate of change of the primary and secondary currents.

Self Energy of Magnetic "Particles"

In Chapter 2, we saw that it cost energy, i.e. one has to do "work," to put together charged particles of the same type. We further saw that this potential or stored energy was reflected in the field caused by the particles.

We will now see a similar effect for "magnetic particles", that is impetus activated charges, and their fields. Since magnetic particles are made from currents, we don't talk about brining in magnetic particles from infinity but turning on currents. This takes work because of the self inductance, the action of one part of the circuit on another as it is turned on. Recall that the potential energy stored in a charged particle is released when the pieces of that particle are allowed to fly apart. The energy stored in the magnetic circuit is released when the currents are switched off. We can calculate the work done in turning on a magnetic circuit in the following way.

Recall that the work per unit charge done by an A-field is

4.30 $$\mathcal{E} = -\frac{1}{c}\frac{\partial}{\partial t}\int_C \vec{A} \cdot d\vec{l} , \quad \text{where: } \vec{A}(\vec{r}') = \frac{1}{c}\oint \frac{I d\vec{l}'}{|\vec{r} - \vec{r}'|} ,$$

This means

4.31
$$\mathcal{E} = -\frac{1}{c^2}\frac{\partial}{\partial t}\int_C \oint \frac{I d\vec{l}' \cdot d\vec{l}}{|\vec{r} - \vec{r}'|} \Rightarrow V = L\dot{I} \text{ with } L = \frac{1}{c^2}\int_C \oint \frac{d\vec{l}' \cdot d\vec{l}}{|\vec{r} - \vec{r}'|}$$

where the minus sign is absorbed into the definitions of the relative sign of I and V

Hence the work per unit time is:

4.32
$$\frac{dU}{dt} = VI = \frac{I}{c^2}\frac{\partial I}{\partial t}\oint\oint \frac{d\vec{l}' \cdot d\vec{l}}{|\vec{r} - \vec{r}'|},$$

Integrating this and using L defined in equation 4.31 gives:

4.33
$$U = \frac{1}{2}I^2\frac{1}{c^2}\oint\oint \frac{d\vec{l}' \cdot d\vec{l}}{|\vec{r} - \vec{r}'|} = \frac{1}{2}LI^2$$

This is comparable to the equation for the energy in a capacitor. Recall in that case: $q = CV$, which implies that $I = C\dot{V}$. Energy is $\dot{U} = IV = C\dot{V}V$, which gives:

4.34
$$U = \frac{1}{2}CV^2$$

Now, we found that the electric field could be used as a sign for the potential energy. Mathematically, we can write:

4.35
$$U_E = \int \frac{E^2}{8\pi}dV$$

To find an equivalent relationship for B, we note that equation 4.27 and equation 4.30 imply that $\frac{1}{c}\frac{\partial}{\partial t}\int_C \vec{A} \cdot d\vec{l} = L\dot{I}$, so that we have:

4.36
$$LI = \frac{1}{c}\oint \vec{A} \cdot d\vec{l}$$

Substitution into equation 4.33 and using the definition of $\vec{J}$ gives:

4.37
$$U = \frac{1}{2}\frac{1}{c}I\oint \vec{A} \cdot d\vec{l} = \frac{1}{2}\frac{1}{c}\oint \vec{A} \cdot \vec{J}\, dV$$

Now using the $\vec{J} = \frac{c}{4\pi}\nabla \times \vec{B}$ and various vector identities you will be asked to show in an end of chapter problem how this gives the pair to equation 4.35:

4.38
$$U_B = \int \frac{B^2}{8\pi}dV$$

We now *seem* to have completed our discussion of the nature of the electricity and magnetism in the static and quasi-static domain.

Summary

Induction refers to three distinct effects: electric current "induced" when:

(1) a conductor moves through an unchanging magnetic field.

(2) the magnetic field generated by a *moving current carrying conductor* sweeps over a stationary conductor.

(3) a changing magnetic field from a *changing current* acts on a stationary conductor.

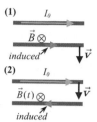

There are two causes for the three inductive effects:

- The "$\vec{v} \times \vec{B}$," which causes the first effect. It arises when the B-field activated plana acts on an impetus-activated charged body. It is described by the Lorentz force law $\vec{F} \propto \vec{v} \times \vec{B}$, hence the name.

- The "$\dot{A}$ **effect**," which causes the second two effects. It arises when A-field activated plana resists changes in its state. It does this by activating impetus in charged particles in contact with it in such a way as to tend to restore the strength of a weakened A-field, diminish the strength of a strengthened A-field or to resist the creation of a new A-field.

   Thus, for example, if a region of A-field activated plana weakens in the x-direction, it resists by causing x-directed impetus in (i.e., applying a force in the x-direction to) a positive charge in contact with it. This impetus activated positive charge then causes A-field in the x-direction thus tending to restore the A-field's strength in that direction.

The $\dot{A}$ *effect*, which is a manifestation of the plana's resistance to changing $\vec{A}$, is so called because of this, and because its magnitude and direction are proportional to $\dfrac{\partial \vec{A}}{\partial t}$. This effect is a new magnetic effect because it is generated by impetus activated charge. Hence, the disposition of the plana whose magnitude and directional type are given by the vector field $\vec{A}(\vec{x})$ has a second action distinct from its B-field effect. However, since, unlike the B-field, the $\dot{A}$ effect acts on charges without any impetus and along the line of the changing A-field, not perpendicular to it, the effect is most conveniently incorporated in our formalism as part of $\vec{E}$. Thus, we write in general:

4.39
$$\boxed{\vec{E} = -\vec{\nabla}\phi - \frac{1}{c}\frac{\partial \vec{A}}{\partial t} \, (cgs)} \qquad \vec{E} = -\vec{\nabla}\phi - \frac{\partial \vec{A}}{\partial t} \, (mks)$$

A key aspect of the nature of the $\dot{A}$ effect that the formalism captures includes the fact that, like $\vec{E}$, the force caused by the $\dot{A}$ effect is along the line of the velocity it causes, not perpendicular to it, as it is with the B-field; thus, like $\vec{E}$, it changes the energy of the particles upon which it acts. However, this does *not* obviate the qualitative difference between the two terms that we have above formally defined as belonging to "$\vec{E}$"; one comes from "magnetized" plana reacting to a changing of its state as quantified by $\vec{A}$, whereas the other is caused by the "electrified" plana as specified by ϕ.

Potential Momentum

In understanding the new term, *mks* units have an advantage, so we write: $\vec{E}_{ind} = -\dfrac{\partial \vec{A}}{\partial t}$, so

that, for a point charge, the force is: $\vec{F} = -q\dfrac{\partial \vec{A}}{\partial t}$, giving: $\Delta \vec{p} = -q\Delta\vec{A}$. The figure below

summarizes the effect for a simple case of a charged bead on a wire. At first, the wire has a steady current generating a steady *A*-field, and the bead is at rest. After the current is switched off, the bead has a momentum of direction and magnitude equal to the former value of $\vec{A}$ at the point of the bead.

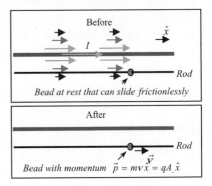

Thus, we say $\vec{A}$ is a potential momentum, in a similar way that ϕ is the potential energy. This is a fundamental reason why we "paint" the plana the same color as the type of impetus that generated it.

Maxwell's Third Equation

The "$\dot{A}$ effect" is formally incorporated into the fundamental formal structure of E&M, Maxwell's equations, by, for example, taking the curl of each side of the analogically generalized definition of the electric field given in equation 4.39:

$$\nabla \times \vec{E} = -\frac{1}{c}\frac{\partial \vec{B}}{\partial t}$$

Faraday's Law
Both types of induction are summarized formally in Faraday's Law:

$$\mathcal{E} = -\frac{1}{c}\frac{d\Phi}{dt}$$

Where $\mathcal{E} = \dfrac{1}{q}\displaystyle\int_C \vec{F}\cdot d\vec{s}$ is the electromagnetic force, EMF, and $\Phi \equiv \displaystyle\oiint \vec{B}\cdot d\vec{a}$ is

flux of *B* through the area bounded by the closed curve C. The relationship between direction of integration around C and the direction of the normal to area element, $\hat{n} = d\hat{A}$, are defined by the right hand rule as shown.

Faraday's law applies to each type of induction for a different reason. It is valid for the $\dot{A}$ effect, because the line integral of the force does equal the flux of *B*-field through the relationship between B and A, $\vec{B} = \nabla \times \vec{A}$ via Stokes Theorem. Mathematically,

$$\mathcal{E} = \frac{1}{q}\int_C \vec{F} \cdot d\vec{s} = -\frac{1}{c}\int_C \frac{\partial \vec{A}}{\partial t} \cdot d\vec{s} = -\frac{1}{c}\int_C \frac{\partial\left(\nabla \times \vec{A}\right)}{\partial t} \cdot d\vec{a} = -\frac{1}{c}\int_S \frac{\partial \vec{B}}{\partial t} \cdot d\vec{a}$$. The law also can be

proven to be valid for the "$\vec{v} \times \vec{B}$ effect" as long as one follows the path of actual currents in defining the surface, S, through which the B fluxes.

Because of its subtlety, Faraday's law should be applied with care and when in doubt one should fall back on the actual physical agencies; namely, calculate using the $\dot{A}$ and $\vec{v} \times \vec{B}$ effects.

Dipole Moment

Because, at the level of abstraction of classical E&M, all magnetic fields are caused by currents, the ideal magnetic dipole is fundamental to the B-field the way that the point charge is fundamental to the electric field. The ideal dipole B-field can be thought of as due to a tiny loop of current, I, of area $\vec{a}$, where the area is allowed to get arbitrarily small while I gets arbitrarily large in such a way that, the dipole moment, $\vec{m} = \dfrac{I\,\vec{a}}{c} = m\,\hat{z}$ remains constant. Its fields are given by:

$$\vec{A} = \frac{\vec{m} \times \hat{r}}{r^2}, \qquad \vec{B} = \frac{m\left(3(\hat{m} \cdot \hat{r})\hat{r} - \hat{m}\right)}{r^3} = -\frac{m\left(3\cos\theta\,\hat{r} - \hat{z}\right)}{r^3}, \text{ where } \theta \text{ is}$$

the angle measured from the z-axis and $\vec{r}$ is the vector to the field point from the center of the source. The A-field makes a circular pattern centered on the z-axis as shown here for the $z = 0$ plane.

Transformers

Transformers can help in visualizing how the principle of (the $\dot{A}$ effect) induction works as well as give an example of the usefulness of induction, as transformers are ubiquitous in the modern world. The basic ideal transformer is composed of two coils, a primary and a secondary wrapped with thin insulated wire (called magnet wire) around a common core. The primary and secondary are wrapped N_p and N_s times respectively. One assumes arbitrarily small resistance, capacitance and radiation. One also assumes that the A-field from one coil is such that it maximally couples to the other coil. In a long cylindrical core, for example, this means that A is in the azimuthal direction ($\vec{A} \propto \hat{\phi}$) throughout the entire cylinder, and there are no gaps in either coil; for this to be true, there must be no curl of A, i.e., no B-field outside the cylindrical core. That is, all of the B-field is confined to the core; we say there is no "leakage."

The ratio of secondary, V_s, to primary voltage, V_p, in an ideal transformer is given by:

$$\frac{V_s}{V_p} = \frac{N_s}{N_p} .$$

Self Inductance
Every loop with a changing current acts one element on another as the A-field generated by one part induces a voltage in another part. This action is due to the $\dot{A}$ effect which resists the change in current. The resistive effect is quantified for a fixed configuration loop in a "lumped" circuit analysis, i.e., treating the loop as one whole component, by its *self-inductance*, L, which is defined by the following relation: $V = L\dfrac{dI}{dt}$.

Mutual Inductance
Every loop with a changing current induces a voltage in any second conducting loop as the virtual plana inside the second conductor resists the change in the state of its A-field. This effect is quantified for a fixed physical configuration in a lumped analysis by the *mutual inductance*, M, which is defined by the following relation: $V_2 = M\dfrac{dI_1}{dt}$, where V_2 is the voltage generated in the second loop and I_1 is the current in the first loop. The parallel relation holds for the reverse situation in which the changing current, I_2, in the second loop induces a voltage, V_1, in the first loop: $V_1 = M\dfrac{dI_2}{dt}$, with the same value of M (hence, the term *mutual* inductance).

More on ideal transformer
The schematic symbol for an ideal transformer is shown on the left below. The solid dots signify that if a voltage is put across the terminals (the unfilled circles) on the left with the positive on the dot, a voltage will appear across the opposite pair of terminals with the positive on the dot on that side. On the right is a schematic model of the ideal transformer.

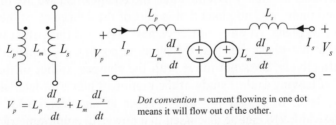

$$V_p = L_p\frac{dI_p}{dt} + L_m\frac{dI_s}{dt}$$

Dot convention = current flowing in one dot means it will flow out of the other.

L_p and L_s are the self inductances of the primary and secondary coils, and $L_m = \sqrt{L_p L_s}$ is the mutual inductance.

The relationship between the voltage in the primary, V_s, and the current in the secondary, I_s, for a long cylinder ($a \ll l$) of radius a and length l is given by $V_s = M\,\dot{I}_p$, where $M = \dfrac{4\pi^2 a^2 N_p N_s}{c^2 l}$. The self inductance of the primary is: $L_p = \dfrac{4\pi^2 a^2 N_p^{\ 2}}{c^2 l}$.

(see next page)

The magnetic *self energy* of a system that generates a B-field is the work it takes to turn on the circuit (whatever currents are responsible for the B-field) to a constant current level, I, (in our quasi-static approximation). The self energy is given by:

$$U_B = \frac{1}{2} I^2 \frac{1}{c^2} \oint \oint \frac{d\vec{l'} \cdot d\vec{l}}{|\vec{r} - \vec{r'}|} = \frac{1}{2} L I^2 \text{, where the integrals are taken over each part of}$$

the circuit acting on every other part and L is the self induction of the circuit.

This formula is analogous to the self energy of an electrical system, which is given by:

$$U_E = \frac{1}{2} C V^2 .$$

Field Energy

As with the E-field, the B-field is a sign of the energy it takes to put the system together. The total self energy can be written: $U_B = \int \frac{B^2}{8\pi} dV$ and the B-field energy density can be written: $\frac{dU_B}{dV} = \frac{B^2}{8\pi}$. Thus, the total field energy density due to electric and magnetic effects can be written: $\frac{dU_{total}}{dV} = \frac{1}{8\pi} \left(E^2 + B^2 \right)$.

Notes on Plana and Empiriometric Abstraction

We have sometimes implicitly assumed that the charges that we are discussing had no virtual plana around them, i.e., no plana that is part of the substance of the charge that has similar powers to free plana. At other times we have assumed (and often will assume) that the charges have virtual plana, but that the plana is continually replaced as the charges move and thus, effectively, does not move with the charges. Either of these assumptions achieves our goal of leaving aside a study of the motion of plana for later studies so that we can remain at the level of abstraction most proper to classical E&M. We will see in Chapter 7 that classical E&M can be discussed in a powerful empiriometric framework that captures a deep aspect of nature but in which (by a being of reason (mental construct) that facilitates the use of the formalism that captures that aspect) we can replace the plana with an experimental measure of the distance between objects.

(Note, by way of tangent, in replacing the plana by a measure of the distance, we replace a substance (plana) with a relation that we have conveyed by a number (e.g., the distance in so many centimeters). This is a characteristic note of the empiriometric method which creates a web of logic and mathematics to facilitate handling important aspects of the physical world, including relating them to each other and to experiment without having, for example, certain substantial natures in the world break the relational and quantitative nature of the system. The relational and quantitative nature of the system allows it to be more easily embodied in symbols, i.e. in a formal system, that, in turn, allows answers to be obtained somewhat more mechanically and thus with more certainty and even also therefore allowing one's study of the world to be much more directly aided by computers.

This casting of the *empirical* into a *formal and symbolic logical system* makes the empiriometric method a special case (though it is the most important case) of a more general *empirio-logical* method.)

Nonetheless, as we are concerned about reality, a thoughtful student will have the following question. If virtual plana moves with the body, which seems natural, will not that virtual plana resist the changing A-field that is caused as it moves through a region of non-uniform A? In this case, since no such $\dot{A}$ effect has been observed, it appears that the virtual plana (which is "trapped" plana, i.e., moving with the body) is differently disposed because it has, for example, an analogical type of impetus or it is acted on by an analogical type of force to which it responds by changing in a certain way, changing qualitatively.[5] This qualitative change disposes it to movement so that changes that arise from being in the new place are already "expected" aspects of its new disposition, so that the plana does not resist these changes in A. However, changes not associated with its movement will be resisted.

Helpful Hints

In reference to A-field activated plana's resistance to change, keep in mind that there is a difference between resisting and nullifying. The plana *resists* change, it does not necessarily succeed (indeed, seldom if ever does so completely) in nullifying the change.

It is helpful to remember this relationship between *cgs* and *SI* charge: $\dfrac{C}{esu} = 10c$ (cf. end of chapter problem on this topic.)

Note the common use of capital "A" for both area and vector potential can cause confusion. To avoid this, we have and will use a, and sometimes S, for area, including surface area. This type of decision is part of the balancing act of symbol assigning that all good physicists must learn. Don't, however, let it take too much of your time; it is, after all, only a name.

Keep in mind that EMF is an integral of a force that is responsible for a potential drop (in the general sense, not just ϕ-field; for example the $\dot{A}$ effect can cause an EMF countered by the EMF of actually separated charges that result in a $\nabla\phi \neq 0$), for it is easy to fall into the habit of thinking of it as simply a voltage, especially because it is measured in *volts*. There is, for example, a difference between the voltage drop across a resistor and the voltage across the battery causing that drop.

---

[5] It would seem that there are two ways for "trapped" virtual plana to move: (1) it can have some power analogous to impetus (which moves massive bodies) or (2) it can be moved by a power ("force") of the massive parts that are part of the substance in such a way that a massive part acts on the parts of the plana in contact with it, which in turn acts on the next part of the plana until all parts of the virtual plana have been acted on so that the virtual plana is moving uniformly along with the massive parts. In either case, we can obtain the result that we posited, i.e. that all the parts of the substance, massive and non-massive (virtual plana), move together. Note also that, in either case, virtual plana has been qualitatively changed.

Problems

1. Assume that you have a bicycle with an aluminum back wheel with no coating so that you can access it electrically. a) Suppose you do the following. Take the tire off to get better access to the edge of the wheel. Turn the bike upside down so that you can turn the pedals and spin the back wheel without the bike moving. Put a magnet (which is approximately 2 *cm* wide and 5 *cm* long and has a field of about 300 *G*) very close to the rim so that its  long side lines up with the tangent to the 1 *cm* wide rim, just covering it. In the figure on the top right, the magnet is shown wrapped in aluminum foil that is used as a handle for holding it near the rim. Next, connect a voltmeter with the positive end to the central axis of the wheel and the negative to a metal brush such as shown in the figure. Rotating the wheel  gives the voltage shown ($-1.9mV$). i) Using the Lorentz force law, explain why there is a voltage. Is it necessary to put the brush directly opposite the magnet? (Assume that the *B*-field is perpendicular to the rim and only pierces the wheel under the magnet, nowhere else.) ii) Give the equation for the voltage in terms of the strength, B_0, of the *B*-field, the radius, R, of the wheel, the width, w, of the magnet, and the frequency, f, in turns completed per second. Use both the Lorentz force law and Faraday's law. iii) If $R \sim .3m$, approximately, how fast is the wheel spinning? What speed of travel of the bike does that translate to? Given the direction of the rim velocity indicated in the figure, what direction is the *B*-field pointing? b) Neglecting all other effects, what voltage would be produced if you are riding down the road at a speed of 20 *mph* with the axis of rotation of your wheels parallel to the Earth's *B*-field, which is about .5*G*?

2. Consider dropping a magnetized ball with a diameter about 4 *mm* down a vertical copper tube with a radius of about 4 *cm*. Assume, for conceptual simplicity, that the ball is dropped with the north end pointing up and that it stays that way during the entire fall, no matter what happens. Also, assume that the field of the ball is generated by currents circulating around the surface of the ball. In terms of the *A*-field, what will happen to the ball and why?

3. Write down the formula for the dipole field given in equation 4.4 in Cartesian coordinates.

4. a) Explain what one means by *EMF*. In particular, explain the connection between the $\dot{A}$ effect, the $\vec{v} \times \vec{B}$ effect and ϕ. b) Explain EMF in the following four cases: a transformer, a generator, a capacitor discharging through a resistor and a battery across a resistor.

5. Consider the following physical system consisting of a rectangular circuit with an initially discharged capacitor. One of the circuit's legs has a length l and is near and parallel to a long straight current-carrying wire. The

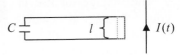

capacitor is arbitrarily far away from the wire so that the effect of the wire's current on it is arbitrarily small. Suppose that at a certain time the current in the long straight wire begins to increase continually, say $I(t) \propto t$. Assume the rate of change of the A-field generated by the rectangular circuit is arbitrarily small (so we, for example, ignore its self-inductance effects). a) Heuristically explain the result of the line integral around the rectangular circuit both right at the moment the current switches on and after the capacitor is charged. b) Also, explain the result in the context of physical EMF (i.e. $\vec{v} \times \vec{B}$ and/or $\dot{\vec{A}}$ effect), and in the context of Faradays law. Include in your answer why the capacitor stops charging. c) Repeat parts "a" and "b" for the integration path shown in orange above; note that the path follows the circuit wire on top and bottom and along the right leg, whereas on the left it takes a shortcut vertical path through "empty space."

6. Take an ordinary "soda can" and wrap it 3 times around the middle with a fairly heavy gauge and then discharge a $400 \mu F$ capacitor that had been pre-charged to 2000volts through the coil. The can will be deeply compressed in the middle. If the coupling is good enough the can will be ripped in half. One can treat the wound wire as a primary and the can as a single turn secondary. Explain qualitatively why the can gets pushed in. Use a drawing to help illustrate your answer.

7. The idea of the flux of the B-field through a surface, $\oiint_S \vec{B} \cdot d\vec{A}$, is weakly analogous to the flux of fluid, $\oiint_S \vec{J} \cdot d\vec{A}$. Given the velocity of a fluid, show how the speed is related to the flux of the fluid and explain the analogy. *Hint:* It is helpful to use the definition of , $\vec{J}$, the current density.

8. A neutron star is a "dead" star whose gravity has collapsed it so that it is more like a giant atomic nucleus (in fact, it is composed of nearly all neutrons) than ordinary matter. Consider a neutron star of radius 10 km, which has a magnetic field of $10^{12} G$ at the surface on one of its poles. Assuming approximately a dipole distribution, how much energy is stored in this B-field? How much energy did it take to make this field? Using $E = mc^2$, how much mass is this equal to? If all this energy was readily available and supposing the US's annual power expenditure is $\sim 10^{20} J$, how long could it power the US?

9. The voltage for use in your house is wired around at above 100kV to save power. Why? It is "stepped down" using a transformer such as shown here to the 20kV range and piped into a neighborhood, then

another transformer steps it down to the voltage to be fed to a group of houses, by a transformer such as shown here. Given the voltage coming into your house is 240V, what transformer ratio is needed? If each transformer feeds 5 houses that expend 10kW of energy, how much current must be in the secondary? How much in the primary?

10. Show that that the A-field (and thus the B-field) of a small loop of current gives the dipole field in the far field limit.

11. Using the B-field given in the text for dipole with moment $\vec{m} = m\,\hat{z}$, calculate its A-field in just the x-y plane.

12. What happened to conservation of linear momentum of massive bodies in **Figure 4-1**? In particular, the "before" figure shows no net momentum of massive bodies and the afterward pictures shows the bead with rightward momentum, suggesting that total momentum went from zero to some finite value. *Hint*: consider how the system got into the state shown in the "before" picture.

13. Using the $\vec{J} = \dfrac{c}{4\pi}\nabla \times \vec{B}$ and various vector identities derive: $U_B = \displaystyle\int \dfrac{B^2}{8\pi}\,dV$.

14. The "dynamo effect" is believed to cause the Earth's magnetic field.
Since the fluid in the center of the earth is conductive, its motion can be modeled by a conductor. Consider the model shown on the right[6] which illustrates the principle of operation of the dynamo effect. The gray regions represent the rotating conductors. The black lines are fixed wires, which are connected by a brush (or other type of connection) that allows them to remain fixed while making electrical contact with the rotating conductors. The pancake-shaped conductor rotates along with the rod shaped ones which are

welded to it. Explain how an initial small ambient B-field in the upward direction can become self sustaining, if the pancake is rotating fast enough.

---

[6] A similar diagram was first used, as far as I can tell, in Purcell's introduction to E&M, 1985 edition.

15. Why is it said that a B-field is effectively frozen in very good conductors? Look up superconductors, which, because of quantum mechanical effects, are more than just perfect conductors. What happens to a B-field inside a material when that material becomes superconducting?

16. Give the equation for the voltage induced in a square loop of wire (with side of length a) by a long straight wire carrying a current I at the moment when the close edge of a square loop is a distance x from the long straight wire and the loop is moving away from the straight wire at a speed v as shown. Given $x = 1\,cm$, $a = 10\,cm$, $v = 30\,mph$, and $I = 1\,A$, what is the induced voltage?

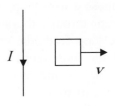

17. Treating an ordinary magnet of area $10\,cm^2$ as a dipole which, 1 cm along its axis, has a field of 300 gauss, what is the approximate effective current flowing in the magnet?

18. A physicist named H. Jehle[7] proposed that the electron did not really have any charge, but only a magnetic field and spin. Its spinning, he claimed, caused an electric-like force. He based his speculation on the idea that, for example, a long cylindrical magnet rotating around its own axis drags its B-field around with it (it seems this means the plana would be dragged with the magnet) causing a force in the wires attached to the voltmeter, not a force in the magnet as it moves in the fixed B-field. In other words, in his hypothesis, there is a "$\left(-\vec{v}\right) \times \vec{B}$" type effect, where

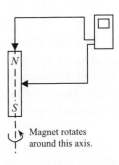
Magnet rotates around this axis.

$\vec{v}$ is the velocity of the B-field rather than the charge. What experiment might be done to distinguish between the two theories? *Hint*: use a cylindrical capacitor structure around the magnet to connect the two points on the magnet to the voltmeter?

19. In the text, a bead is hooked to a rod as shown to constrain it to move without friction parallel to an infinite wire. Suppose one removes the rod, still wanting to do the experiment of watching qA momentum develop in the x-direction. What constraint must be put on T, the approximate time scale during which the current I is switched off, to keep substantial y-axis speed from developing so that the experiment will still work?

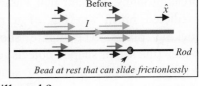
Bead at rest that can slide frictionlessly

20. Prove that, in the quasi-static conserved current case, the action of magnetic field of two arbitrarily shaped current loops in the same plane always conserves the linear momentum of a system of massive bodies. In particular, assume that current is conserved, that changes in current are arbitrarily slow and the propagation speed is taken to be infinite.

Hint: using the vector identity: $\vec{a} \times \left(\vec{b} \times \vec{c}\right) = \left(\vec{c} \cdot \vec{a}\right)\vec{b} - \left(\vec{b} \cdot \vec{a}\right)\vec{c}$

---

[7] Jehle, Phys. Rev. D 3, 306 (1971); 11, 2147 (1975)

21. Consider a very long copper cylindrical shell with a uniform current flowing azimuthally around it, i.e., an ideal solenoidal electromagnet. The solenoid also has plastic spokes in evenly spaced planes along it to support its rotation around its own axis. Suppose a wire loop, which can also rotate around this axis, feeds current through the solenoid as shown. Further suppose that the wire loop feeds the current through the negligibly thin shell and that the loop keeps its rectangular shape as it rotates. Explain, *without* solving explicitly, how the angular momentum of the system of massive bodies is conserved. In particular, explain how the magnet (solenoid) causes the same torque on the wire loop that the loop causes on the magnet (under the assumption of constant current and infinite speed of field propagation). Assume that the outside vertical leg of the wire loop is very close to the cylinder and the bottom leg of the loop is very far from the end of the solenoid. (For extra credit for an advanced student explain in detail (again only qualitatively) the results in the experiment described in the paper titled "The homopolar motor: A true relativistic engine")[8]

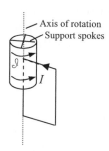

22. Does the electric field in the extended sense always have to be zero in a conductor? Consider a metal made up of a uniform distribution of pair charges, a proton and the electron attached to springs of constant k. Before gravity is applied the spring is in its equilibrium condition. What happens to the field inside our model conductor when it is put into a gravity field?

23. Calculate the approximate self inductance per unit length of two parallel wires of radius a and separation d (that connect at infinity), when $d \gg a$.

24. As testimony to the difficulty of applying Faraday's law in some cases, we noted that Richard Feynman, responsible for so many important physics breakthroughs, thought that Faraday's law predicted much too big of a result for rocking plates. The figure shows such plates (the sector shaped objects) executing half of a rocking motion in snapshots at successive instants of time. The plates are in a uniform *B*-field pointing out of the page. Feynman apparently reasoned that the shortest geometric path was a reasonable choice for use in Faraday's law. Namely, he choose the path from the vertex

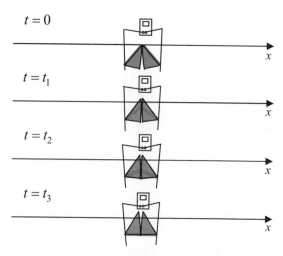

of one plate to the contact point between the plates to the vertex of the second plate, then back around to the voltmeter. The red lines show the circuit path through the plates. What is wrong with his analysis? What is a simple path that *can be* used in Faraday's law? Shade

[8] Am. J. Phys. 70 (10), Oct. 2002, pg 1052

the area that your path through the plates sweeps out and that could be used in computing, from Faraday's law, the EMF induced.

25. Suppose we alter Feynman's "rocking plate" problem (discussed in the previous problem) in such a way that the only current path is the one Feynman used, i.e. consider the configuration shown in the adjoining figure. What voltage appears across the terminals marked " + " and " − "?

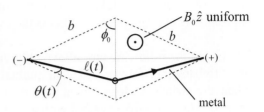

26. We have seen that, in *SI* units, the strength of the *A*-field is measured, like the potential impetus that can be activated in a unit charge that the *A*-field represents, in buridans/coulomb. a) What is the measure of *A* in *cgs*? b) By what factor must one multiply the *cgs* measure of *A* to get the measure of *A* in *SI*? c) How many phors are in a buridan? d) How many *esu*'s are in a coulomb? Show the derivation of all your answers.

27. On the way to Mars, our astronauts will likely want artificial gravity, which can be provided by rotation. Suppose there is a need for a central region of the craft that is not rotating, say to do zero-*g* experiments (so called micro-gravity experiments, but more accurately called free-fall experiments) en route. a) Describe how one might use induction to provide power across the interface between the rotating and non-rotating portion of the spacecraft. Discuss how your solution addresses the problem of coupling and the problem of protecting communication and test instruments from electronic noise. b) Suppose now that one has a thin cylindrical ship with the outside rotating and the inside not. And, suppose that one decides to provide power between the two parts of the ship by nested thin cylinders (here after called "rings") wrapped tightly with wire so as to form a transformer. The outer ring connects to the rotating part of the ship, while the inner ring connects to the fixed part. Assuming that the power circuits drive a 20 *A* current at 60Hz in the primary coil (to allow direct use of standard equipment), use the long wire approximation to estimate how far away a square wire loop with .5 meter sides must be to keep its induced electronic noise below $100 \mu V$. Assume no magnetic shielding is used.

28. Describe how a probe that does not make direct physical contact with a wire might still measure the changing current in that wire. Draw a simple diagram of how such a probe might be constructed.

29. Resistors are sometimes made by wrapping resistive wires around a core many times to get a high resistance. How would you wind such a resister to minimize inductance? Explain your answer.

30. Show *directly* that equation 4.6 gives $\mathcal{E} = -\frac{1}{c}\frac{d\vec{B}}{dt} \cdot \Delta S$.

31. a) Suppose a $1mH$ inductor has a current of $1A$ flowing through it. How much energy is stored in the inductor? The circuit is quickly broken, and the voltage builds up to $10kV$ before it arcs across the switch. Why does the voltage buildup? Assuming a total series circuit resistance of $.1\Omega$, what is the time scale of the energy release? What is the approximate average power? What is the average current?

32. Given the following circuit, which gives the fundamentals of a "flyback transformer," explain qualitatively what happens when the switch is closed for a period of time then abruptly opened. The arrow-like symbol with a vertical line next to it is called "a diode," and it *only* lets current go in the direction of the arrow.

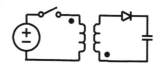

33. The text mentions the importance of the negative feedback effect of Lenz's law, which is fundamentally the resistance of the plana to changes in A. Consider what would happen if there were positive rather than negative feedback by replacing the "$-$" sign with a "$+$" sign to get: $\vec{E} = +\dfrac{1}{c}\dfrac{\partial \vec{A}}{\partial t}$.

34. a) Explain, without solving exactly, what happens in the metal of a rotating Faraday disk in the following cases: i) when no external circuit is connected. ii) when an external current is allowed to flow through a resistor. b) Does gravity, in principle, cause a voltage in a slab of metal, say 1 meter thick? Explain? If yes, how much voltage? Compare and contrast this effect to that of the Faraday disk?

35. Consider a uniform B-field in the y-direction, and a rectangular wire loop of length l and width w hinged to an infinitely massive thin slab that lies along the x-z plane as shown in side view in the figure. The loop is shown with its width into the page and at an angle ϕ from the slab upon which it is hinged along the z-axis. The loop has an electrical resistance R. The following experiment is done in a weightless environment (for example, on the way to Mars) so that gravity can be ignored. You apply a force in the $\hat{\phi}$ direction to the middle of the top wire, such that the loop rotates toward the slab at a constant radial speed ω_0 around the z-axis. What is the equation for this force in *SI* units?

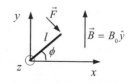

Chapter V

The Last Quasi-static Effect

Introduction

Despite possible appearances, we have not really completed *even* our investigation of slowly changing (quasi-static) electric and magnetic fields. This should be clear since we have not discussed the effect of charge build-up (or depletion), which can happen without quickly changing currents (in first or second derivative), i.e. without violating our generic quasi-static assumption, which fundamentally includes those situations in which we can ignore the finite speed of propagation of the fields.[1]

---

[1] Exactly how fast must the current change before we violate our quasi-static assumption must wait till we have included this last quasi-static effect and completed Maxwell's equations.

Now, charge build-up or depletion implies that some current paths end or begin. Formally, in our empiriometric language, we say $\nabla \cdot \vec{J} \neq 0$, i.e. the divergence of J is nonzero. Before investigating the role of the diverging part of J, we will bring out what we know about the B-field and its source, current, especially the non-diverging part of the current. After probing further the nature of the B-field, we will proceed to examples to illuminate the nature of the "quasi-static effect" that we introduce. Then, we will complete our formal development of Maxwell's equations and end by turning to deepen our understanding of the A and ϕ fields.

The *B*-field and Its Source

The Non-Divergent Part of *J*

Only the Curling part of *J* Can Cause a *B*-field

Recall that we showed in Chapter 3 that the nature of impetus-activated charge is to cause a power in the plana called the A-field, and, wherever there is a curl of that A-field, there is a B-field. Now, only when the magnitude of a component of $\vec{J}$ varies in a direction perpendicular to that component, such as shown below, will $\vec{J}$ have a curl, for this is essentially what one means by a field having a curl.[2] And, it can be shown that only a curling J can cause an A that varies perpendicular to itself, i.e. can cause an A with a curl. Because of the obvious importance of the curl in this discussion, it behooves us to spend a little more time on its detailed nature.

$$\nabla \times \vec{J} \quad \cdots\!\!\longrightarrow\!\!\odot\!\!\longrightarrow J(y)$$
$$\longrightarrow J(y\text{-}dy)$$

The Ways a Curl can Arise

The field illustrated above does *not* change direction. For a field that *does* change direction, one has to include the change of the magnitude of the field in the direction perpendicular in two components to determine the curl in any one direction. Confining oneself to two dimensions, there are two ways in which a field can change perpendicular to its value at any one point; in particular, the component of the field in the x-direction can change in the y-direction and vice versa. (Note well that this includes "acquiring" a component that the field does not have at a given place). For example, consider the field shown in Figure 5-1 for which we have drawn the value of the field at four points labeled "a" through "d" in a small region; the values of the field at each of the four points are indicated by black arrows. We now want to understand the curl at point "c." To do so, we start at the center of the square with the four points as corners. To get the curl at "c," after our analysis, we keep "c" fixed and let the size of the square be arbitrarily small; i.e. we take the limit in which: $\Delta x = \Delta y \rightarrow 0$. In this way, the center moves to "c." The vector at

---

[2] This fact is most evident when one expresses the curl in Cartesian coordinates. This, in turn, is true because every body has height, width and depth, the generic three dimensions, and these are most aptly described by Cartesian coordinates. In other coordinate systems, this same reality is at work but, because such coordinate systems represent a more complex division of the fundamental directions (and in that sense less abstract), the more natural way to approach it is more complex. However, one could untangle the fact pointed at here by use of Cartesian coordinates even when the physical symmetry of the situation does not call for it. Of course, here we discuss extension alone, the province of Euclidean geometry, which excludes all qualification including gravity. In short, we confine ourselves to the level of abstraction proper to classical E&M.

point "b" is drawn below to give a sense of the structure of the field, but, as we will explain after our initial analysis, that vector is *not needed*.

We see from Figure 5-1 that there is a change in the y-component of the field (shown in blue) as we move along the x-axis, i.e. perpendicular to that y-component, from "c" to "d." The y-component decreases from zero at "c" to some small negative value at "d." Such a differential will rotate a curl meter in the clockwise direction, i.e. the minus z-direction $(-\hat{z})$ in our diagram. By contrast, the change in the x-component of the field in the direction perpendicular to it (i.e., along $\hat{y}$) moving from "c" to "a," would cause a rotation in the counter clockwise ($\hat{z}$ direction). Both contributions are needed to get the full curl.

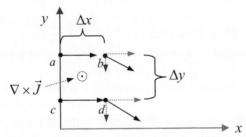

Figure 5-1: The value of a continuous field whose derivative exists at every point is shown at four points in an arbitrarily small region. We use the diagram to understand the curl in the small region around the center of the four points; the center is marked by the circled dot. In the appropriate limit ($\Delta x, \Delta y \rightarrow 0$) with "c" fixed, the center point shrinks to point "c", so that, in that limit, we are actually discussing the curl around "c."

Now, we can see more clearly why the vector at point "b" is not needed. Namely, the value at point d suffices for use in calculating the change in the y-component of the field in the x-direction, *and* the value at point "a" suffices for use in calculating the change in the x-component of the field in the y-direction. As shown in the appendix III, if we were to take an average that included the values at point "b", we would see that the contribution of point "b" becomes arbitrarily small as the size of the box becomes arbitrarily small.

We can make each of the two contributions to the curl even more clear by showing the two contributions in separate diagrams. On the left, we show the contribution of the curl that is due to the change in the y-component of the field along the x-direction. On the right we show the contribution to the curl that is due to the change in the x- component of the field in the y-direction.

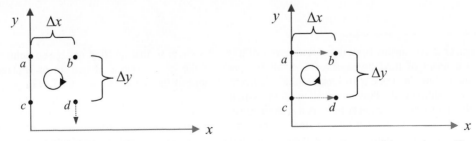

To further digest the import of the two contributions to a curl in a given direction, consider the B-field generated around an "infinite" straight wire. We have seen in Chapter

3 that there is *no* curl of *B* outside the wire. How can this happen? The fall-off of *B* is just such that the curl due to the change in one component of *B* (say, the component in the *y*-direction, B_y) in the

direction perpendicular to it (i.e., the *x*-direction) is *exactly canceled* by the curl due to the change in B_x along the *y*-direction. Note, however, that this curl-less *B*-field arises from a curl of *J*.

With a deeper appreciation of how curl arises, we return to the primary point we want to make about *J*. Clearly if *J*, which represents the current which is the source of the *A*-field, were the same throughout all (unending) space (a being of reason), i.e. a constant field, then there would be no curl of *A*, i.e. no *B*. The *A*-field is constant because any point in the plana is equidistant from all source points. More generally, if there is to be a curl to the *A*-field (a *B*-field) in a region, a component of *J* must change perpendicular to itself. In shorthand phrasing, *only a curl of J can cause a B-field.*

Curling *J* only Generates Circulating *B*

Because the *B*-field is *always* perpendicular to the direction of its propagation,[3] and because of the nature of the fall-off of its strength with distance, only circulating *B*-fields are created, i.e. no *B*-field with a net flux coming in or out of a point is created; empiriometrically: $\nabla \cdot \vec{B} = 0$. Note well that, as we discussed above, *a circulating field does not necessarily have a curl at every point*. In terms of the *A*-field, the fact that *B* must always circulate, never diverge, is immediately evident, for *B* exists only where *A* has a curl (recall formally we write the measure of the direction and strength of *B* as $\vec{B} = \nabla \times \vec{A}$). Namely, since we proved in Chapter 3 that fields whose magnitude and direction are specified by the curl of another field can have no divergence, *B*-fields can have no divergence.

In summary, the presence of a curling $\vec{J}$ (i.e., a *J* that somewhere has a curl[4]), indicates the creation of circulating *B* since every *B*-field circulates. Now, it can be shown that a circulating *B* implies that *B* has a curl somewhere (see end of chapter problem). Hence, a curling *J* results in a *B*-field which somewhere has a curl.

Then, if *J* has a divergence, then, speaking loosely, *J* must be doing something else besides simply creating a curl of *B*.

Splitting *J* into a Non-diverging and Non-curling Parts

As we have seen earlier, a field can be broken into a curl part and a divergence part. As we have said, the curling part of *J* can cause a *B*-field, but the diverging part of *J*

---

[3] Remember *B* propagates from the various *parts* of a given source, so that the *B* at any given point in the plana is a result of field propagation from all the parts of the given source. Hence, to understand the propagation of the field one must look at how a differential piece of the source causes its effect in the plana, e.g. via the solution to the Poisson equation for the *A*-field.

[4] Note that physically, whenever there is a *J*, there must be a variation in *J* in the direction perpendicular to any one of its components, because, at some point, *J* must end in a given direction, as it must end in every direction to avoid the contradiction involved in such a completed infinity. The only exception to this is the spherically symmetric case of a point source (or sink), in which $\vec{J} \propto f(r)\hat{r}$, because in this case, the direction along which *J* changes in order to come to an end can be everywhere parallel to *J* itself.

cannot. To better understand the role of the two parts of J in causing B consider the following.

Think about bending a straight current carrying wire into a circle. In the process, keep in mind that the B-field at any point is caused by *all* the regions throughout the length and breadth of the wire where J has a curl. Also, it is important to note, again, that *while there is circulation* of B, i.e. B coming back on itself, the B-field outside the wire has *no* curl, which is a local property of the field, a property of an arbitrarily small region around a given point. Before bending, the B circulates around the wire along the J direction which is straight. When you bend the (whole) wire, the curl of B inside the wire (where $\nabla \times \vec{B} \neq 0$) bends with it because all of J, the source of that B, bends. Now, this bending of J changes the curl of J, and thus changes B. Indeed, the bending results in a change in the curl of the curl of J and thus, as can be proved in an end of chapter problem, changes the curl of B. This change in the curl of B inside the wire is made evident by the change in circulation of B outside of the wire, i.e. outside of the location of the actual curl of B.

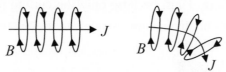

Suppose now that, as shown in Figure 5-2 below, the wire, rather than being bent, comes to an abrupt end; this ending, which of itself introduces a divergence but no new curl, does not affect B, *except* in its ending of the curling of J, thus in its ceasing to generate B.

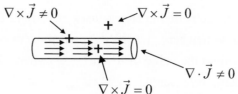

Figure 5-2: The curling part of a uniform current in a wire is on the edges. Think of water moving uniformly along a straight pipe. There is no curl in the water, but there is on the edges. Put a curl meter (shown as crosses) in the middle of a wire (or pipe) with uniform J and it will not move. Put it outside the wire, and, again, it will not move. Put it on the edge and it will. In fact of course, the current does not stop abruptly at the edges but gradually goes to zero, so there is a finite region of curling J.

To bring home the point that the curling part of J causes B, not the diverging part, consider a J that has a divergence but no curl. Consider a point source of J. As we will discuss in detail shortly, such a case in which charge is shooting out from a point *cannot* generate a B-field because the circulating B around each radial element of J cancel each other. In terms of the A-field, one has A that diverges only and thus has no curl.

From another point of view, the curling of B, which every B-field must have somewhere, can be changed by curling that curl, which is, in turn, done by curling the curling of $\vec{J}$. However, one cannot affect the curling of B (which is the only type of B that can be generated) by causing it to diverge, because, as we know, a curl has no divergence.

Thus, to make it diverge would be to make it cease to be a curling B. Or, in terms of the source, one cannot diverge the curl of J without it ceasing to be a curl.

So, again, we see that *only* the curling part of J, $\vec{J}_{curl}$, can cause B; the divergence part of J, $\vec{J}_{div}$ cannot affect B. Formally, we write: $\vec{J} = \vec{J}_{curl} + \vec{J}_{div}$ where $\nabla \times \vec{J}_{curl} \neq 0$, $\nabla \cdot \vec{J}_{curl} = 0$ and $\nabla \cdot \vec{J}_{div} \neq 0, \nabla \times \vec{J}_{div} = 0$.[5]

All this points to the curl of B, $\nabla \times \vec{B}$, as the primary sign of the state of B. Indeed, since every B-field is divergence-less, specifying the curl of B everywhere effectively specifies B everywhere. So, what can we say about the curl of B? It turns out that, because of the nature of the A-field's generation by impetus activated charge, including its fall off with distance, the curl of B is proportional to the value of J at the given point as long as there are no diverging components of J. If the current is zero at a particular place, the curling portions of J throughout the circuit conspire to make the curl of B zero at that point. If, by contrast, J is not zero at a certain point, then the curl of B at that point is proportional to the $\vec{J}$ there. Thus, because only the curling part of J, what we call $\vec{J}_{curl}$, i.e. the part of J that has no divergence, is relevant to the creation of B, we formally write the proportionality described above as:

5.1 $$\vec{J}_{curl} = \frac{c}{4\pi} \nabla \times \vec{B}$$

Note well that $\vec{J}_{curl}$ can have zero curl in some places; the only requirement is that it have no divergence anywhere. However, for it to be non-zero it must have a curl somewhere.

In summary, impetus activated charge causes an A-field that results in a B-field if the curl of J is non-zero somewhere, i.e. if $\vec{J}_{curl}$ is nonzero somewhere. The $\vec{B}$-field produced can be represented by the value of its curl everywhere. The value of the curl of B at a given point is, in turn, represented by $4\pi\vec{J}_{curl}/c$ at the given point, because it represents the net effect of all the curling parts of J in the entire circuit acting on the given point. *Note well* that the *J at the given point is not the sole cause of what it represents*, i.e. the curl of B at that point; the curl of B at that point is, of course, a result of *all* the currents, all the J's, in the source. Equation 5.1 is valid as long as 1) there are no diverging currents and 2) we can neglect propagation, i.e. the quasi-static approximation.

Equation 5.1 is the correct form of the relation that up till now we have written as: $\vec{J} = \frac{c}{4\pi} \nabla \times \vec{B}$. Without the subscript "curl" the relation is actually wrong even quasi-statically, because it equates something that can have a divergence to something that cannot. Physically, given the above analysis, the equality can wrongly suggest that divergent parts of J also cause curls of B, which is impossible. In an end of the chapter problem, you will see that $\vec{B}$ can be formally written as an integral sum of the curls of $\vec{J}$ weighted by the appropriate inverse distance to the field point. It is interesting to note in passing that one can, in principle, add another source of curl of B to the left hand side of

---

[5] We can also write: $\vec{J}_{div} = \vec{\nabla}f$ $\vec{J}_{curl} = \nabla \times \vec{\alpha}$, where f and $\vec{\alpha}$ are respectively a function and a vector field.

equation 5.1, which can be written $\nabla \times \vec{X}$, but no experiment has found evidence of such an X source.

The Role of the Diverging Part of J

What then does the divergence part do and how do we incorporate it into our formalism? A divergence in J means a build-up of charge, which causes a changing electric field. It can be shown that the direction of the time rate of change in E-field is the same as $-\vec{J}_{div}$ and the measure of the rate of its change is proportional to its magnitude, $\left|\vec{J}_{div}\right|$.

Using our uniform radial point source example, radial outward $\vec{J}$ means a decreasing positive charge and thus a decreasing radial E-field; the larger the J the faster the charge changes thus the faster the field changes. In fact, it can be shown in this particular case and, in general, (see end of chapter problem) that the changing E-field at a given point can be used to deduce the divergent part of J, i.e. the part of J that has no curl. In particular, we have:

5.2 $$\vec{J}_{div} = -\frac{1}{4\pi}\frac{\partial \vec{E}}{\partial t}$$

Where the 4π is inserted so that Gauss's law ($\nabla \cdot \vec{E} = 4\pi\rho$) and the conservation of current ($\nabla \cdot \vec{J} = -\dot{\rho}$) are both respected.

The alert reader will immediately notice that we seem to be saying there is current where there is no charge, because we can clearly have $\dot{\vec{E}} \neq 0$, and thus $\vec{J}_{div} \neq 0$, where there is no charge. A careful reader will soon realize the answer. The actual current cannot physically be split into a part that consists of no curls and a part that consists of no divergences. Real current consists of flows that diverge and/or curl. Trying to separate the two will inevitably be artificial, though it does reveal something about the nature of the given flow. Thus, J_{div} can be non-zero even where there is no actual J, because $J = J_{curl} + J_{div}$; all that is necessary is that $J_{div} = -J_{curl}$ so that $J = 0$. Interestingly, *in such places* both the curls and divergences of J are zero so that: $\nabla \cdot J_{div} = 0$ and $\nabla \times J_{curl} = 0$, but, as we have said, they cannot be zero everywhere.

Point to Note: $\vec{J} = \vec{J}_{curl} + \vec{J}_{div}$, where:

$\vec{J}_{curl}$ = the part of J that has a curl and is *distinct* from the curl of J.

$\nabla \times \vec{J} = \nabla \times \vec{J}_{curl}$, but $\vec{J}_{curl} \neq \nabla \times \vec{J}$. It has a curl somewhere, not necessarily everywhere, and a divergence nowhere.

$\vec{J}_{div}$ = the part of J that has a divergence and is *distinct* from the divergence of J. $\nabla \cdot \vec{J} = \nabla \cdot \vec{J}_{div}$, but $\vec{J}_{div} \neq \nabla \cdot \vec{J}$. It has a divergence somewhere, not necessarily everywhere, and a curl nowhere.

What does this mean for our question about what the diverging part does? With diverging J, we can now have curl of B even where there is no charge. This happens because the curling parts of J no longer complete a circuit to make the curl of B sum to zero at every place outside the current path. In other words, some of the cancellation cannot take place because part of the circuit is missing. In the case in which there is no divergence of J anywhere, i.e. $J = J_{curl}$, one can easily make use of equation 5.1. However, when the parts of the circuit are missing, or more precisely when $\nabla \cdot J_{div} \neq 0$ somewhere, equation 5.1 is no longer valid with $J_{curl} = J$. In words, J at a given point can no longer represent the curl of B. To use equation 5.1, we need an expression for J_{curl}.

Given the above, obviously, the answer is: $J_{curl} = J - J_{div} = J + \dfrac{1}{4\pi}\dfrac{\partial \vec{E}_{quasi}}{\partial t}$; note that, because of our quasi-static approximation, at this point, we have simply:[6] $\vec{E} = \vec{E}_{quasi} = -\vec{\nabla}\phi$. So that we can write:

5.3 $$\nabla \times \vec{B} = \frac{4\pi J_{curl}}{c} = \frac{4\pi J}{c} + \frac{1}{c}\frac{\partial \vec{E}_{quasi}}{\partial t}$$

In summary, impetus-activated charges whose measure of magnitude and directional type are captured by J, cause, depending only on the curling parts of J, a curling B even in places where there is no current. This latter happens when there are diverging currents, because such currents signal that there is a break in the current, and thus in the curl of J's that are responsible for B. Again, only in a complete circuit do all the contributions to the curl of B cancel outside the wires. The measure of net magnitude and directional type of that part of the curl of B that is missing due to this diverging J is represented by $\dfrac{1}{4\pi}\dfrac{\partial \vec{E}}{\partial t}$ in the equation.

For a **complete circuit:** $\nabla \times \vec{B} = 4\pi \vec{J}/c$, so that the contribution of all of the various parts of the source are exactly such that their effect in creating the curl of $\vec{B}$ at a given point can be represented by $\vec{J}$ at that point.

If the circuit is **not complete**, i.e. $\nabla \cdot \vec{J} \neq 0$, then one must account for the "missing" currents. This is done by subtracting off the effect of the missing currents, which can be represented by $\vec{J}_{div} = -\dfrac{1}{4\pi}\dfrac{\partial \vec{E}}{\partial t}$, so that the curl of B at a given point is given by $\vec{J}$ at the point minus $\vec{J}_{div}$: $\nabla \times \vec{B} = \vec{J} - \vec{J}_{div}$

To see this in a simple case, consider eliminating half of an "infinite wire" to get the physical configuration sketched below. This ending of the current, i.e. this diverging of J, clearly means that parts of the current distribution that are needed to maintain the $1/r$ fall-off of B that is characteristic of the long wire are missing. B remains in the azimuthal direction, i.e. circulating around the wire, but, for example, becomes weaker as one moves away from the last part of the wire.

---

[6] Not yet analogically extending that definition to include $\partial \vec{A}/\partial t$

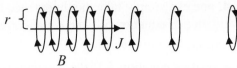

Note: the B-field gets weaker as we proceed from the end of the wire to the right; i.e., B_ϕ decreases.

These changes that result from cutting off half of the wire imply that there is a curl of B in two directions in "empty space." First, the fact that B falls off faster than $1/r$ implies, that the there is a curl of B along the direction of J. Second, the decrease in B along the wire from $1/r$ fall-off to this faster fall-off implies that there is a curl of B along the radial ($\hat{r}$) direction.[7]

In short, divergence in J brings a new possibility because it allows new ways for curls of B to exist.

Rejoining J_{curl} and J_{div}

The above formalism obviously has advantages, not the least of which is its ability to point to the curling part of J, not the diverging part, as the cause of the curl of B. Still, we would like to rejoin the rather unnaturally separated parts of J. One way to accomplish that reunion is to use the forms of equations 5.1 and 5.2 that result from taking the curl of the first and the divergence of the second to yield:

5.1'
$$\nabla \times \vec{J}_{curl} = \frac{c}{4\pi} \nabla \times (\nabla \times \vec{B})$$

5.2'
$$\nabla \cdot \vec{J}_{div} = -\frac{1}{4\pi} \frac{\partial \nabla \cdot \vec{E}}{\partial t}$$

And, in these two equations J_{curl} and J_{div} can, respectively, be then replaced by J. A simpler, more standard way, to accomplish the reunion is to add the two contributions, i.e. add equations 5.1 and 5.2, to get:

5.4
$$\frac{4\pi \vec{J}}{c} = \frac{4\pi \left(\vec{J}_{curl} + \vec{J}_{div} \right)}{c} = \nabla \times \vec{B} - \frac{1}{c} \frac{\partial \vec{E}}{\partial t}$$

Although this equation hides key physical realities that are somewhat manifest in the more explicit forms from which it comes, when coupled with the other Maxwell's equations, it is equivalent to them. However, some things are more evident in this form, as from it we immediately see that even at a point where J is zero, we can have both B and E.

Moreover, at some level, the separation between the two types of J is implicitly maintained even in this reunited J formula. That is, because $\vec{J}_{curl}$ and $\vec{J}_{div}$ do not "overlap," i.e., as we say, they are orthogonal, such a formalism stills keep their distinct effects implicitly separate, and this separation is particularly easy to see for our present limited case:[8] $\vec{E} = -\vec{\nabla}\phi$ since gradients have no curl. For example, it is clear from equation 5.4 that if there is no J divergence anywhere, there is no $\dot{\vec{E}}$ anywhere. Of course, to completely

---

[7] These two new components conspire, in the manner described in Chapter 3, to keep the divergence of the curl of B zero.

[8] Note this separation will not be so clear when we leave our quasi-static construct and allow for finite speed of propagation of the fields. A new term proportional to $\ddot{\vec{A}}$, which can contain both curl *and* div components, will need to be introduced. In addition to radiation effects, this new term will deal with the fact that A-fields corresponding to different states of the source can exist simultaneously in the plana.

verify this separation still exists in equation 5.4, one can simply take the divergence of that equation and note that it is the same as the divergence of equation 5.2, the equation relating how $\vec{J}_{div}$ is represented by changing E-fields caused by the diverging parts of J. And, in similar fashion, taking the curl of equation 5.4 verifies equation 5.1, the equation relating how $\vec{J}_{curl}$ is represented by the curl of B-field caused by *all* the curling parts of J's. Or solving for $\nabla \times \vec{B}$, $\vec{J}_{curl}$ represents *all* the curling parts of J's that cause the curl of B.

Momentary inclusion of Analogically Extended *E*

We now give a peek at the full meaning of equation 5.4 that we will only begin to use in the next chapter when we relax our infinite speed of propagation approximation. Quasi-statically, we formally take: $\vec{E} = -\vec{\nabla}\phi$. However, you will recall that because the magnetic effect of the plana's resistance to a changing A-field is an E-like effect,[9] it is formally useful to extend the definition of E to include this effect, which, as usual, we can

do as long as we are cognizant of its meaning. That is, we usually write: $\vec{E} = -\vec{\nabla}\phi - \dfrac{1}{c}\dfrac{\partial \vec{A}}{\partial t}$.

Thus, the approximation we make in this chapter is that $\left|\vec{\nabla}\phi\right| >> \dfrac{1}{c}\left|\dfrac{\partial \vec{A}}{\partial t}\right|$. When we add this

"$\dot{A}$ effect" back, we will have to say that the curl part of J also represents the effect of the second derivative of curl part of A. This, in turn, ends up implying that the non-diverging part of J at a given point represents the effect of all the curling parts of J in causing the curling of B at that point *only* after one accounts for the effect of the propagation of the A-field.[10] In our empiriometric formalism, in terms of the extended E that incorporates the $\dot{A}$ effect, we write the complete version of equation 5.1:

5.5
$$\vec{J}_{curl} = \frac{c}{4\pi}\nabla \times \vec{B} - \frac{1}{4\pi}\frac{\partial \vec{E}_{curl}}{\partial t},$$

Where $\vec{E}_{curl} \equiv -\dfrac{1}{c}\dfrac{\partial \vec{A}_{curl}}{\partial t}$, so that $\dfrac{1}{4\pi}\dfrac{\partial \vec{E}_{curl}}{\partial t} = -\dfrac{1}{4\pi c}\dfrac{\partial^2 \vec{A}_{curl}}{\partial t^2}$

So we have:

5.5'
$$\vec{J}_{curl} = \frac{c}{4\pi}\nabla \times \vec{B} + \frac{1}{4\pi c}\frac{\partial^2 \vec{A}_{curl}}{\partial t^2}$$

Similarly, equation 5.2 can be extended as:

5.6
$$\vec{J}_{div} = -\frac{1}{4\pi}\frac{\partial \vec{E}_{div}}{\partial t}$$

Where $\vec{E}_{div} = -\vec{\nabla}\phi - \dfrac{1}{c}\dfrac{\partial \vec{A}_{div}}{\partial t}$, so that $\dfrac{1}{4\pi}\dfrac{\partial \vec{E}_{div}}{\partial t} = -\dfrac{1}{4\pi}\dfrac{\partial \vec{\nabla}\phi}{\partial t} - \dfrac{1}{4\pi c}\dfrac{\partial^2 \vec{A}_{div}}{\partial t^2}$

So we can also have:

---

[9] Recall the cause is magnetic, coming from impetus-activated charge, though the effect is like the electric effect.

[10] Using equation 5.6 below, at a point with no current, we can write: $\vec{J}_{curl} = 1/(4\pi)\dot{\vec{E}}_{div}$.

5.6'
$$\vec{J}_{div} = \frac{1}{4\pi}\frac{\partial \vec{\nabla}\phi}{\partial t} + \frac{1}{4\pi c}\frac{\partial^2 \vec{A}_{div}}{\partial t^2}$$

Using equation 5.5' and 5.6', we see that we cannot deduce the value of either part of J without taking into account the finite speed of propagation of the A-field, since its second time derivative will, in general, be changed by such propagation. But, further discussion of this, since it is outside the domain of quasi-statics, properly belongs in the next chapter. You can investigate the conditions under which one can simply take $\vec{E} \sim -\vec{\nabla}\phi$ (i.e., $\vec{\nabla}\phi \gg (1/c)\dot{\vec{A}}$) in a couple of the end of chapter problems.

With this analysis, we now have the complete Maxwell's fourth equation which is usually written as:

5.7

> **4th Maxwell's Equation** $\nabla \times \vec{B} = \frac{4\pi\vec{J}}{c} + \frac{1}{c}\frac{\partial \vec{E}}{\partial t}$

Where: $\vec{E} = -\vec{\nabla}\phi - \frac{1}{c}\frac{\partial \vec{A}}{\partial t}$

Again, without the context provided in this section, this equation can be misleading from a causal point of view, since it can imply, *to the unwary*, that the *full* $\vec{J}$ (i.e. with both J_{curl} and J_{div}) is the cause of curl in B by, for example, leading one to think that the J on the right hand "side stands in" for the contribution from all the *full J* of the source. A similar mistake can also be made with the $\frac{\partial \vec{E}}{\partial t}$ term; both mistakes, of course, arise from not making the *div/curl* decomposition explicit. Sometimes, indeed, in working formally with the theory, it may be convenient to ignore these and other distinctions; this is fine and good as long as we do *not* ultimately forget those realities, recognizing, as always, that convenience does not define reality.

And, we have danced around another much deeper difficulty with the above Maxwell equation which is shared by all of Maxwell's equations. None of them are written as a field in terms of its source; for example, they do not give: $\vec{B} = f(\vec{J})$, (equations that do will be introduced in the next chapter). Even our more explicit versions do not give this direct relation; however, if one brings the correct context, especially through the solutions for the ϕ-field and the A-field in terms of their sources, charge and impetus activated charge, one can avoid significant physical misinterpretations of the equations.

Now, before we dive further into Maxwell's equations in the next chapter, we need to better understand the nature of $\frac{1}{c}\frac{\partial \vec{E}}{\partial t}$ term, which is often called the displacement current, in the quasi-static limit by investigating several examples. These examples will also give us an opportunity to lay out the causal origins of A and B more fully. After some calculational work for each example, we will explain the import of the quasi-static term (displacement term) that we've seen represents the divergent part of J, J_{div}.

Parallel Plate Capacitor

Exacerbating the just mentioned problem of confusing the mathematical formalism with the physics, equation 5.7 is often written as:

5.8
$$\nabla \times \vec{B} = \frac{4\pi \left(\vec{J} + \vec{J}_{displacement} \right)}{c},$$

$$\text{with } \vec{J}_{displacment} \equiv \frac{1}{4\pi} \frac{\partial \vec{E}}{\partial t}.$$

Not only does this ignore the distinction between J_{curl} and J_{div}, it is written as if the changing field is actually a current. Indeed, this term is often treated as if it were a source of *B*-field, i.e., as if it were an impetus activated charge. This goes back to James Clerk Maxwell (1831-1879AD), who thought real currents were responsible for this "displacement current" term, currents in the vacuum, what we call the plana. [11] To dispel this idea and explain the full meaning of this term, we now turn to the parallel plate capacitor for the concreteness and simplicity of principle that this example can provide.

We take the case of a disc-shaped parallel-plate capacitor (see Figure 5-3) fed by long straight wires, called "leads," that are charging the capacitor. This means the left plate is becoming more positive as the current from the left lead dumps charge on it,[12] and the right plate is becoming more negative, as the current going into the right lead takes positive charge from it.

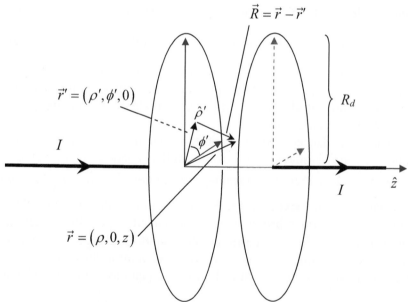

---

[11] Starting with ideas that are not abstract enough (not generic enough because overly specialized and too closely hemmed in by sensorial particulars) is a problem that is hard to avoid if one has not firmly laid out the foundational physica. When one does not have these true starting points of thought (the ten categories, the four types of causality etc. discussed in Chapter 1 of *PFR-M*) articulated in a clear rigorous manner, it is inevitable that one will make analogies that are not well understood and thus also not well thought out.

[12] Here as in other places, we assume for simplicity that we have positive mobile charges, rather than the negative electrons which are the mobile charges in standard wires.

Figure 5-3: Parallel plate capacitor with disc shaped plates fed by long wires. A current I flows into and out of a circular plate capacitor. $\vec{r}\,'$ is the vector to a differential current element in the disk, while $\vec{r}$ is the field point vector.

To understand the fields that are produced, we have to first determine what they are. To do this, we, in turn, first determine the current distribution which is the source of the fields we are interested in.

Current Distribution of Charging Capacitor

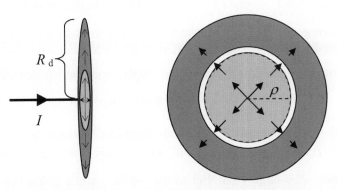

Figure 5-4 Side and face view of the left circular plate showing the current flow pattern. ρ is the arbitrary radius at which one calculates the current that enters and leaves the white annulus on the right.

Using the Figure 5-4 above, we can calculate the current I_{disk} at a given radius, for the steady state, quasi-static (instantaneous propagation) case. The current I moves along the wire perpendicular to the plate. It charges the circular plate uniformly starting from the center. It deposits one annulus of charge after another until the remaining current is used to fill the last annulus whose outer boundary is the boundary of the disk itself. In a given time Δt, charge Δq is put down uniformly on the annulus to give a surface charge of σ everywhere on the surface of the disk, where:

5.9 $\qquad \sigma = \dfrac{\Delta Q}{\pi R_d^{\,2}}$ with $\Delta Q = I\Delta t$, the charge put down on the whole disk in a

time Δt and where R_d is the radius of the disk.

An annulus of thickness $d\rho$ at radius ρ has area $da = 2\pi\rho d\rho$, and the amount of charge put down in this annulus is:

5.10 $\qquad \Delta q = \dfrac{da}{\pi R_d^{\,2}}\Delta Q\,.$

Hence, since the charge accumulated per Δt in the annulus is the current coming in from the inner boundary (radius ρ) minus that leaving the outer boundary of the annulus (radius $\rho + d\rho$), we get, by taking the limit as $\Delta t \to 0$:

5.11 $\qquad I_{disk}\left(\rho\right) - I_{disk}\left(\rho + d\rho\right) = I\dfrac{2\rho}{R_d^{\,2}}d\rho$

Notice that the current is less for the greater radii $(\rho + d\rho)$, because it no longer has to fill the annulus just filled. Integration from ρ to 0, using $I(0) = I$, gives:

$$5.12 \qquad I_{disk}(\rho) = I\left(1 - \left(\frac{\rho}{R_d}\right)^2\right)$$

B-field of Charging Capacitor

From this, one can get the magnetic field using Ampère's law in path integral form:

$$5.13 \qquad \int_C \vec{B} \cdot d\vec{l} = \int_S \frac{4\pi\vec{J}}{c} \cdot d\vec{S} = \frac{4\pi I}{c} \qquad \text{(Recall the differential form is: } \nabla \times \vec{B} = \frac{4\pi\vec{J}}{c}\text{)}$$

Remember the divergence part of J doesn't contribute to B so we can always include it; i.e., we do *not* need to separate out the curl part of J which alone contributes. Also, note that we have omitted the $\dot{\vec{E}}$ term (the so-called displacement term) from the right hand side of equation 5.13. We can do this by picking a surface, S, on which the component of $\dot{\vec{E}}$ perpendicular to the surface is arbitrarily small.

In particular, we choose the cylindrical "open can" surface, S_2, shown in Figure 5-5, taking the radius, ρ, to be arbitrarily small, i.e. $\rho \ll R_d$. Along the sides of the "can" inside the capacitor, $\dot{\vec{E}}$ is largely parallel to the surface, while along the sides outside the capacitor and on the bottom of the can, $\dot{\vec{E}}$ can be made arbitrarily small by decreasing the radius of the can. Applying Ampère's law (in path form) to S_2, we then get:

$$5.14 \qquad \int_C \vec{B} \cdot d\vec{l} = B_\phi 2\pi\rho = \frac{4\pi}{c}\left(I - I_{disk}(\rho)\right),$$

where we take ϕ in the direction shown in the figure. Note that our definition of the circulation in the $\hat{\phi}$ means that the inward normal for S_2 is the positive direction for representing its area.

Hence, we get the standard result:[13]

$$5.15 \qquad B_\phi = \frac{2I\rho}{cR_d^{\,2}}, \qquad \text{for } \rho \ll R_d, d \ll R_d$$

We will see later that this result can be obtained using the displacement current $\vec{J}_D = \frac{1}{4\pi}\frac{\partial\vec{E}}{\partial t}$ as if it were a real current. Before we do, we complete our understanding of the magnetic effect by calculating A.

---

[13] Note that symmetry shows us that there is no ϕ dependence and no z dependence to first order in the idealized limit in which the diameter of the plates is very large relative to the gap between them, d, (i.e., $d/R_d \ll 1$).

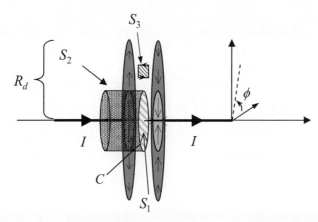

Figure 5-5: Parallel circular plate capacitor shown with cylindrical surface composed of surfaces S_2 and S_1 each of which can be used in Ampère's extended path law. Surface S_3 can be used for calculating the flux of the B-field from the curl of the A-field.

A-field for Capacitor

To understand the particulars of the magnetic field, i.e. the A-field, as well as its causal production, we calculate $\vec{A}$ inside the capacitor near the center, using $\vec{A} = \frac{1}{c}\int \frac{\vec{J}}{|\vec{r}-\vec{r}'|}d^3r'$, which in differential form is $d\vec{A} = \frac{I\,d\vec{l}}{c\,R}$. To do the integration completely, we need, as we did in Chapter 3, a more complete circuit configuration. This time we use the configuration shown in Figure 5-6, which respects, as did the one used in Chapter 3, the cylindrical symmetry.

Figure 5-6: Cylindrically symmetric experimental setup for charging a two parallel circular plate capacitor in the thin plate limit with the return-path coaxial cylinder much further away than the ends[14] ($R_0 \gg L$, $L \gg R_d$, $R_d \gg 2d$, where R_d = radius of disk, while R_0 = radius of hemispherical cap). Note the shift of the x-y plane to the center of the disks which helps in the integrations that follow. We can make the influence of the caps as small as we like by making L and R_0 proportionally larger).

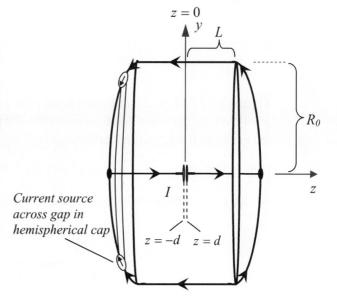

Current source across gap in hemispherical cap

---

[14] One could simply use a plain flat plate at one end and at the other end a flat plate with a concentric thin annulus-shaped current source inserted somewhere between the wire and the coaxial return, but it would be harder to illustrate.

We start by calculating the contribution to the A-field from the long wires, i.e., the leads, on either side of the capacitor. By inspection of the figure, we see that near the capacitor, but outside of it (i.e. outside of the region $-d < z < d$), the current is I along the z-axis. We first calculate the contribution from the current in the right lead:

5.16
$$\vec{A}^{Right} = \int \frac{I\,\vec{dl}}{c\left|\vec{r}-\vec{r}'\right|} = \frac{\hat{z}\,I}{c}\int_d^L \frac{dz'}{\sqrt{\rho^2 + (z-z')^2}},$$

This gives, solving by substitution:

5.17
$$A_z^{Right} = \frac{I}{c}\left(sinh^{-1}\frac{L-z}{\rho} - sinh^{-1}\frac{d-z}{\rho} \right)$$

The contribution of the left lead is then easily found by taking $z \to -z$

5.18
$$A_z^{Left} = \frac{I}{c}\left(sinh^{-1}\frac{L+z}{\rho} - sinh^{-1}\frac{d+z}{\rho} \right)$$

From the graph shown in Figure 5-7 below, which gives the total A-field due to the leads as a function of radius for $z = 0$, we can see that the A-field intensity decreases with radius, as we might expect.

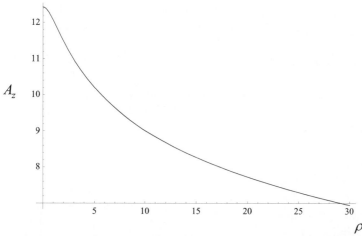

Figure 5-7: Plot of A_z (in-between the plates of a circular plate capacitor, at $z = 0$) due to the leads that feed current to the plates versus ρ, the radius from the center of the capacitor. (arbitrary units)

Noting that $B_\phi = \dfrac{\partial A_\rho}{\partial z} - \dfrac{\partial A_z}{\partial \rho}$, and that the left and right leads contribute the same near $z = 0$, we see that the B-field contributed by the leads then is given by:

5.19
$$B_\phi^{Leads} = -\frac{\partial A_z}{\partial \rho}\bigg|_{z=0} = \frac{2}{\rho^2}\frac{I}{c}\left(\frac{L}{\sqrt{1+\left(\frac{L}{\rho}\right)^2}} - \frac{d}{\sqrt{1+\left(\frac{d}{\rho}\right)^2}} \right),$$

If we stay sufficiently inside the capacitor as we've decided, i.e. $\rho << R_d$, so that (recalling, $L >> R_d$), $\rho << L$, and expand to first order, the *first term in equation 5.19* gives $\frac{2I}{\rho c}$. The *second term in equation 5.19* is more complicated but we will only investigate it in the region $\rho >> d$ where the leading term in the expansion for the second term is $-\frac{2I\,d}{\rho^2 c}$ which yields, using both terms, the approximate field from the leads near $z = 0$:

$$5.20 \qquad B_\phi^{\,Leads} = \frac{2I}{\rho c}\left(1 - \frac{d}{\rho}\right) \sim \frac{2I}{\rho c}. \quad \text{For } \rho >> d \text{ (the largest region).}$$

The calculation of the A-field inside the capacitor due to the currents in the *disks* will yield an A-field with only radial components and will be a function of z and ρ; i.e. it will yield only $A_\rho(\rho, z)$. The exact calculation of this field (see end of chapter problem) is more complex than that for the leads. However, we already know that the B-field due to the disk must cancel those due to the leads so that the total B-field will vary according to equation 5.15. Notice that this total field is much smaller than the field due to the leads alone:

$$5.21 \qquad \frac{B_\phi^{\,Leads}}{B_\phi^{\,Total}} = \frac{\dfrac{2I}{c\rho}}{\dfrac{2I}{cR_d}\left(\dfrac{\rho}{R_d}\right)} = \left(\frac{R_d}{\rho}\right)^2 >> 1 \qquad \text{for } \rho >> d$$

This means the job of the *disk's* A-field is to cancel out enough of the curling of the A-field caused by the *leads*, i.e. cancel out enough of the B-field due to the leads, to leave only the total $B_\phi^{\,Total}$. Mathematically, we write $\left|B^{Leads}\right| - \left|B^{Disks}\right| = \left|B^{Total}\right|$.

Notice, that the circulation of A, i.e. the B-field, due to the *leads* is accomplished by gradients of the z component of A in the ρ direction, while the circulation of A due to the *disks* is accomplished by gradients of the ρ component of A in the z-direction. Mathematically, in the ϕ component of the curl of A given by $B_\phi = \dfrac{\partial A_\rho}{\partial z} - \dfrac{\partial A_z}{\partial \rho}$, the first is the "*disks*" term, and the second is the "*leads*" term. This brings out again the analogical abstraction involved in the B-field, which, by itself, cannot specify some basic aspects of the source; as we've said, there are multiple ways to make the same *local* B-field. In this sense, the A-field carries more detail about the source. Again, in this capacitor example, the A-field carries information about the disks and the leads that the local B-field alone does not tell us.[15] We need the A-field to understand the causal situation.

---

[15] Using 1) the Lorenz gauge that we will discuss shortly, 2) knowledge of B in *all* of the plana, and 3) $\nabla \cdot \vec{B} = 0$, for localized sources and fields, is equivalent to specifying A within a constant; however, once we have A we have the "extra" information *explicitly*.

A-field and the Meaning of Displacement Current
Parallel Plate Capacitor

The analysis above manifests that the A-fields are being generated by the currents, charges activated by impetus, and not by changing electric fields as might be thought if one interpreted the displacement current, $\vec{J}_D = \dfrac{1}{4\pi}\dfrac{\partial \vec{E}}{\partial t}$, as a literal or even an analogical current in Ampère's law. The changing electric field is caused by a changing distribution of ability to activate electric field, i.e. a changing charge distribution. The magnetic field is caused by charges with impetus. Now, because the total charge (the measure of the intensity of the capacity for activating impetus), as given by Gauss's law,[16] is conserved, a changing distribution of a charge, and hence a changing electric field, can only occur when impetus is activated in a charged massive body, which means that body produces an A-field in the plana that, in general, has a curl, i.e. creates a B-field.

We now show that the changing charge distribution in the parallel plate capacitor results in a changing electric field that is exactly proportional to the diverging current that serves as a stand in for the currents that cause the B-field inside the capacitor ($B_\phi = \dfrac{2I\rho}{cR_d^{\,2}}$).

Referencing the geometry given in Figure 5-5, using Gauss's law and symmetry, we can write the electric field inside the capacitor as:

5.22 $$\vec{E} = 4\pi \frac{Q}{\pi R_d^{\,2}}\hat{z} = \frac{4Q}{R_d^{\,2}}\hat{z}$$

We then get $\vec{J}_D = \dfrac{1}{4\pi}\dfrac{\partial \vec{E}}{\partial t} = \dfrac{1}{\pi R_d^{\,2}}\dfrac{\partial Q}{\partial t}\hat{z}$, which yields:

5.23 $$I_{Displacement} = J_D \pi \rho^2 = I\left(\frac{\rho}{R_d}\right)^2$$

This is the same as the net current ($I - I_{disk}$) referenced in equation 5.14. That is, we get the same result for the right hand side of equation 5.14 if we use surface S_1 with J_d instead of S_2 with real currents. From Figure 5-5, one can see that both surfaces are bounded by the same curve C inside the capacitor and thus applying equation 5.14 yields, as it must, the same B-field for both cases.

Thus, the effect of the currents causing the A-field (and thus the B-field) can be calculated either directly from the charge motion or from its effect on the electric field. Again, we see that, because of charge conservation, a changing electric field means there must be moving charged particles, i.e. currents. This, in turn, means there will be an A-field and generally also a B-field, because only exceptional current distributions have no curling components. In our capacitor example, which can be loosely characterized as a long wire with a gap in it, there will be a magnetic field in the gap from the currents outside the gap. Now, the gap in the current means that there will be charge build-up on either side of the gap, and so, as a result, there will be an electric field in the gap. Again, clearly, the

---

[16] It is an experimental fact at our level of analogical abstraction that Gauss's law works also for moving charges; i.e., the charge calculated via a surface integral around the charge does not depend on the surface over which one integrates the electric field.

azimuthal *B*-fields around a current element will not end abruptly at such a gap because the currents that cause the *A*-fields are close by.

We could also quickly determine *B* for the capacitor interior by immediately invoking the results of the analysis in the introduction to this chapter, i.e., the split of *J* into curl and divergence components. Recall that the displacement current, in the quasi-static case, is more properly thought of as the negative of the divergent part of *J*, i.e. $-J_{div}$. This is manifest in our capacitor case as the changing electric field pattern has no curl, and it is generally manifest, as we pointed out earlier, when one writes the term as a time rate of change of the gradient of the potential. Now, though $\dfrac{1}{4\pi}\dfrac{\partial \vec{E}}{\partial t}$ is not an actual current, it does represents the effects of an aspect of the current; in a generic way, it represents the sum total of the effects of the currents that are missing. Recall that if there were no divergences in *J*, the curl of *B* at a given point would be given by a quantity proportional to *J* at that point, so that there would be no curl of *B* in empty space.[17] As we also saw previously, in empty space, the curl of *B* is generally given by the negative of the divergence term, i.e. exactly what we've called the displacement term. Again, when we move beyond quasi-static sources and fields in the next chapter, we will see a further source of curl of *B* in empty space arises from the analogically defined "part" of *E,* i.e., $\vec{E} \sim -\dot{\vec{A}}$, which does not represent the effect of currents (or absence of currents) but, represents a separate cause of curling *B* beyond what the curl of *J*'s themselves cause. We will see this cause is the finite speed of propagation of changing *A*-field.

The parallel plate capacitor is a simple example that helps us understand *A* and *B* as well as this last quasi-static effect, i.e. the effect due to charge build-up. Because this effect helps us unify all we've learned up to this point by giving deeper access to the full Maxwell equations, we consider two more examples which focus even more closely on the "displacement current."

A Current Element and an Exploding Ball

Finite Current Element

We first consider the case of a finite current element such as shown in Figure 5-8. It satisfies charge conservation by starting with an accumulation of positive charge at the source on the left end and an accumulation of negative charge at the sink at the other end of a long straight wire.

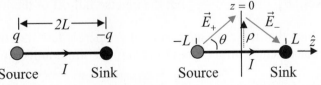

Figure 5-8a. (*left*) Single wire current feed by a source of positive charge on the left and sink on the right. **b.** (*right*) This diagram is for use when integrating contribution of *E* field over the center plane; one needs to project *z*-component out, for the other component cancels due to the symmetrical nature of the fields.

---

[17] Remember that by "empty" we simply mean empty of all but plana *only* qualified by the very *B* and *E*-fields of which we speak.

Using Ampère's extended law that includes both conduction and "displacement current," we here show that (1) the B-field approaches that of a long wire as the charge separation approaches infinity and (2) the total "current" (displacement + conduction), with fixed separation, approaches zero as the radius of observation approaches infinity, unlike for the infinite long wire construct.

The electric field on the plane passing through the center of the two charges and perpendicular to the line connecting them, which we take to be the z-axis, is:

$$\vec{E} = \frac{2q\cos\theta}{r^2}\hat{z} = \frac{2q}{r^2}\frac{L}{r}\hat{z} = \frac{2qL}{\left(\sqrt{L^2+\rho^2}\right)^3}\hat{z}\,.$$

Integrating the electric flux over a disk of radius ρ_0 then gives:

$$\Phi_E = \int_{disk}\vec{E}\cdot d\vec{a} = \int_0^{\rho_0}\frac{2qL}{\left(\sqrt{L^2+\rho^2}\right)^3}2\pi\rho d\rho = 4\pi Lq\int_0^{\rho_0}\frac{\rho}{\left(L^2+\rho^2\right)^{3/2}}d\rho = 4L\pi q\left(\frac{1}{L}-\frac{1}{\sqrt{L^2+\rho_0^2}}\right)$$

Thus, the extended Ampère's path law: $\displaystyle\oiint\nabla\times\vec{B}\cdot d\vec{A} = \oint\vec{B}\cdot d\vec{l} = \frac{4\pi}{c}\oiint(\vec{J}-\vec{J}_{div})\cdot d\vec{A}$

yields for:

Case (1):
The limit in which the separation between the charges, L, approaches infinity yields:

$$Lim_{L\to\infty}\left(B_\phi\cdot 2\pi\rho_0 = \frac{4\pi}{c}\left(I - \frac{1}{4\pi}\frac{\partial\Phi_e}{\partial t}\right)\right) = \frac{4\pi}{c}I\,,$$

$$\text{because } I_d \equiv \frac{1}{4\pi}\frac{\partial\Phi_e}{\partial t}\to 0$$

So, we get the standard result for a long wire B.

Case (2):
To *first order*, when the distance, ρ_0, from which we measure the field approaches infinity we get:

$$Lim_{\rho_0\to\infty}\left(B_\phi\cdot 2\pi\rho_0 = \frac{4\pi}{c}(I-I_d)\right)\sim\frac{4\pi}{c}\left(\frac{L}{\rho}\right),\quad\text{since}\quad I_d\equiv\frac{1}{4\pi}\frac{\partial\Phi_e}{\partial t}\to I\left(1-\frac{L}{\rho}\right)\quad\text{with}$$

$I = \frac{\partial q}{\partial t}$. This gives: $B_\phi\propto\frac{2I}{\rho c}\frac{L}{\rho}\propto\frac{1}{\rho^2}$, *not* the inverse fall-off of the long wire, but a much

faster one. Note also that the right hand side of Ampère's law, i.e. the sum of the actual current plus "displacement" current, goes to zero in the limit, unlike in the case of the long wire.

To better understand these results, draw circles centered at various places along the source, as shown in the figure below, and determine what the divergence is at each point by asking whether or not there is more current coming into a given circle than comes out. Only at the ends, i.e. regions A and E, is there a divergence in $\vec{J}$.

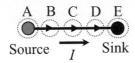

Thus, in terms of the decomposition of J into J_{div} and J_{curl}, we see the following.

The *first* limit obtains, because as we push the ends apart, the divergence part, for which the changing E, the "displacement current," is used as a stand-in, moves to infinity where it cannot play a role, we are left with only $\vec{J} = \dfrac{c}{4\pi}\nabla \times \vec{B}$. That is, within a distance that can affect the field point, there are no missing circuit elements so that the curl of B becomes closely proportional to the value of J at the given point, i.e. zero in empty space.

In terms of Ampere's law, $\oint \vec{B} \cdot d\vec{l} = \dfrac{4\pi}{c} \oiint (\vec{J} - \vec{J}_{div}) \cdot d\vec{A}$, in which the line integral on the right measures the total circulation of the B-field by adding up all the contributions of the curl of B within the area under consideration, in the limit only J contributes to the total. This is because the divergent part of J at each point in the area (represented by the changing E-field, which in turn represent the part of J that over estimates the effect of all the curls of J in the source) is zero in the limit.

In the *second* case, we move so far away that the short current cannot contribute much, so that we have minimal net B-field. Said another way, we cease to see the effect of the current because, at great distance, the source and the sink appear nearly on top of one another. By contrast, if we remove the divergent parts (whose effect on the curl of B at a point are represented by the changing E-field at that point), the distant edges of the wire would exist and thus would continue to contribute, making a larger B-field at the given distance (quantitatively, the fall-off would be $1/\rho$ instead of $1/\rho^2$). In short, when we are close to the finite wire, the missing curling elements of the "infinite" current are not "felt," but as we move away, they are felt more and more.

In terms of Ampere's law, close to the wire, the contributions to the net circulation given by the line integral of B are closely represented by the current that pierces the planar surface contained by the line integral. Only a little correction needs to be made due to the missing ends, and this appears as the small value surface integral of $\dot{\vec{E}}$. However, the correction, i.e. what is subtracted off in that surface integral of $\dot{\vec{E}}$, approaches the value of the current piercing the surface as we move away, expanding the area of integration, thereby adding more and more of the non-zero curl of B in empty space. Of course, for the infinite wire construct, the curl of B would zero in empty space, so that expanding the area of integration, in that case, should and does add nothing.

Exploding Ball

Lastly, we consider the thought experiment introduced into a freshman textbook by Feynman, i.e., an exploding ball of charge. Assume a positively charged ball initially has a size R and then (quasi-statically) expands outward isotropically.

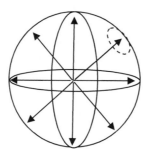

Figure 5-9: Isotropically exploding ball of positive charge.

To see the role of the displacement current as "pointing" to real currents, first consider the A-field at any point around the ball by using the formula: $\vec{A} = \int \frac{I\, d\vec{l}}{c\,R}$. From the symmetry, the A-field must be radially directed, and it can only vary in the radial direction. That is, A can vary neither in azimuth nor altitude angle (ϕ nor θ), because all points at the same radius are causally influenced in the same way. Such a uniform radial A-field cannot produce a B-field, because it has no circulation. Hence, under our approximations, B must be zero.

But, you might say, doesn't Ampère's path law predict that there must be a non-zero B-field, because every circle, such as drawn in Figure 5-9, is threaded by a real non-zero current? Of course, the answer is, as we saw in the capacitor example, that there are nearby currents that also contribute to the curling part of the A-field; so, the effect of these currents must also be taken into account. These currents can be accounted for (or "pointed at") through the electric fields they cause by the use of $\vec{J}_D \equiv \frac{1}{4\pi} \frac{\partial \vec{E}}{\partial t}$. To see this, take a spherical Gaussian surface, S, of radius R and take Q as the charge inside S at a given time, to get:[18]

5.24 $\qquad \frac{\partial Q}{\partial t} = \frac{\partial}{\partial t}\left(\frac{1}{4\pi} \int_S \vec{E} \cdot d\vec{a} \right) = \int_V \left(\frac{1}{4\pi} \frac{\partial \vec{E}}{\partial t} \right) \cdot d\vec{a}$,

Substituting the definition of displacement current above, gives:

5.25 $\qquad \frac{\partial Q}{\partial t} = \int_V J_D \cdot d\vec{a}$

In other words, the displacement current is the rate of change of charge in the region; this rate is negative, since it is decreasing. Obviously, the current, I, is equal in magnitude but opposite in sign to the rate of change of the charge in the region. Since the current is uniform across the surface, we can easily see that: $J_{Displacement} + J_{conduction} = 0$; hence, the expanded Ampère's law correctly predicts that there is no B-field.

In terms of the decomposition of J into J_{div} and J_{curl}, we see that there are no real currents that have a curl, so that $\vec{J}_{curl} = 0$, which means, via equation 5.1, $\nabla \times \vec{B} = 0$, so

---

[18] Remember, that the surface integral of the field, through Gauss's Law, is our definition of charge; hence, again, the functional equivalence of a changing field to a changing charge should not be such a surprise.

that $B = 0$. The only currents are the divergence currents which can be represented by $\dfrac{1}{4\pi}\dfrac{\partial E}{\partial t}$.

More Formal Derivations of the Displacement Current

Displacement Current from Biot-Savart and Conservation of Charge

There are several ways to formally introduce the displacement current. While the empiriometric consistency of Maxwell's equation is usually emphasized in the discussion of displacement current, since we are doing physics in the full sense, we start with the causal statement implicit in the equation for the A-field as a function of the current:

$\vec{A} = \dfrac{1}{c}\displaystyle\int \dfrac{\vec{J}}{|\vec{r} - \vec{r}'|}d^3r'$. Taking the curl of that equation while keeping in mind our quasi-

static condition that we neglect speed of travel of the fields, we can get the standard Biot-Savart law (see problem at end of chapter):

5.26 $$\vec{B} = \int \dfrac{\vec{J}\times\hat{R}}{cR^2}dV$$

Then, you can prove the generalized Ampère's law (in differential form) known also as 4th Maxwell equation (equation 5.7) by taking of the curl of equation 5.26 and incorporating

$\nabla\cdot\vec{J} = -\dfrac{\partial\rho}{\partial t}$, rather than, as we do in the purely static case, $\dfrac{\partial\rho}{\partial t} = 0$, (see end of chapter

problem).

Displacement Current from Ampère's Force Law and Conservation of Charge

A second approach to the "displacement term" that begins more directly from experiment will give further insight. We start, with the result of the Ampère two wire experiments (i.e., Ampère's force law) and use the force law for the magnetic field to deduce the magnetic field of a long wire. We then, following the arguments of Chapter 3, deduce that any path integral of the magnetic field around such a wire is related to the current threading the path in the following way:

5.27 $$\int_C \vec{B}\cdot d\vec{s} = \dfrac{4\pi}{c}I$$

As we mentioned in Chapter 3, analysis verified by experimental evidence reveals that we can extend this law to arbitrary shaped wires. In Chapter 3, we assumed that there was no charge build-up, i.e., $\nabla\cdot\vec{J} = 0$, and this meant we could replace the above equation with:

5.28 $$\int_C \vec{B}\cdot d\vec{s} = \dfrac{4\pi}{c}\int_S \vec{J}\cdot d\vec{a}$$, where S is *any* surface bounded by C.

However, if there is charge build-up, then we need to use the full conservation

equation, $\nabla\cdot\vec{J} = -\dfrac{\partial\rho}{\partial t}$, if we want the equation to work for *any* surface bounded by C, for,

in this case, certain surfaces will not thread a current. The only way equation 5.28 above

can be maintained is if we extend "$\vec{J}$" in equation 5.28 to somehow account for $\dfrac{\partial\rho}{\partial t} \neq 0$.

To do this, we rewrite the integral form of Ampere's path law as follows. We imagine starting with any surface, S, bounded by a curve C, and then we supply a second surface, called a "closing surface," which when coupled with the surface S, creates a *closed* volume of space. We use the convention that the normal to the area points *inward*:

5.29 $\quad \int_C \vec{B} \cdot d\vec{s} = \frac{4\pi}{c} \int_S \vec{J} \cdot d\vec{a} = \frac{4\pi}{c} \left(\int_{S+\text{Closing Surface}} \vec{J} \cdot d\vec{a} - \int_{\text{Closing Surface}} \vec{J} \cdot d\vec{a} \right).$

$\qquad = \frac{4\pi}{c} \left(-\int_V \nabla \cdot \vec{J} \, dV - \int_{\text{Closing Surface}} \vec{J} \cdot d\vec{a} \right)$

Where, in this last line, we have used the divergence theorem and the minus sign arises in the first term because of our inward definition of $\widehat{da}$.

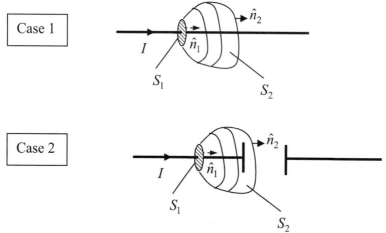

Figure 5-10: ***a.*** (case 1) Arbitrarily shaped closed surface (which is composed of two surfaces, S_1 and S_2) around a continuous wire of constant current. ***b.*** (case 2) Arbitrarily shaped closed surface around a wire that breaks, feeding a capacitor plate.

Using Figure 5-10a and b above with $S = S_1$ and S_2 is the "closing surface" of S, we see that in *case 1*, where $\nabla \cdot \vec{J} = 0$, equation 5.29 gives:

5.30 $\quad \int_C \vec{B} \cdot d\vec{s} = \frac{4\pi}{c} \int_{S_1} \vec{J} \cdot \hat{n}_1 d\vec{a} = \frac{4\pi}{c} \int_{S_2} \vec{J} \cdot \hat{n}_2 da$, where, in the last equality, we have

used the fact that the normals to the differential surfaces in S_1 are defined positive inward, but those in S_2 are defined positive outward, so as to agree with the right handed rule applied to the actual B-field circling around the wires. For *case 2*, where $\nabla \cdot \vec{J} = -\dfrac{\partial \rho}{\partial t}$,

again taking into account the sign of the area normal, equation 5.29 becomes:

$$\int_C \vec{B} \cdot d\vec{s} = \frac{4\pi}{c} \int_{S_1} \vec{J} \cdot \hat{n}_1 da$$

$$= \frac{4\pi}{c} \left(\int_V \frac{\partial \rho}{\partial t} dV + \int_{S_2} \vec{J} \cdot \hat{n}_2 da \right)$$

5.31

$$= \frac{4\pi}{c} \left(\frac{1}{4\pi} \frac{\partial}{\partial t} \int_{S_1 + S_2} \vec{E} \cdot \hat{n} da + \int_{S_2} \vec{J} \cdot \hat{n}_2 da \right)$$

$$= \frac{4\pi}{c} \left(\int_{S_2} \frac{\partial \vec{E}}{\partial t} \cdot \hat{n}_2 da \right)$$

Here, in the third equality, $\hat{n}$ is the normal pointing *outward* from a differential element of the closed surface formed by S_1 and S_2.

In the second equality, we used the conservation equation; in the third equality, we used Gauss's law, $\nabla \cdot \vec{E} = 4\pi\rho$ and the divergence theorem; in the last, we substituted the particular values for our capacitor problem, i.e. $\vec{E} \sim 0$ everywhere on S_1 and $\vec{J} = 0$ everywhere on S_2. This shows that, unlike *case 1*, in *case 2* we cannot freely switch surfaces without a revision in our formula. Namely, we need to add the "displacement current" term if we are to be able to switch between surfaces in the way we wish; in particular we need:

5.32
$$\int_C \vec{B} \cdot d\vec{s} = \frac{4\pi}{c} \int_S \left(\vec{J} + \frac{1}{4\pi} \frac{\partial \vec{E}}{\partial t} \right) \cdot d\vec{a}$$

Indeed, we need this surface independence to apply Stokes theorem to move from the integral form to the more standard differential form:

5.33
$$\nabla \times \vec{B} = \frac{4\pi \vec{J}}{c} + \frac{1}{c} \frac{\partial \vec{E}}{\partial t} \, .$$

We can physically understand the particular result in equation 5.31 by recalling that, in empty space, the $\dot{\vec{E}}$ term represents the effect of the missing curling currents that would be present were there a complete circuit. That is, because they are "missing," there is a curl of B in empty space at each point that must be integrated (over the entire area surrounded by the closed curve, C), i.e. $\int_S \left(\nabla \times \vec{B} \right) \cdot d\vec{S}$, to get the net "circulation" of the B-field around C, i.e. $\int \vec{B} \cdot d\vec{l}$, in the finite region.

A Simple Formal Derivation of the Displacement Term

Lastly, we come to the direct formal consistency of the equations. We could, following the logic developed in our introductory section, see the need for the displacement term by simply looking at the static form:

5.34
$$\nabla \times \vec{B} = \frac{4\pi \vec{J}}{c}$$

Clearly, J cannot have a divergence, but must be written as in equation 5.1. Thus, if J has a divergence, i.e. if $\nabla \cdot \vec{J} = -\dot{\rho} \neq 0$, we must add another term for this equation to continue to be valid. This term must by Gauss's law equal the displacement term given above.

Thus, we have made our way again back to the fourth Maxwell's equation, first discovered by the great physicists James Clerk Maxwell (1831 -1879AD), see history box below. These equations are all together:

$$I. \quad \nabla \cdot \vec{E} = 4\pi\rho \qquad\qquad II. \quad \nabla \cdot \vec{B} = 0$$

5.35 I-IV

$$III. \quad \nabla \times \vec{E} = -\frac{1}{c}\frac{\partial \vec{B}}{\partial t} \qquad IV. \quad \nabla \times \vec{B} = \frac{4\pi\vec{J}}{c} + \frac{1}{c}\frac{\partial \vec{E}}{\partial t}$$

with the force law: $\quad \vec{F} = q\vec{E} + q\dfrac{\vec{v}}{c} \times \vec{B}$

Displacement and Divergent Current Summary

In the course of our examples and analysis, we have seen the role of the new ("displacement") term and its importance as a sign of more complex currents that arise as charges are built up. We have seen in each example that the changing divergence part of the E-field, $\dfrac{1}{4\pi}\dfrac{\partial \vec{E}_{div}}{\partial t}$, which is the only important term quasi-statically, represents, stands in the place of, real currents. Thus, it makes sense that this displacement term is the negative of divergence part of $\vec{J}$, i.e. $\vec{J}_{div}$, and therefore cannot *directly* represent $\nabla \times \vec{B}$ per equation 5.1: $\nabla \times \vec{B} = \dfrac{4\pi}{c}\vec{J}_{curl}$. However, as we have said, $4\pi\vec{J}_{div}/c$ at a given point does represent the correction to the case in which one imagines completing the circuit (i.e. so that there is no diverging part of J) such as shown in the figure below. For any complete (closed) circuit, $4\pi\vec{J}/c$ at any given point gives the magnitude and direction of the curl of B at that point, and that curl of B is caused by all the curls of J in the circuit. $\vec{J}_{div}$ reveals how much of J needs to be subtracted off to yield the effect of all of the curling parts of J as it actually is, i.e., without the imagined parts that close the circuit. Now, you may ask: Doesn't the effect of the various curls of J that are missing depend on the details of what's missing? That is, doesn't the correction, $\vec{J}_{div}$, depend on how we imagine that we've closed the current path? Concretely wouldn't it depend on whether we start with the circuit shown below on the far right or the one in the middle? No, it does not. It only depends, as it must, on the source configuration as it actually is (far left below).

Two possible imagined completions of the "circuit" shown on the left

This is where we see the amazing effect of the nature of the A-field coupled with conservation of charge. Namely, as we have seen, a circuit can be closed in *any* way; as long as it is closed, the current at a given point will be proportional to (thus in someway represent) the net curl of B due to *all* of the J's in the source. Nevertheless, the value of the current, and hence the curl of B, obviously does depend on the circuit. If we consider eliminating part of a given closed circuit, to create source with a divergent J, it *does* matter what part of a completed circuit we leave; obviously, what is left, not what was taken away, will determine B. But, because part has been eliminated, the fancy cancellation accomplished by those parts, which was responsible for $\nabla \times \vec{B} = \dfrac{4\pi}{c}\vec{J}$ at any given point, will no longer occur. Moreover, there are an infinite number of possible source configurations besides the one that actually did close the circuit. This shows that *only* some generic information about what is missing is needed to find out what part of the curl of B is no longer generated so that one may subtract it off from that which would be generated were the circuit complete. It depends on the "end points" of the current flow, i.e. on the places where $\nabla \cdot \vec{J} \neq 0$, whose consequences are reflected in the mathematical device: J_{div}. Formally, we write: $\nabla \times \vec{B} = \dfrac{4\pi}{c}\left(\vec{J} - \vec{J}_{div}\right)$. The correction effect, written as $(4\pi / c)\vec{J}_{div}$, is given by the time rate of change of the electric field at the point of interest. As explained, this makes intuitive sense as one expects the E-field that accompanies the build up in charge to give some generic indication of the parts of the circuit that are "missing."

We should also note the important role of symmetry in our calculations in this and other chapters and in E&M generally. In particular, for the sources we've discussed in this chapter (which all are the "remaining" part of an imagined completed circuit) are handled implicitly by incorporating the symmetry of the source, especially through the use of Gauss's and Stokes theorems.

Now, as long as we consider the E field in the above quasi-static way, the full "displacement" term, $\dfrac{1}{4\pi}\dfrac{\partial \vec{E}}{\partial t}$ really is important in some cases, but, as we have seen it doesn't introduce any radically new phenomena. It's when we leave this quasi-static world and take into account the finite speed of travel of the field that the real import of the new term is evident. In the next chapter, we will see how the addition of the displacement current term naturally leads us to the finite speed of propagation of electric and magnetic field that we have till this point ignored. Before we can move to this and other related issues, we need to better understand A and ϕ, especially their interrelation.

James Clerk Maxwell (AD 1831 -1879) was born June 13 in Edinburgh Scotland. He is the true "Newton of Electricity and Magnetism" in the sense of bringing E&M to empiriometric unity with his four Maxwell's equations. Einstein considered this the model of a good theory. It surpassed Newton's theory in its empiriometric level. Like Newton, he "stood on the shoulders of giants." His "Maxwell's equations" broadened and deepened the use of equations in physics, thereby doing the same for the empiriometric method itself. He published that work in *A Treatise on Electricity and Magnetism* in 1873. And, as if this massive achievement were not enough, he had also contributed fundamentally to thermodynamics in 1871, establishing what are now called the four "Maxwell's relations." He made ground breaking contributions to statistical physics and other areas, including presenting the world's first color photo in 1861 and writing on color perception and color blindness.

Maxwell had a keen mathematical/scientific interest at a young age and discovered the regular polygon on his own. He won multiple academic prizes while still growing up.

In a time when empiriometric physics was more quickly than ever before separating from its natural philosophical origin and aims, Maxwell kept some of the ideal of Newton and his medieval predecessors that began with Aristotle that science was about understanding the world, not just describing it.[1] And, he thought our knowledge really advanced in this regard. He said "It was a great step in science when men became convinced that, in order to understand the nature of things, they must begin by asking, not whether a thing is good or bad, noxious or beneficial, but of what kind it is? And how much is there of it? Quality and Quantity were then first recognised as the primary features to be observed in scientific inquiry."

Maxwell also shared the medieval interest in the Bible verse "But by measure and number and weight, Thou didst order all things" [Wisdom 11:20] (one of the most quoted verse in medieval times), when he linked the important idea of atoms to that verse. Maxwell's motivation in his work can be seen in his various statements such as: "I believe with the Westminster Divines and their predecessors *ad Infinitum* that 'Man's Chief End is to glorify God and to enjoy him for ever.'"

Maxwell was a devout Christian, taking each Sunday to study theology. He also said, "Each individual man should do all he can to impress his mind with the extent, the order, and the unity of the universe, and should carry these ideas with him as he reads [the Bible]." Maxwell considered his scientific study of the world as an act of worship. He respected the Bible, treating it as calling for and confirming the good of rigorous, exact experimental science, not substituting for it, as some in his day tried to get him to do.

[1]Note, Maxwell did overspecialize too quickly to a mechanical analog for the plana, as seen with his interpretation of the displacement current, and this tendency to not start generic enough, letting nature tell us the details, has plagued physics and driven some number further into the belief that mathematical description is sufficient for physics. Despite this, concrete models are often the road to deeper *physical* insight if their limits and unproven aspects are not allowed to recede from sight.

The A and ϕ Potentials and Their Interrelation

Gauge Fixing: the Lorenz Condition

In Chapters 2 and 3, we discussed the potentials as local specifications of the electric and magnetic fields. In Chapter 3, we found that to obtain an equation for A we needed another condition that empiriometrically connects A and ϕ, a so called gauge condition. This word "gauge" has a somewhat complex history that does relate to "gauge's" meaning as "measured length," but since further amplification would bring in topics beyond the scope of this textbook, we will leave the history there. We turn now to (1) understand the Lorenz gauge condition that we introduced earlier and (2) to further penetrate the related issue of the evidence for taking A and ϕ as representing real aspects of the physical world.

We start with an investigation of the Lorenz condition itself. Recall that the Lorenz gauge, in *SI* (*mks*) is, in differential form:

5.36a,b $\qquad \vec{\nabla} \cdot \vec{A} + \dfrac{1}{c^2}\dfrac{\partial \phi}{\partial t} = 0 \qquad\qquad$ (in *cgs*, $\nabla \cdot \vec{A} + \dfrac{1}{c}\dfrac{\partial \phi}{\partial t} = 0$)

In this section, we show the analogical conservation law that this entails. In particular, note that equation 5.36 is very much like the equation of charge conservation:

5.37a $\qquad \nabla \cdot \vec{J} = -\dfrac{\partial \rho}{\partial t}$,

which in integral form is:

5.38b $\qquad \displaystyle\int \vec{J} \cdot d\vec{a} = I_{exiting} = -\dfrac{\partial}{\partial t}Q$.

It simply describes the fact that as charge flows out of a given region, the amount of charge in that region must decrease by exactly that much. The Lorenz gauge can be viewed in this light if the proper analogies are assumed. In integral form, the Lorenz gauge in *SI* (which bests manifests the potential momentum meaning of A), i.e. equation 5.36a, is:

5.39 $\qquad \displaystyle\int_V \nabla \cdot \vec{A}\, dV = \int_A \vec{A}\cdot d\vec{a} = -\dfrac{1}{c^2}\int_V \dfrac{\partial \phi}{\partial t} dV$.

We saw in Chapter 3 that A and ϕ are, respectively the potential momentum and energy per unit charge. Hence, for a positive test charge of *unit* magnitude, A is the potential momentum and ϕ is the potential energy, and we can introduce a more suggestive notation: $\vec{A} = \vec{p}_{pot}, \phi = E_{pot}$. Note that to make the *SI* units work out correctly one should imagine that there is a $q = 1C$ multiplying the left hand side of each equation.[19]

To proceed with our interpretation of the Lorenz condition, we need an actual speed of propagation, not our being-of-reason "infinite speed."[20] For this, we, in turn, need a brief digression.

In the next chapter, we will see that including the Lorenz condition leads to the conclusion that A and ϕ (and hence E and B) propagate at the same speed c, the speed of light. This fact that the Lorenz gauge does not introduce any other speeds can be taken as

---

[19] Of course, by altering the units, we could make the unit charge as small as we want, e.g., many orders of magnitudes smaller than a coulomb.

[20] One can articulate the functional equivalent of the infinite speed assumption in the language of approximation by saying that the speed is "so fast it doesn't matter;" yet, this just avoids the issue by pushing off the question: "what is its speed?"

an indication of the correctness of our choice, for all other (finite speed) gauge conditions (which are, indeed, mathematically possible, in the sense of consistent with Maxwell's equations) give multiple speeds of propagations. Now, we should not assume a proliferation of speeds without evidence. In addition, evidence codified in special relativity indicates a special role for c. In short, there is no reason to hypothesize a speed other than c unless evidence forces it. Furthermore, in the same vein, and which is a reason why many physicists prefer the Lorenz condition, it is the only gauge condition that is of a special type of equation called Lorentz (note the "t") invariant, which means its mathematical form is the same in every uniformly moving frame, independent of its velocity (under the conditions required for special relativity).

Before we continue our main thread, recall that even statically the field propagates, in a different sense, because one part of the field has to be maintained in existence by the part next to it nearer to the source. Also, recall that for fields that move at speed c only, such as light, the energy, the rate of transfer of the impetus, is: $E = pc$.[21]

Now we will show, using the analogy to equation 5.37, that the Lorenz condition is reasonably interpreted as a kind of conservation of field energy, which, in turn, is reflective of the potential energy of the system of charges.

Remembering that the potential momentum is understood by what actual momentum it can cause, to determine the flow of impetus (i.e., to get the energy), we have to decide the mass of the test particle that will receive the impetus. Now, the A field and the ϕ field are powers (i.e. abilities or capacities) of the plana and the plana has no mass. Furthermore, a body of *any* mass can be activated by A-field or ϕ-field activated plana. Hence, no mass can be singled out without specializing the plana or its action beyond its own generic nature. Thus, we reasonably take the case of activating the plana in some way with an analogical impetus similar to (but not the same) as we did with light, i.e. we effectively take the case of zero mass to probe the nature of ϕ and A. More specifically, we allow the A-field to cause some aspect of the plana to change so as to carry the given amount of potential impetus A (per unit charge) through the plana at speed c. (We pick c for the reasons given above.) We can concretely picture the effect of such a propagation by imagining that the plana activates impetus in a *test* particle of arbitrarily small mass, so that its energy relation approaches that of light, $E = pc$. In this model, we would also take the charge on the test particles to be arbitrarily small thereby keeping their effect on the A- and ϕ-fields that we are probing arbitrarily small. (To avoid the issue of charge, you can think of the immediate discussion that follows as being in terms of momentum and energy *per unit charge* even though we will say simply momentum and energy to avoid multiplying words.)

Now, remember that when the A-field source is switched off, the plana causes a test particle at position z to gain an impetus of intensity $\left|\vec{A}(z)\right|$ in the direction of $\vec{A}(z)$. As we've said, the speed of our particle of infinitesimally small mass (and charge) will be c. To probe the nature of the A-field activated plana, we could consider the effect of A on a

---

[21] Note that the relationship $E = pc$ assumes $p = |\overline{p}| = \sqrt{p_x^2 + p_y^2 + p_z^2}$. Note that, for systems of light *alone*, because $E = |p|c$, we get conservation of $|p|$, as well as conservation of momentum, along each perpendicular axis.

distribution of such test particles spaced equally along a line, say in the z-direction. And, in order to not introduce an extraneous constant, we take there to be precisely one per unit length and assume our system's features are much bigger than this scale, so that we have sufficiently sampled the system by our series of particles.

With this analytical tools in hand, we use Figure 5-11, which shows a finite current element source of A, to present our interpretation of the Lorenz condition and thus the relation between the two potentials. The arrow labeled p^L_z on the left represents the average momentum that leaves the dashed box and enters the solid orange box in time Δt. The arrow labeled p^R_z on the right represents the average momentum that leaves the solid orange box in time Δt. This means there is an analogical activity (energy) of $p^L_z c$ entering the box and $p^R_z c$ leaving so that, for the figure shown, there is a net increase in energy in the box of $\left(p^R_z - p^L_z\right)c$ in time Δt.[22] This suggests then that there should be local increase (one that is not traveling, i.e., one like ϕ) in potential energy in this time Δt.

And, this is exactly what the Lorenz condition predicts. The Lorenz condition obviously deals with the divergence part of A, not the curling part (i.e., the part associated with B), and cries out to be treated as a conservation law. To understand the prediction of the Lorenz condition, we use Figure 5-11 again for concreteness. We pick the box to be lined up with the A-field, but the mathematics will, of course, cover any more complicated situation and the general Gauss's law type argument will remain valid.

Figure 5-11: Current element causes decreasing A-field in z-direction, which is necessarily accompanied by a time varying electric potential ϕ, because of the charge build-up. The box drawn with solid orange lines represents differential volume of cross sectional area $\Delta x\Delta y$ and height (Δz) defined by the distance light travels in time Δt.

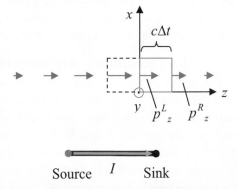

Thus, using differential areas and taking $\vec{A} = \vec{p}_{pot}, \phi = E_{pot}$, and defining $\widehat{da}$ as positive inward, equation 5.39 gives:

$$c\left(\vec{p}_{pot}{}^L \cdot d\vec{a} + \vec{p}_{pot}{}^R \cdot d\vec{a}\right) \sim \frac{\partial}{c\,\partial t}E_{pot}\,dxdy\,(cdt)$$

5.40
$$c\left(p_{pot}{}^L \cdot dxdy - p_{pot}{}^R \cdot dxdy\right) \sim \frac{\partial}{c\,\partial t}E_{pot}\,dxdy\,(cdt)$$

$$\Rightarrow c\,\Delta p_{pot} \sim \Delta E_{pot}$$

---

[22] In terms of A, these left and right averages would be $A_z\left(\Delta x\,/\,2, 0, \Delta z\,/\,2 \pm \Delta z\,/\,2\right)$, where $\Delta z = c\Delta t$.

This equation says that the field energy crossing the boundary of the region must equal the field energy lost from the region. Hence, we see that the Lorenz condition can be interpreted as related to the deep principle of conservation of energy. Again, the condition indicates an increase in potential energy locally because of the state of the potential around that locality, thus linking the state of the two key fields, ϕ and A, through a simple energy relation. Remembering that ϕ and A are already linked in that ϕ comes from charge and A comes from *impetus-activated* charge, it is natural for the maximum speed that impetus can move a massive body, i.e. c, to, in some way, link A and ϕ. We can bring out part of the principled connection between A and ϕ and their sources with the following diagram.

$$\nabla \cdot \vec{J} = -\frac{\partial \rho}{\partial t} \qquad\qquad \nabla \cdot \vec{A} = -\frac{1}{c^2}\frac{\partial \phi}{\partial t}$$

$\sim$ Charge

Caused by charge

$\sim$ Impetus activated charge

Caused by impetus activated charge

The analogy between the two equations outlined in this diagram naturally suggests that A divergence should cause ϕ build-up like J divergence causes q build-up. Still, to make sense of the build-up of ϕ from A, one needs to introduce the speed of travel, c, in the manner described above.

Thus, it is reasonable to take the Lorenz condition as valid. However, it should be said that at the level of abstraction of classical E&M, no firm conclusion can be drawn, and it is thus best to consider the question of gauge condition an open matter. Note also that the gauge condition is of central importance in E&M since it manifests the relation between the two aspects of the subject, the electric and magnetic potential.

Summary Arguments for Reality of ϕ and A

We can now briefly summarize our reasons for thinking ϕ and A represent measures of real strengths (and, in the case of A, directional types) of qualities of the plana. We begin by noting a key reason, why one could argue that A and ϕ are not real. Namely, the Maxwell's equations for E and B given in equations 5.35I-V along with the Lorentz force equation will describe any situation. And, these equations remain the same no matter what gauge condition we pick. However, the magnetic field and the electric field do naturally call for the existence of the potentials A and ϕ that describe those fields. Furthermore, each of those fields has a natural physical meaning relating, respectively, to the potential momentum and potential energy that they can cause per unit charge. Causing a certain amount of energy in a certain distance are indeed essential properties of the electric and magnetic fields. And, directly related to this fact, there must be some specific relation between one part of field activated plana and another; they cannot remain abstract but must be this or that. The potentials provide a natural specification. The fact that we have not completely pinned down their value does not obviate the need for them or something like them. Furthermore, without them, more physical causes must be posited and the more complex fields, E and B, become primary. By contrast with A and ϕ, fundamentally, one has all of E&M. The coup-de-grace is outside electromagnetism, in advanced area of physics called quantum field theory; it is called the Aharanov-Bohm effect, where A affects particles where there is no B or E. However, though we answer multiple issues by accepting A and ϕ, we still do have the so-called gauge freedom that we started this paragraph with, what of it?

To answer this question, we have to understand the full known empiriometric (and probably things we don't yet know) meaning of the gauge condition and that is beyond the scope of classical *E&M*, so we leave this problem for those more advanced topics. We can, however, argue as we did in the previous section for the plausibility of the Lorenz gauge. Still, even with given the Lorenz gauge, some further freedom remains that needs to be explored; however, this freedom gives solutions that are disconnected from sources and thus can be treated separately. And, indeed, we will ignore them in this course.

The student does not need to worry about how all these details work out or even try to memorize them but only: (1) note that we will continue to treat the A and ϕ field as representing real aspects of the plana, specifying them through the use of the Lorenz gauge and (2) notice that Maxwell's equations have exciting depths, beyond what we will reach in this text or even in advanced classical E&M, that still remain to be plunged.

Having exhausted the fundamental principles relating to static and quasi-static fields, we will turn to discuss the propagation of E and B and their specifications, ϕ and A.

Summary

Impetus-activated charge, whose measure of "density" and directional type are represented by the current density J, causes an A-field. An A-field that curls shows the plana has a B-field; formally, we write: $\vec{B} = \nabla \times \vec{A}$.

Only the curling part of J contributes to B. It can be shown that $\vec{B} = \frac{1}{c} \int \frac{\nabla \times \vec{J}}{R} dV$.

The curl of B is a sign of B.
The curl is thus crucial to our understanding of B and A.

We can say that the curl is the _net_ change of a vector field *perpendicular to itself*, if we take proper account of the two distinct ways that a field can change perpendicular to itself: (1) *the strength of the field* can change as one moves perpendicular to its direction (left figure), or (2) the field can acquire a

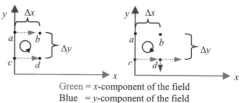

Green = x-component of the field
Blue = y-component of the field

component perpendicular to itself as one moves along its initial direction (right figure).[23] Said another way, there are two ways that a curl around a given axis, which we can take to be the z-direction, can arise: (1) changes in the x-component of the field in the y-direction (left figure), or (2) changes in the y-component of the field in the x-direction (right figure).

Because the B-field is *always* perpendicular to the direction of its propagation, and because of the nature of the fall-off of its strength with distance, only circulating B-fields are created, i.e. no B-field with a net flux coming in or out of a point is created; empiriometrically: $\nabla \cdot \vec{B} = 0$. A "circulating" field connects back to itself, such as shown around the wire here. (This use of "circulating" refers to a large region (*not* a "point"), but the word can also be used at a point to contrast "circulation" with a *generic* curling B, which does not *necessarily* connect back on itself). A circulating field does not necessarily have a curl at every point. That B has no divergence implies that, if there is a B-field, it must have a curl; this is why curl of B is used as a sign of B, i.e. is often used in place of it, (indeed, the curl of B specifies B when it is given everywhere).

Thus, the curling part of J creates a circulating B.

---

[23] We could also say a vector field has a curl in a region whenever a change in the field *perpendicular to itself* (left figure) is not compensated by a change in the field *along itself* that introduces a component perpendicular to itself (right figure). Where, in the first clause "changing perpendicular to itself," we mean that there is a difference between parallel components of the field that are separated by a slight distance. And, "introduces a component perpendicular to itself" means that a point close to the point of interest has a component that is perpendicular to the direction of the field at the point of interest.

One cannot diverge the curl of J, but one can curl the curl of J. This happens, for example, when one bends a wire thereby changing the curls of J that exist at least at the edges. It can be shown that the curl of B is only influenced by the curls of the curl of J from each source point; namely: $\nabla \times \vec{B} = \frac{1}{c} \int \frac{\nabla \times (\nabla \times \vec{J})}{R} dV$.

We split J into its two parts formally writing: $\vec{J} = \vec{J}_{curl} + \vec{J}_{div}$ where $\nabla \times \vec{J}_{curl} \neq 0$, $\nabla \cdot \vec{J}_{curl} = 0$ and $\nabla \cdot \vec{J}_{div} \neq 0, \nabla \times \vec{J}_{div} = 0$. $\vec{J}_{curl}$ and $\vec{J}_{div}$ do not "overlap;" i.e., they are orthogonal.

The diverging part of J cannot cause a B; an example of this is found in the exploding uniform charge discussed below.

$$\boxed{\vec{J}_{curl} = \frac{c}{4\pi} \nabla \times \vec{B}}$$ (equation 5.1) is valid as long as 1) there are no diverging currents and 2) we can neglect propagation, i.e., the quasi-static approximation. *Note well* that the J *at a given point is not the sole cause of the curl of B at that point*; the curl of B at a point is, of course, a result of *all* the currents, all the J's, in the source.

$$\boxed{\vec{J}_{div} = -\frac{1}{4\pi} \frac{\partial \vec{E}}{\partial t}}$$ (equation 5.2), where quasi-statically $\vec{E} = \vec{E}_{quasi} = -\vec{\nabla}\phi$. A divergence in J implies a build-up of charge, which causes a changing electric field. The field $\vec{J}_{div}$ reveals the degree to which a source is not a *closed* circuit.

Because of the non-local nature of the div/curl decomposition of J (i.e. the decomposition at one place depends on far away places), J_{div} and J_{curl} can each be non-zero even at points where there is no current. No current at a given point only means $J = J_{curl} + J_{div} = 0$, so all that is required is: $J_{div} = -J_{curl}$. Still, in such places, both the curls and divergences of J are zero, i.e., $\nabla \cdot J_{div} = 0$ and $\nabla \times J_{curl} = 0$.

Maxwell's Fourth Equation (Quasi-static form)
Adding together equations 5.1 and 5.2 (see above) re-unites J giving:

(5.3) $$\boxed{\nabla \times \vec{B} = \frac{4\pi J}{c} + \frac{1}{c} \frac{\partial \vec{E}_{quasi}}{\partial t}}$$

This equation thus has a physical J, but is not a directly causal equation, for the right hand side does not cause the curl of B at the given point but only represents the effect of all the curling J. One can also reunite J, by using the following equations:

$$\nabla \times \vec{J} = \frac{c}{4\pi} \nabla \times (\nabla \times \vec{B}), \nabla \cdot \vec{J} = -\frac{1}{4\pi} \frac{\partial \nabla \cdot \vec{E}}{\partial t}.$$

For a *complete circuit:* $\nabla \times \vec{B} = 4\pi \vec{J} / c$; i.e., the contribution of all of the various parts of the source are exactly such that the measure of their effect in creating the curl of

$\vec{B}$ at a given point is given by $4\pi\vec{J}/c$ at that point. For an *incomplete circuit*, i.e. $\nabla\cdot\vec{J}\neq 0$, the "missing" currents are accounted for by subtracting off the measure of their effect, whose measure is represented by $\vec{J}_{div} = -\dfrac{1}{4\pi}\dfrac{\partial\vec{E}_{quasi}}{\partial t}$, so that the curl of B at a given point is given by $\vec{J}$ at the point minus $\vec{J}_{div}$ at that point: $\nabla\times\vec{B} = \vec{J} - \vec{J}_{div}$, i.e., equation 5.3 above.

Quasi-statically, one can only get a curl of B in "empty space" (i.e. free plana having no powers but A and B) by introducing charge paths that end. However, leaving quasi-statics by introducing the finite speed of field propagation, provides a third way, which is represented by the equations: $\vec{J}_{curl} = \dfrac{c}{4\pi}\nabla\times\vec{B} - \dfrac{1}{4\pi}\dfrac{\partial\vec{E}_{curl}}{\partial t}$, where $\vec{E}_{curl} \equiv -\dfrac{1}{c}\dfrac{\partial\vec{A}_{curl}}{\partial t}$;

$\vec{J}_{div} = -\dfrac{1}{4\pi}\dfrac{\partial\vec{E}_{div}}{\partial t} = \dfrac{1}{4\pi}\dfrac{\partial\vec{\nabla}\phi}{\partial t} + \dfrac{1}{4\pi c}\dfrac{\partial^2\vec{A}_{div}}{\partial t^2}$. These will be studied more in the next chapter.

Maxwell's 4th Equation (full form)

(5.7)
$$\boxed{\nabla\times\vec{B} = \frac{4\pi\vec{J}}{c} + \frac{1}{c}\frac{\partial\vec{E}}{\partial t}}$$
where: $\vec{E} = -\vec{\nabla}\phi - \dfrac{1}{c}\dfrac{\partial\vec{A}}{\partial t}$

This equation, like all of Maxwell's equations, does not directly represent the physical causes at play, i.e. it does not give the fields as a function of their sources. It relates measures of fields and sources *all at one point*, rather than giving a measure of a single field *at a given point* as the sum of the measures of the effect of *all the parts of all of the sources*. Note also that the split between the *div* and *curl* parts of the equation is not explicit, so should be kept in mind.

The term $\dfrac{1}{c}\dfrac{\partial\vec{E}}{\partial t}$ is often called the displacement term, $\vec{J}_D = -\vec{J}_{div}$, since Maxwell erroneously thought that plana was closely analogous to a massive media with charge (i.e., he thought of it as a dielectric, see Chapter 8).

Three Examples which Involve Charge Build-up (and thus the $\dot{E}$ term)
1. Parallel Plate Capacitor

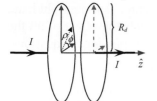

The current in the disks can be shown to be:

$$I_{disk}(\rho) = I\left(1 - \left(\frac{\rho}{R_d}\right)^2\right).$$

The *A*-field due to right and left leads are:

$$A_z{}^{Right} = \frac{I}{c}\left(sinh^{-1}\frac{L-z}{\rho} - sinh^{-1}\frac{d-z}{\rho} \right),\ A_z{}^{Left} = \frac{I}{c}\left(sinh^{-1}\frac{L+z}{\rho} - sinh^{-1}\frac{d+z}{\rho} \right)$$

The *B*-field inside the capacitor due to the leads is:

$$B_\phi{}^{Leads} = \frac{2I}{\rho c}\left(1 - \frac{d}{\rho} \right) \sim \frac{2I}{\rho c} \quad \text{for } d << \rho << R_d \text{ in the } z = 0 \text{ plane.}$$

The linear lead currents and the radial disk currents both contribute to the *B*-field inside the capacitor:

The only component of *B* is the ϕ component which is given by: $B_\phi = \dfrac{\partial A_\rho}{\partial z} - \dfrac{\partial A_z}{\partial \rho}$. The first term is the "*disks*" term, and the second is the "*leads*" term.

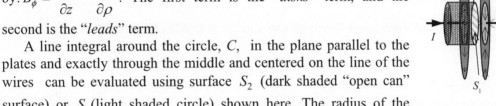

A line integral around the circle, *C*, in the plane parallel to the plates and exactly through the middle and centered on the line of the wires can be evaluated using surface S_2 (dark shaded "open can" surface) or S_1 (light shaded circle) shown here. The radius of the circle is taken to be much smaller than the diameter of the disks ($\rho << R_d$). With S_1, one must use the $\dot{E}$ term to represent the "missing" currents, whereas with S_2, one only deals with real currents, since, in the region of interest, $\vec{\dot{E}}$ is perpendicular to the disks inside the capacitor and is arbitrarily small outside the capacitor.

2. Finite Current Element

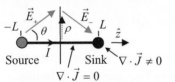

As the current flows at rate *I* from the positive to the negative charge it generates a *B*-field in the plana, which we investigate by looking at the $\vec{\dot{E}}$ along the plane $z = 0$. The *E*-field (where we take the primal definition: $\vec{E} = -\nabla\phi$) in this plane is: $\vec{E} = \dfrac{2qL}{\left(\sqrt{L^2 + \rho^2}\right)^3}\hat{z}$. Note that the divergence of

J is only non-zero on the ends, so that the ends define how the *B*-field is different from an infinite line of current. As the element gets infinitely long, they recede into the distance, and, thus, the *B*-field approaches that of the infinite line. By contrast, as one moves outward along the $z = 0$ plane (i.e., using figure above, $\rho \to \infty$), the distance between the two ends shrinks relative to the view distance so that the source and the sink become more and more *effectively* on top of each other, decreasing the effective current flow thus more quickly diminishing the *B*-field. More precisely, because of this effect, the *B*-field falls off as $1/\rho^2$, rather than the $1/\rho$ of the infinite line of current.

3. Exploding Ball

A ball of charge that expands uniformly in all directions causes only radial *A*-field, which has no curl and hence no *B*-field. Using a small circle (shown dotted to the left) in a plane perpendicular to a radially

outgoing element of current in Ampère's path law without the $\dot{E}$ term implies that there is B-field. However, this only accounts for the current piercing the disk bounded by the circle (and it does so as if that current were an infinite line). When we account for all the other currents in the source by integrating $\dot{E}$ over that disk, the calculation correctly predicts no B-field for the exploding ball.

One can formally deduce Maxwell's fourth equation (equation 5.7), by taking the curl of the Biot-Savart formula: $\vec{B} = \int \dfrac{\vec{J} \times \hat{R}}{cR^2} dV$.

One can also deduce it by requiring that the definition of "$\vec{J}$" be expanded so that: $\int_C \vec{B} \cdot d\vec{s} = \dfrac{4\pi}{c} \int_S \vec{J} \cdot d\vec{a}$ for *any* surface S bounded by C and *any* source. By using

$\nabla \cdot \vec{J} = -\dfrac{\partial \rho}{\partial t}$ and the divergence theorem, one can show that the integral form must be:

$\int_C \vec{B} \cdot d\vec{s} = \dfrac{4\pi}{c} \int_S \left(\vec{J} + \dfrac{1}{4\pi} \dfrac{\partial \vec{E}}{\partial t} \right) \cdot d\vec{a}$, so that $\vec{J} \to \vec{J} + \dfrac{1}{4\pi} \dfrac{\partial \vec{E}}{\partial t}$. The very surface

independence just allowed then allows one to move from this to the differential form given in equation 5.7.

Every source with impetus-activated charges causes an A-field, and, in general, a curl of A, i.e., a B-field. If there is no charge build-up, there is no relation between $E = \nabla \phi$ and B, and

the relation between $\vec{B}$ and $\vec{J}$ can be written: $\nabla \times \vec{B} = \dfrac{4\pi \vec{J}}{c}$. This means that the curl of B

at a given point of the plana is equal to the $\vec{J}$ at that same point. All of the curling parts of J add to make this result true whenever all charges follow *complete* paths.

However, whenever a part of a path is "missing," i.e., whenever a path ends, it has to be accounted for through the use of the $\dot{E}$ term, which is non-zero because charge builds up at such places. The $\dot{E}$ term is added on the right hand side of Maxwell's fourth equation (equation 5.7) and is equivalent to subtracting off the non-curling part of $\vec{J}$, $\vec{J}_{div}$. The changing E-field ($\dot{E}$) gives a generic representation of the ending points, the effect of which need to be subtracted off the first term because that first term is only validly used alone on the right hand side when there are no breaks in the charge paths.

In order to move beyond quasi-static fields, we have to allow a finite speed of propagation of the fields. To do this for A and ϕ, we need a constraint connecting them. The Lorenz gauge means we apply the following constraint, called the Lorenz condition:

$$\nabla \cdot \vec{A} + \frac{1}{c} \frac{\partial \phi}{\partial t} = 0 \ (cgs) \qquad\qquad \left(\nabla \cdot \vec{A} + \frac{1}{c^2} \frac{\partial \phi}{\partial t} = 0 \ (SI) \right)$$

Recalling that $\vec{A}$ is the potential momentum per unit charge and ϕ is the potential energy

per unit charge and by making an analogy with the conservation equation: $\nabla \cdot \vec{J} = -\dfrac{\partial \rho}{\partial t}$, we can argue that the Lorenz condition is natural. In the analogy, a divergence of $\vec{A}$ implies a build up of analogous energy (per unit charge) due to an imagined property of the plana that carries (analogous) momentum (per unit charge) and propagates at speed c, which, in turn, represents the action of the A-field on certain nearly speed c particles (so $E \sim pc$). This means that a divergence of the analogous potential momentum (per unit charge) symbolized by $\vec{A}$, results in a *local* build up of analogous potential energy (per unit charge) symbolized by ϕ.

The availability of such an apparently natural interpretation, along with the fact that it is the only gauge that does not introduce further speeds (i.e., in addition to c), gives support for our choice. More advanced disciplines will give qualified further support, while they raise further and deeper questions.

Maxwell's Equations:

$I.$ $\nabla \cdot \vec{E} = 4\pi\rho$ $II.$ $\nabla \cdot \vec{B} = 0$

$III.$ $\nabla \times \vec{E} = -\dfrac{1}{c}\dfrac{\partial \vec{B}}{\partial t}$ $IV.$ $\nabla \times \vec{B} = \dfrac{4\pi \vec{J}}{c} + \dfrac{1}{c}\dfrac{\partial \vec{E}}{\partial t}$

with the force law: $\vec{F} = q\vec{E} + q\dfrac{\vec{v}}{c} \times \vec{B}$

Helpful Hints

Note the word "static" is often used to describe systems with constant currents, though clearly they are not unmoving, not "static," but "stationary," which means not increasing or decreasing, having reached some kind of steady state. A similar caveat applies for the term quasi-static.

The word "circle" can be a point of confusion. Technically, it means the locus of points in two dimensions that are a fixed distance from a point called the center; this means the circle itself is one dimensional. However, in ordinary speech "a circle" often means a two dimensional plane figure that has this as its boundary. Here we will use the word "disk" to mean this latter (two dimensional) shape, keeping the word shape for the first (one dimensional) shape.

The *divergence* of a field is related to how the field changes along its own direction, while the *curl* of the field is related to how the field changes perpendicular to itself. In both cases, however, one must keep in mind what all components of the field are doing. For example, in the case of the flux of the E-field from a point particle at the origin, although the field strength decreases as one moves with the field direction upward along the z-axis, that loss of field strength is made up for in the strength "given to" the x and y-components. And, in the case of the line integral of the B-field in the "empty plana"[24] near a long straight current-carrying wire lying on the x-axis, the field strength decreases as one moves upward along the z-axis, i.e., perpendicular to the field. However, in the limit of an infinitely long line, this contribution to the line integral is exactly compensated for by the change in B_z in the y-direction (see end of chapter problem).

---

[24] We can also call it "empty space," if we are mindful of its meaning that we simply mean that we have a region consisting of plana with no massive bodies in it.

Problems

1. Consider removing half of an infinite wire of constant current, I, to leave a diverging part of the current exactly at the coordinate origin as shown to the right. a) Using Maxwell's fourth equation in integral form applied to the planar region, S, which is bounded by the circular orange path (which is slightly in front of the end of the wire), calculate the B-field along that circle. b) Taking the quasi-static approximation and assuming the radius of the path is a and the distance from the end is z, such that $z \ll a$, calculate the B-field using the Biot-Savart law. c) Compare the results and explain how a part "a" type analysis accounts for the actual currents. In particular, there will be an instantaneous jump in the current piercing S, when C is moved slightly to the left. This seems to imply a large jump in the B-field which is not consistent with the smoothly varying B-field that seems to be implied by the smoothly varying distance from the currents that cause the B-field, which is formally manifest in the Biot-Savart law. Resolve this paradox. d) Using your expression for B, calculate the curl of B, for $z > 0, z \ll \rho$, showing that, as mentioned in the text, there is a non-zero component of the curl of B arising from the change in B_ϕ in both the axial direction ($\hat{z}$) and the radial direction ($\hat{\rho}$).

2. Consider the cylinder shown on the right, which has a positive point charge source at the center of its left base and a negative point charge source at the center of its right base. Assume that the current flows from the positive charge to the negative, but that the cylinder is neutral and is not a conductor so that the ϕ-field produced by the two point sources is not changed by the current flow. A steady total current I flows uniformly from the positive to the negative source. a) Using the displacement current, calculate the B-field inside the cylinder in the plane that passes perpendicular through the axial center. Give and explain, the result for $L \to \infty$. b) Setup (but do not solve) the integrals for calculating the A-field directly from the current distribution. *Hint*: This involves setting up one of the integrals for the parallel capacitor not done in the text.

3. We will here investigate how the various current elements in a closed source ($\nabla \cdot \vec{J} = 0$) act together to make the curl of B zero in empty space (free plana). Consider the arbitrarily thin (x-direction) rectangular flow, with height l and width b, of positive charge shown to the right. The orange *solid* lines mark the boundary of the region of current inside of which there is only empty space, whereas the orange *dotted* lines mark the end of 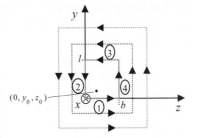 the current flow, outside of which there is only empty space. A current I flows around the circuit. Assume that the flow is everywhere uniform except at the turning points and that the width of the flow, i.e. the distance between the dotted and solid orange lines, is arbitrarily small. By noting that once we understand the generation of the B-field due to the

inside boundary, the outside boundary contributes in a parallel way, we can focus on just the inside boundary. a) Draw $\nabla \times \vec{J}$ at the boundary, where $\vec{J}$ is the current density of the flow. Draw your answer inside the rectangle for each of its four sides, which are labeled 1, 2, 3 and 4. Also draw the value of $\nabla \times \left(\nabla \times \vec{J} \right)$ on that boundary; since space, is tight use another figure for this drawing. Explain any contributions that cannot be easily drawn. b) Using the advanced techniques given at the end of Chapter 4, show: $\nabla \times \left(\nabla \times \vec{J} \right) = -\nabla^2 \vec{J}$, and explain how your results are in concordance with this fact. c) For the case in which l and b are arbitrarily large, so that the influence of edges 3 and 4 can be ignored, explain qualitatively how edge 1 and edge 2 act together to make $\nabla \times \vec{B} = 0$ at the point labeled $(0, y_0, z_0)$. In particular, how can contributions from a horizontal and a vertical wire cancel, when those contributions *appear* to be everywhere perpendicularly to each other.

Hint: Note $\nabla \times \vec{B} = \int \dfrac{\nabla \times \left(\nabla \times \vec{J} \right)}{R} dV$, and treat each wire separately in its *entirety*, neglecting only the region at infinity.

The following part is only for advanced students who know how to use the unit step function, $u(x)$, and its derivatives (the Dirac-delta function, $\delta(x)$, and its derivatives $\delta^n(x)$). d) Consider compressing the width of the flow so that the limit to zero width is completed giving the rectangular 1-dimensional wire circuit shown to the right. Write the formal expression for $\vec{J}$ for wires 1 and 2, calculate $\nabla \times \vec{J}$ and $\nabla \times \left(\nabla \times \vec{J} \right)$ and use part (b) to check your work. Using these results, repeat part (c).

4. The figure to the right shows a parallel plate capacitor being charged by a constant current I. a) Draw the E-field. b) Explain heuristically, using Maxwell's fourth equation, what the value of the azimuthal B-field is at the point P situated as shown in the figure to the right. In your explanation, use the surfaces S_1 and S_2, both of which are bounded by the circular path, C, shown inside the capacitor. In particular, how does the B-field at that point compare to that of an uninterrupted "infinite" line of current? Physically and in terms of the displacement term, explain what happens if we make the radius of C arbitrarily big?

5. Suppose we have a fully charged parallel plate capacitor that is shorted as shown in the figure to the right and is draining via a constant current I. a) Draw the currents that flow in the disks and the E-field inside the capacitor. b) Using Maxwell's fourth equation in integral form applied to the circle whose cross section is shown dotted in the figure, give a rough estimate of the value of the B-field at the point, P, just outside the edge of the capacitor. Qualitatively explain that value in terms of Maxwell's equation and in terms of its physical origin.

6. Explain why a circulating B-field, i.e. one that eventually circles back on itself in some finite region, for example around a long straight current, always implies that there is a non-zero curl of B inside that region.

7. As mentioned in a footnote in the text, we can write the two components of $\vec{J}$ as $\vec{J}_{div} = \vec{\nabla} f$ and $\vec{J}_{curl} = \nabla \times \vec{\alpha}$. Making an obvious analogy between $\vec{E}$ and ϕ and $\vec{J}$ and f, give an example of how $\vec{J}_{div}$ can be non-zero even where there is no source.

8. Using Maxwell first equation and conservation of charge show that $\vec{J}_{div} = -\dfrac{1}{4\pi} \dfrac{\partial \vec{E}}{\partial t}$.

9. The following two problems help circumscribe the limits of our quasi-static approximation. In particular, if the E-field can simply be taken to be its primary part, so that $\vec{E} \sim -\vec{\nabla}\phi$, which means $\left| \vec{\nabla}\phi \right| >> \dfrac{1}{c}\left| \dfrac{\partial \vec{A}}{\partial t} \right|$, the quasi-static approximation applies. Here

we examine this inequality for the case of a *point capacitor* of width $2d$ driven by a voltage $V = V_0 \sin \omega t$. Such a point capacitor is shown schematically to the right, but we here consider the coaxial design and parameters shown in Figure 5-6 (for concreteness take $L \sim 10d$). At the mid
point of the capacitor, what relationship between d and ω is required by the above inequality? How is your answer related to the infinite speed of propagation of the fields approximation? *Hint*: Make use of the general formula for the relationship between angular frequency and wavelength: $\omega = 2\pi \dfrac{c}{\lambda}$.

10. Similar to the above problem, here we will investigate the conditions under which $\left| \vec{\nabla}\phi \right| >> \dfrac{1}{c}\left| \dfrac{\partial \vec{A}}{\partial t} \right|$ is true for a concrete physical configuration. Consider a line of charge of uniform linear density $\lambda << c$, and length $2L$, which is moving at speed $v << c$ along its own axis. Allow the wire to begin accelerating perpendicular to its own length accelerating line of charge and consider the two parts of (the analogically extended) E-field at a radius ρ_0 from the wire in the plane perpendicular to the line of charge that cuts through the wire's center point. Calculate, for $L \sim \rho_0$, write the above inequality. Now, assuming a typical time scale, T, that is paired with the distance scale L, for both the time scale of the acceleration and the velocity v, relate this inequality with the propagation speed of the field.

11. Show that if E and B are solutions to Maxwell's equations (equations 5.35) then, as alluded to in Chapter 3, $\vec{E}' = \vec{E}\cos\theta + \vec{B}\sin\theta$ and $\vec{B}' = -\vec{E}\sin\theta + \vec{B}\cos\theta$ are solutions to a revised system of Maxwell's equations that includes magnetic charge (magnetic monopoles) and magnetic current. Explicitly write the form for the new charge and current densities, $\rho'_e, \rho'_m, \vec{J}'_e, \vec{J}'_m$ in terms of the old $\rho_e, \rho_m, \vec{J}_e, \vec{J}_m$.

12. Show that the Lorenz condition also makes sense from the following point of view. In the static case, it gives $\nabla \cdot \vec{A} = 0$. Prove directly from this, using a Gaussian surface, that an "infinitely" long straight current cannot produce a radial A-field.

13. By taking the curl of $\vec{A} = \frac{1}{c} \int \frac{\vec{J}}{|\vec{r} - \vec{r}'|} d^3 r'$, derive the Biot-Savart law:

$\vec{B} = \int \frac{\vec{J} \times \hat{R}}{cR^2} dV$. *Hint*: Use $\nabla \times (f\vec{V}) = \vec{\nabla} f \times \vec{V} + f \nabla \times \vec{V}$ and the theorem derived from

the divergence theorem: $\int_V (\nabla \times \vec{V}) dV = \int_S \hat{n} \times \vec{V} dS$, where $\hat{n}$ is the unit normal to the surface S that bounds the volume V.

14. Show that B only depends on the curl of J. In particular, show that, in the quasi-static approximation where we neglect propagation, $\vec{B} = \frac{1}{c} \int \frac{\nabla \times \vec{J}}{R} dV$ by taking the curl of Maxwell's 4th equation. *Hint*: Use the appropriate vector identity and the already given solution for Poisson's equation.

15. Using the vector identity in the problem above: a) Derive $\vec{B} = \frac{1}{c} \int \frac{\nabla \times \vec{J}}{R} dV$ by taking

the curl of $\vec{A} = \frac{1}{c} \int \frac{\vec{J}(\vec{r}')}{R} d^3 r'$; recall: $R \equiv |\vec{r} - \vec{r}'|$. b) Derive the following by taking

further curls: $\nabla \times \vec{B} = \frac{1}{c} \int \frac{\nabla' \times (\nabla' \times \vec{J})}{R} dV'$, $\nabla \times \vec{B} = \frac{1}{c} \int \frac{\nabla' \times (\nabla' \times \vec{J})}{R} dV'$,

$(\nabla \times)^n \vec{B} = \frac{1}{c} \int \frac{(\nabla' \times)^{n+1} \vec{J}}{R} dV'$ and $(\nabla \times)^n \vec{A} = \frac{1}{c} \int \frac{(\nabla' \times)^n \vec{J}}{R} dV'$, for all $n \in \{0, 1, 2..\}$

Hint: Transition from field point coordinates to source point integration coordinates, i.e., from ∇ to ∇', by proving and using the following: $\vec{\nabla} R = -\vec{\nabla}' R$.

16. Take the curl of the Biot-Savart formula: $\vec{B} = \int \frac{\vec{J} \times \hat{R}}{cR^2} dV$ and use $\nabla \cdot \vec{J} = -\frac{\partial \rho}{\partial t}$, rather

than $\frac{\partial \rho}{\partial t} = 0$, to get the generalized quasi-static Ampère's law (in differential form), the 4th Maxwell equation, equation 5.7. *Hint*: Make use of the advanced techniques given at the end of Chapter 4 and/or appropriate vector identities.

17. If we say, in a shorthand speak, that "a vector field that does not change perpendicular to itself, does not have a curl" we mean "does not have a *net* change perpendicular to itself." That is, if the two contributions to a curl in a given direction, i.e. the two different "components" of the change of the field perpendicular to itself, cancel each other, there is clearly no *net* change though there is a change in each component. The 2-dimensional vector field given by: $\vec{V} = \dfrac{\hat{\phi}}{r}$ is an example of such a cancellation. Show this explicitly. In particular, explain how it can have no curl when the *y*-component of $\vec{V}$ decreases as one moves outward along the *x*-axis.

18. Explain how the distinction between part and whole allows us to talk about the field changing perpendicular to itself when we describe the meaning of curl. *Hint:* Consider free plana which is a single substance.

19. In the simplest way, draw the two different ways a vector field can, in the sense described in the summary, change perpendicular to itself. Use arrows at two nearby points for each drawing, and to make the simplest possible drawings, assume the field changes as quickly as needed between the two chosen points.

Chapter VI

Maxwell's Equations and the Propagation of Fields

Introduction

With the completion of the last chapter, we have incorporated into our *formalism* all aspects relevant to the subject of classical E&M, the last being our accounting for the different effects that arise from curl and the diverging parts of the current. In short, we have introduced the complete Maxwell's equations:

6.1 *I-IV*

$$I. \quad \nabla \cdot \vec{E} = 4\pi\rho \qquad II. \quad \nabla \cdot \vec{B} = 0$$

$$III. \quad \nabla \times \vec{E} = -\frac{1}{c}\frac{\partial \vec{B}}{\partial t} \qquad IV. \quad \nabla \times \vec{B} = \frac{4\pi\vec{J}}{c} + \frac{1}{c}\frac{\partial \vec{E}}{\partial t}$$

$$\text{with the force law:} \quad \vec{F} = q\vec{E} + q\frac{\vec{v}}{c}\times\vec{B}$$

However, we have not yet drawn out a key implication of the powers of the plana that we have discussed. In particular, we have not discussed the propagation of the electric and magnetic fields or their specializations, the *A* and *ϕ* fields. In this chapter, we will do both, starting with the potentials and then moving to energy propagation or radiation in terms of potentials bringing out its meaning in terms of *E* and *B*.

Maxwell's Equations Coupled with Lorenz Condition

To introduce complete equations for *A* and *ϕ*, we recall the relations that connect them to the *E* and *B* fields. In particular, we formally take $\vec{E}$, as we have already introduced in previous chapters, to include the magnetic term $-\dfrac{1}{c}\dfrac{\partial \vec{A}}{\partial t}$ because it is an effect like that of a

true electric effect, so that we have $\vec{E} = -\nabla\phi - \dfrac{1}{c}\dfrac{\partial \vec{A}}{\partial t}$. In this way, we can write the

Maxwell equations 6.1, in terms of A and ϕ to get:

6.2 *I-IV* $\nabla \cdot \vec{E} = 4\pi\rho$ ---------- $-\nabla^2\phi - \dfrac{1}{c}\dfrac{\partial \nabla \cdot \vec{A}}{\partial t} = 4\pi\rho$

$$\left.\begin{array}{l} \nabla \cdot \vec{B} = 0 \\[2ex] \nabla \times \vec{E} = -\dfrac{1}{c}\dfrac{\partial \vec{B}}{\partial t} \end{array}\right\} \quad \text{----------} \quad \left\{\begin{array}{l} \vec{B} = \nabla \times \vec{A} \\[2ex] \vec{E} = -\nabla\phi - \dfrac{1}{c}\dfrac{\partial \vec{A}}{\partial t} \end{array}\right.$$

$\nabla \times \vec{B} = \dfrac{4\pi\vec{J}}{c} + \dfrac{1}{c}\dfrac{\partial \vec{E}}{\partial t}$ ---------- $-\nabla^2\vec{A} + \vec{\nabla}\left(\nabla \cdot \vec{A}\right) = \dfrac{4\pi\vec{J}}{c} - \dfrac{1}{c}\dfrac{\partial \vec{\nabla}\phi}{\partial t} - \dfrac{1}{c^2}\dfrac{\partial^2 \vec{A}}{\partial t^2}$

Note that the last equation uses the vector identity, $\nabla \times \left(\nabla \times \vec{V}\right) = \nabla\left(\nabla \cdot \vec{V}\right) - \nabla^2\vec{V}$. As mentioned earlier, there should be no anxiety about this equation, for at the level of this course, it can be treated as a given. The student does not need to be able to derive it, but only needs to know that this formula is helpful in such derivations.

Note that the definitions of the fields in terms of the potentials (given on the right side of equations 6.2 *II-III*, i.e. the middle two lines of the above list of four equations) suffice to give the middle two Maxwell's equations "automatically." Making these kind of simplifying definitions, which ultimately of course arose from the desire to succinctly capture some real aspect of the physical world, is a hallmark of the empiriometric method.

We now formally supplement the Maxwell's equations with the Lorenz condition:

6.3 $\nabla \cdot \vec{A} + \dfrac{1}{c}\dfrac{\partial \phi}{\partial t} = 0$ *(cgs)*

Hence we can write the last equation above (equation 6.2 *IV*) as:

6.4 $-\nabla^2\vec{A} + \dfrac{1}{c^2}\dfrac{\partial^2 \vec{A}}{\partial t^2} = \dfrac{4\pi\vec{J}}{c}$,   (which is also written: $\Box\vec{A} = \dfrac{4\pi}{c}\vec{J}$)[1]

This is the so-called (inhomogeneous) wave equation or propagation equation with a

source term $\dfrac{4\pi}{c}\vec{J}$ for the A-field. The solution of this equation is:[2]

6.5 $\vec{A}(\vec{r},t) = \dfrac{1}{c}\displaystyle\int \dfrac{\vec{J}\left(\vec{r}',t - R/c\right)}{R}dV'$,

Where, as earlier, $R \equiv |\vec{r} - \vec{r}'|$ and the primed coordinate refers to the vector pointing from the origin to the source current (cf. Figure 6-1).

---

[1] This box symbol is called a D'Alembertian.

[2] This is often written: $\vec{A} = \dfrac{1}{c}\displaystyle\int \dfrac{[\vec{J}]}{R}dV'$ where the square brackets mean that the quantity is evaluated at the retarded time, $t - R/c$.

Note that this equation is different from the quasi-static current case we introduced in Chapter 3:

6.6
$$\vec{A}(\vec{r},t) = \frac{1}{c}\int \frac{\vec{J}(\vec{r}',t)}{R}dV'$$

This equation is the special case of equation 6.5 in which we take the current to vary slowly enough that the propagation time can be neglected. It does not have the shifted-time, $t - R/c$.

The shifted-time term simply accounts for the delay introduced by propagation. As we saw in the beginning of Chapter 2 (for an electric field), a source acts on the part of the plana nearest to it; that part acts on the plana part next to it, and this continues out to the point under consideration, the so-called field point or view point. This speed of travel is c, the speed of light.

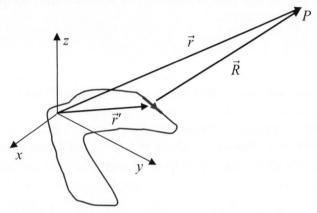

Figure 6-1: An arbitrary source is shown as a brown loop. We discuss the A-field caused at a point P in the plana by a small current element at $\vec{r}'$ which is a displacement $\vec{R}$ from that element. The element of the source under consideration is highlighted by making the line wider and drawing a blue arrow to represent the current.

Hence, if we have the situation shown in Figure 6-1, it is not hard to see that, at time, t, when the small element at $\vec{r}'$ has a current density $\vec{J}(\vec{r}',t)$, the A-field that it causes will take R/c seconds to reach the point P. Thus, the field now being caused by $\vec{J}(\vec{r}',t)$ will not be felt at point P for R/c seconds. And, the A-field activated at point P at time t will be the field "sent out" R/c seconds earlier than time t, i.e. $t - R/c$, so that the field be $\frac{\vec{J}(\vec{r}',t - R/c)}{cR}dV'$. Of course, integrating this over the volume of the entire source to account for all of its parts gives equation 6.5.

Note here that it is the displacement current term, $\sim \partial \vec{E}/\partial t$, that we added that "makes" the A propagate. Well, not really. The 'makes' is in quotations because, of course, *terms* do not make *things* propagate. What the term really does is incorporate into our formalism the fact of field propagation. Note well that it does this only when, as we stated above, we analogically extend the definition of E, by including the changing A; i.e.,

$$\vec{E} \equiv -\nabla \phi - \frac{1}{c}\frac{\partial \vec{A}}{\partial t}.$$

Now, equation 6.5 is one of two of the fundamental equations that we need in this chapter. The second fundamental equation of this chapter comes from applying this analogical extension to Maxwell's first equation. We see the formal result explicitly in the right hand equation of the equations 6.2 *I*. After applying the Lorenz condition (equation 6.3) which we view as an analogical conservation of energy condition that reveals the way ϕ and A are related, we get:

6.7 $\dfrac{1}{c^2}\dfrac{\partial^2 \phi}{\partial t^2} - \nabla^2 \phi = 4\pi\rho$ (which is also written: $\square\phi = 4\pi\rho$)

Note this equation is just like one component of equation 6.4, so you should not be surprised to learn that its solution is:

6.8 $\phi(\vec{r},t) = \displaystyle\int \frac{\rho(\vec{r}',t-R/c)}{R}dV'$

Equations 6.6 and 6.8 are the full solutions of Maxwell's equations coupled with the Lorenz condition. They are often called the "retarded solutions" because they include the fact that any effect takes a certain time to move or propagate from the source to some place in space, so that there is a delay or "retardation" between cause and effect. Like equation 6.6, each piece of source, in this case each $dq = \rho dV'$, contributes its delayed effect at a given distance, and only adding the effect from all such pieces of the source do we get the total effect at the given field point located by the displacement vector $\vec{r}$ from the coordinate origin.

Introduction to Radiation

Equations 6.5 and 6.8 also imply that there is a type of propagation that we have not studied; namely, electromagnetic radiation, which is sometimes called "light." It is called light after the most obvious example of electromagnetic radiation, visible light. We show below the visible spectrum of pure colors, which is often memorized through the acronym, which can be thought of as a name: ROY G BIV, which stands for Red, Orange, Yellow, Green, Blue, Indigo, and Violet. As illustrated below, there is actually a whole "spectrum" of moving dispositions of the plana, known as electromagnetic radiation, only a small subset of which can be seen or sensed in any way by us. As shown below, visible light is only a small part of the electromagnetic spectrum. Radio waves, of which there are a wide spectrum, are another commonly-known category of electromagnetic radiation.

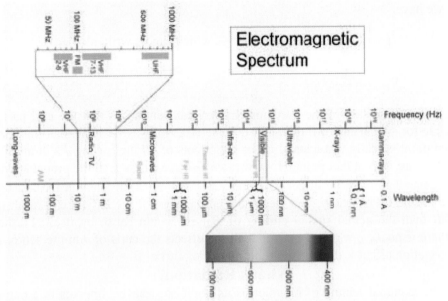

In electromagnetic radiation, E and B propagate in such a way that they carry analogical momentum; that is, they for example, activate impetus in their direction of travel upon being absorbed by a massive body. Formally, this new effect comes about because of the double derivative term, $\ddot{\vec{A}}$, seen in equation 6.2 *IV* which results in the wave equation, equation 6.4, rather than the Poisson equation given in Chapter 3. Physically, it comes about because, with the inclusion of propagation, the A-field can exist in the plana in a new way. What about ϕ? ϕ is not necessary for this effect but can be included. Why so?

Remember any propagation that results in a change in the field, as opposed to just maintenance of the field, always requires a changing source (outside of the relativistic perspective which we will discuss later). That is, since the net charge cannot be made or destroyed, to change the field a charge must move, which means we must have impetus activated charge and hence the *A-field* (of the same analogical type as that impetus) that it causes. Thus, magnetic effects, i.e., effects due to plana-activated A-field, must always be associated with electromagnetic radiation. By contrast, we can eliminate the ϕ term by just considering a neutral wire with current flowing in a circuit that does not allow charge buildup. Thus, we will consider this essential case first, while cases including ϕ will be considered afterward. For concreteness, we will shortly take the case of a long wire with a current that we can change all along its length at the same time. This is not a standard choice of antenna, because it would be very difficult to make. However, its simplicity lets us more quickly approach the key concepts. Later we will move to a short wire antenna, which will introduce ϕ but still keep the current at one instant the same all along the wire.

To start this analysis, recall the equations we introduced in Chapter 5:

6.9 $\qquad \vec{J}_{curl} = \dfrac{c}{4\pi}\nabla\times\vec{B} - \dfrac{1}{4\pi}\dfrac{\partial\vec{E}_{curl}}{\partial t}$

$\qquad$ where $\vec{E}_{curl} \equiv -\dfrac{1}{c}\dfrac{\partial\vec{A}_{curl}}{\partial t}$, so that $\dfrac{1}{4\pi}\dfrac{\partial\vec{E}_{curl}}{\partial t} = -\dfrac{1}{4\pi c}\dfrac{\partial^2\vec{A}_{curl}}{\partial t^2}$

So that we have:

6.10a $\qquad \vec{J}_{curl} = \dfrac{c}{4\pi} \nabla \times \vec{B} + \dfrac{1}{4\pi c} \dfrac{\partial^2 \vec{A}_{curl}}{\partial t^2}$, which can be written:

6.10b $\qquad \nabla \times \vec{B} = \dfrac{4\pi}{c} \vec{J}_{curl} - \dfrac{1}{c^2} \dfrac{\partial^2 \vec{A}_{curl}}{\partial t^2}$

Which, as we have said in the previous chapter, can remind us that the curl part of J is responsible for the curling of B (and hence B itself), as long as we remember that the J_{curl} at the given point is really just a stand in for the effects of *all* the curling J's in the source in causing the curl of B at that point. However, we also see a new fact; the propagation results in some cancellation of the effect of J. Indeed, if the $\ddot{A}$ is big enough the last term can overtake the first. Physically, this means the propagation of the A-field can result in more curl of B than the actual spatial curling of J itself would cause by itself. Furthermore, where there is no J_{curl}, which signals that all the effects the curls of J in the source cancel, this propagation effect is the only possible source of curl of B.

What's Radiation?

The essential feature of radiation is its ability to transfer impetus to a point *along* the direction of propagation *in such a way* that we can assign the radiation itself an analogical type of impetus. Because of this, we will, as we mentioned in *PFR-M*, speak of the momentum of light.

Unlike the *force* that can be exerted by a given part of an E or B-field-activated plana, radiation is a feature of the plana that does not need to be maintained by the source. If charge could be created, i.e. if conservation of charge could be violated, radiation could be likened to the front of the propagation of the static E or B-field in that once started, it keeps going. Indeed, such a front exists within any already established field, but has no visible effects other than the very fact that field is still there. However it is unlike such an effect in that a certain flux of radiation can cause, no matter how it spreads out, a fixed total amount of momentum.

Thus, the key empiriometric sign of radiation is that the net impetus (and energy) that crosses a given small area in a given small time falls off as the inverse square of the distance (from the distant confined source), so that the total (analogical) momentum flowing through a spherical surface around the source is the same (within the given level of approximation) no matter how big the sphere is. This is important because it indicates that the radiation is a traveling power of the plana since no matter how far one gets away from the source the radiation loses none of its power to act, but rather spreads it out.

In this sense, radiation is a distinct feature. Though, as we have indicated, radiation can occur with various dispositions of the qualities of A and ϕ, a certain disposition of A is necessary for radiation. For example, A must be such that $\ddot{A} \neq 0$.

Radiation is also distinct from induction effects, which also involve changes in A. For example, consider two identical circular conducting loops threading the same axis but in different planes as shown below.

Initially   *Finally*

If one loop is driven by a sinusoidal voltage with a period much longer than the time it takes light to travel to the second loop, the changing A-field caused by the first will cause a current in the second as the plana tries to resist the change in A by activating impetus in the charges in the second loop via the magnetic E-like effect given by $-\dfrac{1}{c}\dfrac{\partial \vec{A}}{\partial t}$.

Assume that the wires are completely non-resistive wires, so that no net voltage can be sustained; for example, a gradient of ϕ must be compensated for by an inductive effect (i.e. the $\dot{A}$ effect). And, *inside the wire* of the secondary loop, where there is no ϕ-field, the change in current must be exactly such as to cancel A; it can be shown that this, in turn, means that there can be no net B-flux through the secondary loop (see end of chapter problem). With the non-resistive wire assumption, we do not have to worry about losing energy to heat. However, the current flowing in the second loop acts back (after propagation time during which the current, according to our assumption, changes very little) on the first loop causing an A-field that acts to try to decrease the A-field caused by the current flowing in the first loop. This means that $\partial A / \partial t$ in the first loop is now smaller than necessary to cancel to fixed applied voltage (which is of the type represented by $-\nabla \phi$). Hence, the increasing A-field from the second loop causes a net force on the free carriers in the direction of the already existing current, thus increasing the current until the net force on the carriers is again zero. Thus, effectively, i.e. in terms of the end result, the energy input into the first loop is transferred to the second loop. There is now a current, I_2, in the second loop as a result of voltage that is applied to the first loop. That is, an energy of $\dfrac{1}{2}LI_2{}^2$ (where L is the inductance of the second loop) appears in the second loop from power applied to the first loop, so that we have an energy transfer, and, thus, in a certain way, impetus is transferred. However, note that the effect is not localized but requires consideration of both loops. There is no single localizable region of the field that can be identified as moving outward "carrying" a fraction of the impetus caused; it's only the net effect that gives us a sort of energy transfer or impetus transfer. It is important to notice that the A-field, which is a potential momentum, does indeed propagate outward from each differential element of impetus activated charge. But, it is the interaction of these various A-fields within each loop as well as one loop on another that is responsible for the *complete* effect that we are discussing.

A second reason that we do not look for an analogical impetus propagation in induction fields is that such fields cannot be said to have a (closely) analogical impetus as they propagate forward, since the impetus they cause is *perpendicular* to the direction of their propagation and impetus, by definition, moves along its direction of travel for it is the very cause of the motion of the body of which it is a property.

Of course, there is a B-field type interaction occurring along with the properly inductive ($\dot{A}$) effects, which can cause impetus in the direction of its travel, but we have already treated above the difference between radiation and B generically (along with E proper, i.e., $-\nabla \phi$).

In contrast to inductive effects, a radiative field can locally cause impetus in a body (e.g., a small piece of metal) in the direction of the field's propagation, when it "hits" the body. Furthermore, as an accelerating charge is causing radiation in the plana *outside* the charge, the action of accelerating the charge has already resulted (due to interactions *inside* the charge) in a force on the charge that acts in the opposite direction of the emitted radiation. That is, we say a "back reaction" force acts on a body that is emitting radiation. The radiation emitted has the capacity to cause impetus of exactly the same magnitude but of the opposing type as that caused in the radiating massive body, thus conserving momentum in an analogical sense. We will see in a rough general way how this comes about after we investigate the cause of electromagnetic radiation. At that point, we also will give the more refined answer to our title question: "What is radiation?"

Radiation from an Infinite Wire

The simplicity of our long wire example allows us to calculate its A-field rather quickly in some cases. If, for example, we allow the wire's current to everywhere change quadratically in time, namely we take: $I(\vec{r},t) = I_0 \left(\dfrac{t}{\tau}\right)^2$, then equation 6.5 gives:

$$6.11 \qquad \vec{A}(\vec{r},t) = \frac{\hat{z}}{c} \int_{-b}^{b} \frac{I(\vec{r}'',t-R/c)}{R} dz' = \hat{z}\frac{I_0}{c\tau^2} \int_{-b}^{b} \frac{(t-R/c)^2}{R} dz'$$

Where $R = \sqrt{x^2 + y^2 + z'^2}$, because the current is on the z-axis, (i.e., $x' = 0$, $y' = 0$), and the long wire should be independent of z so we take $z = 0$. Note also we eventually must let b get very large to make our long wire approximation.

After substituting and taking the limit (see end of chapter problem), we get: $\ddot{A} = \dfrac{4I_0}{c\tau^2} sinh^{-1}(b/\rho) \rightarrow \dfrac{4I_0}{c\tau^2} \ln\dfrac{2b}{\rho}$, where $\rho \equiv \sqrt{x^2 + y^2}$ and we take $b \gg \rho$ to get the second form. Hence, by equation 6.10b, we have a curl of B even in the free plana (i.e. no current at the viewpoint) with a *closed* (at infinity) source. Note that this would not be the case if the current increased linearly, i.e. $I(t) = I_0 \dfrac{t}{\tau}$, for then it is clear upon inspection of

$$\vec{A}(\vec{r},t) = \frac{I_0}{c\tau} \int_{-b}^{b} \frac{(t-R/c)}{R} dz' \text{ that } \ddot{A} = 0.$$

However, we emphasize that $\ddot{A} \neq 0$ does not necessarily mean radiation is produced. It is a necessary condition for radiation in the usual sense, but not sufficient. This can be illustrated by our case of a quadratically increasing current in a wire. For example, in the infinite line limit, calculation (see end of chapter problem) shows that the net momentum (and energy) does not fall off as $1/\rho$, which is necessary to conserve momentum (and energy) in cylindrical symmetry, thus apparently ruling out radiation in this limit. Now, because the energy and momentum of radiation are conserved, radiation can reach "infinity;" so, it is natural to ask if this conclusion holds up when we look at infinity, i.e. arbitrarily far from the source. This means we must face, for example, the closure of the circuit, making say a rectangle. We can close the circuit using the more symmetrical circular loop discussed in Chapter 3. In that case, we can see that our quadratic-in-time current leads to a momentum that falls off faster than the inverse square of the distance from the source; hence, the loop does not emit radiation in the usual sense.

We can already begin to sense the complexity involved in radiating sources. In order to avoid the complication of detailed sources while still capturing the nature of radiation in its simplest form, radiation is usually discussed in the "plane wave approximation" in which the fields oscillate in a sinusoidal manner at a fixed frequency with fixed peak magnitude and propagate in a fixed direction. We will introduce and explain plane waves in a later section.

For now, we return to understanding how the *propagation* of changes of A (which implies $\ddot{A} \neq 0$) can produce qualitatively new states of A. In particular, we can see physically how the propagation of A can result in curl of the curl of A, i.e. curl of B, through the following conceptually transparent case. Consider our long wire antenna with a current that is increasing then decreasing then increasing so that it causes an A-field that is first large then small then large again. The resulting pattern at some point in space will be as shown here.

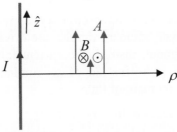

Such a pattern breaks the $1/r$ dependence of B that is responsible, as we saw in Chapter 3, for curl B being zero everywhere but at the wire (in the limit of an infinitely thin wire). In short, because the finite speed of propagation one can have a curling B in a region that does not include any current.

More on Origin of B

To better understand the origin of B, and thus the nature of radiation which necessarily has a B-field, it is helpful to derive the equation for B as a function of J. This is done very directly by taking the curl of equation 6.9, then using Maxwell's 2nd and 3rd equations ($\nabla \cdot \vec{B} = 0$, $\nabla \times \vec{E} = -\dfrac{1}{c}\dfrac{\partial \vec{B}}{\partial t}$), the vector identity given below equations 6.2, and the solution to $\Box \vec{V} = \overrightarrow{source}$ gleaned, for example from equations 6.4 and 6.5, we get:

6.12
$$\vec{B}(\vec{r},t) = \nabla \times \vec{A}(\vec{r},t) = \frac{1}{c} \int \frac{[\nabla' \times \vec{J}(\vec{r}',t)]_{t \to t-R/c}}{R} dV'$$

$$\text{where: } \vec{R} = \vec{r} - \vec{r}' \text{ and } \vec{\nabla}' \equiv \hat{x}\frac{\partial}{\partial x'} + \hat{y}\frac{\partial}{\partial y'} + \hat{z}\frac{\partial}{\partial z'}$$

Note well the square brackets in the integrand which indicate retarded time, the signature of finite speed of propagation.

This equation clearly shows that what matters is the curling of J as it is propagated to the given part of the plana being affected. As discussed for the previous similar integrals (e.g. equation 6.8), the quality (in this case the power or ability represented by $\vec{B}$) that is caused at the given spot is the result of the action of all the different pieces of the source, and this fact is manifested formally in the sum (integral) in the equation.

Finite Current Element

If we next consider a finite current element such as discussed in Chapter 5 and shown again below, we automatically introduce a ϕ, indeed a changing ϕ. And, for example, if the current has a second time derivative, such as the current with the t^2 dependence mentioned above, there will be a non-zero $\ddot{\phi}$ as well as an $\ddot{\vec{A}}$ component in the field. This is obvious because of the charge build up shown in the diagram below.[3]

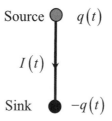

Radiation results when we force an oscillatory current. This is fundamentally an electric dipole antenna. However, standard dipoles are driven from the center and do **not** have a constant current throughout; instead, the current falls to zero at the ends and the charge is distributed along the wire instead of accumulating just at the ends. Such an antenna is shown below at one moment of time.

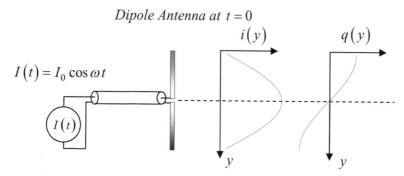

The spring mounted antennae that you sometimes see on cars or small trucks, especially on police cars, are fundamentally dipole antennas. In such "whip" antennas, the lower half of the dipole is provided by "reflection" off the metal of the car.

Uniformly Moving and Accelerated Point Charge

An even simpler example that generates both ϕ and A radiation is an accelerating small charged particle, the so-called accelerated "point" charge. The arbitrarily moving point charge generates an E-field and a B-field with two distinguishable parts, the "velocity" component and the acceleration or radiation component of the $\vec{E}$ and $\vec{B}$ fields. We will not derive the equations for the fields, leaving that for more advanced courses, but will state the results:

6.13a,b $$\phi = \left[\frac{q}{\kappa R}\right], \qquad \vec{A} = \left[\frac{q\vec{v}}{c\,\kappa R}\right]$$

6.14 $$\vec{B} = \left[\hat{R} \times \vec{E}\right]$$

---

[3] It's a helpful exercise to convince yourself that only a non-zero first derivative in the current is needed to yield a $\ddot{\phi} \neq 0$.

$$6.15 \quad \vec{E}(\vec{x},t) = e\left[\frac{\hat{R} - \dfrac{\vec{v}}{c}}{\gamma^2 \kappa^3 R^2}\right] + \frac{e}{c}\left[\frac{\hat{R} \times \left\{(\hat{R} - \dfrac{\vec{v}}{c}) \times \dfrac{\dot{\vec{v}}}{c}\right\}}{\kappa^3 R}\right]$$

Where, as earlier: $\vec{R} = \vec{r} - \vec{r}'$, $\vec{r}$ is the field point, $\vec{r}'$ is the position of the charge,

$\kappa = \left[1 - \hat{R} \cdot \dfrac{\vec{v}}{c}\right]$, and the square brackets indicate a quantity that is to be evaluated at

the retarded time, i.e., one replaces t with $t - R/c$.

The first term in equation 6.15 is the velocity term, the second is the radiation term. Once again, don't be afraid of these two rather imposing equations. You will not be expected to memorize or to be able to manipulate and use them for this course (though there are some simple end-of-chapter problems on part of them). All that is required at this point is to recognize that the finite propagation time of ϕ and A introduces significantly new effects as is suggested by the much greater complexity of the E and B relative to ϕ and A.

To get further insight into the effect of a finite speed of propagation, we will first do a quick study of the ϕ-field of a *uniformly* moving point charge. Then we will move to a discussion of the fields produced by an *accelerated* charge.

Potential of a Uniformly Moving *Part-less* Charge

We start very simple by ignoring the extension of the charge; that is, we treat it as one discrete part with no extension.[4] For simplicity and to insure consistency with this assumption, we take the plana to be composed of large discrete cells, taking the particle to be the same size and shape as the plana parts. We then consider the particle to jump into the place of one such part of the plana after another every so often, so that it moves, on the average, at a uniform speed, v. (We assume a slow speed of travel so that special relativistic effects can be ignored.) What ϕ results? The values of ϕ in the forward and backward directions are the easier to understand since the propagation is simpler in those directions. If, at a given time, we look at a distance d from the particle along the line of travel (i.e. in *front* of it), the field is reduced by a factor of $1 - v/c$ relative to what it would be if the charge were at rest. And, if we look a distance d *behind* the particle, the field is enhanced by a factor of $1 + v/c$. Figure 6-1 shows the field as it travels in such a grid. Because the particle is shown traveling only at a fraction of the speed of light the backward enhancement and forward suppression are not very obvious, but they would become increasingly so as the speed approached c.

---

[4] We make a mental construct (being of reason) by ignoring any distinction of parts, treating it as a part-less substance.

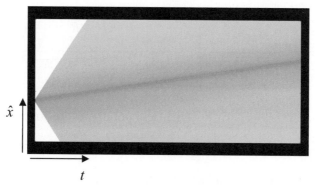

$\hat{x}$

t

Figure 6-2: Evolution of ϕ-field due to a uniformly moving charge in a discrete plana. Darker regions indicate higher potential. The dark line represents the path of particle. Note it moves at angle at $1/10\ c$. The lines that mark the boundaries between the yellowish region and the white represent the position as a function of time for objects traveling at the speed of light in the plus and minus x-direction.

The suppression/enhancement effect is easily understood in the following way by considering the propagation. The particle activates a certain intensity ϕ in the plana cell next to it. This cell then activates the next, moving along at speed c. However, the particle catches up somewhat to the field that it emitted forward, so that at any time after the steady state field is reached, the value of ϕ so many cells in front of the particle looks like it is weaker by a factor $1 - v/c$. It is, of course, *not* emitted weaker, it is only that the field at, say, the 20th cell in front of the particle was not emitted when the particle was 20 cells away (as it might naively appear since the particle at the current moment is 20 cells away), but the field was actually emitted $\dfrac{20}{1-v/c} \sim 22$ cells ago (taking $v = 1/10c$ as used in Figure 6-1, so that $1 - v/c \sim .9$); hence, it is down in amplitude by $\dfrac{1-v/c}{20}$ rather than simply $1/20$. For the backward emission case, the particle catches up to the stronger, more recently emitted, field and thus gets an apparent enhancement by $1 + v/c$. Thus, the field is stronger in the back, weaker in the front. The detailed proof of these two effects is left as an end of chapter problem.

But, what about the other directions? They can be treated, in some respects, similarly by re-introducing the forward angle θ measured from the forward pointing velocity vector (say in the counterclockwise direction, but it makes no real difference). However, the calculation is much more complicated, and we will reserve this treatment for the problems section.

Potential of a Moving Particle *with* Parts

As we should have expected since we assumed, contra-reality, that the particle had no parts, the above result is not the correct result. In fact, the particle's ϕ-field follows it as if it were rigidly attached to it. We now show the effect of accounting for the parts of the particle and their true extension. We start with an examination of the field in front of the particle (on the line travel). Concretely, consider a particle moving toward the viewpoint which we take to be the origin.

We treat the particle as a line of length l (ignoring the dimensions perpendicular to the motion as not directly involved in the motion) of uniform charge distribution, λ, along the direction of motion. We assume the length of the line of charge is rigidly maintained during the motion.[5] Each part of the particle will contribute to the field at the viewpoint at a different retarded time. We need to take the sum of all the contributions of the finite number of parts of length Δx and then take the limit of infinitesimally small parts ($\Delta x \to 0$), which gives an integral.[6] Note the key point will be that each part will *not appear* to have the same charge at the viewpoint as it actually does. We proceed now to study this difference between apparent and actual in detail.

In actual fact, each part of the charge of length Δx has a uniform charge per unit length, λ, and so each part has a charge $\lambda \Delta x$; hence, the total charge is, in the limit of arbitrarily small parts: $q = \int_0^L \lambda dx$. However, at the origin, the length of each part of the particle will *appear* to be increased by a factor $\alpha = \dfrac{1}{1 - v/c}$. More generally the relationship between the apparent length, l', and the actual length, l, is: $l' = \alpha l$.[7] By "apparent distance (or length)," we here mean the distance that we deduce from a signal transmitted through the plana from a point of the particle to the viewpoint. This α factor arises because it takes a finite time for a signal traveling at speed c to get from the given part to the origin (see end of chapter problem).

With this in hand, we take the charge per unit length as a primitive. That is, we assume that it is the very nature of what it means to be a charged particle to have the quality of (active and passive) charge distributed uniformly according to length with measure λ. This means that the charge on each particle is *apparently* increased by α; namely, $q^{retarded} = \int_0^L \lambda[dx] = \lambda \int_0^L \dfrac{dx}{1 - v/c} = \dfrac{q}{1 - v/c}$, where here we use the square brackets around dx to indicate the apparent length of the little extension dx because the length expansion is due to the retarded time, i.e., due to the finite time it takes for the field to travel. Thus, combining the increased apparent charge with the increased apparent length, the potential as a function of the distance, r, in front of the particle can be written:

$\phi = \left[\dfrac{q}{r}\right] = \dfrac{q/(1 - v/c)}{r/(1 - v/c)} = \dfrac{q}{r}$. Thus, as can be seen by their common cause, the two effects cancel. A similar result obtains for the viewpoints along the trajectory behind the particle. Hence, the ϕ-field in front of and behind the charged particle looks the same as it does at rest.

---

[5] We ignore special relativistic effects at this point.

[6] Note that it is part of the nature of the "continuum" (i.e., extension, the first accident of all physical substances) that one cannot treat it as the infinite limit of part-less parts (i.e. "parts" that have no extension in some secondary sense). Note even a countably infinite number of such "parts," which are more like points than parts, can not make even a part of the continuum, because each new such "part" one adds is radically unlike the continuum one is trying to make.

[7] For those somewhat familiar with special relativity, note that this is *not* a special relativistic effect. In fact, recall that we have deliberately assumed a slow moving particle ($v \ll c$) so as to avoid special relativistic considerations.

In fact, as long as the particle is traveling sufficiently slowly relative to the speed of light ($v/c \ll 1$), the particle looks as if it is carrying its whole ϕ-field with it like a turtle carries its shell. Furthermore, because of this and a similar result for A, the E-field travels with the particle, and the particle with its E-field is more like a perfectly round porcupine flying through space with its extended quills pointing radially outward. As one approaches the speed of light, ϕ changes such that the E-field lines (the "quills") compress along the direction of travel.

A, ϕ and E, B for an Accelerated Charge in the Point Limit

Now that we have some feel of how to go about analyzing the finite propagation time, we will leave the detailed calculation for a more advanced course and simply show the results one gets if one uses the program by Dr. Joseph Haller (which we will make available when ready[8]) to do the work for us. In particular, we give the results for a particle that starts at rest and then for a short time, T, accelerates at a constant rate a and then continues on at a uniform speed, aT.

We will illustrate the results for ϕ and A using flip animations. The instructions and perforated pages for creating your own "flip books" of ϕ and A for the briefly accelerated point charge are at the back of this book. Those pages also include details of the physical situation being modeled. The flip books use a color scheme in which a color scale represents the magnitude of those fields, which is also explained there. The flip books are created for a particle with a final speed $v \sim .4c$, so that there are slight relativistic effects; most prominently, notice the slight ovalness in the field patterns. Each image shows the state of the plana in the x-z plane, where the particle moves along the z-direction.

Notice the sharp contrast between ϕ and A. Namely, the first image in the flip book represents the field generated by the particle, which has up to this point been at rest. In particular, it has produced a symmetrical ϕ-field, leading to the multicolor band structure you see. Whereas, since the particle has no impetus before the time shown in the first frame, the particle produces no A-field, and the first image is thus black. At the other end of the motion, once the particle is traveling at a uniform speed, it generates an approximately spherically symmetric field, the so called velocity fields, that travel with the particle. During the brief acceleration, both ϕ and A become compressed in front. The band that travels out in a circle like a ripple in a pond is the radiated portion of the field. It is due to the fact that the accelerating particle gets closer to the previously emitted field and thus "bunches up". An analogous spreading occurs behind the particle.

The flip books reveal that in the velocity-field domain, i.e., that part of the field that is due to the charge's uniform motion, A is pretty much the same as ϕ. This is to be expected as equations 6.13 show that, in that domain, A only differs from ϕ by a multiplicative factor of $\vec{\beta} = \vec{v}/c$.

Though there are 3 components of the E and 2 components of B,[9] of these only E_z includes contributions from both ϕ and A. A 20-frame set is given in the pages at the end of the book for the $\nabla\phi$ "half" of the E_z-field as well as the $\dot{A}_z$ "half." Study carefully the various frames of ϕ and $\vec{A}$ and their derivatives to learn how they work together to

---

[8] See iapweb.org/pfrem/software

[9] $E_x = -\nabla_x\phi$, $E_y = -\nabla_y\phi$, $E_z = -\nabla_z\phi - (1/c)\dot{A}_z$, $B_x = \partial A_z / \partial y$, $B_y = -\partial A_z / \partial x$

cause the effects we see. The last set of 20-frames shows $\partial \phi / \partial x$. Note in all three sets, the negative sign properly belonging to the E-field definition is omitted for simplicity.

Summary of the Causes of B and E and Their Relation to Radiation

From these examples and from the calculation given in equation 6.12, we see that there are fundamentally two causes of B: (1) impetus activated charge which causes a curling B and (2) force(s) that change the impetus of an extended charge, causing second time derivative of A, $\ddot{A}$, in the plana, which results in curl of B as represented in equation 6.10b. In the case of radiation, these forces, as we have mentioned and will see below, will have to do extra work, work beyond just giving energy to a particle of a given mass, to create a curl of B in regions far from the sources.

Another way to approach the generation of B as well as E is by use of the relatively easily derived, but only recently discovered, Jefimenko equations (in *cgs*):

6.16a,b
$$\vec{E} = \int \left(\frac{[\rho]}{R^2} \hat{R} + \frac{[\partial \rho / \partial t] \hat{R}}{Rc} - \frac{[\partial \vec{J} / \partial t]}{Rc^2} \right) dV$$

$$\vec{B} = \int \left(\frac{[\vec{J}] \times \hat{R}}{c R^2} + \frac{[\partial \vec{J} / \partial t] \times \hat{R}}{c^2 R} \right) dV$$

Note that since these equations use E and B, they tend to obscure the further specification of A and ϕ.

Again, you are not expected to memorize or to be able to use these equations to solve problems (except in very simple cases); they are introduced so that, at least once in your study of E&M, you will have seen the formal relation between E and B and their sources. That there is such a relation should have already been evident from the things you already know. In particular, charge causes the qualification of the plana known as ϕ and impetus activated charge causes A, and their particular form is set by the relative location of the parts of the source along with the effects of propagation that arise as the state of the source changes. And, the B-field arises when the plana is disposed in such a way that there is a curling part of A, while what we call E is due to two different agents, the plana resisting the changes in A ($-\frac{1}{c}\frac{\partial \vec{A}}{\partial t}$) and true electric field whose structure is indicated by the divergence of ϕ ($-\vec{\nabla}\phi$). Hence, in empiriological terms, since ρ and J respectively cause ϕ and A, and E can be expressed in terms of ϕ and A, there is a mathematical relation between E and ρ and J. Similarly, because B can be expressed in terms of A, there is a relationship between B and J.

In terms of Maxwell's equations (equations 6.1), we have seen that the "displacement" term, $\frac{1}{c}\frac{\partial \vec{E}}{\partial t}$, contributes $-\frac{1}{c}\frac{\partial \vec{\nabla}\phi}{\partial t} - \frac{1}{c^2}\frac{\partial^2 \vec{A}}{\partial t^2}$, without which we would not have the variation in second time derivative of the A-field that "makes" A-propagation, and hence radiation, possible. Recall that the role of the displacement term is to incorporate $\nabla \cdot \vec{J} = -\dot{\rho}$. That is, it incorporates the fact that charge can build up and currents can change, but charge can only build up and current can only change when there are currents (impetus-activated charge) to build the charge in the first place and to be the subject of change in the second, and with currents come magnetic fields, i.e. an A-field, which usually means a B-field.

Now, from the more full perspective of this chapter, we see that this necessarily bring us to propagation. Namely, as we've seen, when the current is changing, the A-field changes; this means the B-field (if there is one) and the electric field (at least, in the extended sense, i.e., $E_i \sim \partial A_i / \partial t$) change. This, in turn, implies that usually a changing E is accompanied by a changing B and vice-versa. Thus, it is often thought that a changing E causes changing B and vice versa. Note carefully and remember that, though this may sometimes be a useful fiction, it is indeed a *purely* mental construct, since E does *not cause* B nor vice versa.[10]

Indeed, even within the equational structure of the theory in terms of E and B, one needs to be careful of this construct. For example, consider a steady current causing a linear build up in charge on a capacitor, so that i.e. $\rho(t) \propto t$, such as considered in Chapter 5. In such a case, there is no changing $\vec{J}$,[11] thus no changing A and thus no changing B, but there is a changing E. Hence, in this case, it is clear then that a changing E does not cause a changing B.[12]

Electromagnetic Plane Waves

Now that we have seen something of the sources of fields, in particular radiation fields, we will now simplify our approach to radiation by looking far enough away from the source that we can neglect the non-radiative components of the field. That is, we neglect what are called the near and the intermediate fields such as those of the long wire approximation (see the summary for a precise definition of these fields). This leads us via Maxwell's equations (6.1 *I-IV*) to the two wave equations for E and B:

6.17a,b $\qquad \dfrac{1}{c^2}\dfrac{\partial^2 \vec{E}}{\partial t^2} - \nabla^2 \vec{E} = 0 \qquad\qquad \dfrac{1}{c^2}\dfrac{\partial^2 \vec{B}}{\partial t^2} - \nabla^2 \vec{B} = 0$

The general *physical* solution to these equations (within a phase angle, i.e., within the addition of a constant angle to the argument of each cosine below), coupled with the second two Maxwell's equations, is:

6.18a,b $\qquad \vec{E} = E_0 \cos(\omega t - kz)\,\hat{x}, \qquad\qquad \vec{B} = B_0 \cos(\omega t - kz)\,\hat{y}$

$$\text{where: } E_0 = B_0,\ \omega = k\,c$$

These are the equations for plane waves traveling in the z-direction. They are so called because at a given time each of the fields is the same at every point in a plane. Figure 6-3 shows what the field looks like on the $z = 0$ plane at $t = 0$, one eighth period later and then finally one half period later. Figure 6-4, shows the way the plane wave qualifies the plana at one particular moment of time ($t = 0$), showing the field for only the $z = 0$ plane and one half wavelength away from that $z = \lambda / 2$, though of course, such field diagrams

---

[10] There is no reason to say that B causes E, but there are reasons to say it does not.

[11] In fact, the current source will have some changing currents, because it will not be a perfect current source (consider, for example, a basic current source consisting of a single transistor with a Zener diode at its base).

[12] Note, however, it seems in the real world, as opposed to the world of beings of reason in which charges are points that densely and uniformly fill space, there will be no perfect cancellation of the E-fields from the various charges that are decelerated, for example, as they stop to build the charge of a given region. This simply means we have to respect, as we must with the full classical E&M theory itself, the limits of validity of our approximations.

could in principle be drawn for any plane $z = z'$. Such plane waves are, of course, clearly a mental construct (being of reason) that keeps us focused on the most generic properties of radiation, and as such we have to be careful about the limits of its validity.

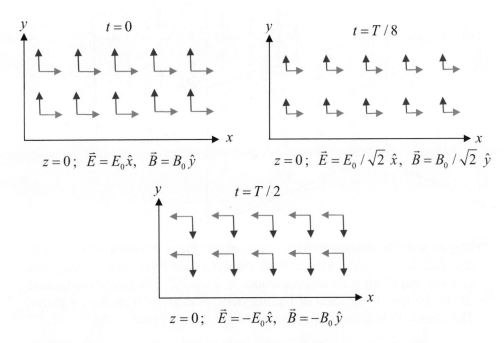

Figure 6-3: A series of snap shots of a plane wave in the plane perpendicular to its direction of travel, note that for such a wave $\vec{E} = E_0 \cos(kz - \omega t)\hat{x}$ and $\vec{B} = B_0 \cos(kz - \omega t)\hat{y}$: **a.** (*top left*) at $t = 0$, $z = 0$, showing $\vec{E} = E_0\hat{x}$, $\vec{B} = B_0\hat{y}$. b. (*top right*) $t = T/8$, $z = 0$ showing $\vec{E} = E_0/\sqrt{2}\,\hat{x}$, $\vec{B} = B_0/\sqrt{2}\,\hat{y}$ **c.** (*bottom*) at $t = T/2$, $z = 0$, showing $\vec{E} = -E_0\hat{x}$, $\vec{B} = -B_0\hat{y}$. The electric field is shown in red and the magnetic field in blue.

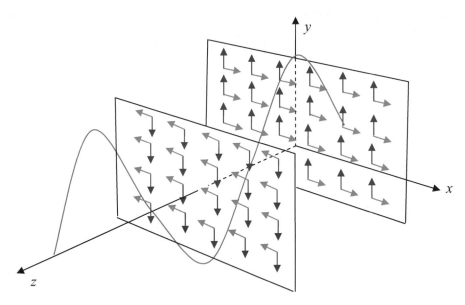

Figure 6-4: Snap shot at $t = 0$, of a plane wave traveling along the z-direction shows the electric (red) and magnetic field (blue) that fill the plane at $z = 0$ and a half wavelength distance at $z = \lambda / 2$. The sine wave (green) shows the spatial variation of the magnetic field at $t = 0$ in the y-z plane. The electric field follows this same pattern in the x-z plane.

A more realistic picture can be drawn, for example, in the dipole antenna approximation; it is shown in Figure 6-5.

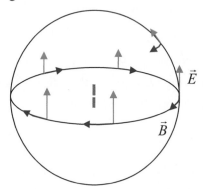

Figure 6-5: Electric and magnetic field for radiation emitted from an ideal dipole. Note that far away from the dipole and for a small region of the sphere, i.e. locally, the E and B pattern approach that of the plane wave.

Since ϕ and A are specifications of E and B, it is helpful to look at radiation in terms of them. ϕ and A also satisfy wave equations, as can be seen by taking equations 6.7 and 6.4 with $\rho = 0$ and $\vec{J} = 0$:

6.19 $\qquad \dfrac{1}{c^2}\dfrac{\partial^2 \phi}{\partial t^2} - \nabla^2 \phi = 0 \qquad \dfrac{1}{c^2}\dfrac{\partial^2 \vec{A}}{\partial t^2} - \nabla^2 \vec{A} = 0$

The simplest plane wave solution that results in the E and B plane waves described in equations 6.18 is:

6.20 $\qquad \vec{A} = A_0 \sin\left(\omega t - kz\right)\hat{x}$

This wave moves along the z-direction. The figure below illustrates how it implies the existence of E and B described by equations 6.18a,b. Physically, various parts of the source (which is some great distance away) cause an A-field of a certain magnitude that then propagates to the point of interest. The source varies in a sinusoidal manner so that the intensity of A at the same point in the plana a moment later is smaller (or larger). The plana resists this change (the "$\dot{\vec{A}}$ effect"), so that we have an electric field in the extended sense in the x-direction. Furthermore, at any moment in the plana, the source has caused an A-field that varies along the z-direction; this means there is a B-field in the y-direction. Hence, we have sinusoidally varying E and B-fields, each of which is perpendicular to the direction of the travel as described by equations 6.18a,b.

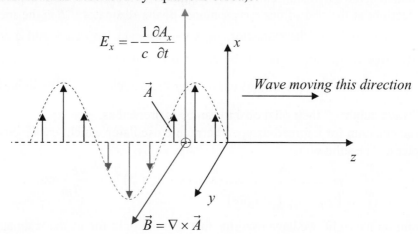

Clearly, there is no corresponding way to reproduce that same E and B-field pattern using only the ϕ-field. Thus, the simplest way to think of plane wave radiation is as an A-field traveling outward from the source in sine-wave form in the manner shown above.

Further Answer to "What is Radiation?"

To further our understanding of radiation (standard meaning), we need to analyze its distinguishing feature, its ability to cause impetus along the direction of its propagation at large distances from the source with no diminution in its total power along its *entire* wavefront as it travels outward in all directions. How does it cause impetus along its own direction?

Consider a portion of the plana that has radiation moving towards an object as shown. For simplicity, take the radiation to be resonant with the atomic structure of the material it hits.

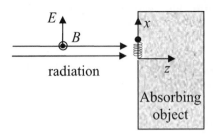

radiation

Thus, modeling the resonator in the material as a damped harmonic oscillator, at the surface of impact which we take to be $z = 0$, and assuming a plane wave incident on the surface of the form given by equation 6.18, we can write:

6.21 $\ddot{x} + \beta \dot{x} + \omega_0^2 x = \dfrac{q}{m} E_{0x} \cos(\omega t - kz) = \dfrac{q}{m} E_{0x} \cos \omega t$

Where x is the displacement from equilibrium of the charged particle, say an electron, at the end of our spring model of the resonator, β is the measure of the resistive ability of the resonator, ω_0 is the resonant frequency and q and m are the effective charge and mass of the electron in our model. ω is the frequency of the light and $k = \dfrac{2\pi}{\lambda}$, is a way of expressing the wavelength of the light, λ, called the "wave number;" their relationship can be expressed as: $\omega = kc$.

This is the equation for a forced damped harmonic oscillator which was discussed in *PFR-M*, Chapter 6. The solution is:

6.22 $v_x = \dfrac{qE_{0x}}{m} \dfrac{\omega}{\sqrt{\left(\omega_0^2 - \omega^2\right)^2 + \left(\beta\omega\right)^2}} \sin(\omega t - kz + \phi) \xrightarrow{\omega \to \omega_0} \dfrac{qE_{0x}}{\beta m} \cos \omega t$

In the limit on the right, we have used the fact that the light forces the resonator at its own resonant frequency, so that the phase: $\phi = \tan^{-1} \dfrac{\beta\omega}{\left(\omega^2 - \omega_0^2\right)} \xrightarrow{\omega \to \omega_0} \dfrac{\pi}{2}$.

Taking the equation for the fields of the radiation:

6.18 $\vec{E} = E_{0x} \cos(\omega t - kz) \hat{x}, \quad \vec{B} = B_{0y} \cos(\omega t - kz) \hat{y}$

with set $z \to 0$, and using the Lorentz force law $\vec{F} = q\left(\vec{E} + \dfrac{\vec{v}}{c} \times \vec{B}\right)$, we can write the following force equation:

6.23 $F_z = \dfrac{dp_z}{dt} = q\dfrac{v_x}{c} B_y = \dfrac{q}{c}\left(\dfrac{qE_{0x}}{\beta m} \cos \omega t\right)\left(B_{0y} \cos \omega t\right) = \dfrac{q^2 E_{0x} B_{0x}}{c\beta m} \cos^2 \omega t$

Hence, we see that the radiation does indeed exert a force in the z-direction and thus does cause impetus along its direction of travel.

Poynting Vector: Flow of Momentum and Energy in Electromagnetic Wave

Now, we can determine the relationship between momentum and energy. Using the Lorentz force law, we can write the following power equation:

6.24 $\dfrac{dE}{dt} = \vec{F} \cdot \vec{v} = q\left(\vec{E} + \dfrac{\vec{v}}{c} \times \vec{B}\right) \cdot \vec{v} = qE_x v_x = \dfrac{q^2 E_{0x}^2}{\beta m} \cos^2 \omega t$

In the last, equality we have used equation 6.22 in the requisite limit.
Hence, we have, using equation 6.24 along with equation 6.23:

$\dfrac{\dfrac{dp_z}{dt}}{\dfrac{dE}{dt}} = \dfrac{1}{c}\dfrac{B_{0x}}{E_{0x}}$, so we can write: $F_z = \dfrac{dp_z}{dt} = \dfrac{1}{c}\dfrac{dE}{dt}\left(\dfrac{B_{0x}}{E_{0x}}\right)$, and since $E_0 = B_0$, we have an

equality independent of the fields, and we can write simply:

6.25 $$p_z = \dfrac{E}{c}$$

Thus, we have momentum associated with the outgoing electromagnetic wave that is related to energy in the way we discussed in Chapter 10 of *PFR-M*. In particular, the radiation has the power to cause a certain amount of impetus per unit time, and that power always moves through the plana at the same speed, c. Hence, there is no factor of a ½ in our relation between momentum and energy, rather we get: $E = pc$ as the rate of transfer of the momentum.

We now turn to use this relation between momentum and energy to bring out a few important equations. A simple, yet useful, one is easily obtained. We divide equation 6.25 by a field volume, and obtain the momentum per unit volume, g, in terms of the energy density, $U \equiv \dfrac{dE}{dV}$:

6.26 $$g = U / c$$

Since the electromagnetic wave is continuous, the momentum comes as a certain amount per unit time. The wave applies a certain number of phor per second per cm^2, i.e., a force per unit area to an absorbing surface. This, in turn, can, via equation 6.25, be expressed in terms of energy per unit time per unit area. To say more, we need to introduce a couple of results that can be deduced from Maxwell's equations. First, the magnitude of the instantaneous power (dE / dt) per unit area, as well as the direction of the radiation is given by "Poynting's vector:"

6.27 $$\vec{S} = \dfrac{c}{4\pi}\vec{E} \times \vec{B}$$

Second, it can be shown that

6.28 $$U = \dfrac{1}{4\pi}\left|\vec{E} \times \vec{B}\right|, \text{ so that}$$

6.29 $$\left|\vec{S}\right| = cU$$

This means we can write the momentum, including its direction, per unit time per unit volume as:

6.30 $$\vec{g} = \dfrac{\vec{E} \times \vec{B}}{4\pi c}$$

In these ways, as alluded to in Chapter 2, we can talk about the radiation fields as having an energy density and as carrying an analogical momentum and energy.

Radiation Back Reaction: The Force Caused on the Source as Radiation is Emitted

The previous explains how and in what sense radiation carries potential momentum in the direction of its propagation. But, how does radiation causes a force as it is generated by a charged body? It does *not* accomplish it by literally acting back but in the following way.

We treat the particle as made up of a truly extended charge, so that even within minimal parts the particle will have some analogical or effective divisions of its intensity with respect to position. Then, to see the essential causality of the effect, we consider a charged body that consists of two isolated small charged parts, say in the point particle limit, held together by a stiff rod. The charged parts are taken to be a distance d from each other as shown below. We will eventually let this distance d (and thus the rod) become arbitrarily small to get our single point particle approximation that includes the fact that the particle is actually extended.

Now, one charged part of this body will act on the other via the A and ϕ fields, and thus you should immediately see a vehicle for a net force. In particular, the fields have to propagate from one part to another, and if there is an asymmetry between the way the two effects propagate, there is a possibility of a violation of Newton's third law. In other words, the force that the first part exerts on the second part can be different than the second part exerts on the first. It can be shown (end of chapter problem) that this does not happen when the particles are in uniform motion[13], but only happens when the body is accelerated. Newton's third law is then recovered analogically because of the momentum that can by analogy be attributed to the emitted radiation as described previously.

There is much more detail to discuss that is beyond the scope of this book. People like Max Abraham, Hendrik Lorentz, Paul Dirac and many others investigated the import of radiation and you are encouraged to look at their work, in particular the so called Abraham-Lorentz-Dirac equation for the motion of a radiating charge in the point limit.

Finally, we can give the summary definition of radiation. Radiation is a certain disposition of A and ϕ activated plana, such that, upon being caused in the plana, leaves the source with a certain momentum that it did not have before, and acts to change successive parts of the plana so that it is maintained in such a way that upon being absorbed it can cause impetus in the direction of its motion. It has E and B perpendicular to each other and to the direction of travel, and the fields diminish according to $1/r$ as it moves away from the source. Extra energy is required to emit radiation. In particular, to accelerate a charge you have to give energy beyond what it takes to give impetus to the massive particle to get it up to the given speed. The so-called radiative force acts against you causing you to have to do extra work.

We have now introduced all the key physical properties of the plana that are packed into the Maxwell formalism. We will now step back to look at the integrity of the whole system of primal causes involved in E&M as revealed by special relativity. Because of its deep special relativistic nature, classical E&M can be called the first relativistic theory.

---

[13] As long as the particle does not move faster than the speed of light, which cannot happen in empty plana, but can happen in water and other media. This later effect is not relevant to this chapter, but properly belongs under the heading of dielectric and magnetic materials.

Summary

The electric and magnetic powers of the plana (to which Maxwell's equations are relevant) can be schematized in the following way.

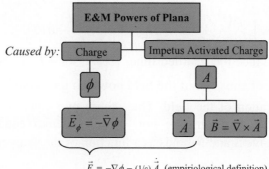

$\vec{E} \equiv -\nabla\phi - (1/c)\,\dot{\vec{A}}$ (empiriological definition)

Maxwell's equations coupled with the equations for fields in terms of the potentials,

$\vec{E} = -\nabla\phi - \dfrac{1}{c}\dfrac{\partial\vec{A}}{\partial t}$, $\vec{B} = \nabla \times \vec{A}$, give (in *cgs*):

6.2 *I-IV* $\nabla \cdot \vec{E} = 4\pi\rho$ --------- $-\nabla^2\phi - \dfrac{1}{c}\dfrac{\partial \nabla \cdot \vec{A}}{\partial t} = 4\pi\rho$

$\nabla \cdot \vec{B} = 0$

$\left. \begin{array}{l} \\ \nabla \times \vec{E} = -\dfrac{1}{c}\dfrac{\partial\vec{B}}{\partial t} \end{array} \right\}$ --------- $\left\{ \begin{array}{l} \vec{B} = \nabla \times \vec{A} \\[2mm] \vec{E} = -\nabla\phi - \dfrac{1}{c}\dfrac{\partial\vec{A}}{\partial t} \end{array} \right.$

$\nabla \times \vec{B} = \dfrac{4\pi\vec{J}}{c} + \dfrac{1}{c}\dfrac{\partial\vec{E}}{\partial t}$ --------- $-\nabla^2\vec{A} + \vec{\nabla}\left(\nabla \cdot \vec{A}\right) = \dfrac{4\pi\vec{J}}{c} - \dfrac{1}{c}\dfrac{\partial\vec{\nabla}\phi}{\partial t} - \dfrac{1}{c^2}\dfrac{\partial^2\vec{A}}{\partial t^2}$

Applying the *Lorenz gauge condition*, $\nabla \cdot \vec{A} + \dfrac{1}{c}\dfrac{\partial\phi}{\partial t} = 0$ (cgs), (or in *SI*: $\nabla \cdot \vec{A} + \dfrac{1}{c^2}\dfrac{\partial\phi}{\partial t} = 0$),

to the last equation on the right in equations 6.2 gives:

6.4 $\dfrac{1}{c^2}\dfrac{\partial^2\vec{A}}{\partial t^2} - \nabla^2\vec{A} = \dfrac{4\pi\vec{J}}{c}$, which can be written: $\Box\vec{A} = \dfrac{4\pi\vec{J}}{c}$

This has the solution:

6.5 $\vec{A}(\vec{r},t) = \dfrac{1}{c}\displaystyle\int \dfrac{\vec{J}\left(\vec{r}\,',t - R/c\right)}{R}dV'$,

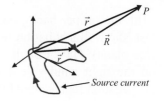

Where $R \equiv |\vec{r} - \vec{r}\,'|$ is as defined in the figure.

The primed coordinate, $\vec{r}\,'$, refers to the vector pointing from the origin to the source current. The unprimed coordinate, $\vec{r}$, points to the field point, P.

Applying the gauge condition to the first equation on the right in equations 6.2 gives:

6.7 $\dfrac{1}{c^2}\dfrac{\partial^2\phi}{\partial t^2} - \nabla^2\phi = 4\pi\rho$, which can be written: $\Box\phi = 4\pi\rho$

> *Note that it is through the mediation of the A-field (via the Lorenz condition) that Gauss's law remains satisfied even for arbitrarily moving charges.*

This has the solution:

6.8 $\phi(\vec{r},t) = \displaystyle\int \frac{\rho(\vec{r}',t - R/c)}{R}dV' = \int \frac{\left[\rho(\vec{r}',t)\right]}{R}dV'$

Both 6.5 and 6.8, introduce the retarded time, $[t] \equiv t_r = t - R/c$, which shifts the current time, t, to the time, t_r , of emission of the field now being received, thus accounting for the finite speed of propagation of the field. Thus, these two equations are explicitly causal, whereas the Maxwell's equations are not. The Jefimenko equations give the causal equations in that they give E and B-fields as a function of the sources J and ρ ; viz,

6.16a,b $\vec{E} = \displaystyle\int \left(\frac{[\rho]}{R^2}\hat{R} + \frac{[\partial\rho/\partial t]\hat{R}}{Rc} - \frac{[\partial\vec{J}/\partial t]}{Rc^2} \right) dV$

$\vec{B} = \displaystyle\int \left(\frac{[\vec{J}]\times\hat{R}}{cR^2} + \frac{[\partial\vec{J}/\partial t]\times\hat{R}}{c^2R} \right) dV$

Maxwell's full equations (6.2 *I-IV*) imply the existence of *electromagnetic radiation*, which is a new moving disposition of the qualities (i.e., the A and ϕ-fields) of plana that arises from the finite speed of travel of the fields. Equations 6.5 and 6.8 contain in principle all such effects. Electromagnetic radiation is sometimes simply called "light" after the most obvious example of electromagnetic radiation, visible light, i.e., the light that makes objects we see visible. The visible spectrum of pure colors can be memorized through the acronym, ROY G BIV, which stands for Red, Orange, Yellow, Green, Blue, Indigo, and Violet. The term "electromagnetic radiation" is used to designate a whole "spectrum" of moving dispositions of the plana, only a small subset of which can be seen or sensed in any way by us (see schematic illustration in text). We categorize these dispositions by their frequency and wavelength in the plane wave approximation summarized below. These moving dispositions of the plana carry information about their sources.

With the proper background understanding given in Chapters 3-5, the curl portion of the last Maxwell's equation reveals the *two ways a curling B-field can exist in a vacuum* (i.e., free plana): (1) Through incomplete closure of current paths represented by the first term on the right hand side and (2) time variation of the source through the finite speed of propagation represented by the second term on the right hand side.

$$\nabla\times\vec{B} = \frac{4\pi}{c}\vec{J}_{curl} - \frac{1}{c^2}\frac{\partial^2\vec{A}_{curl}}{\partial t^2}$$

Physically, there are fundamentally *two causes of B*: (1) impetus activated charge which causes a curling B and (2) force(s) that change the impetus of an extended charge, (causing second time derivative of A, $\ddot{A}$, in the plana), which results in curl of B. In the case of radiation, these forces have to do extra work, work beyond just giving energy to a particle of a given mass, to create a curl of B in regions far from the sources.

The *essential feature of radiation* is its ability to transfer impetus to a distance *along* the direction of propagation *in such a way* that we can assign the radiation itself an analogical type of momentum that is conserved in conjunction with mechanical momentum[14] because of the back-reaction summarized below. Thus, for example, as the radiation spreads out from its source, its total momentum does not change.

In light of the above, we can see that radiation is fundamentally different than "static and qausi-static" E and B-fields which also move outward from their sources.

Induction between two loops is fundamentally different from radiation because no single localizable region of the field can be identified as carrying a fraction of the impetus *and* because it *only* causes motion perpendicular to the direction of its propagation. When induction occurs via an A-field that *does* propagate along its own direction, it is fundamentally like the static E-field mentioned above, which cannot be assigned a fixed moving and conserved momentum.

The two contributions to the generation of B are manifest in the following equation.

6.31
$$\vec{B}(\vec{r},t) = \nabla \times \vec{A}(\vec{r},t) = \frac{1}{c}\int \frac{[\nabla' \times \vec{J}(\vec{r}',t)]_{t \to t-R/c}}{R}dV'$$

Note well the square brackets in the integrands which indicate retarded time, the signature of finite speed of propagation.

An "infinite" length straight wire with a current $I(t) = I_0\left(\dfrac{t}{\tau}\right)^2$ will not

generate radiation, but will cause a curl of B in the plana by breaking the $1/r$ fall-off of B which is characteristic of the long wire static field. This follows from the fact that $\ddot{A} \neq 0$ and the 4th Maxwell's equation. This result shows that, though $\ddot{A} \neq 0$ is a necessary condition for radiation, it is not sufficient.

Cutting off the ends of our "infinite" wire yields the finite current element shown to the right and brings us a step closer to the standard dipole antenna. If the current, $I(t)$, is oscillatory, it will emit radiation. Since the current effectively piles up charge at both ends, it generates radiation that contains a non-zero ϕ as well as a non-zero A. However, only A is *necessary* for radiation. The standard dipole antenna is as shown to the lower right.

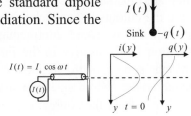

---

[14] That is, the term "momentum" as applied to massive bodies, which is the first meaning of the term and means the measure of the intensity (strength) of the impetus of a massive body.

The fields generated by *an arbitrarily moving point charge* are helpful in investigating the nature of radiative fields. The potentials caused by such a charge (called the Liénard-Wiechert potentials) can be written:

6.13a,b
$$\phi = \left[\frac{q}{\kappa R} \right], \qquad \vec{A} = \left[\frac{q\vec{v}}{c\,\kappa R} \right]$$

Where q and $\vec{v}$ are its charge and velocity, and $\kappa \equiv \left[1 - \hat{R} \cdot \dfrac{\vec{v}}{c} \right]$ and R and the brackets are as defined above.

These imply the E and B-fields given in equations 6.14 and 6.15.

ϕ-field of a uniformly and slowly ($v \ll c$) moving point charge

An analysis of the propagation of the ϕ-field caused by an (idealized) *part-less* particle yields the following. At some fixed distance in front of the particle (i.e., ahead of it along its direction of motion), the field is reduced by a factor of $1 - v/c$ relative to what it would be at rest. This is because the particle catches up to fields that have become weaker through spreading themselves into the plana around the particle. By contrast, directly in back of the particle (i.e., behind it in the direction opposite to its direction of motion) the field is increased by a factor of $1 + v/c$. This is because the particle recedes from the just emitted radiative fields, making them appear to have been emitted sooner ago then they were, thus making them appear stronger relative to the at rest case.

By considering a particle with extended parts, thus giving the particle real extension, we see that, at some distance in front of the particle, the apparent length is augmented by a factor $1/(1 - v/c)$. However, since the particle's length and charge are directly related through the linear charge density, λ, the charge is also augmented by the same factor. A complete calculation shows that a similar result obtains for all angles so that the net result is that the ϕ-field appears to exactly follow the particle. A similar result obtains for the A-field so that the particle with its E-field is like a perfectly round porcupine flying through space with its extended quills pointing radially outward. As one approaches the speed of light, ϕ changes such that the E-field lines (the "quills") compress along the direction of travel.

Fields of an accelerated slowly ($v \ll c$) moving point charge

These fields teach us more about the propagation of ϕ and A and about the radiation fields. They are illustrated in the figures in the chapter and in the "flip books" that one can make using the cutouts at the back of the book. They show how the radiation fields start "mixed in" with the near and intermediate fields and begin to emerge in the far field. Only the radiation fields survive in the far zone. These zones can be identified for a sinusoidally varying source with frequency v (and thus wavelength $\lambda = c/v$) for which the characteristic source size a is such that: $a \ll \lambda$. The zones are specified by their radial distance r from the source: (1) *near zone*: $a \ll r \ll \lambda$, (2) *intermediate zone*: $r \sim \lambda$, (3) *far zone*: $r \gg \lambda$.

Though changing E almost always appears with a changing B, a changing E does not cause a changing B or vice-versa, but instead have a common cause, the changing source. (See the equations for the potentials (6.13a,b) and/or the equations for the fields (Jefimenko equations, 6.16a,b) as a function of their sources). This is important to note as the opposite is often implied, and even sometimes said directly.

Plane Waves

In the far field, we can write, using Maxwell's equations, the *wave equations*:

6.17a,b $\dfrac{1}{c^2}\dfrac{\partial^2 \vec{E}}{\partial t^2} - \nabla^2 \vec{E} = 0, \quad \dfrac{1}{c^2}\dfrac{\partial^2 \vec{B}}{\partial t^2} - \nabla^2 \vec{B} = 0$

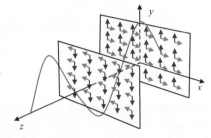

The general *physical* solution to these equations (within a phase angle, i.e., within the addition of a constant angle to the argument of each cosine below), coupled with the second two Maxwell's equation's, is:

6.18a,b $\vec{E} = E_0 \cos(\omega t - kz)\,\hat{x}, \quad \vec{B} = B_0 \cos(\omega t - kz)\,\hat{y}$

(where: $E_0 = B_0$, $\omega = k\,c$)

This wave travels at the speed of light along the *z*-direction and manifests a further *characteristic of radiation: E is perpendicular to B and to the direction of its travel.*

Note surprisingly, the ϕ and *A*-fields also satisfy their own wave equations. As a result, for example, *A* can be written as: $\vec{A} = A_0 \sin(\omega t - kz)\,\hat{x}$, which is a wave of amplitude A_0 traveling at speed *c* along the *z*-direction. This wave implies the existence of an *E* and *B* exactly as given above in equations 6.18. This is so because the plana resists the changing in time of *A* in the *x*-direction causing an analogical *E*-field in

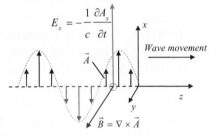

that direction, and, at a given instant of time, there is a curl of A_x along the *z*-direction, which means there is a *B* along the *y*-direction.

We can understand the key property of radiation, its analogical impetus, by investigating how light interacts locally with a small object, say a small piece of metal. We do this by modeling the surface as a damped harmonic oscillator, in particular a damped spring with one end fixed and the other attached to a charged particle. The light impacting our modeled surface is shown to the right. The *E*-field, for example, the plana resisting the changing A in the *x*-direction, causes impetus to be activated in the charge so that the *B*-field exerts a force on the charge in the *z*-direction. The net result is that the potential momentum of the *A*-field, being half of the time upward and half of the time downward cancels out, while the forward impetus given to the particle as a result of the secondary action of the *B*-field survives giving a net, though small, impetus to the absorbing body. Using the above specified model, the measure of the impetus (i.e. the momentum) given to the body can be shown to be: $p_z = \dfrac{E}{c}$. This is the standard relation between the momentum of light and its energy (see for example *PFR-M*). Light traveling along the *z*-direction hits the metal and activates z-type impetus of measure p in it and deposits, in that same process, a total energy E. Hence, using this fact, we can assign an analogical momentum and energy to "light."

The momentum per unit time per unit area, is given by: $\vec{g} = \dfrac{\vec{E} \times \vec{B}}{4\pi c}$, while Poynting's vector, $\vec{S} = \dfrac{c}{4\pi}\vec{E} \times \vec{B}$, describes the power per unit area deposited. We also have: $\left|\vec{S}\right| = cU$, where U is the field energy density.

The momentum and energy of radiation fall-offs as $1/r^2$ from the source, and the E and B-fields each fall-off as $1/r^2$.

Radiation back reaction
Because of the finite speed of travel, Newton's third law can, in a strict sense, be violated inside a particle. Modeling a particle as two separated points such as shown to the right, it can be shown that when the particle is accelerating, the front point exerts (via the plana) a force on the back point that is different from the force caused by the back on the front. This results in a net force on the particle. However, in this process, radiation is emitted. The radiation has the just discussed ability to cause impetus of the same strength but opposite type as caused by the excess force on the particle. In this way, Newton's 3rd law is recovered by analogically extending its meaning. Namely, the particle causes an analogical impetus in the plana (radiation) at a certain rate but the plana *apparently* (actually it's the internal parts of the particle including its virtual plana) acts back causing an equal but opposite impetus per unit time in it.[15]

Radiation is most easily understood far from all sources. Far from the sources, in classical E&M, no further radiation can occur; hence, the total radiated momentum and energy remains the same. Because of this conservation law, radiation is often empiriologically defined as that which falls-off in such a way that its total energy "makes it to infinity." In the roughly spherical symmetry that obtains for sources confined to a distant region, the energy should fall-off as $1/r^2$ and thus the E and B-fields should fall off as $1/r$.

---

[15] Note the shorthand way of speaking. For example, when we say that the virtual plana effectively causes "an equal but opposite impetus" in the particle, we mean to include, for instance, the possibility that the particle already has impetus of the requisite type (or its opposite) and, in that case, the back reaction will increase (or decrease) the strength (intensity) of that type to the same degree that it caused analogical impetus in the radiation.

Helpful Hints

There are various levels at which we can understand and talk about radiation in particular and E&M in general. Of course, our goal is to understand physical reality and thus we list that goal first. However, it is often convenient, as we have said, in the process of (1) *understanding* and (2) *applying* the concepts to employ equations and schema to investigate and manipulate the concepts. As we proceed to the more purely practical, our way of speaking becomes increasingly dense and can, if we are not careful, lead to confusion and even outright misunderstanding of the concepts. However, these various modes of approach have their important and even essential place. We list them in the order in which they proceed from discussing the reality itself to the most practical bent. Note that some of the levels are identified simply by the people who use the given level. The key point is that each level has its own language and mode of approach.

Five Levels of Description of Reality
I. Fully physical understanding at a given level of abstraction
II. Empiriometric
III. Electrical Engineering
IV. Hams and Radio Enthusiasts
V. Technicians

We identify five different levels but there may be more or less depending on the given subdivision. Furthermore, note that these divisions are not hard and fast (for example, some technicians may have a more fundamental empiriometric approach (level II) than a given Ham), but are meant to give insight into the various modes of approach that are indeed distinct and can be seen such by the language they use. In reading other material, it is important to know what approach, it will be taking.

As with our comment about integrals in Chapter 3, when applying the wave equation and like in curvilinear coordinates, one must be careful because in general the curvilinear unit vectors depend on the curvilinear coordinates. In particular, in curvilinear coordinates, one cannot, in general, write: $\nabla^2 F_i = \dfrac{1}{c^2}\dfrac{\partial^2}{\partial t^2} F_i$, though one can in Cartesian coordinates.

Note that often times in this text that we have repeated (and will repeat) a formalism in order to look at it once again from a different angle, while introducing a new aspect of it. This is helpful in rounding out ones understanding and in digesting the concepts as well as in gaining facility with using the formalism itself. It is recommended that you follow this procedure in your study of the material, especially in areas in which you find yourself weak.

In solving problems, it can sometimes be helpful to remember that E and B are the most analogically general fields. Remember, forces can be expressed in terms of them alone. This is the reason that, in introductory physics texts, Maxwell's equations are nearly always formulated in terms of E and B alone.

Problems

1. Show, by graphing, that $\vec{A} = A_0 \sin(\omega t - kz)\hat{x}$ travels in the z-direction and thus is transverse, i.e., perpendicular-to-direction of travel.

2. Given the fact that radiation can cause impetus along the direction of its travel, i.e. that there is a kind of potential momentum along the direction of travel, one might suppose that radiation consists of a longitudinal A-field wave, i.e. an A-field that points and waves along the direction of travel (e.g., $\vec{A} = A_0 \sin(kz - \omega t)\hat{z}$). Explain what is wrong with this theory.

3. Explain why the propagation of "static" E and B-fields (e.g. see the discussion at the beginning of Chapter 2) are each different from radiation. Explain separately the case of the analogical E-field which is due to the $\dot{A}$ effect.

4. a) Using the spherically symmetric wave equation written in spherical coordinates, i.e. $\nabla^2 \psi = \frac{1}{r}\frac{\partial^2 (r\psi)}{\partial r^2}$, show that $\psi = \psi_0 \frac{e^{i(kz - \omega t)}}{r}$ is a solution to the wave equation ($\frac{1}{c^2}\frac{\partial^2 \psi}{\partial t^2} - \nabla^2 \psi = 0$) and give the resulting dispersion relation. Write down the form of E and B that are perpendicular to each other and the direction of travel, thus giving the equations for radiation fields far from a uniformly radiating source. b) Using Poynting's formulas, calculate the energy and momentum emitted per unit time per unit area. Explain why the momentum and energy fall off in the way they do. c) Repeat (b) for the plane waves given in the text.

5. Consider radiation of the form $\vec{A} = A_0 \cos(kz - \omega t)\hat{y}$ impinging on a square loop of wire of length l lying in the y-z plane with one edge parallel to the y-axis as shown. a) Explain why such a loop is called a magnetic dipole antenna if the wavelength, λ, is such that $l \ll \lambda$, but can be called an electrical antenna if $\lambda \sim l$. b) Calculate the voltage as a function of time "received" by i) a small rectangular loop ($kl \ll 1$) ii) an "electrical" antenna loop in which $l = \lambda/2$ (make the simplest assumptions) c) Calculate the peak signal for each.

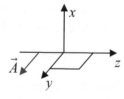

6. Explain how the second time derivative of the A-field results in a curl of a curl of A, i.e. $\nabla \times B \neq 0$, by drawing arrows for a field that has a non-zero $\ddot{A}$; use a field different from the one given in the text.

7. The accelerations fields given in equations 6.14 and 6.15 are the radiation fields for an arbitrarily moving "point" particle of charge q. a) Taking $v \ll c$, write the formal equations for the radiation fields (i.e., E and B) for a particle that has a uniform acceleration a along the z-axis. b) Write the magnitude of each field in terms of the angle θ

measured from the z-axis. Using equation 6.27, calculate the amount of power emitted in per unit area in the θ direction by the particle. c) Show that, in *cgs*, the total power radiated in all directions is: $P = \dfrac{dE}{dt} = \dfrac{2}{3}\dfrac{q^2}{c^3}a^2$.

8. Recall the following physical situation discussed in the text of Chapter 4. A bead is hooked to a rod as shown to constrain it to move without friction parallel to an infinite wire. Suppose, as in the problem at the end of Chapter 4, one removes the rod, still wanting to do the experiment of watching qA momentum develop in the x-

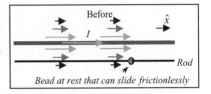
Bead at rest that can slide frictionlessly

direction. a) What intuitive constraint can be put on T, the approximate time scale during which the current I is switched off, to avoid substantial radiation acting on the bead? b) Using the equation for total power radiated by a uniformly accelerated point charge given in the problem above, roughly, what constraint can be put on the bead to assure it does not emit a significant amount of radiation? Explain your result intuitively. Assume for both questions that the current in the wire is not affected by the bead.

9. Use the instructions given in the text to prove equation 6.12.

10. Consider the loop with current $I(t)$ shown to the right. a) Assuming $d \gg b$, b is arbitrarily large, and $I(t) = \begin{cases} t>0 & I_0\left(\dfrac{t}{\tau}\right)^2 \\ t<0 & 0 \end{cases}$, calculate the "steady state" E and B-fields near the wire along the negative x-axis, i.e. give the equations for E and B in the "infinite" straight wire limit. b) Using Poynting's vector ($\vec{S} = \dfrac{c}{4\pi}\vec{E} \times \vec{B}$), calculate the energy per unit time per unit area. Comment on the arbitrarily far in the future (i.e. $t \to \infty$) form of $\vec{S}$. Does it fall off as $1/\rho$? Explain why a $1/\rho$ fall-off would conserve energy in a cylindrically symmetrical system in which the energy travels outward along the cylindrical radial direction ($\hat{\rho}$). c) Using the damped spring model of an absorbing surface given in the text, argue why a field that constantly increases with time, such as the one in this problem, will run into trouble if one attempts to described it as carrying potential momentum in the sense defined by this model.

11. As mentioned in the text, give the A-field for the infinite straight wire with a current $I(t) = \begin{cases} t>0 & I_0\left(\dfrac{t}{\tau}\right)^2 \\ t<0 & 0 \end{cases}$. Show that: $\ddot{A} \approx \dfrac{4I_0}{c\tau^2}\ln\dfrac{2b}{\rho}$.

12. Calculate the leading terms in far field E and B-fields for a circular loop of current such as discussed in Chapter 3, assuming a current $I(t) = \begin{cases} t > 0 & I_0 \left(\dfrac{t}{\tau} \right)^2 \\ t < 0 & 0 \end{cases}$. Does it radiate? Explain.

13. Show that a loop made out of perfectly conducting material that starts with no B-field currently threading it can never have a B-field threading it, i.e., the flux through the loop must always remain zero. Formally, we write: $\Phi = \int \vec{B} \cdot d\vec{S} \equiv 0$.

14. Calculate the radiation produced by a rotating disk of uniform charge, such as used in Rowland's experiment mentioned in Chapter 3.

15. Using the Jefimenko equations 6.16, show that for a current I_0 coming in along the z-axis and ending at the origin so that $q = I_0 t$ is deposited in an arbitrarily small region at the origin, B does not change in time but that E does, thus verifying that E cannot be the generic cause of changing B. Neglect the retarded time.

16. Consider a point charge that starts at rest at the origin, undergoes uniform acceleration along the z-axis for a short period of time, T, then continues along at speed aT forever afterward. In order to avoid relativistic issues, suppose $aT \ll c$. Using Maxwell's equations and a figure like shown to the right: a) Show that this action generates an E-field that, at great distances from the particle, is perpendicular to the direction of travel. b) Show that E falls off as $1/r$ and varies as $\sin\theta$. c) Give 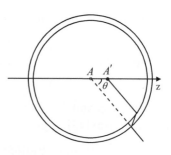 the *cgs* formula for the components of E parallel and perpendicular to the direction of travel, $E_\parallel$ and $E_\perp$, and use Poynting's vector to calculate the power radiated per unit area in a given direction. d) In terms of A and ϕ, hypothesize about what effects result in absence of the component of the field along the direction of travel?

17. Consider an arbitrarily short "line of charge" with linear density λ moving uniformly at speed v along the x-axis as shown to the right. Making use of the propagation speed, c, of the ϕ-field, calculate ϕ as a function of the angle, θ and the distance of the particle from the viewpoint at the present, rather than the retarded time. Verify that your answer agrees with the result given in the text for ϕ directly in front and in back of the particle.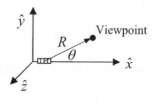

18. Derive the formula given in the text for the apparent length enhancement that occurs for a line particle moving toward the origin along the x-axis, such as shown on left. (Ignore special relativistic effects.) Also calculate the suppression factor for the case when the particle is moving away from the viewpoint. *Hint*: the apparent enhancement (or suppression) occurs due to the finite speed of travel of the signal that informs the observer of the presence of each end of the line particle.

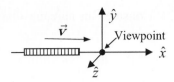

19. Starting with Maxwell's fourth equation, and using the vector identity, $\nabla \times (\nabla \times \vec{V}) = \nabla (\nabla \cdot \vec{V}) - \nabla^2 \vec{V}$, derive an equation for the electric field as a function of the retarded $\dot{\vec{J}}$ and $\vec{\nabla}\rho$. Then, use $\nabla'[f(\vec{x},\vec{x}',t)] = [\nabla'f(\vec{x},\vec{x}',t)] + \dfrac{\hat{R}}{c}\dfrac{\partial f(\vec{x},\vec{x}',t)}{\partial t}$, and the fact that $\int_V \vec{\nabla}f\,dV = \int_S f\,\vec{dS}$ to recover the electric field given by Jefimenko.

20. There are two possible linear polarizations of radiation. If, for simplicity, we assume plane wave radiation and take the direction of travel along the z-direction, then these polarizations correspond to the E-field along the x or y-directions. a) Assuming no ϕ-field, write an equation and draw the A-field for the case of radiation with: i) x-polarization only, ii) y-polarization only, and iii) a polarization with the E-field at angle θ_0 from the x-direction. b) Given that light is polarized only in the horizontal direction when coming off at a glancing angle from the surface of water, say a lake, explain how polarized lenses work. c) Given the model used in the text for incident light (radiation) against a surface, explain why horizontal light would tend to be preferred.

21. Starting with plane waves given in equations 6.18, give a form of a radiation field that is a combination of both A and ϕ. In particular, write the equations for A and ϕ as a function of rectangular space coordinates and time. *Hint*: Start with $\vec{A} = (A_x, 0, 0)$ and then introduce ϕ and an $A_z \neq 0$. Note also that A and ϕ, like E and B, have to satisfy their wave equations.

22. In Chapter 3, we discussed the store and give back role of E and the shepherding role of B. In these terms and using the damped spring model of the interaction of radiation with a massive body, discuss the role of E and B in radiation pressure, i.e., radiation causing impetus in a massive body.

23. Show that a particle in uniform motion does not experience a radiative back reaction force by using the dumbbell model at the end of this chapter to show that Newton's third law can be interpreted without need of discussing field momentum.

(see next page)

24. Consider an antenna that has field amplitudes: $E = V_0 \dfrac{\sin \theta}{R}$, $B = \dfrac{V_0}{c} \dfrac{\sin \theta}{R}$ and thus

power per unit area emitted according to (in *SI*): $S = \dfrac{1}{\mu_0} V_0^2 \dfrac{\sin^2 \theta}{cR^2}$. Assuming the antenna

is directed so that the peak of its pattern is directed at the observer: a) Given a 50kW television transmitter, what is the strength of the electric field in V/m at 20km from the antenna. b) Given a 25watt transmitter on Mars, what is the magnitude of the electric field back at Earth at the point of its closest approach?

Chapter VII

Introduction to
the Relativistic Nature of Electricity and Magnetism

Introduction

We have seen that the electric field is caused by charge and the magnetic field by impetus-activated charge. We have also seen that, for example, a small massive body with the property of charge activates the electric potential (the ϕ-field) in the plana starting near it and moving radially outward. Since ϕ decreases in strength as it propagates outward, it is clear that an *E*-field is generated as well. That *E*-field acts to cause impetus (in a positive test charge) in that radial direction. By contrast, impetus-activated charge causes a power that acts in a plane defined by the perpendicular to that radial direction and the impetus' directional type. In particular, it activates an *A*-field with a directional type that is the same as that of the impetus of the charge that caused it, and that *A*-field decreases in strength as it activates parts of the plana further from the source. Maxwell's equations encapsulate this as well as further details of how this all happens. Thus, we already have a profound understanding of the nature of two of the generic powers of nature, which we call electric and magnetic "forces."

However, one could still ask further questions. *Given* that there is a new power associated with impetus-activated charge, it makes sense for that power to have the direction and fall-off that we see that it does. But, we can ask: what is the purpose (what's called its intrinsic final cause) of having this extra force? Why does impetus-activated charge have to cause a fundamentally different type of power to exist? Can we dare to answer such a question?

Yes, the answer is encapsulated in Einstein's special theory of relativity. However, like all empiriometric theories, it must be unpacked. For a complete unpacking of the primary principles, read Chapter 10 of *PFR-M*. In this chapter, we assume a basic understanding of the material given there.

Thus, we start by giving the summary answer based on the essentials given in *PFR-M*, which do apply at the level of abstraction of classical E&M.[1] In *PFR-M*, we establish that a substance with impetus contracts in the direction of its motion (i.e., the direction specified by the impetus's type) and it behaves more sluggishly because its mass increases with impetus, resulting in a slowing of the time it takes things to happen. These effects after being lumped with the appearance of bodies measured from the uniformly moving frames (using an operational definition of simultaneity appropriate to the given frame), are called, respectively, length contraction, mass increase, and time dilation.[2] These effects can be viewed as consequences of the principle of integrity that every substance acts in such a way as to maintain itself, keep its same integrity, in uniform motion as it has at rest.

The principle of integrity imposes certain constraints on the forces, i.e. that which changes the impetus. These constraints require the existence of the magnetic field given the existence of the electric field, or in terms of potentials given ϕ requires A. Said still another way, if charge causes a power in the plana, impetus activated charge must cause a related power in order to avoid violation of the principle of integrity, which is apparently a deep principle of nature, which experiment shows in some way, extends beyond electric and magnetic forces to all other known primary generic powers (gravity, strong and weak as well as E&M).

Special Relativity via the *Setup* and *Dual-Setup*

To better understand the deep interrelation between the electric and magnetic fields revealed by special relativity as understood through the integrity principle, we will now analyze several examples using a symmetrical pair of physical configurations or setups called the "setup" and the "dual setup."

We begin by briefly summarize the reasoning (given in detail in Chapter 10 of PFR-M) that leads to the fundamental relativistic principle that, in turn, leads to the importance of the "setup" and "dual-setup." We recall first that the principle of integrity is formalized in the internal relativity principle that no experiment done within a given house can distinguish one house from another and that only the absolute speed of a given house determines whatever differences there may be between houses. Internal relativity, along with the existence of a maximum measured speed, implies the fundamental principles of special relativity:

---

[1] For example, properly including the further specification of gravity would require us to bring in general relativity.

[2] This combination can be called, for short, *LMOST*, which stands for *L*ength contraction, *M*ass dilation, *O*perational *S*imultaneity and *T*ime dilation.

1) Any experiment done in one inertial frame will yield the same in any other inertial frame

2) Measurement of the speed of light will yield the same result in every inertial frame.

This, in turn, implies that, at the appropriate level of abstraction, (for example leaving out gravity so that we can have objects that serve as our network of clocks, rulers etc. that continue in their state of uniform motion without being affected by their environment, e.g. clocks and rulers of other observers) *one cannot determine whether one is at rest or in uniform motion*. This consequence of the integrity principle results from the interior of each house, including ones inside others, having to, operationally, behave the same in motion as at rest.

Furthermore, in order for the integrity principle to be satisfied we saw that mass and time dilation and length contraction must occur. This places constraints on the forces. We can investigate these constraints through the frame symmetry implied by the relativity of motion through the use of the setup and its dual.

In particular, a given source, say a line of charge, moving uniformly away from me and my measuring instruments (including, in the case to be discussed first, my test particle) at speed *v* should yield the same *measured* results as when the same line charge is at rest and my measuring instruments and I are moving away from it at speed *v*. Recall a frame simply means a uniformly moving system of measuring instruments (rulers and clocks…etc.); an observer is typically present as well, but the instruments can be set to record the data and be observed later. Thus, using this language, we can say that observing a uniformly moving line charge from an absolute rest frame should give the same results as observing a stationary (in the absolute sense) line charge from a frame moving uniformly away from it at the same speed. In this sense, then it does not matter whether the line charge moves at speed *v* away from me or I move at speed *v* away from it. Indeed, all that matters at this operational level is the relative speed between them; that is, neither has to be in at absolute rest, and, indeed, one cannot determine whether either is at rest or not. Hence, we come to the idea of the setup and dual setup to understand the physical origin of this result.

To better understand the physical realities behind this relativity of motion, we consider a setup that looks at the actual physical state of the object under consideration, which is revealed in the absolute rest frame, because there my measuring instruments are not changed by having impetus. Thus, in the **setup**, we let the physical object under consideration move at speed *v* away from us, and we see what happens. While in the ***dual setup***, we let the object under consideration remain stationary, and we measure it or its effects from a frame moving at speed *v*. So, in the *setup*, we see how the physics of the object itself changes, while in the *dual-setup* our measuring instruments are changed by impetus, and the effects that result come from that rather than the object under consideration. It is the impetus changing the way the bodies behave that is responsible for the mass and time dilation and length contraction that result, in turn, in getting the same experimental results in the setup and dual setup. Again, in the first case, it is due to changes in the object, in the second it is due to changes in the measurement instruments. To understand this further, we proceed to a sequence of concrete examples.

Note that once we understand the physical origin of the frame invariance, we can often ignore the distinction between the setup and its dual in order to get the experimental predictions of new situations. Indeed, in ordinary empiriometric relativity, one identifies the setup and dual setup *because* of the very fact that, at the level of abstraction appropriate to special relativity, they would yield the same experimental results. However, as we've said, ignoring a reality, such as the real distinction between the setup and its dual, does not make it cease to be. Hence, we move on to our concrete examples to better understand the physical aspects revealed in the special relativistic character of E&M.

A Line Charge

Now, we return to the simple concrete example of a line charge because, in motion along its own length, it has more symmetry than a point particle in motion and thus generates a simpler field. The *setup* shown in Figure 7-1a has the observer in the (absolute) rest frame with a long line of positive charge moving at speed v along the negative x-axis. The *dual-setup* in Figure 7-1b has the observer in the frame moving along the positive x-axis and the line charge at rest. Each observer needs a measuring device that incorporates a test charge, which is shown as a small red dot.

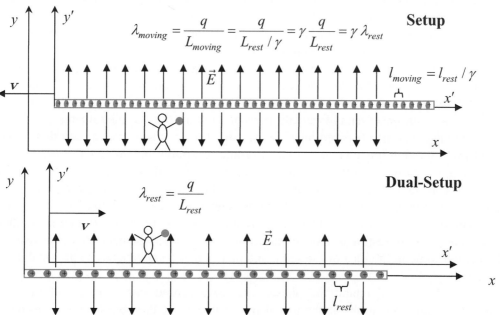

Figure 7-1: a. *(top) Setup* in which a long line charge moves at speed v to the left and is observed from the rest frame. Note the primed coordinate system, which moves with the line charge is moving the opposite direction from which it does in the standard configuration used in special relativity. **b.** *(bottom) Dual-setup* in which the line charge is at rest and is viewed from a frame moving at speed v. q is the total charge of the line charge. L_{rest} and L_{moving} are, respectively, the total length of the line charge when the line is at rest and when it is in motion. The line is taken to be arbitrarily thin and arbitrarily long so that cylindrical symmetry can be applied to Gauss's law to get $E = \dfrac{2\lambda}{r}$, where r is the perpendicular distance from the wire to the observation point, i.e., the field point. In *both* figures, the actual state of line charge, not the state as viewed from the moving frame, is shown.

Setup: **Field of Moving Line Charge**

The actual change in the line charge arises as the impetus modifies the length and mass. Of relevance to us is the shrinkage of the line, which causes a higher line charge density in the moving rod than in the rest rod. Using the approximation of an arbitrarily long and thin line of charge, one can use cylindrical symmetry and Gauss's law to get: $\vec{E} = \dfrac{2\lambda}{r}\hat{r}$, where λ is the density of charge per unit length (in esu/cm) and r is the radial distance from the line (in *cm*). Since λ is γ times bigger in motion, the electric field, which is perpendicular to the rod's direction of motion, is γ times bigger when the rod is moving, i.e. mathematically we have:

7.1 $E_{\perp moving} = \gamma E_{\perp rest}$

Now, only the electric field acts on our test charge which is at rest, so this completes the explanation of the measured results expected in the setup. However, there is a magnetic field because the line charge does have impetus. The magnetic field is given by Ampère's force law or by application of cylindrical symmetry to Ampère's circuital law to get: $\vec{B} = \dfrac{2I}{rc}\hat{\phi} = \dfrac{2\lambda v}{rc}\hat{\phi} = \dfrac{2\lambda\beta}{r}\hat{\phi}$. Thus, in terms of the charge density of the line charge at rest, we write:

7.2 $B_\phi = \dfrac{2\gamma\beta\,\lambda_{rest}}{r}$

Again, this magnetic field causes no force in this case in which the test particle has no impetus but we will return to the case in which it does have impetus later.

Dual-Setup: **Field of Line Charge at Rest Measured from Moving Frame**

Now, in the *dual-setup* of Figure 7-1b, in which the line charge is at rest but the observer and his test particle is moving, the length contraction and mass and time dilation of the moving observer and his measuring devices work together to make operational simultaneity natural in that moving frame. These effects, which can go by the acronym *LMOST*,[2] work together in such a way to make it appear that, in the moving frame, the rod is also contracted by a factor of γ so that the electric field is also γ times greater in the perpendicular direction. You may surmise that there must also be an *apparent B*-field, i.e. the effects on an apparently moving charge should *appear* just as they would if there were a true *B*-field acting. This apparent field will be discussed briefly later in this chapter when we allow our test charge to have an apparent motion in the moving frame. It is also analogically like what we will do later in a *setup*.

With the two line charge experiments, that are the *setup* and *dual-setup*, in mind, we can re-explain how the integrity principle leads to the equivalence of the measurements made in the setup and the dual setup. Namely, if the line charge moving away from me "looks" different than when I am moving away from it, then the effects on me are dependent on whether I am in motion or not and thus the integrity principle would be violated. Thus, in terms of terse logical propositions, we can say: If no equivalence, then no integrity, and, thus, if integrity, then equivalence.

Current in a Wire: Parallel Relative Motion

We next consider the slightly more complicated example of a long wire, which of necessity has both positive and negative charges, with a test charge.

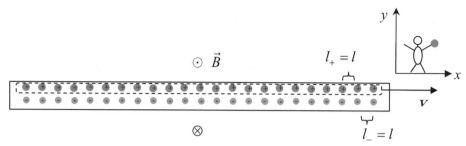

Figure 7-2: Current-carrying wire at rest. For conceptual simplicity, we take the ions to be negatively charged and the outer "orbiting" particle to be positively charged. Think of an anti-matter wire, in which positrons are flowing rather than electrons so that current flows in the direction of the flow of the positrons. Note that since the observer is in the absolute rest frame (for which, here, we use unprimed coordinates) and the wire is at rest, **this is *not* a setup or a dual-setup**.

For conceptual simplicity, against the ordinary reality but as Franklin generically[3] thought, take the fixed ions to be negatively charged and the "electrons" moving at average speed v to be positively charged. Perhaps one might like to think of an anti-matter wire. Such a wire will produce a static azimuthally-directed B-field; Figure 7-2 shows its direction for the slice that the paper makes through the wire. The moving positive charges, say positrons, arrange themselves so that they are spaced in the same way as the fixed negative charges; given a distance l between the ions, the average distance between the positrons is l, thereby minimizing the potential by making the mean (at the appropriate scale) charge density in the metal zero.

Now, we are ready to analyze how, in this case, the bodies and their fields implement the integrity principle. To do so, we analyze: (1) the *setup* to see how the moving wire looks from a fixed frame and (2) the *dual-setup* shown in detail in Figure 7-2 to see how a fixed wire looks from a symmetrically related moving frame. The dual configurations are summarized here in Figure 7-3.

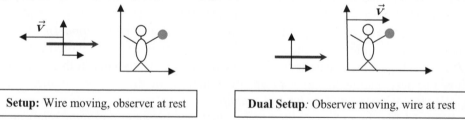

| **Setup:** Wire moving, observer at rest | **Dual Setup:** Observer moving, wire at rest |

Figure 7-3: *Setup* (left) and *dual-setup* (right) for a test charge (red ball) and a current carrying wire (brown arrow) in relative motion.

Setup: Electric Field of a Moving Wire

In the *setup* shown in detail in Figure 7-4, we consider the wire to be moving in the $-\hat{x}$ direction at speed v, against the direction of motion of the positrons. For simplicity, we take the wire to be composed of positrons evenly distributed along a one dimensional body bundled with a second one dimensional body composed of evenly distributed ions;

---

[3] Obviously, Franklin did not know about atoms but simply hypothesized moving charges that were electro-statically balanced by some fixed charges.

we will call the first the positron string and the second the ion string. Of course, in all this, we have, as in earlier chapters, ignored many qualities that are essential to the nature of a standard metal wire so as to focus on certain key features.[4] Moving the wire at speed v to the left then means: (1) to deactivate the rightward impetus of positron string, so that the positrons are at rest and (2) to give leftward impetus to the negatively charged ion string so that the ions move with speed v to the left (see Figure 7-4). Remember impetus, a quality of a massive body, qualifies or determines, to some level, the quantity (extension) of that massive body; in particular, the length (extension) of a massive body with impetus decreases as the impetus increases, according to $l \to \dfrac{l}{\gamma}$, where $\gamma = \sqrt{1 + \left(\dfrac{p}{m_0 c}\right)^2}$ (or in terms of the effectiveness of that impetus in causing motion, i.e. the resulting speed, we get the standard form: $\gamma = 1 / \sqrt{1 - \beta^2}$).

Considering the wire to be effectively infinite in length by confining our observations close enough to the wire that the end effects are not significant, we ask about the line charge density of the wire of an infinitely long wire that is moving as described. Now, since the positron string no longer has impetus, the positrons are now further apart ($l_+ = \gamma l$), while the ions have impetus and thus are moving; that impetus determines the ion string length to be shorter ($l_- = l / \gamma$).

For concreteness, take $v = \sqrt{3} / 2$ and $c = 1$, so that $\beta = \sqrt{3} / 2$ and $\gamma = 2$ as shown in Figure 7-4 below. When the wire is at rest, the total line charge density is: $\lambda_{rest} \equiv \lambda_0 = \lambda_+ + \lambda_- = 0$, because $\lambda_+ = q / l \equiv \lambda_0$, $\lambda_- = -q / l$, where q is the magnitude of the total charge on each string. However, when the wire is moving, we get $\lambda_+ = \dfrac{q}{\gamma l}$, $\lambda_- = -\gamma \dfrac{q}{l}$, which gives:

$$\lambda = \lambda_+ + \lambda_- = -\frac{q}{l}\left(\gamma - \frac{1}{\gamma}\right) = -\lambda_0\left(\gamma - \frac{1}{\gamma}\right) = -\lambda_0\left(\frac{1 - 1/\gamma^2}{1/\gamma}\right) = -\lambda_0 \gamma \beta^2 = -\frac{3}{2}\lambda_0$$

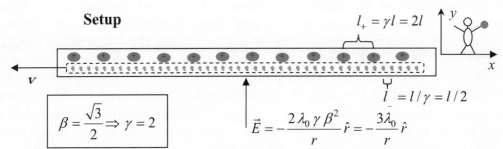

Figure 7-4: A portion of the "infinite length" wire moving at speed v to the left; this means the positrons are at rest and the ions in the antimatter wire are moving to the left as shown. The electric field results from application of Gauss's Law using a cylindrical surface. Note for an ordinary (not antimatter) wire, the field would be outward rather than inward.

---

[4] Again, this is typical of the empiriological method that focuses on one aspect in detail before incorporating another.

This means, the general formula for electric field around the moving wire, which is obtained using a cylindrical Gaussian surface, is:

7.3
$$\vec{E} = -\frac{2\lambda_0 \gamma \beta^2}{r} \hat{r}$$

This is the only force a charged particle at rest will experience. The moving wire will produce a magnetic field in the plana, as it would at rest, but it cannot act on a charge that has no impetus.

Dual-Setup: *Apparent* **Electric Field of a Wire Viewed from a Frame Moving Uniformly along the Wire**

Next, consider the *dual-setup* shown in Figure 7-5.

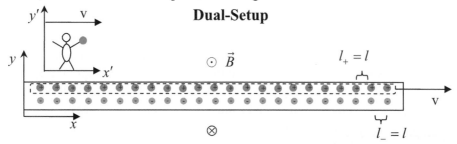

Figure 7-5: Current carry wire *at rest* viewed (in unprimed coordinate system) from moving frame (measured in primed coordinates). The actual state of the wire, not as it appears in the moving frame, is shown.

We now observe the wire, which stays at rest, from a vehicle moving along the wire at speed v along the direction of motion of the positrons. In this moving frame, the positrons *appear* to be at rest and the ions *appear* to be moving to the left at speed v. Remember also that as we move faster, all activity becomes sluggish because mass increases with impetus intensity; this means all of the time standards we carry with us slow down. Also, we adopt operational simultaneity to set all of our clocks. Furthermore, the length of all rods moving with the frame shrink in the direction of the motion, as those rods that we carry with us are qualified by impetus. As we said, the use of operational simultaneity coupled with mass increase and length dilation results in the *apparent* shrinkage of all bodies in the absolute rest (no impetus) frame. This means the entire length of the resting wire *appears* to shrink in the moving frame. A methodical application of the transformation laws of special relativity gives the new values for the spacing of the positive and negative charges. In particular, the positrons *appear* further apart ($l_+ = \gamma l$) because they appear to be at rest and thus *appear* to no longer be shrunk because of impetus they no longer *appear* to have, while the ions are *apparently* moving now so they appear closer together ($l_- = l/\gamma$).[5] Hence, as expected given the principle of relativity, we get the same

---

[5] The apparent shrinkage and expansion are largely due to operational simultaneity, which causes us to use measurements of the field and positrons at two different actual times and thus two different *actual* states of the field in the plana and two different *actual* positions of the positrons. Recall also that the use of operational simultaneity also means the actual distance between the measurements *in the plana* will not be the actual distance between the two measurements (even accounting for length contraction) *in the frame* (say as

expression for the field as obtained above: $\vec{E} = -\dfrac{2\lambda_0 \gamma \beta^2}{r}\hat{r}$, but with a different physical reality causing the appearances.

To understand the physical difference in the causes operating in the two setups, note that we determine the veracity of the above equation for the electric field in the moving frame by using a charged test particle moving along with the frame. This means the test particle is *actually* moving, actually has impetus and thus is able to be acted on by the magnetic field of the wire. However, in the moving frame, this magnetic field *acts like* an electric field in that it causes a force towards an apparent charge distribution, on an *apparently* at rest particle; thus it *appears* to act in the direction of the field rather than perpendicular as does a *B*-field.

Again, referencing Figure 7-3, whether in *the setup* in which the wire moves and I stay still with my test particle, or in *the dual-setup* in which the wire stays still and I and my test instruments (including my test particle) move together, the *apparent* motions, as measured in the way natural to each frame, are the same.

The Necessity of the Magnetic Field

Thus, we clearly see that the principle of integrity, as it appears in the empiriological system of special relativity, can only be satisfied if there is such thing as a magnetic field. Again, if there were no magnetic field caused by positrons with impetus then the test particle in the moving frame would not experience a force, for remember that the net charge in the region is only an *apparent* net charge resulting from our use of operational simultaneity in the moving frame, not an actual net charge, so that line charge does not itself attract the test charge. If there were no attraction, the system would not act the same in motion as it does at rest, and hence the principle of integrity would not obtain. In particular, a current carrying wire at rest would not attract a positron moving parallel to it (which, by itself, would seem fine). However, an observer moving along with a moving wire would measure a net charge that should attract the positron, yet it would not be attracted. This is, of course, despite the fact that the same charge arrangement at rest would indeed attract a moving positron. It is an important testimony to the intelligible unity of purpose of nature (as seen through her most generic powers) that charges which cause electric fields of their nature need to move and, in turn, need impetus to move and that very impetus empowers such bodies to cause the very magnetic field which is necessary to maintain integrity during uniform motion of bodies that have charge.

Current in a Wire: Perpendicular Relative Motion

To understand the setup and its dual for perpendicular motion, we need to know what the field is like around a uniformly moving "point" charge, since in the appropriate approximation, one can take the wire as composed of point charges.

Field of a Uniformly Moving Point Charge

As usual, the point approximation means we consider the field on a large enough scale that we can neglect the size of the particle, i.e. we work, in the limit, with the being of reason of a point charge. We know the field of a charge at (absolute) rest. A moving charge measured in the frame moving with it must appear to have that same rest field distribution.

This means the electric field must appear to travel with the particle. We also know

---

measured on a board carried with the frame), because the frame will have moved in the time between the measurements.

that bodies that have impetus contract, the so-called Lorentz contraction in the direction of motion. If we follow this line of thinking, we expect the field lines to contract in the direction of motion. That is, we consider the field lines as like stiff spikes that stick out from the point particle, like, for example, those on a porcupine (see Figure 7-6a). We then note that such spikes contract in the direction of motion, but not perpendicular to the direction of motion. From this reasoning, we can calculate (see end of chapter problem) the electric field to get

7.4 $$\vec{E}_{moving} = \frac{Q}{r^2}\frac{1}{\gamma^2\left(1-\beta^2\sin^2\theta\right)^{3/2}}\hat{r}$$

Where θ is the angle measured from the x-axis in the lab frame as shown in Figure 7-6b; r is the radius from the charge to the field point, Q is the charge of the particle; β is the velocity of the uniformly moving charge as a fraction of the speed of light, and $\gamma = 1/\sqrt{1-\beta^2}$. Note the field is radially directed.

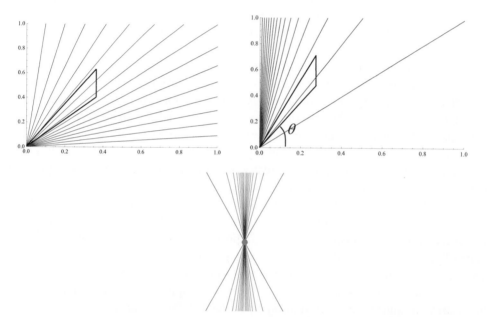

Figure 7-6: *a.* (*top left*) A 2-dimensional slice of the field lines[6] of particle at *rest **b.*** (*top right*)[7] A slice of the field lines of a moving particle with $\gamma = 10$.   ***c.*** (*bottom*) Full two dimensional slice of the E-field lines of the charged particle moving to the right.

---

[6] Note for these type of diagrams to be accurate they must be drawn in three dimensions rather than two so that the $1/r^2$ decrease in intensity will obtain (because of the surface area of the sphere increases as r^2).

[7] The first figure is created as follows: each line is written as $y = \tan\left(j\,\theta_0\right)x$, where $j \in \{1,2,3..\}$ with $\theta_0 = \left(\pi/2\right)/16$. The second figure works as follows: each line is $y = \gamma\tan\left(j\,\theta_0\right)x$, since $\tan\theta = \Delta y/\Delta x$ and contracting the Δx by γ is the same as multiplying the tangent by γ.

To better understand the angular dependence of the field strength, we graph it below in Figure 7-7.

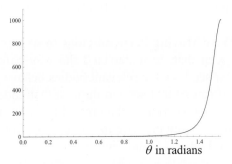

Figure 7-7: Graph of the electric field strength of a moving charge particle ($\beta = .995$) as function of angle; here the field strength is normalized to one for small angles.

To understand the physics of this effect, one needs to break the E-field into its two analogical parts, represented by the measures $-\vec{\nabla}\phi$ and $-\dfrac{1}{c}\dfrac{\partial \vec{A}}{\partial t}$. The contraction of the E-field of a uniformly moving particle relative to that of a rest particle is generally caused by the propagation of ϕ and by the newly existing (i.e., not present for a particle at rest) A-field and the effect of its propagation. To understand how it works, one needs to do the type of analysis that we did in Chapter 6 for the ϕ-field directly in front and behind the uniformly moving particle for all angles around the particle and for both the ϕ-field and the A-field. Upon doing that analysis, one will see, as we saw in Chapter 6, the importance of the retarded time, i.e., the time at which the source point caused the effect in the plana that is just now reaching the field point under observation. Also, as we saw for the case discussed in Chapter 6, the actual extension of the particle is crucial to that analysis; that is, one cannot consider a particle as composed of discrete parts and get a result that agrees with experiment. The A-field will cause a certain E-field like effect because, behind the particle, there will be an A-field caused by the particle that is not maintained at the same level. Thus, the A-field, which points along the direction of the motion of the particle will decrease, thus creating an $\partial A / \partial t$ effect that acts in the direction of the particle's motion. Recall that effect arises because the plana reacts to the falling A-field, trying to maintain the A-field by exerting a force on any body receptive to such action, i.e., to any charged particle. The effective E-field due to this $\dot{A}$ effect will be exactly the opposite in front of the particle. Hence, the net result is that, in both the front and the back of the particle, the $\dot{A}$ effect will tend to decrease what would be the static "$\vec{\nabla}\phi$ contribution" to the E-field (recall that the static E points along the direction of the motion in front of the particle and against the direction of motion in back of it).

To see the different effects that ϕ and A have in this context in the equations, we note that along the trajectory the ϕ looks, as we said in Chapter 6, the same in front and back as a rest particle does. However, for this same $\theta = 0$ or $\theta = \pi$ case, the time rate of change of the A-field contributes another term just like the ϕ-field term but with a factor of $-\beta^2$, giving the net $1/\gamma^2$ factor relative to a particle at rest that we can see in the formula for the electric field given in Equation 7.4 by taking $\sin\theta = 0$.

The derivation of the equations for ϕ, and A in general is a little complex and the calculation from those potentials to E is more complex; we will do neither here. Instead, we will continue on to understand the physical effects for relative motion perpendicular to wire.

Setup: **Wire Moving Perpendicular to its Length**

Generally, we have seen that, to understand the way nature acts to preserve the principle of integrity, we must see how the relevant bodies behave when in motion. In this way, we pinpoint the physical causes that act and through that understand the nature of the powers and hence the substances involved. In the present case, we are interested in how the wire acts on a test charge, so we allow the wire to move away from the initially at rest test charge in the fashion shown in Figure 7-8 and Figure 7-9.

Figure 7-8: *Setup*-- Simplified drawing of motion of current carrying wire and test charge shown in detail in Figure 7-9.

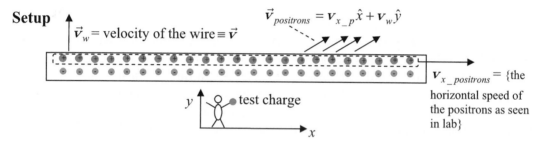

Figure 7-9: *Setup* detail-- Current carrying wire moving perpendicular to the direction of current flow.

Since the wire is moving at velocity $\vec{v}$ and the ions are at rest with respect to the wire, the ions are simply moving speed v in the $\hat{y}$ direction. These moving ions create a magnetic field with which we are not currently interested, because the test charge has no impetus and thus cannot receive the action of a magnetic field.

The motion of the positrons is slightly more complex, for they have motion both in the y-direction, due to the impetus of the wire, and in the x-direction, due to the impetus that causes the original current. Figure 7-9 shows the direction of motion of the positrons. If we draw in the field of a couple of the positrons (using our previously found form for the field), we get the standard result pictured in Figure 7-10 below.[8]

---

[8] Note that the decreasing of the E-field along and against the direction of motion of the point charge (with increasing speed) illustrated in the figure is clearly related to the length contraction of a rod, but how they are related is a complex problem that you might want to ponder. Note that, currently, length contraction seems to obtain, at the appropriate level of approximation (which is related to our level of abstraction), for all classes of forces not just E&M.

Setup

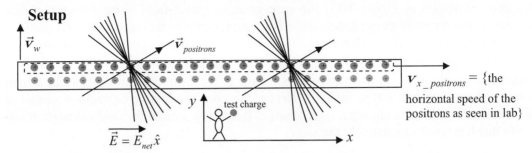

Figure 7-10: The electric field caused by the positrons squashes toward the axis perpendicular to the motion causing a net "electric" field in the direction shown. Test charge is at rest.

In words, the electric field (in the extended sense that is common empiriological usage) from the positrons on the right of the test particle push less to the left than the field from positrons on the left push to the right, because of the nature of the fields and the way they propagate as described earlier. This can be seen in the figure above by noting the field lines are denser on the left than the right. Hence, there is a net force acting in the x-direction on the test particle. This same fact can be seen in an even more direct way.

Namely, the wire moving away neither contracts the positron line nor the ion line, so that no net electric charge appears in any region. The two line charges share the same uniform vertical motion, so the A-field caused by those vertical-impetus-activated charges cancel, since they are of opposite sign, as long as one is far enough away (and assuming a steady state condition of the plana). However, the positron motion to the right is not shared by the ions, and the rate of this motion, i.e. the positron's rightward speed, is reduced by a factor of γ because of mass dilation (note that one can show that if one gives vertical speed v to a mass that initially only has a horizontal speed u, then u will decrease to u/γ because of mass increase[9] because of the more intense net impetus--see end of chapter problem). Of course, the reduced positron speed still causes a rightward A-field. Now, unlike in the unmoving case, this A-field at a given point in the plana decreases as the wire moves away, and thus, the plana, acting against this decrease, causes the analogical E-field in the x-direction. Here we see clearly that it is not the ϕ-field but the A-field that is responsible for the effect.

From the principle of relativity, we know that an observer (experimenter) moving away from the wire at rest at speed v must "see" the same apparent effects (observe the same experimental results) as when, as in the above case, he is *truly* at rest observing the wire actually moving at speed v away from him. Now, our goal, as always in this context, is to understand, via the final causal principle of integrity, what actual physical causes are involved in producing this apparent sameness. Having seen the generic causality of the primal case in which one views the moving wire (which is changed by its impetus) from rest, we investigate the causality of the dual case.

Dual-Setup: Fixed Wire Viewed by Observer Moving Perpendicular to it

That is, we consider the causes involved when the observer and his test charge are, under the action of their respective impeti, moving at speed v away from a fixed wire. In

---

[9] $\gamma = 1 / \sqrt{1 - (v/c)^2}$

this case, as shown in Figure 7-11, the wire generates an A-field along the current that decreases as one moves away; hence, there is a circulation into the page at the test particle, i.e., a B-field into the page. Hence, as discussed in the static magnetic field chapter, the wire tries to pull the downward ($-\hat{y}$ direction) moving test particle back by activating impetus of x-ward type, i.e. the plana exerts a force in the x-direction, for recall like impeti attract and the wire has x–type impetus. Hence, the same qualitative behavior is observed as in the case of the moving wire, fixed charge. Relatively straightforward calculation can verify that it is also quantitatively the same.

Dual-Setup

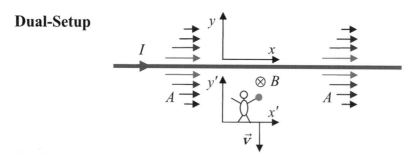

Figure 7-11: Simplified *dual-setup* for case of motion perpendicular to wire: test charge moving near *fixed* wire. Note non-standard configuration for Lorentz transformation: the primed frame is moving *against* the y-direction *instead* of *along* the x-direction. As usual, we show the actual state of the wire and plana, not the state as seen in the moving frame.

Yet, we can say a little more about why this observer moving away from the wire (*dual-setup*) looks the same experimentally as the wire moving away from the observer (*setup*) (which are, for that reason, empiriologically, treated as the same). The observer has impetus so that all of his rulers contract in the y-direction and all of his mass standards "gain" mass. This contraction does nothing to the apparent net charge, so we see no apparent ϕ-field, thus agreeing with the measurements in the setup. On the other hand, this contraction and dilation together with operational simultaneity change the apparent speed of the positrons from the simple Galilean result so as to *apparently* give the same current as we saw in the *setup*. The velocity-addition law for special relativity that encapsulates these effects reveals that a body moving along the x-axis at speed u will appear, in an inertial frame moving at speed v in the y-direction, to be moving at speed u/γ in the x-direction (in addition to apparently moving at speed v in the negative y-direction). Note that this is the same horizontal speed that we calculated earlier would result if we gave a body of the same mass that was initially moving at speed u in the horizontal direction enough impetus to move at speed v in the vertical direction. This is as it must be according to relativity. Namely, moving vertically away at speed v from a body of horizontal speed u (*dual-setup*) must look the same as the following situation. Observing that same body of initial horizontal speed u *after* it has been given, in addition, a speed v in the vertical direction away from me (*setup*), i.e. so that it has, as just stated, a horizontal speed u/γ (see end of chapter problem). Each of these situations is pictured here.

Setup: Body moving at vertical speed v.   **Dual-Setup:** Body with no vertical speed.

The upshot of all this is that, in the dual setup, there *appears* to be a changing A-field in the plana of the same type as seen in the setup.

Lastly, in our point charge description, length contraction, mass and time dilation and operational simultaneity play the part of making the fixed charge look like its field lines were contracted in the direction of motion.

In this analysis we have touched on the role of A and ϕ. In a later section, we will discuss the general physical causality reflected in the empiriological transformation of the A and the ϕ in more detail. For now, we are ready to summarize what we have learned about the interrelation between A and ϕ.

The Principle of Integrity and the Fields

As we have said, the principle of integrity (as formalized in the internal relativity principle) that implies the special relativistic laws is a deep principle inscribed deeply in physical nature, i.e. in the constellation of contiguous substances that is her reality. Electrically charged massive bodies produce an electric field (ϕ field), and, because of their physical nature, they must also be able to have impetus; hence, they can also generate and receive the action of an A-field with a circulation (at our current level of abstraction). Furthermore, because they are charged, they can already, of their nature, receive the action of a changing A-field. These various qualities behave in precisely the way necessary to keep a uniform moving body in the same effective state as the same body when at rest. The most restrictive aspect of this is seen in the way A and ϕ work together so that any experiment done by an observer moving away from an object will give the same results if that object is moving away from him. Indeed, only the relative motion between the two matters. Of course, this assumes that one can ignore the environment in which the object and observer exist. Clearly, for example, a field in the environment that is not associated with the observer or the object, would allow one to distinguish more than just the relative motion between the observer and the object. This simply means one has to be careful to apply the dual configuration in a way that is consistent with its usage as a probe of the natures of the physical changes involved in special relativity (see end of chapter problem).

Again, classical E&M is a special relativistic theory, and hence, there is no experimental way, within that domain, to distinguish between the *setup* and the *dual-setup*, between motion and rest. The *setup* and the *dual-setup* are *empiriometrically* (but not really) equivalent situations. By remembering that they are *actually* distinct, we can, as we have done in a few cases here, more fully understand the profound interrelation between electricity (ϕ) and magnetism (A) that makes that experimental equality so.

Transformation of Electric and Magnetic Fields

We now would like to obtain a general formula for, as well as a general understanding of, the transformations of E and B and later do the same for A and ϕ. That is, given the electric field and the magnetic field at a point in the plana (viewed from rest), we would like to predict what the apparent E and B are in the moving frame. In particular, we would like a formula for E' and B', i.e. the values of E and B as measured in the moving frame, in terms of the actual E and B, i.e. as they would be measured in the rest frame. Note that we use primed symbols for the values in the moving frame and unprimed symbols for values in the rest frame. Because the principle of special relativity is valid at the level of abstraction of E&M, such a formula will then relate the E and B in one inertial frame to any other frame moving uniformly relative to it; we will typically call one the lab (unprimed) frame and the other the moving (primed) frame. Rather than derive the general formulas in complete detail (which can be done by various transformation arguments based on the relativity of motion and symmetry), we show how, in a special case, a simple E-field can appear as a B-field in a second frame and how to generate the relevant transformation law for that case. This will give us a general idea of the meaning of the transformation laws.

Now, it is clear how a magnetic field can appear as an electric field; the impetus of a moving test charge is not seen in the particle's own rest frame (i.e. in the moving frame), but the B-field nonetheless acts on it. Take, for example, the field around our fixed current carrying wire. As we saw earlier, moving parallel to the current will cause an inward force that will be seen in the moving frame as an electric field since the particle is not apparently moving in that frame.

But, how does an apparent magnetic field arise from a pure electric field? For concreteness, take the case of a fixed positive charge. This charge has an electric field but no magnetic field. When viewed from the moving frame it is clear that we will see the previously described Lorentz contracted field. However, there will also be an apparent magnetic field; where does that come from? At the simplest level where we side-step "trapped" plana,[10,11] it comes from the contraction of the length and the dilation of the mass coupled with the use of operational simultaneity that change the *apparent* behavior of a force viewed from the moving frame.

Apparent Force and Fields of a Line Charge as Seen from a Moving Frame

We now investigate this in more detail with an eye towards developing a transformation law. To do this, we consider measuring the effect of the E-field, say in the

---

[10] Consider the case, for example, of a plana trapping box (a closed house in the language of *PFR-M*) moving uniformly (with its internal plana) with a charged particle at absolute rest within it, so that the plana is moving over it. We could then ask about the nature of that physical situation; in particular, we would ask how the principle of integrity operates in such a situation.

[11] Properly speaking, starting with what is most primary in itself, we should start with plana that is a part of a substance that stays with a moving substance (cf. also the note at the end of Chapter 4). Plana that is part of a substance we call virtual plana. From there we can talk about the possibility, which seems highly unlikely, of plana being literally trapped by certain kinds of material. Trapped plana then would be the analogical generalization that would include any plana that moves along with the frame, whether it is virtual plana, trapped plana or whatever other scenario that might be possible. Plana will be discussed shortly, but only briefly as plana motion, as already mentioned, is not, of itself, a topic that belongs at the level of classical E&M.

y-direction, which is perpendicular to a *line* of uniform positive charge density.[12] That is, we want to look at how E_y appears in a frame that is moving at speed v along the wire.

Clearly, there is a gradient of the ϕ-field in the y-direction and that gradient does not change *along the wire*, i.e. the $\hat{x}$ direction. Note that we cannot use a test charge moving along with us in the moving frame to "test" the *apparent* magnetic field, so we use one that is not moving at all. Such a particle will have an *apparent* velocity of $-v\hat{x}$ in the moving frame. When we measure the apparent effect of the E-field (i.e., $-\partial\phi/\partial y$) in the moving frame, we are asking about the rate of change of the particle's speed in the y-direction. This means we have to measure how much speed, $\Delta v'$, it picks up in the y-direction in a given interval of time, $\Delta t'$, as measured in the moving frame.

For simplicity, we assume the particle starts from rest, so, in the absolute rest frame (or, what is experimentally equivalent, the lab frame), we have: (1) the speed is v, an arbitrarily small quantity used to calculate the force, (2) the mass is, to requisite order, m_0, the rest mass, and (3) the time interval is Δt. Hence, the measure of the magnitude of the force, f_0, which is manifest by the rate of the change of the impetus, is $\dfrac{m_0 v}{\Delta t}$.

In the moving frame, the measured speed is a factor of $\dfrac{1}{\gamma}$ times less than the actual speed of the particle because of combination of operational simultaneity, length contraction and mass dilation (*LMOST*). For the same reasons, the mass and time elapsed each appear γ times larger. This means that the apparent force in the moving frame, f', which is manifest by the rate of change of impetus it apparently causes, is γ times less than it would be in the fixed frame, i.e. $f' = \dfrac{\gamma m_0 v/\gamma}{\gamma \Delta t} = \dfrac{f_0}{\gamma}$. This, then, is the transformation law for a force whose direction is perpendicular to the direction of motion of the moving frame. Thus, we can say, given the invariance of charge with impetus (and thus apparent impetus): when the test particle is apparently moving at velocity $-v\hat{x}$, the force per unit charge $= f/q$, is:

7.5 $\dfrac{f'_y}{q}(v' = -v) = \dfrac{E_0(y)}{\gamma}$, where $E_0(y)$ is the actual value of the field at a distance y from the line charge.

Furthermore, it can be shown (also using a similar force transformation argument--see end of chapter problem), and experiment verifies, that when the test particle is at rest in the moving frame, i.e. moving with the frame, the force per unit charge acting on it is simply:

7.6 $\dfrac{\vec{f}'}{q}(v' = 0) = \vec{E}' = \gamma E_0(y)\hat{y}$.

From these facts, we can conclude that, when the particle *appears* to move, a force from a second type of field comes into play, a force that has the following form: $-\gamma \beta^2 E_0(y)\hat{y}$. We can, in turn, interpret this field as an apparent magnetic field in the negative z-direction, because with:

---

[12] As already mentioned, a line charge in motion has more symmetry than a point in motion and will allow us to better focus on the generic principles at stake.

7.7 $\vec{B}' = -\gamma\beta\, E_0(y)\hat{z}$,

and a test particle of velocity $-v\hat{x}$, we get the apparent force per unit charge as:

$$\vec{f}'/q = \left(\gamma E_0(y) + \beta\cdot B_z'\right)\hat{y} = E_0(y)\left(\gamma - \gamma\beta^2\right)\hat{y}$$

7.8

$$\vec{f}'/q = \frac{E_0(y)}{\gamma}\hat{y}$$

And, this is indeed the value one obtains experimentally.

A similar analysis of the force on a test charge with an arbitrary apparent speed in the x'-direction, say $\beta' = v'/c$, in the moving frame results in the following equation for the force seen in the moving frame:

7.9 $\vec{f}' = q\left(\overbrace{E_y'}^{\gamma E_0(y)} - \beta'\cdot B_z' \right)\hat{y}$, with B_z' defined as in equation 7.7.

This is precisely the form needed for the second term to be a magnetic-field-like force, i.e. proportional to speed and perpendicular to $\vec{B}'$ *and* $\vec{v}'$; in particular, it is a particular case of $\vec{f}' \propto \vec{v}' \times \vec{B}'$.

Now, having established the magnetic-field like character of part of the apparent force in the moving frame, we can write the transformation law to a frame moving perpendicular to an *E*-field. Namely, using equations 7.6 and 7.7, respectively we have:

7.10a,b $E_y' = \gamma E_y$ $B_z' = -\gamma\beta E_y$

Notice that these can be verified by actually doing the calculations for the case of the line charge. In the rest frame of the line charge, we have $E_\perp = \dfrac{2\lambda}{r}$. While, in the moving frame, the charges appear to be moving at speed v in the negative *x*-direction and appear to be γ times closer together so that the linear density $\lambda \to \lambda\gamma$; hence we get: $E_\perp' = \dfrac{2\gamma\lambda}{r} = \gamma E_\perp$, where we have used the fact that $r = r'$, which is true because the motion is perpendicular to *r*. Similarly, for *B*, we have: $B_\phi' = -\dfrac{2I'}{r'} = -\dfrac{2\gamma\lambda\beta}{r} = -\gamma\beta E_\perp$.

In addition to more completely uncovering how the effects of an electric field *appear* in a moving frame, we have also seen something of the way forces transform and the correlative empiriological force transformation laws. Note that the latter is done in detail in the appendix to this chapter. At this point, rather than developing each transformation law in detail, we will summarize the general transformation laws.

Transformation Equations for *E* and *B*-fields

The transformation laws for the fields from the rest frame to the moving frame in standard configuration (i.e., a "boost") are given by Equations 7.11a-f. Prototypical examples are illustrated in Figure 7-12.

7.11 (For a boost in the *x*-direction)

a. $E'_x = E_x$ **b.** $B'_x = B_x$

c. $E'_y = \gamma(E_y - \beta B_z)$ **d.** $B'_y = \gamma(B_y + \beta E_z)$

e. $E'_z = \gamma(E_z + \beta B_y)$ **f.** $B'_z = \gamma(B_z - \beta E_y)$

Figure 7-12a gives standard configuration in which we define the coordinate systems for the lab frame (unprimed) and the moving frame (primed frame). Note in our analysis, we always start with an absolute rest frame (unprimed) and then transition to a uniformly moving frame (primed). Then, using the principle of relativity of motion, we leave behind whether the unprimed frame is moving or not and just take it to be our "lab" frame; the second (primed) frame we define only as moving at some uniform speed *v* relative to the lab frame, leaving behind what the actual state of motion of the frames are.

In Figure 7-12b, a *B*-field in the *z*-direction is shown in blue (since *B*lue begins with *B*). In the natural time keeping and measuring standards of the moving frame, plana disposed in this way appears to have an *E*-field *and* a *B*-field, which are represented, respectively, by the red and blue arrows. Using the transformation equations 7.11, and assuming a *γ*-factor of 2, the apparent *B*-field is shown twice as big as the actual *B*-field. Or said in special relativistic terms, the *B*-field in the moving frame (i.e., any frame moving uniformly relative to a given inertial frame called the lab frame) is in the same direction as, but twice the magnitude of, the *B*-field in the lab frame. And, in addition, an *apparent E*-field in the negative *y*-direction shows up in the moving frame and is shown in red in the figure.

Figure 7-12c shows an *E*-field, like discussed at the end of the last section, pointing in this case along the *z*-direction. In the primed frame, the *E*-field (in red) is in the same direction but apparently increased by a factor of two, while an apparent *B*-field (in blue) in the *y*-direction also appears.

Figure 7-12d, the last figure, illustrates the transformation of *B*-field along the *y*-direction. However, this transformation can be considered superfluous, if we recall the implicitly assumed symmetry of the two frames. Namely, one can rotate the fields in the second figure, Figure 7-12b, around the x axis by 90 degrees in the left handed sense to get this last figure. In terms of our transformation laws equations 7.11, this is equivalent to taking $\hat{z}(\hat{z}') \rightarrow \hat{y}(\hat{y}')$ and $\hat{y}(\hat{y}') \rightarrow -\hat{z}(-\hat{z}')$. In this way, one can generate equations 7.11d and e from 7.12f and c respectively.

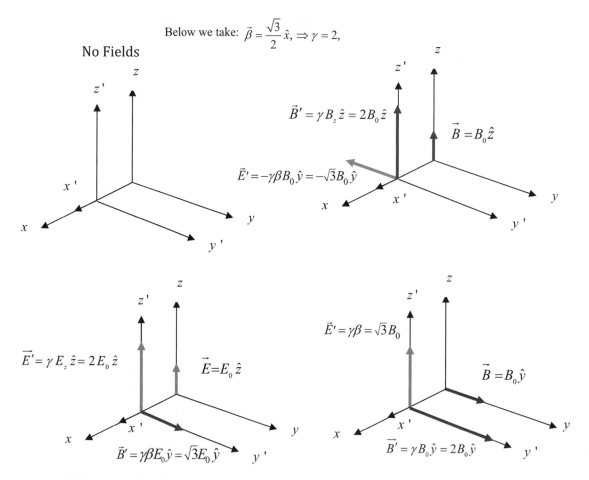

Figure 7-12: *a.* (*top left*) Standard boost configuration showing an unprimed coordinate system for the frame at absolute (or in the general case relative) rest and a primed coordinate system for the frame moving at uniform speed in the *x*-direction. ***b.*** (*top right*) *B*-field (blue) along *z*-axis in the rest frame appears as a *E*-field (red) and a *B*-field as shown in the moving frame. ***c.*** (*bottom left*) *E*-field (red) along *z*-axis in the rest frame appears as an *E*-field (red) and *B*-field as shown in the moving frame. ***d.*** (*bottom right*) *B*-field (blue) along *y*-axis in the rest frame appears as an *E*-field (red) and *B*-field as shown in the moving frame. This last is obtained by rotating the fields in figure *b* by 90 degrees clockwise around *x*-axis.

Note that these transformation laws abstract from the particular sources and from the potentials that specify the plana and reflect something more of the particular sources. We now move to incorporate A and ϕ.

Transformation Equations of A and ϕ Fields

It is convenient to write the A and ϕ fields as a single "4-vector," because, as we will see, such a form expresses a fundamental reality about the magnetic and electric potentials. In particular, we can write A as a column vector with ϕ in the first row and the three components of the A-field in the last three rows. Then, the transformation law between two inertial frames in standard configuration (see Figure 7-12), the so-called Lorentz transformation, for A and ϕ can be written succinctly as:

$$7.13 \quad \begin{pmatrix} A^{t\,'} \\ A^{x\,'} \\ A^{y\,'} \\ A^{z\,'} \end{pmatrix} = \begin{pmatrix} \gamma & -\gamma\beta & 0 & 0 \\ -\gamma\beta & \gamma & 0 & 0 \\ 0 & 0 & 1 & 0 \\ 0 & 0 & 0 & 1 \end{pmatrix} \begin{pmatrix} A^{0} \\ A^{x} \\ A^{y} \\ A^{z} \end{pmatrix} = \begin{pmatrix} \gamma A^{0} - \gamma\beta A^{x} \\ -\gamma\beta A^{0} + \gamma A^{x} \\ A^{y} \\ A^{z} \end{pmatrix} = \begin{pmatrix} \gamma\phi - \gamma\beta A^{x} \\ -\gamma\beta\phi + \gamma A^{x} \\ A^{y} \\ A^{z} \end{pmatrix}$$

To understand the transformation, let's return to our line charge, for which there is obviously no A-field. Or, said in the analogical general way typical in special relativity, a line charge presents no A-field in its own rest frame. However, equation 7.13 shows that in the moving frame, we will "see" an apparent A-field: $A^{x\,'} = -\gamma\beta\phi$.

In for example the case of the line charge, we have a ϕ such as shown in Figure 7-13, which changes in the y-direction. Focusing on the apparent B-field in the y'-direction at a fixed value of y' (note y will be the same in the moving frame, i.e., $y' = y$), we can write:

$$7.14 \qquad B_z' = \left(\nabla \times \vec{A}'\right) = \frac{\partial A^{y'}}{\partial x'} - \frac{\partial A^{x'}}{\partial y'} = -\frac{\partial A^{x'}}{\partial y'} = \gamma\beta\frac{\partial \phi}{\partial y}.$$

We would now like to see how this comes about in terms of the potentials. As we have pointed out before, if there is an effect seen in the moving frame, it is only because there is first, in principle, a qualitative change in a substance (or substances) when it moves, because it must have impetus to move. Our key task in understanding the nature of these substances with respect to their participation in the principle of integrity is to find out the nature of this change, especially its causal structure. We are here concerned with the nature of fields in moving plana. Hence, we are studying "trapped" plana, plana that moves along with the moving frame. What then happens in such plana? We will see that this question is best answered by keeping the trapped plana associated with the charges that cause the fields.[13] To see this, we return our focus to the case of relative motion along a line charge, which generates a static field in the plana if we allow enough time for the field to reach steady state.[14]

If we treat the plana as moving *without* the charges, the fields are the same in the moving plana (i.e., in a moving closed house) as in the rest plana (i.e., a fixed open house). Recall the fact that the measured results will not change when we open a closed house was shown in *PFR-M* and is a consequence of the relativity principle that follows from the principle of integrity. Now, there is no reason at (or near) our level of abstraction to think that plana can move without massive bodies, so we consider plana moving with the particles, the iconic case being virtual plana, plana that is part of a single substance.[15]

---

[13] We here will treat the movement of plana generically without the goal of deciding the issue of how it actually moves but only with the goal of bringing out key issues at stake in its movement and resolving them.

[14] That is, we consider the system only after the fields have had a chance to be established in the plana, so that no further change in the plana, at least in the region of interest, is occurring.

[15] Furthermore, in terms of the dual configuration, it would make no sense to have the unqualified plana (i.e. plana with no fields *anywhere* or *ever* activated in it, which means no sources anywhere) move away, for example, in the *setup* because it exercises, at our level of abstraction, no influence on test bodies. In the first

If the charges are allowed to move with the plana, the *actual* fields, as opposed to the apparent fields, are different in the moving closed (or open) house that contains the line charge moving with it, than they are in the *dual-setup* in which the line charge is not moving, i.e. is at absolute rest. In particular, in the *setup*, the line charge actually moves, so that there *is really* an *A*-field (and a *B*-field) in the trapped plana, though it does not *appear* as such in the frame moving with the line charge. Figure 7-13 below illustrates the *setup* and *dual-setup* for this case.

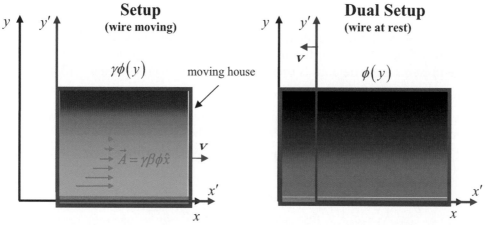

Figure 7-13: a. (*left*) *Setup*: Moving closed house with trapped plana and line charge inside moving with it. In the *setup*, one measures from the absolute rest frame so one gets an understanding of the physical state relative to undilated clocks, masses and lengths and without operational simultaneity. **b.** (*right*) *Dual-setup*: Stationary closed house with trapped plana and a line charge at rest, but viewed from moving frame, so the relative velocity is same as setup. All fields are drawn to indicate their *actual* state, not how they appear in a moving frame. The ϕ-field is shown as green, with brighter green indicating more intense ϕ. A group of red arrows indicate the intensity and direction of the *A*-field in the region close to the wire. The fields, of course, extend outside the walls of the house, but are not shown to simplify the drawing.

---

form of the dual configuration, all bodies that exercise an influence (which, by definition, excludes test particles or measurement instruments) must be moving at the same speed in the *setup* and must be at rest in the *dual-setup* so that the observer can move away from them all. The point of this is (1) in the setup, to learn how moving bodies change, (2) in the dual setup, to learn how bodies at rest appear in the moving frame, and (3) to explore their deep interrelation. However, at some point, one also wants to investigate how a moving body appears in a moving frame and what that says about the deep interrelation between the setup and its dual. In this second more complex form of the dual configuration, the observer in the *dual-setup* still views from a uniformly moving frame; however, now, instead of being at (absolute) rest, the bodies must be given appropriate (in general non-zero) velocities in accordance with the *setup* and the physical implications of relativity (e.g., see the example in the section "An 'Atomic' Clock made out of Rotating Charges" and secondarily see "Conclusion and Last Example").

To better understand the transformation law, we make an analogy to the case of a massive body at rest. In such a case, we have $p = \begin{pmatrix} m_0 \\ 0 \end{pmatrix}$. In our case with the potentials, we have: $p_{pot} \equiv A^{\mu} = \begin{pmatrix} \phi \\ 0 \end{pmatrix}$. Recall that the rest mass energy, m_0 is a kind of potential energy that can, under the right causal agency, be made to cause impetus with a certain level of activity, i.e. it can cause energy, and in so doing it expends itself in proportion to the energy it makes.[16] Hence, in summarizing this in the packaged-to-go language of the empiriometric system, we say mass is converted to energy.

The resting mass has no impetus. However, when viewed from the frame moving at $-v\hat{x}$, it *appears* to have an impetus of intensity $\gamma m_0 v$ in the x-direction, and its energy also appears thereby increased, for it has an apparent kinetic energy $= (\gamma - 1) m_0$. Now, when such a massive body is actually moving, it *actually* does have more energy and more mass. From the viewpoint of the principle of integrity, this then is the primal source of the apparent increase in mass-energy in the moving frame. Empiriometrically, we write the transformed "momentum-mass 4-vector" as:

7.15 $\qquad p' = \begin{pmatrix} \gamma m_0 \\ \gamma m_0 \beta \end{pmatrix}$ $\qquad$ which we get from: $p' = \begin{pmatrix} \gamma & \gamma\beta \\ \gamma\beta & \gamma \end{pmatrix} \begin{pmatrix} m_0 \\ 0 \end{pmatrix} = \begin{pmatrix} \gamma m_0 \\ \gamma m_0 \beta \end{pmatrix}$

In a similar way, ϕ is a potential energy of the different type already described. Now, this potential energy "boosted" into a new frame (which means, as we saw above, the charges must also be boosted), seems to act like the rest mass. Empiriometrically, this is reflected in the "4-vector potential," A, transforming in the same way as p:

7.16 $\qquad A' = \begin{pmatrix} \gamma \phi \\ \gamma \phi \beta \end{pmatrix}$

Hence, we are led again to the mass-energy equivalence.[17] Namely, this potential for the plana to cause energy itself has an effect that is analogically like mass. This is manifest, at one indirect level, if we consider first converting the potential energy in the field to rest mass energy to get us exactly to the rest mass case.

More directly, in terms of the moving plana, say virtual plana which is part of a substance, we expect the body to gain more potential for causing energy when it is moving. This is reflected in the top term in equation 7.16. We also expect that if we boost to a moving frame, we should see an increase (or decrease) in potential momentum as well. This, in turn, implies that moving plana changes (because of the moving charge) so that it can cause greater (or lesser) momentum. Indeed, the potential momentum that results, which we saw is given by A, is precisely what is predicted.

---

[16] Recall the two meanings of mass needed here: inertial mass, the resistance to the action of the impetus, and what we can call "potential energy mass," or simply the rest mass, which is the fact that the amount of initial inertial mass (i.e. the body's mass when only an "infinitesimal" impetus is activated) is proportional to an ability to cause energy, a certain rate of transfer of impetus, in another body.

[17] We also have to account for the fact that, unlike the massive particle, the field is a property of the plana in some region and thus defined on a grid or a space rather than at a single place. When we transform the field, we also have to transform the coordinates which label the points of the space; there are thus two things to transform. This is similar to the massive particle in that it has a location as well as a momentum, except it only has one location.

In particular, first, the potential in the *x-y* plane (for $y > 0$) when the line is at rest is: $\phi = 2\lambda \ln \dfrac{y}{k}$, where k is a constant picked for convenience in defining the zero of ϕ. This yields the expected field: $E_y = \dfrac{2\lambda}{y}$. When the line charge was moving, we saw earlier that $E_y^{\ moving} = \dfrac{2\gamma\lambda}{y}$, which verifies the top term in equation 7.16, using the equivalence of the setup to the dual-setup (which gives $E_y{'} = E_y^{\ moving}$) and the form of ϕ just given. Second, by similar reasoning, we verify that $A_x{'} = 2\gamma\lambda\beta \ln \dfrac{y}{k} = \gamma\,\beta\,\phi$.

Physically, we can see that the potential energy (ϕ) and the potential momentum (A) represent real aspects of a moving line charge in that those fields, as we saw earlier, are signs of the energy and momentum "stored" in the act of creating those fields. Said another way, if we stop the line charge, stored "self-inductive" energy and momentum will be released. And, if the forces holding the line charge together are "shut off," the line charge will break into (effectively) arbitrarily small pieces that move to infinity, thus also releasing a certain amount of "self-capacitive" energy.

Hence, we see the trapped plana and the associated charged massive body change in such a way as to produce the behavior described by the transformation laws. The impetus acting in a massive substance which includes virtual plana modifies the charge and/or virtual plana so that the fields change in such a way that the substance behaves in uniform motion as it did when at rest.

Other important questions can be asked, but we leave them for other courses because dealing with them properly requires information outside of classical E&M. So we shift to summarize the qualitative points of this subsection.

Note that here we have, for the first time, begun to see the virtual plana as an integrated part of the whole of a given massive substance, as it must be if it is indeed a substance about which we speak. In particular, a massive substance (formally) causes its property of inertial mass, which specifies its receptivity to the action of impetus. Ordinary massive bodies are composed of virtual charge as well as virtual plana; the charge causes fields in the virtual plana. The field can cause energy and thus can be thought of as having potential energy. As we have seen, the fields caused in the plana also depend on whether impetus is activated in the substance or not. If these statements generalize to other fields, then it may be that the fields are, in some analogical sense, the source of both types of rest mass: the inertial rest mass and the potential energy rest mass.[18] More accurately said, since the core of the essence of an ordinary massive substance is to emanate both its ability to receive impetus, i.e. its inertial mass, and its charge (and the fields from them), it makes sense that they should be proportioned in some specific way. We could then say, if this is verified by further experiment and analysis, that the fields are, in some sense, responsible for the convertibility of mass and energy.

---

[18] Some of what we call the potential energy of the rest mass should be considered actual energy as it is due to actual impetus of the parts of the substance. Note that such a use of the word "impetus" is an analogical use, because in it "impetus" belongs to a part, not the whole in the way we usually speak; furthermore, because of this, the word energy is also here used in an analogical sense.

Extra--for Advanced Students

In more advanced courses, you will learn to be proficient in the formal mathematics of tensor analysis. This mathematics expresses the special relativistic results we have given in a way that, if understood properly, brings out its essentials. There you will learn that one can write the so-called Maxwell tensor, which connects the potentials (ϕ and A-fields) with the E and B-fields:[19]

7.17
$$F_{\mu v} = \frac{\partial A_v}{\partial x_\mu} - \frac{\partial A_\mu}{\partial x_v}$$

In this notation, the indices μ and v can be t, x, y or z which represent, in order, the time, x, y and z components, where $A_t \equiv \phi$. In matrix notation, we make the following identification: [20,21]

7.18
$$F_{\mu v} = \begin{array}{c} \quad\; t \quad\; x \quad\; y \quad\; z \\ \left. \begin{pmatrix} 0 & E_x & E_y & E_z \\ -E_x & 0 & -B_z & B_y \\ -E_y & B_z & 0 & -B_x \\ -E_z & -B_y & B_x & 0 \end{pmatrix} \right\} \begin{array}{l} t \\ x \\ y \\ z \end{array} \end{array}$$

Equation 7.17 is written in so-called covariant form (which is written with subscripts) and is simply related to the contravariant form (which is written with superscripts) that we have used up till this point. Contravariant vectors are vectors that transform the way the space-time coordinates transform, i.e. the way "ordinary" vectors do.

Now, as an example in using the Maxwell tensor, we make the equation explicit for the $\mu, v \to t, x$. We take the first index as the row and the second as the column giving: $F_{tx} \equiv E_x$ and $F_{yz} \equiv B_z$, which coupling with equation 7.17 and taking $c = 1$, gives:

$$F_{tx} \equiv E_x = \frac{\partial A_x}{\partial t} - \frac{\partial A_t}{\partial x} = \frac{\partial A_x}{\partial t} - \frac{\partial \phi}{\partial x} = -\frac{\partial A^x}{\partial t} - \frac{\partial \phi}{\partial x}$$

In the last line, we have used the special relativistic relationship between the covariant and contravariant components of the A-field; $A^\mu = \sum_\sigma \eta^{\sigma\mu} A_\sigma$, which yields: $A^t = A_t$, $A_i = -A^i$, where $i \in \{x, y, z\}$.

The advantage of this tensor notation is that in the moving (primed) frame the Maxwell tensor (equation 7.17) is written in exactly the same way except all quantities are marked with a prime to indicate that they are measured in the primed frame. One can show

---

[19] You will also learn a short hand notation so that this formula can be written: $F_{\mu v} = \partial_\mu A_v - \partial_v A_\mu$ (and in the moving frame: $F'_{\mu v} = \partial'_\mu A'_v - \partial'_v A'_\mu$).

[20] Note that we use the plus sign for the temporal component and minus signs for the spatial components of the Minkowski metric.

[21] One can get the covariant form of F in the other metric signature by changing the signs of all the field components in this covariant matrix. To get the contravariant form of F, simply change the signs on the E-field components only.

that this is the case using the fact that both the derivative $\dfrac{\partial}{\partial x_\mu}$ and A_ν, being covariant,

transform in the manner shown in equation 7.19. (Whereas, contravariant "4-vectors" such as the time-space vector $\begin{pmatrix} t & x & y & z \end{pmatrix}$ transform as shown in equation 7.13 (see also *PFR-M*)).

Hence, for the covariant 4-vector A, which consists of $A_0 = A^0 = \phi$ in the first row and the three components, $A_i = -A^i$ in the last three rows, we have:[22]

7.19
$$\begin{pmatrix} A_0' \\ A_x' \\ A_y' \\ A_z' \end{pmatrix} = \begin{pmatrix} \gamma & \gamma\beta & 0 & 0 \\ \gamma\beta & \gamma & 0 & 0 \\ 0 & 0 & 1 & 0 \\ 0 & 0 & 0 & 1 \end{pmatrix} \begin{pmatrix} A_0 \\ A_x \\ A_y \\ A_z \end{pmatrix} = \begin{pmatrix} \gamma A_0 + \gamma\beta A_x \\ \gamma\beta A_0 + \gamma A_x \\ A_y \\ A_z \end{pmatrix}$$

Returning to the case of a uniformly charged line, we now calculate B_z with this formalism. We get, using the conversion from covariant to contravariant given earlier:

7.20
$$F_{yx}' = B_z' = \partial_y' A_x' - \partial_x' A_y'$$
$$= \partial_y \left(\gamma\beta A_0 + \gamma A_x \right) - \left(\gamma\beta\partial_0 + \gamma\partial_x \right) A_y$$
$$= \gamma \left(\partial_y A_x - \partial_x A_y \right) + \gamma\beta \left(\partial_y A_0 - \partial_0 A_y \right)$$
$$= \gamma \left(-\partial_y A^x + \partial_x A^y \right) + \gamma\beta \left(\partial_y A^0 + \partial_0 A^y \right) \quad \Rightarrow B_z' = \gamma\, B_z - \gamma\,\beta E_y$$

Where we use the following short hands: $\partial_0 = \dfrac{\partial}{\partial t}, \partial_x = \dfrac{\partial}{\partial x}, \partial_y = \dfrac{\partial}{\partial y}$

For the line charge, since there is no A-field, we can get as $B_z' = +\gamma\beta \left(\partial_y \phi \right)$, the form already obtained in equation 7.14.

Charge

The discussion and analysis in this chapter so far bring home the essential dependence of all of E&M on charge. We will thus take this opportunity to bring together what we know about charge and fill in a few holes that we have left.

Charge Invariance

Up to this point, we have tacitly assumed that charge is invariant under Lorentz transformation, i.e. when a charged body moves uniformly its charge, in some sense, stays the same. Furthermore, implicit in this statement is the empiriological lumping of active and passive charge into a single mental construct "charge," which is only possible when we assume the universal ratio of active to passive charge.[23] Thus, we have also assumed, as

---

[22] To get the transformation matrix for such a covariant vector, one simply takes $\beta \to -\beta$ in the matrix for the contravariant transformation.

[23] Note once we establish the invariance at our current level of analogical abstraction, we again leave behind distinction between active and passive charge as well as the new active and passive nature of charge as modified by impetus. In this way, we can enter the empiriometric domain of special relativity and focus on the parts of the reality that can be simply represented in our formalism, which captures the principles as they

experiment has verified to good accuracy, that this universal ratio continues to hold for charged massive bodies with impetus.

In particular, charged (positive and negative) massive bodies can cause two[24] different types of dispositions of the plana, A and ϕ, which means there are *two types of active charge*. The effect of these active charges is written in terms of current distributions and charge distributions as follows:

7.21
$$\frac{1}{c^2}\frac{\partial^2 \vec{A}}{\partial t^2} - \left(\frac{\partial^2 \vec{A}}{\partial x^2} + \frac{\partial^2 \vec{A}}{\partial y^2} + \frac{\partial^2 \vec{A}}{\partial z^2}\right) = \frac{4\pi}{c}\vec{J} \qquad \Rightarrow \vec{A} = \int \frac{\vec{J}\left(\vec{r}',t - \frac{|\vec{r}-\vec{r}'|}{c}\right)}{|\vec{r}-\vec{r}'|}d^3r',$$

where $\vec{J} = nq\vec{v}$, *n is the number of charges with charge q per unit volume.*

7.22
$$\frac{1}{c^2}\frac{\partial^2 \phi}{\partial t^2} - \left(\frac{\partial^2 \phi}{\partial x^2} + \frac{\partial^2 \phi}{\partial y^2} + \frac{\partial^2 \phi}{\partial z^2}\right) = 4\pi\rho \qquad \Rightarrow \phi = \int \frac{\rho\left(\vec{r}',t - \frac{|\vec{r}-\vec{r}'|}{c}\right)}{|\vec{r}-\vec{r}'|}d^3r'$$

where $\rho = nq$

However, there are three types of passive (receptive) charge, three ways charged bodies can receive impetus from E&M fields: (1) The ϕ-field, i.e. ϕ-activated plana, can cause impetus in charged massive bodies according to the gradient of the field ($\vec{E} = -\nabla\phi$, so that). (2) The A-field, i.e. A-activated plana, can similarly cause impetus in *impetus-activated* charged massive bodies according to the curl of the A ($\vec{B} = \nabla \times \vec{A}$). (3) When the source does not support the A-field in a constant way, the plana resists the changes by trying to activate impetus in charged massive bodies ($\vec{E} = -\frac{1}{c}\frac{\partial \vec{A}}{\partial t}$). So, we have:

7.23a-c
$$\vec{F}_1 = q_1\left(-\nabla\phi\right), \quad \vec{F}_2 = q_2\frac{\vec{v}}{c}\times\vec{B}, \quad \vec{F}_3 = q_3\left(-\frac{1}{c}\frac{\partial \vec{A}}{\partial t}\right), \text{ where } q = q_1 = q_2 = q_3$$

Much analysis in terms of the nine categories of properties could be applied to these equations and to the relations for the fields in terms of the sources (equations 7.21 and 7.22), but it is beyond the scope of this course to do so. Instead, beyond noting that future physics may hinge on (or even lie hidden in) the simplifying assumptions and mental constructs[25] packed into those equations, we confine ourselves to a few important points.

First, note that the receptivity of a body to the first and last type of force (equations 7.23a and c) are *not* dependent on the impetus of the test body, but only on the same measure of its receptivity that we saw in the static case. By contrast, since the second type of force is proportional to the velocity of charged body, the receptivity is dependent on the effectiveness of the impetus of which velocity is a measure. Hence, by not requiring

---

can be used to predict the results of quantitative measurements. Some day, as the equivalence principle of gravitational theory did with respect to mass, it may even be important empiriometrically to come back to these distinctions which remain real even when we don't focus on them in our highly successfully theories.

[24] In the full sense, then, there are four types of active charge.

[25] For example, in a certain (probably untenable, but interesting) theory of elementary charge, impetus changes the way an elementary charge acts. Namely, in that theory, in order to produce the field pattern seen for moving charges, one supposes that the charge acts differently in causing a field in the plana, depending, for example, on whether it is acting in the forward or backward direction.

ourselves to write a single term for the receptivity, we make the separate dependence on q and $\vec{v}$ more clear. Namely, we write the force in terms of the measure of the velocity and the measure of the receptivity of the charged body *before* any impetus is activated in it. In this way, the receptivity to the action of these diverse forces can be treated as proportional, indeed equal so that we have $q = q_1 = q_2 = q_3$, so that by a mental construct (being of reason), we treat, by analogical generalization, the various passive charges as if they were one.

Second, we note that the passive charge defined in this way, coupled with the analogical definition of active charge implied in the equations 7.21 and 7.22, means that passive and active charge are proportional and can be treated as having equal measures independent of the motion of the particle. This then is the fuller meaning of the universal ratio of passive to active charge.

We can express the key aspect for relativity, i.e. the invariance of charge with uniform motion, in a more results-oriented, though less precise, language in the following way. Experimentally, the force a charged particle experiences in a given electric field is independent of its motion, which means that it is independent of the speed, the measure of effectiveness of its impetus. Furthermore, in all our discussions of charges moving in free plana, we have seen that the field propagation accounts for the state of the plana around such particles *without* needing to consider a change in the effectiveness of the charge's production of the field. Thus, we can say that, in some sense, the motion of the charge neither affects its active or passive charge.

Lastly we note that within moving trapped plana, as opposed to within free plana, there may be reason to suspect that active and/or passive charge do physically change. In particular, because the virtual plana case is primary, we focus on it. Ordinary substances have virtual electrons, protons and virtual plana. Each of these is present by its power, so when the substance has impetus, that impetus qualifies the whole substance in a different[26] way. Of course, if the active and/or passive charge (and/or the correlative receptivity in the virtual plana) change because of this, they, at our level of abstraction, must change in such a way that they produce a field that behaves in the way described by the transformation laws. Such an observation is our reminder to leave the issue here. Though the issue is of some importance it is, again, beyond the scope of this course, so we leave it by saying that however it may be, it is irrelevant at the level of abstraction of E&M as is evident in the fact that Maxwell potential equations (coupled with the Lorenz condition) are special relativistic, i.e. that they are in invariant under the Lorentz transformations.

Other Important Features of Charge

A couple of other important empiriological and physical interpretation issues relative to charge need to be discussed.

Firstly, we note we need to re-examine the Gauss's Law definition of charge given in Chapter 2 in light of the possibility of moving charges. Recall that in differential form of Gauss's law says that the measure of the charge, q, per unit volume is proportional to the divergence of the E-field at that point: $\rho = \dfrac{1}{4\pi} \nabla \cdot \vec{E}$. This is a definition of active charge in the static case, where $\vec{E}_{static} = -\nabla \phi$. However, when there are moving charges, we get:

---

[26] Yet obviously consistent with its nature, otherwise it would destroy it.

$\rho = \dfrac{1}{4\pi} \nabla \cdot \left(-\nabla\phi - \dfrac{1}{c}\dfrac{\partial \vec{A}}{\partial t} \right)$, so that using the Lorenz condition, $\nabla \cdot \vec{A} = -\dfrac{1}{c}\dfrac{\partial \phi}{\partial t}$, that says

that the ϕ-field and A-field is always produced in such a way that the divergence of the A-field out of a region indicates the rate of decrease of ϕ in the region. Mathematically, we get, the equation below, from which we can get the previously given solution for ϕ.

7.24 $\rho = \dfrac{1}{4\pi}\left(-\nabla^2\phi - \dfrac{1}{c}\dfrac{\partial \nabla \cdot \vec{A}}{\partial t} \right) = \rho = \dfrac{1}{4\pi}\left(-\nabla^2\phi + \dfrac{1}{c^2}\dfrac{\partial^2 \phi}{\partial t^2} \right)$

We re-write this as:

7.25 $-\nabla^2\phi = \nabla \cdot \vec{E}_{static} = 4\pi\rho - \dfrac{1}{c^2}\dfrac{\partial^2 \phi}{\partial t^2}$

From which, we see that the, divergence of the E-field proper, i.e. that specified by ϕ, is still determined by the charge, as in the static case, except there is a correction due to the acceleration of the magnitude of ϕ, which can be related back to the finite propagation speed of that field. To begin to see this in a heuristic way, think of trying to determine the charge of a source inside of a sphere centered at a certain point by using the strength of the ϕ-field on the surface of that sphere, i.e. by doing an integral of ϕ over the sphere. In this way of looking at the issue, we can see, in the following way, that we need a term to account for the change in the ϕ-field's distribution around the charge due to its motion. If the speed of propagation of the field were infinite, there would be no such change because the field distribution would always exactly reflect the charge that is causing it at the given instant. In other words, a changing source *at a given moment* would produce the exact same ϕ-field as it would if it were not changing. However, because the speed of propagation is actually finite, the ϕ-field can, for example for a moving particle, bunch up in some places. Thus, we cannot take the value of ϕ to reflect the charge of the particle in the same way in motion as it does at rest. Someone who does will, in the case of bunching, conclude the charge is stronger than it actually is. In very rough terms, the acceleration term accounts for "bunching" and "stretching" that arise in a changing source due to the finite propagation time.

Secondly, we know, experimentally, that charge comes only in discrete values and is conserved. Much is made of the discrete nature of charge, though it fundamentally follows at some level from the fact that all material bodies have extension, which means actual parts (recall quantity is the first property of all physical substances). Hence, we expect that there will be a smallest particle, which if we are successful in breaking will no longer have the property of charge. Now given conservation of charge, this means that breaking such a smallest charged particle will not be possible. The only way to destroy it would be interaction with a particle of the opposite charge of the same magnitude, so, that in the process, charge is locally conserved. Thus, for example, an electron and a positron can interact with each other and the plana in an "annihilation" event that ends with gamma rays activated in the plana. Keep in mind that the extension of such particles may be very small and, indeed, may not even be in principle *directly* experimentally probe-able. Note again that this understanding of the discreteness of charge depends on conservation of charge.[27] So this leads us to ask: Why is there conservation of charge?[28]

Charge conservation seems to be part of nature's maintenance of conservation of energy. Given mass-energy equivalence as we currently know it, if charge were not conserved, then two like charges could "annihilate," destroying all the electrical potential energy stored in the pair of particles. For example, if you had brought them together, it took you a certain amount of energy to do so. Thus, the maintenance of conservation of energy provided by conservation of charge gives a handle on the internal reasons, the reasons for conservation of charge having to do with the proper interaction of the bodies themselves. As for extrinsic reasons, i.e. the reasons why discreteness and conservation of charge are needed for higher substances beyond the simple particles, we simply note there is a need for primitive powers, such as charge, to be invested in an element-like nature from which those more complex substances, including plants and animals, can draw certain fundamental powers. Thus, for example, quarks, which have charge, form protons and protons become part of sodium, which become part of table salt, which can become part of you.

There is of course more to learn about charge. For one, the nature of charge, including a further understanding of the above aspects, cannot be had without further investigation into the principle of integrity via the special-relativistic type analysis such as we have done. And, as always, further specification of our understanding comes by inclusion of more advanced subject matter (e.g., QFT and general relativity) and, finally, by each new piece of physics research.

An "Atomic" Clock made out of Rotating Charges

We have talked about many of the effects on moving bodies, but we have not yet discussed time. With the transformation laws understood and the key parts of Maxwell's theory in place, we can now look at one type of clock and ask how it looks in a moving frame to understand how it works with other correlative changes to preserve the integrity of moving substances.

The clock in the moving frame is shown in Figure 14b below. The clock consists of a very massive core (like an atomic nucleus) with a very light charged particle moving in a circular orbit around it. The orbital trajectory is in the y-z plane when at rest (shown in Figure 14a; when the observer is in the moving frame, this is a *dual-setup*), and in the y'-z' plane when moving (shown in Figure 14b; when the observer is at rest, this is a *setup*). We take the charge of the nucleus and the orbiting charge to be equal and opposite and of magnitude q. If we take the mass of the nucleus and orbiting particle to be M and m respectively, we have $M \gg m$. We also take the size of each particle to be negligible. Because of its close analogy to an atom, we will call the orbiting particle an electron to make for a more simple sentence structure. For simplicity, we take the moving frame speed to be much greater than the orbital speed of the electron: $v \gg v_e$ and $v_e \ll c$, so that we

---

[27] Note physics is open to discovering all kinds of new types of things, but they need to be based on evidence for many obvious reasons not the least of which is, without evidence, we do not really know what we are talking about.

[28] Or said another way, what is the final causal reason for it?

can neglect relativistic effects of the orbital velocity. We also assume the acceleration is small enough that we can neglect radiative effects.

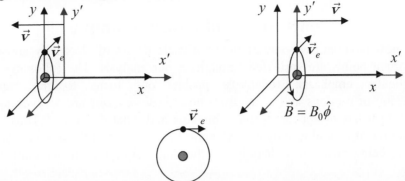

Figure 7-14: Analysis of an "Atomic" Clock in two physically different states. *a.* (*left*) Clock is at rest; we are interested in viewing it mostly from the absolute rest frame (or lab frame) but also from the moving (primed) frame (*dual-setup* relative to setup called out in *b*), because it is from the moving frame that the dynamical symmetry comes into play. *b.* (*right*) Clock is moving; we view from unmoving (unprimed) frame (a *setup*), for we are interested in actual situation, not as viewed empiriologically. Notice for such dual-configurations the key objects are in different physical states. *c.* (*bottom*) Front view of the "atom."

In the absolute rest frame (or lab frame), assuming the electron orbits at a fixed circular radius of r_0 from the point-like nucleus, we have:

$$F = eE_0 = m\omega^2 r_0 \Rightarrow \omega = \sqrt{\frac{eE_0}{m\,r_0}} \equiv \omega_0, \; T_0 = \frac{2\pi}{\omega_0} = 2\pi\sqrt{\frac{m\,r_0}{eE_0}} \; .$$

When the "atom" is moving, we have two forces acting on the electron: a magnetic and an electric force. The electric force in the plane of rotation of the electron is increased by a factor of γ. The magnetic force acts only on the non-azimuthal component of the velocity namely, $v\hat{x}$. Since $\vec{B} = B_0\hat{\phi}$, where $\hat{\phi}$ is defined in Figure 14b above and $B_0 = \gamma\,\beta E_0$, we get the total force on the electron from the nucleus for the moving atom is: [29]

$$\vec{F}_{moving} = (-e)\left(\gamma E_0\hat{r} + \beta\hat{x} \times \gamma\beta E_0\hat{\phi}\right)$$

$$= -e\gamma E_0\hat{r}\left(1-\beta^2\right) = -\frac{eE_0}{\gamma}\hat{r}.$$

Thus, we get, since lengths do not shrink in the perpendicular direction, but mass still dilates: $\gamma m\omega'^2 r_0 = \dfrac{eE_0}{\gamma}$; hence, $\omega' = \sqrt{\dfrac{eE_0}{\gamma^2 mr_0}} = \dfrac{\omega_0}{\gamma}$, so that $T = \gamma T_0$. Hence, the moving clock takes γ times longer to make one revolution than does a non-moving clock of the same type.[30] This is exactly the time dilation we sought to verify. Notice again the

---

[29] Note here we use: $\hat{x} \times \hat{\phi} = -\hat{r}$.

[30] See end of chapter problem that drops the assumption of small orbital velocity.

importance of the existence and particular nature of the magnetic field as an efficient cause in implementing the principle of integrity.

Conclusion and a Last Example

We now have an overview of the relativistic nature of electricity and magnetism. Many important points are sprinkled throughout our analysis. The following is one that needs particular emphasis. Relativity pushes our focus to the "appearances" (measurements) of motions and away from how those *motions* are or even appear to be caused (i.e. what has or appears to have impetus and what doesn't).[31] Thus, to see how relative motion that is produced by a *combination* of a motion of the observer and a motion of the object being observed is handled in the dual configuration, we consider one last simple case.

We look at the case of a positive, uniformly charged line (our line charge), with a positive point charge speeding along it. In this case, as we know, the line charge actually, not just apparently, causes an electric field to be activated in the plana, which, in turn, acts on the point charge. This can be considered from a frame moving at some speed v which is not equal to the speed of the particle, so that, in this moving frame, the particle does not appear to be at rest (see Figure 15a). For simplicity of description, consider that, from this frame, the particle only moves slightly. In this frame, the fields that act *appear* to be different than they were when viewed from absolute rest, but the actual state of the particle, the free plana, and the line charge are clearly unchanged.

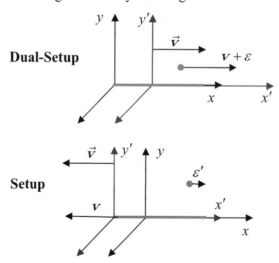

Figure 7-15: *Dual-Setup* and *Setup,* two physically different situations, are illustrated along with natural frames. *a.* (*top*) "Viewing" the stationary line charge and a moving point charge (moving at absolute[32] speed $v + \varepsilon$) from the

---

[31] As a consequence, it also does not pay attention to how impetus (real or apparent) is caused. Hence, as we have seen, from the point of view of empiriometric relativity, what matters is not, for example, whether there is an actual E-field or not, but whether whatever causes the effect produces measured results as if it were an E-field, i.e. as would an E-field of the chosen strength and direction.

[32] Keep in mind one might not be able to determine what the actual speeds are because we might not, in principle, be able to determine what the absolute frame is, but this does not obviate the need for saying, in principle, what is really happening (see *PFR-M* for discussion on place and motion). As we have said before,

primed (blue) moving frame, i.e. the *dual-setup*, gives the same apparent relative motion as **b.** *(bottom)* "Viewing" the line and the point charge from the unprimed (black) absolute rest frame, i.e. the *setup*.[33] Note that all speeds in both diagrams are the *actual* speeds, not speeds as they appear in moving frames.   In the bottom configuration (the *setup*), the line charge generates a magnetic field, whereas in the top configuration (the *dual-setup*), it does not.

However, consider now the two physical situations shown above *along with* the specified observers (i.e., the dual-configurations, the *setup* and its dual), which "look" equivalent, because the relative motion is the same. Recall that, loosely speaking, you cannot tell whether you are moving away from something or it is moving away from you as long as, whatever uniform motion you and the body actually undergo, the relative velocity is the same. In the *setup* in which the particle is moving slightly, and the line charge is moving along its own length (cf. Figure 15b), the line charge generates a magnetic field, an *A*-field, in the plana which then acts, along with a modified electric field, on the point charge. See how physically different this is from the *dual-setup* in which the line is at rest, the charge is moving quickly, and the observer is moving (cf. Figure 15a); yet the *apparent* motion produced in the respective frames is exactly the same.

Now, the most important general principle we have found and explored through such examples as the above is as follows. The nature of electrical substances, which includes ordinary substances because they all contain virtual charge (if not always actual charge),[34] is to have magnetic fields that result necessarily when such bodies have impetus. And, it is only because of this that the principle of integrity and, hence, empiriological special relativity can be true. As always, we see that the intelligibility that we encapsulate in our empiriological systems resides in the natures of substances.

In previous chapters, we established the fundamental causal and empiriological structure of electricity and magnetism. In this chapter, we have elucidated something of the fundamental unity of electricity and magnetism[35] as revealed by the integrity principle whose fundamental consequence is the principle of relativity, which states that the results of any experiment will be the same in any inertial frame.[36] Having thus established the

---

refusing to make real distinctions, such as between rest and motion, to avoid pragmatic difficulties (as good as that may be), even in principle pragmatic difficulties does not obliterate those distinctions. On the other hand, avoiding such difficulties by ignoring such distinctions is usually to be encouraged in our formalism; after all we should focus problem solving and our research on what, at present we know and can know more about.

[33] Note, the primed frames in the two setups, though shown using the same notation, are *different* frames. The primed frame in the right setup is the rest frame of the line charge, while it is not in the left setup. *Primed* here simply means a moving frame.

[34] Because of this, we can also say, in our analogically generally empiriological way: "ordinary *bodies* are electrical *bodies* because they all contain charge, even if they don't always have a net charge." Note that we often even ignore the electrical nature of "macroscopic" parts of ordinary bodies. Many times we consider the parts of a body without considering their composition of positive nuclei and negative electrons, but just consider the larger parts as neutral material on a scale of many atoms.

[35] This is part of what is called the intrinsic finality of electricity and magnetism, i.e. the interior reasons that electricity and magnetism is the way it is rather than some other way.

[36] Recall an inertial frame implies any frame that moves at a uniform (absolute) speed in a non-interactive environment other than the bodies of the experiment itself. This means that (absent the experiment) the plana has no qualifications that will make the experiment depend on the orientation of the experiment in the plana or the place of the experiment or the time at which one starts the experiment. We can say in an empiriological

fundamentals, we now move to use some of these fundamentals: (1) to understand how electric and magnetic fields affect certain materials, (2) to learn more about circuits and (3) to design and build a radio frequency transmitter, our practical finale, where we will make use, in some way, of everything we have learned.

Chapter Appendix: Force Transformation Laws

Force, a power to cause impetus in a body,[37] is revealed by the rate of change of impetus of a body. This is, in turn, found by measuring the rest mass and measuring the body's speed to get the total momentum at two arbitrarily close times from which we calculate the force using the equation: $\vec{F} = \dfrac{d\vec{p}}{dt} = \dfrac{d\left(\gamma m_0 \vec{v}\right)}{dt}$. This is not, however, a four vector, so it will not transform according to the standard relativistic transformation (which is developed in *PFR-M* using the integrity principle). We easily generalize by taking $\vec{p} \to p^\mu$ (taking p^μ from *PFR-M*) and $dt \to d\tau$, where the later is the proper time, the time as read off in the frame moving along with (i.e., at rest with) the particle. Hence, we have:[38]

$$F^\mu \equiv \frac{dp^\mu}{d\tau} = \gamma \begin{pmatrix} \dfrac{dE}{dt} \\ \dfrac{dp^i}{dt} \end{pmatrix} = \gamma \begin{pmatrix} P \\ f^i \end{pmatrix}$$

Note that this equation is formally just like that for the four-velocity:

$$U^\mu \equiv \frac{dx^\mu}{d\tau} = \begin{pmatrix} \dfrac{dt}{d\tau} \\ \dfrac{dx^i}{d\tau} \end{pmatrix} = \gamma \begin{pmatrix} 1 \\ \dfrac{dx^i}{dt} \end{pmatrix} = \gamma \begin{pmatrix} 1 \\ u^i \end{pmatrix}$$

One can calculate the velocity transformation law, for standard transformation along the *x*-axis, from this and get, taking $c = 1$, the following standard relations for a particle moving at speed $\vec{u} = u^i \hat{x}_i$:

$$u_x' = \frac{u_x - v}{1 - u_x v},$$

$$u_y' = \frac{u_y}{\gamma\left(1 - u_x v\right)}, \quad u_z' = \frac{u_z}{\gamma\left(1 - u_x v\right)}$$

---

shorthand, "space" is homogeneous and isotropic and time is homogeneous. Since the results are the same in any frame moving uniformly with respect to any given inertial frame, we can add the final empiriological statement that the space-time metric is invariant under a Lorentz boost in any of the three orthogonal directions.

[37] See "Helpful Hints" for a discussion of the use of the word force.

[38] For completeness, note it also can be written: $F^\mu = \gamma \dfrac{d}{dt}\begin{pmatrix} \gamma m_0 \\ \gamma m_0 u_i \end{pmatrix}$

From these, we naturally get the following relationship among the relevant γ factors:

$$\frac{\gamma(u')}{\gamma(u)} = \gamma(1 - u_x v) \text{ where } \gamma \equiv \gamma(v)$$

We can then read off, by analogy, *part* of the transformation law of the spatial components of F^μ, which are the standard 3-forces; namely:

$$f_x' = ?, \ f_y' = \frac{f_y}{\gamma(1 - u_x v)}, \ f_z' = \frac{f_z}{\gamma(1 - u_x v)}$$

We see immediately the transformation law that we used in the main part of the text. Namely, for the force perpendicular to the particle motion, the force one sees exerted on a test particle is always greatest in the *test particle's* own rest frame and it is reduced by a factor of γ in other frames: $\vec{F} = \frac{1}{\gamma}\vec{F}'$, where the primed frame is the rest frame of the particle.

Now, we expect the first component of the force transformation law to be different because it involves the time component which is not completely analogous to the 4-velocity. If we actually do the transformation, we get (verifying the two components found above):

$$f_x' = \frac{f_1 - v\, dE/dt}{1 - u_x v},$$

$$f_y' = \frac{f_y}{\gamma(1 - u_x v)}, \ f_z' = \frac{f_z}{\gamma(1 - u_x v)}$$

Note that if the rest mass is constant (forces like this W. Rindler calls *pure* forces while he calls those that do change the rest mass *heat-like* forces), we can reduce this to a simpler form.

Indeed, we can write a general expression for dE/dt. Taking our two vectors and looking for invariants we write, in the lab frame:

7.26 $$U^\lambda F_\lambda = \eta_{\lambda\sigma} U^\lambda F^\sigma = \gamma^2 \left(\frac{dE}{dt} - f^i u^i\right).$$

Such an invariant is most easily evaluated in the rest frame of a particle being described; in that frame, we get:

7.27 $$\eta_{\lambda\sigma} U^\lambda F^\sigma = \frac{dE}{d\tau} = \gamma \frac{dm_0}{dt},$$

Where in the last equality, we express the answer in terms of the lab frame time instead of the proper time.

Equating equation 7.26 to equation 7.27 gives: $\dfrac{dE}{dt} = \vec{f} \cdot \vec{u} + \dfrac{1}{\gamma}\dfrac{dm_0}{dt}$ and gives the

special case for pure forces:[39] $\dfrac{dE}{dt} = \vec{f} \cdot \vec{u}$. Hence, the force transformation laws in the

general case are:

$$f_x{}' = \frac{f_1 - v\left(\vec{f} \cdot \vec{u} + \dfrac{1}{\gamma}\dfrac{dm_0}{dt} \right)}{1 - u_x v},$$

$$f_y{}' = \frac{f_y}{\gamma\left(1 - u_x v\right)}, \quad f_z{}' = \frac{f_z}{\gamma\left(1 - u_x v\right)}$$

Note that *only* when the force is pure (m_0 constant) and the force is parallel to the motion of the particle is the force invariant under Lorentz transformation.

Now, again bringing in the definition of force as change of momentum per unit time ($d\left(\gamma m_0 u^i\right)/dt$), we can write the force in terms of acceleration and velocity. Namely, for a pure force:

$$f^i = \gamma m_0 a^i + \frac{d\left(\gamma m_0\right)}{dt}u^i = \gamma m_0 a^i + \left(\vec{f} \cdot \vec{u}\right)u^i$$

From this, we see that all three 3-vectors in this equation are in one plane, but they are only in a line when: (1) the force is parallel to the motion of the particle or (2) the force is perpendicular to the particle motion. These two cases give:

$$f_{\parallel} = \gamma^3 m_0 a_{\parallel}, \qquad f_{\perp} = \gamma m_0 a_{\perp}$$

---

[39] From this, we can write, for a pure force: $F^{\mu} = m_0 \dfrac{dU^{\mu}}{d\tau} = m_0 a^{\mu} = \gamma\left(u\right)\left(\vec{f} \cdot \vec{u}, f^i\right)$

Summary

Einstein's *special theory of relativity* reveals the profound unity of electrical and magnetic forces.

The empiriometric theory of special relativity is unpacked in Chapter 10 of the first volume of the *Physics for Realists* series: *PFR-M*. The current chapter assumes a basic understanding of the material covered there. The essentials given there, which *do* apply at the level of abstraction of classical E&M,[40] are summarized as follows.

A body with impetus (1) contracts in the direction of its motion, (2) behaves more sluggishly because its mass increases with its momentum, resulting in (3) a slowing of the time it takes things to happen. These effects after being lumped with the appearance of bodies measured from the uniformly moving frames (using an operational definition of simultaneity appropriate to the given frame), are called, respectively, length contraction, mass increase, and time dilation. This combination can be called, for short, *LMOST*, which stands for *L*ength contraction, *M*ass dilation, *O*perational *S*imultaneity and *T*ime dilation.

These effects can be viewed as consequences of the principle of integrity that every substance acts in such a way as to maintain itself, keep its same integrity, in uniform motion as it has at rest. For this to be true, the qualification of the plana called the ϕ-field, which is due to the property of charge, must be complemented by a second qualification of the plana called the *A*-field, which is due to impetus activated charge. The principle of integrity is apparently a deep principle of nature, which experiment shows, in some way, extends beyond electric and magnetic forces to all other known primary generic powers (gravity, strong and weak as well as E&M).

The principle of integrity is formalized in the internal relativity principle that no experiment done within a given house can distinguish one house from another and that only the absolute speed of a given house determines whatever differences there may be between houses. Internal relativity, along with the existence of a maximum measured speed, implies the fundamental principles of special relativity:

    1) Any experiment done in one inertial frame will yield the same in any other inertial frame

    2) Measurement of the speed of light will yield the same result in every inertial frame.

This, in turn, implies that, at the appropriate level of abstraction, (for example leaving out gravity so that we can have objects that serve as our network of clocks, rulers etc. that continue in their state of uniform motion without being affected by their environment, e.g. clocks and rulers of other observers) *one cannot determine whether one is at rest or in uniform motion*. This consequence of the integrity principle results from the interior of each house, including ones inside others, having to, operationally, behave the same in motion as at rest.

---

[40] For example, properly including the further specification of gravity would require us to bring in general relativity.

Using the **dual-configuration** consisting of a *setup* and a *dual-setup*, we investigate the deep complementarity between the electric and magnetic field that results in the effective indistinguishability between a body in motion and motion away from that body.

A. The *simplest* form of the dual configuration utilizes the following thought experiments:

  1. A *setup* in which the body under consideration is given impetus to move at speed v so that it can be examined (from the absolute rest frame) to understand what new properties it acquires due to its impetus. Note that in the absolute rest frame the body appears unfiltered by *LMOST* effects on our instruments.

  2. A *dual-setup* in which the given body is kept at absolute rest so that the observer can move away from it at speed v, creating the same *relative* velocity as in the setup. In this way, we learn using *LMOST* how the fields act so that the measurements made here in the *dual-setup* give the same results as those made by the observer in the *setup*.

B. At some point, one also wants to investigate how a *moving* body appears in a moving frame as well and what that analysis reveals about the deep interrelation between the setup and its dual. Here, the observer in the *dual-setup* still views from a uniformly moving frame; however, now, instead of being at (absolute) rest, the bodies are given appropriate (in general, but not always, non-zero) velocities in accordance with the *setup* and the physical implications of relativity (as seen in the examples of the "atomic" clock and the moving line and point charges).

Three examples of dual configurations: **1.** line charge, **2.** a wire moving parallel to its length and **3.** a wire moving perpendicular to its length.

1. *Line Charge*

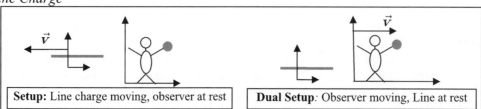

| **Setup:** Line charge moving, observer at rest | **Dual Setup**: Observer moving, Line at rest |

Setup: Line charge moves to the right, viewed by fixed observer, and has: $\vec{E} = \dfrac{2\gamma\lambda_0}{r}\hat{r}$, i.e. pointing radially outward from the line charge, with charge density per unit length λ_0 and $\vec{B} = \dfrac{2\gamma\beta\lambda_0}{r}\hat{\phi}$, i.e. azimuthally in right hand fashion around the line of the current where $\beta = v/c, \gamma = 1/\sqrt{1-(v/c)^2}$ (or $\gamma = \sqrt{1+(p/m_0 c)^2}$). The E-field is γ times bigger for a rod in motion, because the impetus activated in the rod qualifies it in such a way that it contracts by a factor of γ along the direction of its motion, thus increasing its linear density according to: $\lambda' = \gamma\,\lambda_0$.

Dual-Setup: Line charge is at absolute rest and has: $\vec{E} = \dfrac{2\lambda_0}{r}\hat{r}$, $\vec{B} = 0$, but *appears* to have, due to *LMOST*: $\vec{E}' = \dfrac{2\gamma\lambda_0}{r'}\hat{r}'$, $\vec{B}' = \dfrac{2\gamma\beta\lambda_0}{r'}\hat{\phi}'$, and we also have; $I' = \lambda'v$.

2. *Wire moving parallel to its length*

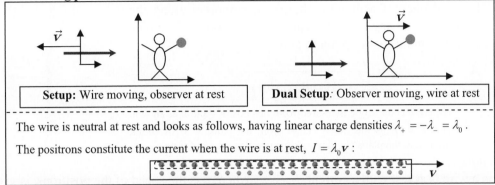

Setup: Wire moving, observer at rest
Dual Setup: Observer moving, wire at rest

The wire is neutral at rest and looks as follows, having linear charge densities $\lambda_+ = -\lambda_- = \lambda_0$.

The positrons constitute the current when the wire is at rest, $I = \lambda_0 v$:

Setup: Current-carrying wire moves to the left and is viewed from (absolute) rest, and

has: $\vec{E} = -\dfrac{2\lambda_0 \gamma \beta^2}{r}\hat{r}$; this results from negative line of charge contracting by γ as it

moves at speed v and positive line of charge expands by a factor γ as it is stopped giving:

$\lambda' = \lambda_+' + \lambda_-' = -\lambda_0\left(\gamma - \dfrac{1}{\gamma}\right) = -\lambda_0 \gamma \beta^2$ and the current being $I' = \lambda' v$.

(There is also a *B*-field which we do not address here but which cannot affect a charge with no impetus like our test charge.)

Dual-Setup: Current-carrying wire is at rest and has: $\vec{E} = 0$, $\vec{B} = \dfrac{2I}{cr}\hat{\phi}$, with $I = \lambda_0 v$.

In the moving frame, the speeds of the charges and the distance between the charges within the line of positive charges and the line of negative charge *appear*, due to *LMOST*, the same as in the setup; however, they are not actually the same, for they have not actually

changed. In the moving frame, we do apparently have: $\vec{E}' = -\dfrac{2\lambda_0 \gamma \beta^2}{r'}\hat{r}'$, which arises

from the actual *B*-field acting on the actually moving test charge as well as from *LMOST*. Note that if impetus-activated charge did not cause its own unique type of field, i.e. did not cause the magnetic field, we would not have this result and relativity would fail.

3. *Wire moving perpendicular to its length*

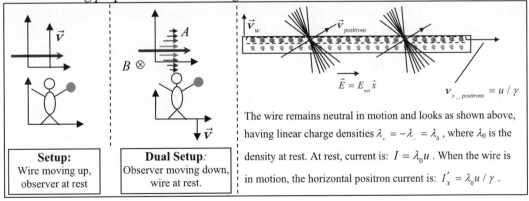

Setup: Wire moving up, observer at rest	**Dual Setup:** Observer moving down, wire at rest.

$\vec{E} = E_{net}\hat{x}$

$v_{x\_positrons} = u / \gamma$

The wire remains neutral in motion and looks as shown above, having linear charge densities $\lambda_+ = -\lambda_- = \lambda_0$, where λ_0 is the density at rest. At rest, current is: $I = \lambda_0 u$. When the wire is in motion, the horizontal positron current is: $I_x' = \lambda_0 u / \gamma$.

To approach this we can make use of:

Field of a moving point charge

The field lines of a moving charge, q, compress along the direction of its travel in Lorentz like fashion and can be derived from this assumption to give:

7.4 $$\vec{E}_{moving} = \frac{q}{r^2} \frac{1}{\gamma^2 \left(1 - \beta^2 \sin^2 \theta\right)^{3/2}} \hat{r}$$

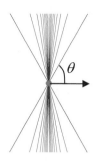

(where θ is the angle measured from the axis of motion as shown, and $\vec{r} = r\hat{r}$ is the radial vector from the particle to the point interest in the plana (field point) at a given moment of time).

Setup:

When a wire is at rest with a current $I = \lambda_0 u$, where u is the speed of the positrons, is given a *perpendicular* speed v, no contraction results along the length of the wire, so the wire remains neutral, thus *not* producing a ϕ-field. Because both negative and positive charges share the same vertical-impetus no A-field is generated. However, only the positrons have horizontal impetus, so an (horizontal) A-field *is* generated by them. Because of the increase in mass that results from the vertical impetus and because the horizontal impetus moving the positrons is unchanged, the original horizontal speed u of the positrons drops to u/γ. This results in a decreasing x-directed A-field at the position of our test charge. The plana resists this decrease ($\dot{A}$-effect) causing an x-directed (analogical) E-field such as shown in the top right figure above.

This E-field can also be understood in terms of the individual positrons shown in that same figure. Considering the region between the two positrons that have their field lines drawn on them, we see the rightward acting field from the left positron is much stronger than the leftward acting field from the right one, thus, as above, resulting in a net rightward E-field.

Dual-setup

Due to *LMOST*, the resting wire as measured from the moving frame does appear to contract along its width but not along its length. Hence, there is no apparent generation of ϕ field. In addition, *LMOST* results in the positrons looking like they are going $1/\gamma$ times slower, which, as above, appears to generate an x-directed E-field. Thus, the dual-setup is operationally equivalent to the setup. However, the actual physical cause in this dual case, unlike the setup, is the B-field. As the test charge has downward impetus moving it away from the wire, the A-field-activated plana acts to give impetus to the test charge along the wire's direction of motion to attract the charge back, i.e. along the A-gradient, to it. Since once the charge is given impetus of its same type, the wire will attract the charge towards it. *LMOST* also results in the individual charges appearing as they did in the setup.

Force and Field Transformation Laws for particular cases (using a line charge)

The force transformation law for motion perpendicular to the force is as follows. Given arbitrarily small speed v and time difference Δt, the force acting on a particle can be calculated as : $f_0 = \dfrac{m_0 v}{\Delta t}$. In the moving frame, *LMOST* giving an apparent new mass and time of γm_0 and $\gamma \Delta t$ results in an equation for the apparent force that yields the transformation law to a frame moving perpendicular to the force: $f' = \dfrac{\gamma m_0 v / \gamma}{\gamma \Delta t} = \dfrac{f_0}{\gamma}$.

Comparing this force ($\vec{f}_y'(v' = -v) = \dfrac{qE_0}{\gamma}\hat{y}$) that is acting on an apparently moving particle with the apparent electric field ($\vec{E}'(v' = 0) = \gamma E_0 \hat{y}$) acting on an apparently at rest particle (i.e. one moving with the frame, i.e., we see that the apparent motion of the particle introduces an extra term in the force given by: $f_B' = \dfrac{qE_0}{\gamma} - \gamma q E_0 = -\gamma \beta^2 q E_0$. We interpret this as due to an apparent B-field: $\vec{B}' = -\gamma\beta E_0 \hat{z}$.

Hence, we have the following field transformation laws: $E_y' = \gamma E_y \quad B_z' = -\gamma\beta E_y$

E and B Transformation Laws
In terms of the fields $\vec{E}$ and $\vec{B}$ measured in a given uniformly moving frame (i.e., an inertial frame), the measured fields, $\vec{E}'$ and $\vec{B}'$ in a second inertial frame moving relative to it as shown below (i.e. moving according to the standard configuration) are given by:

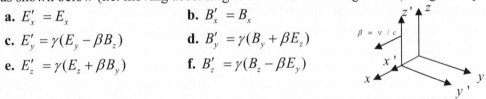

a. $E_x' = E_x$ **b.** $B_x' = B_x$

c. $E_y' = \gamma(E_y - \beta B_z)$ **d.** $B_y' = \gamma(B_y + \beta E_z)$

e. $E_z' = \gamma(E_z + \beta B_y)$ **f.** $B_z' = \gamma(B_z - \beta E_y)$

Operational simultaneity is applied in both frames, and, of course, if one frame happens to be at absolute rest (which cannot be determined at our current level of abstraction) then operational simultaneity reduces to actual time with no intervening mental construct. These figures show how a B-field (blue) and an E-field (red) in the first frame appear in the second frame.

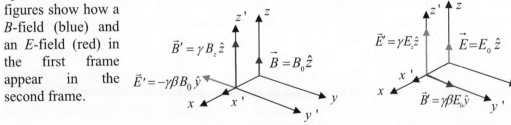
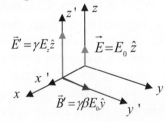

In the first figure, the B-field *apparently* gets larger and develops an apparent E-field to the right when facing the direction of motion. In the second figure, the E-field apparently gets larger and develops an apparent B-field to the left. This latter can be remembered by using the first two letters of *el*ectric to think of *e* field to the *l*eft. The first then is remembered as the opposite.

Transformation of A and ϕ
ϕ and A, like energy and momentum introduced in Chapter 10 of *PFR-M*, can be put into a single special relativistically invariant "4-vector" in that they transform in the same way as the time space 4-vector (t, x, y, z):

$$\begin{pmatrix} A^{t\,'} \\ A^{x\,'} \\ A^{y\,'} \\ A^{z\,'} \end{pmatrix} = \begin{pmatrix} \gamma & -\gamma\beta & 0 & 0 \\ -\gamma\beta & \gamma & 0 & 0 \\ 0 & 0 & 1 & 0 \\ 0 & 0 & 0 & 1 \end{pmatrix} \begin{pmatrix} A^{0} \\ A^{x} \\ A^{y} \\ A^{z} \end{pmatrix} = \begin{pmatrix} \gamma\phi - \gamma\beta A^{x} \\ -\gamma\beta\phi + \gamma A^{x} \\ A^{y} \\ A^{z} \end{pmatrix} \qquad \text{(where } A^{0} = \phi\text{)}$$

A particle at absolute rest has no impetus, but has rest mass energy; this is written as: $p = \begin{pmatrix} m_0 \\ 0 \end{pmatrix}$. Similarly, a particle at rest causes no A-field, but does cause a ϕ-field; this is written: $p_{pot} \equiv A^{\mu} = \begin{pmatrix} \phi \\ 0 \end{pmatrix}$. From a moving frame, these appear, respectively as: $p' = \begin{pmatrix} \gamma m_0 \\ \gamma m_0 \beta \end{pmatrix}$ and $A' = \begin{pmatrix} \gamma\phi \\ \gamma\phi\beta \end{pmatrix}$. In the dual configuration language, this is called the *dual-setup* and results from the relative motion and *LMOST*. It implies, through the relativity principle, that there exists a *setup* in which the physical situation actually changes in this way. In particular, when the particle is moving at speed β it has rest energy and momentum γm_0 and $\gamma m_0 \beta$, while the ϕ-field is amplified by γ and there is an A-field of amplitude $\gamma\phi\beta$ due to the fact that the moving charge has impetus.

To give more concreteness to our thinking about the field, we consider the plana motion, but only briefly as its motion (if at all) is not properly part of E&M. Since there is no reason to think that plana can move without massive bodies, we consider the plana as moving with the charge, for example, a line charge; the iconic case is virtual plana that is part of a single substance that moves all of whose parts massive and nonmassive (i.e. virtual plana) move together. In this way, we also keep the fields, which are properties of the plana, more clearly associated with their source.

When in motion, there really is an A-field, as well as a ϕ-field, in the "trapped" plana in moving substance. Since the potential field energy (ϕ) and potential field momentum (A) can represent how much energy it takes to put the charge together and how much momentum was given away to give it the momentum it has, the fields are closely related to the massive aspects of the particle. For example, the rest mass might be more closely related to fields than some currently think. In any case, the fact that they transform in the same way shows a deep unity exists between the ϕ and A-fields and basic characteristics of their source, i.e., the rest mass and the momentum.

Types of Charge
There are actually multiple types of active and passive charge, which because of their nature, we can treat by an analogical generalization as being equal. Hence, we only need to specify a single charge q, though in fact it represents many properties. Furthermore, because of this equivalence, q is independent of the motion of the particle. We say the charge is invariant under Lorentz boost or charge is a Lorentz scalar.

Gauss's Law in terms of the potentials can be written: $-\nabla^2\phi = \nabla \cdot \vec{E}_{static} = 4\pi\rho - \dfrac{1}{c^2}\dfrac{\partial^2\phi}{\partial t^2}$.

The last term, the "acceleration" term, accounts for the propagation effects that cause bunching up and stretching out of the ϕ-field as the particle moves.

The *discrete nature of charge* fundamentally follows from the fact that material bodies[41] have extension (quantity), which means actual parts and conservation of charge. Hence, we expect that there will be a smallest particle, which if we are successful in breaking will no longer have the property of charge, which is not possible given conservation of charge. The only way to destroy it would be interaction with a particle of the opposite charge of the same magnitude, so, that charge is locally conserved. Thus, for example, an electron and a positron can interact with each other and the plana in an "annihilation" event that ends with gamma rays activated in the plana. Keep in mind that the extension of such particles may be very small and, indeed, may not even be in principle *directly* experimentally probe-able.

The nature of electrical bodies, which includes ordinary bodies because they all contain virtual charge (if not always actual charge), is to have magnetic fields that result necessarily when such bodies have impetus. Because of this fact, charge is the primary property in our study of electricity and magnetism. After mass (and gravity to which mass is inseparably related), *charge is,* indeed, of the properties we are studying, *the closest to the essence of most simple physical substances.*

Reasons for Charge Conservation
Intrinsic reason (proper to the interaction of simple bodies themselves): It seems to be part of nature's maintenance of conservation of energy. Given mass-energy equivalence as we currently know it, if charge were not conserved, then two like charges could "annihilate," destroying all the electrical potential energy stored in the pair of particles. For example, bringing them together takes a certain amount of energy.
Extrinsic reason: There is a need for primitive powers, such as charge, to be invested in an element-like nature from which more complex substances, including plants and animals, can draw certain fundamental powers. Thus, for example, quarks, which have charge, form protons and protons become part of sodium, which become part of table salt, which can become part of you.

Two further examples explored via dual configurations: 4. *an "atomic" clock* and 5. *a line charge moving parallel to itself with test charge moving slightly faster next to it.*
    A clock modeled after a hydrogen atom with a light negative particle orbiting a heavy positive particle ("nucleus") such as shown below will slow down when put in uniform motion due to mass dilation and the changed shape of the *E*-field pattern of the nucleus.

---

[41] Recall that we always start with the things we know directly through our senses, i.e. physical substances that have the nine categories of properties. Other types of things (i.e. non-physical things, things that are essentially different from physical substances in that first (primary) sense of the term) may exist. However, we only introduce them if our study of those things that we can see forces us to conclude that there are things that we cannot see that lack this physical nature in the first sense of the word. Hence, we here start with physical things in the first sense of the word, which do indeed have extension, the first category of property.

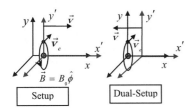

A line charge moving at nearly the same speed as a test charge along side it will generate an A-field as well as a ϕ-field. In the frame that is moving along with the wire, the wire will appear at rest but the test charge will appear to be moving slightly. This example, like the "atomic" clock but more simply, shows that various objects in the setup can have more than one speed of motion, and there will still exist a dual setup in which things looks exactly the same. This, in turn, helps us see further that the focus of special relativity is relative motion, not the particular state of motion of the observer and the objects observed.

Again, despite the fact that in both the "atomic" clock and the moving line example (as in all dual configurations), the actual physical situation is radically different in the setup compared to the dual setup, empiriometric special relativity prescinds from actual motion, handling only relative motion.

Dual configurations are about the physical state not *just* the measured appearances, whereas frames generally are. Once we understand the physical origin of frame invariance, we can often ignore the distinction between the setup and its dual in order to get the experimental predictions of new situation. Indeed, in ordinary empiriometric relativity, one identifies the setup and dual setup *because* of the very fact that, at the level of abstraction appropriate to special relativity, they would yield the same experimental results.

Helpful Hints

A prototypical example of a dual-configuration is:

(1) an observer at rest and a massive body moving at a speed v under the action of its impetus, called the "setup" and

(2) an observer moving at speed v in the opposite direction under the action of his impetus viewing a massive body at rest, called the "dual setup."

In this case, we see that the key physical change is the massive body's actual increase in resistance to the action of its impetus, i.e. its increase in mass by a factor of γ.

Note that the dual-configurations or dual-physical situations are distinguished from dual frames, for in dual frames one leaves the physical object under consideration unchanged and just views it from a different state of uniform motion. This is an important distinction that is not made empiriologically, at least at the present state of our empiriological theory.

The word "force" is usually used only when the power under consideration is actually causing or attempting to cause impetus. Outside simple Newtonian physics, physicists tend to use the word "field" for the power (i.e., a power that is able to cause impetus) *per se* and force when that power is acting. Thus, for instance, we say the force of gravity causes the ball to accelerate downward or the force of gravity is canceled by the upward force of the floor, but we say that the gravitational field is all around the Earth.

Note that a primed frame is specified in the setup and a *different* primed frame is specified in the dual setup. Note well that though they are both marked with the same primed symbols, they *are different* frames. For example, if, in the setup, the frame of the moving body of interest, i.e. the primed frame, is moving along the *positive x*-axis, then, in the dual setup, the moving (primed) frame from which one views is moving in the *negative x*-direction. Thus, the *prime* in this context simply means *a* moving frame.

Problems

1. a) Draw the setup and the dual setup for the case of a body at rest. b) Draw the setup and its dual for the case of a uniformly moving body. c) Explain the violation of the principle of integrity that would result if measurements done in the setups in parts *a* and *b* were different from those done in the dual setups. d) Take part *c* a step further and consider two parts of a body, one that is at rest the other that is moving uniformly. Explain the violation of the principle of integrity that would result in this context by considering this body (which has these two parts) in uniform motion.

2. Draw and explain the setup and dual setup for the case of a rod moving along the direction of its own motion.

3. a) Explain what would happen to charge conservation if, all else being equal, the charge, q, did change under Lorentz transformation. b) How would this affect say the field of a line of charge?

4. The rest mass-energy of an electron, which is the same as that of a positron, is about 511keV. a) How much energy is released in a positron electron annihilation event in which both the positron and the electron are at rest at the moment of annihilation? b) Suppose, in violation of conservation of charge, similar to what is discussed in the "Other Important Features of Charge" section, that two positrons can annihilate. What new factor not currently needed might one try to introduced to preserve conservation of energy? c) Consider a third positron at a distance that is not part of the annihilation, does this affect your answer?

5. Consider the situation illustrated here in which one has three charged bodies: one positive test charge, a positive line charge and a negative plane of charge. The plane of charge can be thought of as generating an ambient field in which the test charge and the line charge "live." a) Ignoring the plane of charge, draw setup and 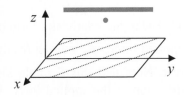 dual setup for the case in which point and line charges each move at speed v. b) Reintroduce the unmoving plane of charge into both the setup and its dual. Will the setup and its dual now still yield the same measured results? Why or Why not? If not, how do you rectify the situation? In particular, draw the proper dual configurations.

6. Show explicitly the E and B-field transformation equations 7.11d and e can be generated from equations 7.11f and c, respectively, from the rotation mentioned in the text.

7. In Chapter 6, we saw that, for a plane wave, a longitudinal A is consistent with the wave equations for ϕ, A, E and B. Approach this fact from a relativistic point of view in the following way. Using a relativistic transformation (i.e. a Lorentz transformation) of the standard (linearly polarized) plane wave, $\vec{A} = A_0 \sin(\omega t - kz)\hat{x}$, show that there must be an actual physical wave that has a ϕ-field *and* a longitudinal A-field. *Hint:* In addition to the Lorentz boost, a rotation is needed; there is no need to formally perform this rotation but only to carefully explain it. (N.B.: advanced students may be asked to do this rotation formally for extra credit. In this case, you should also show that the field with the longitudinal A still has neither a longitudinal E nor a longitudinal B-field.)

8. Derive the formula for the electric field of a charged point particle moving at speed v, i.e. show $\vec{E}_{moving} = \dfrac{Q}{r^2} \dfrac{1}{\gamma^2 \left(1 - \beta^2 \sin^2 \theta\right)^{3/2}} \hat{r}$, where $\beta = v/c$ and θ is the angle measured from the direction of motion. Use the method mentioned in the text, i.e. start with the form of the electric field of a particle at rest and Lorentz contract its field lines by a factor of $\gamma = 1/\sqrt{1 - \beta^2}$ along the direction of motion as if they were stiff rods.

9. To get an idea of the angular dependence of the E-field of a point charge that moves uniformly near the speed of light, make a rough plot of the magnitude of the E-field given in the previous problem as a function of θ for $\beta = .995$.

10. Consider a uniformly moving positive line of charge. Assume its charge density at rest is: λ_0, yielding: $\vec{E} = \dfrac{2\lambda_0}{r}\hat{r}$. a) Show that, in the frame moving with the line charge, the force acting on a test particle that is also moving with the line charge (i.e., at rest in the moving frame) is apparently due to a field: $\vec{E}' = \gamma E_0(y)\hat{y}$. Use a similar technique to that given in the section titled "*Apparent* Force and Field of a Line Charge as Seen from a Moving Frame." *Hint:* Use both the transformation law for velocities and the one for time and space.

11. Using the relativistic form of the momentum, $\gamma m_0 v$, a) Derive the relation between an applied force, $f_{\parallel}$, and the resulting acceleration, $a_{\perp}$, for a particle that is accelerated from rest. Note that this formula also applies to a particle with an initial velocity along the direction of the applied force. b) Derive the relation between: (1) an instantaneous force, $f_{\perp}$, applied perpendicular to the motion of a particle already moving at speed β and (2) its resulting instantaneous acceleration, $a_{\parallel}$?

12. a) In the absence of gravity, consider a body moving initially at speed u in the x-direction. Apply enough impetus to the particle in the y-direction for it to gain a speed v in that direction. Prove without using the velocity transformation laws directly, that, after this operation is complete, the x-component of the velocity is u/γ. b) Considering this final state of motion of the body as the setup, draw the dual setup and explain, using velocity transformation laws, why the actual state of the body you have drawn in your dual setup will yield, when measured from the moving frame, an apparent x-speed of u/γ, i.e., the same measured value as calculated for the setup.

13. a) Explain, in detail, the equations given in the "Atomic" clock section and give the equation for the rotational speed, ω, of the "atom" in terms of the e, m and r. b) Show that the spin angular momentum of the "atom" is unchanged after it is put in uniform motion. c) Show that the spin angular momentum is gives the same measured results in the dual setup as it does in the setup.

14. Show that our "atomic" clock still undergoes the proper time dilation if we drop our assumption of small orbital speed, i.e. allow $v_e \sim c$.

15. $\vec{E} \cdot \vec{B}$ and $E^2 - B^2$ are each invariant under Lorentz transform; we say they are Lorentz scalars. a) Explain what this means and verify by direct calculation that each of these is indeed a Lorentz scalar for the case shown in Figure 7-12b. Do *not* assume the specific speeds given in the figure. b) These are the only two independent field invariants in E&M. Comment on the simple physical implication of the existence of these two invariants. c) Comment on the implication of the invariance of $\vec{E} \cdot \vec{B}$ for radiation.

16. Given that the field at the interplanetary field at the radius of the Earth's orbit from the sun is about $5 \times 10^{-5} G$ (i.e. $5nT$), ignoring all other effects, calculate the maximum apparent E-field seen by our Mars craft if it were moving at $30km/s$. How much voltage across a $1cm$ stretch of vacuum (or air) does this result in?

17. By allowing the plane of positive charge with surface density, σ, shown here to move in the x-direction, derive the transformation law illustrated in Figure 7-12c (and given in equations 7.11d and e with $B_y = 0$). Use the setup and its dual to make your physical argument. Give all E and B-fields in terms of appropriate parameters.

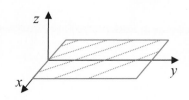

18. Recall the force transformation law $f'_\perp = \dfrac{f_\perp}{\gamma}$ given in the text above equation 7.5 and

the transformation law of electric field given in that same section: i.e. $E_\perp' = \gamma E_\perp$ (for example, take $B_z = 0$ in equation 7.11c). How is it that the apparent force is reduced by a factor of γ but the apparent E-field is increased by a factor of γ. More generally, we say that the force on a test particle is strongest in the rest frame of that test particle, whereas the actual E-field of a given source (assuming there is no actual B-field) always looks bigger in the moving frame. Explain.

19. In frame, Σ, one is moving with absolute speed u_1 along the positive x-direction and measures: $\vec{E} = E_0\hat{z}$, $\vec{B} = B_0\hat{y}$. a) What E and B will one measure in a frame Σ' that sees Σ as moving in the negative x-direction at speed u? How does it depend on u_1? b) What E and B will one measure if one is moving at absolute speed u_2 along the positive x-direction?

20. Explain, in rough physical terms, why the rest mass appears in the time component of the energy-momentum 4-vector. Likewise, explain why ϕ is placed in the time component of the A^μ 4-vector.

Chapter VIII

Dielectrics:
Static Electric Fields Inside Massive Bodies

Introduction

In the first three chapters, we discussed the attraction of various objects: how paper and water are attracted by static charge; how a magnet attracts metal and other magnets, and how it can even attract right through some materials. In this chapter and the next, we will uncover the generic principles that explain these electric and magnetic behaviors and more.

For the electrical case which we discuss in this chapter, we start with the following experiment. Go to a sink and turn on the faucet very slowly until a narrow continuous stream of water comes out. Then, take a comb run it through your hair and put it near the middle of this stream and watch the stream bend toward the comb as shown in Figure 8-1a. Why does it do this? The answer is that water is a polar molecule; i.e., it has an electric dipole moment that defines two poles, a positive and a negative. An electric field, such as generated by the charges on the comb can cause the dipoles to align with the field. Or, said another more generic way, water is a type of *dielectric*, the subject of this chapter.

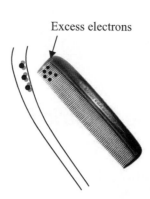

Excess electrons

Figure 8-1: a. (*left*) A thin stream of water flowing out of a sink is attracted towards a comb charged by running it through one's hair. **b.** (*right*) Representational drawing of how a negatively charged comb affects the permanently polar H_2O molecules of the stream, thus leading to bending of the water column.

There are actually two types of dielectrics: (1) those with molecules that have permanent dipole moments, i.e., dipole moments that exist without the imposition of an external electric field and (2) those with molecules that have a dipole induced by the application of a field.

Dipole and other Moments of Charge Distributions

To begin to understand both types of dielectrics, we first establish some general mathematical tools. Then, we can specialize to the charge distributions of atoms and molecules. Just as no actual charge distribution is a point charge, no actual charge distribution is an ideal dipole. This is clear once we give the precise definition of the ideal dipole. An ideal dipole is defined as two point particles of charge q separated by a distance s, where s is made arbitrarily small while the q gets arbitrarily big to maintain a constant product $p = qs$. That product, p, is called the *dipole moment* which is most naturally handled by the formalism introduced in the next section. Through that formalism, we will see that, as some charge distributions can be approximated by a point charge, others are best approximated by a simple dipole in a certain regime.

Multipole Moments

We introduce the relevant formalism by recalling the ϕ-field form of the extended Coulomb's law:

8.1
$$\phi(\vec{r}) = \int \frac{\rho(\vec{r}\,')}{|\vec{r} - \vec{r}\,'|} d^3 r'$$

In more advanced courses, you will learn how to wield a more complex formalism that allows you to expand the right hand side into a *general* series of "multi-poles" for the case in which the source is small compared to the distance to the field point. Here, we only want to emphasize the principles, so we will not seek an explicitly general form. Instead, noting that every direction is generically like every other direction, we can, with no principled loss of generality (though with a formal loss) write down field for a charge distribution as it is seen along the z-axis as:

8.2
$$\phi(r) = \int \frac{\rho(\vec{r}\,')}{\sqrt{(x-x')^2 + (y-y')^2 + (z-z')^2}} d^3r'$$

$$= \int \frac{\rho(\vec{r}\,')}{\sqrt{x'^2 + y'^2 + z^2 - 2zz' + z'^2}} d^3r'$$

$$= \int \frac{\rho(\vec{r}\,')}{r\sqrt{1 + \left(-2\dfrac{z'}{r} + \left(\dfrac{r'}{r}\right)^2\right)}} d^3r'$$

where we have used: $\vec{r} = r\hat{r} = z\hat{r}$, $\vec{r}' = r'\hat{r} = x'\hat{x} + y'\hat{y} + z'\hat{z}$.

Assuming that we are sufficiently far away, i.e. assuming that $r \gg r'$, we can expand using the Taylor's series $(1+\varepsilon)^{-1/2} = 1 - \dfrac{1}{2}\varepsilon + \dfrac{3}{8}\varepsilon^2 + O(\varepsilon^3)$, we get:

8.3
$$\phi(r) = \frac{1}{r}\int \rho(\vec{r}\,')\left(1 + \frac{z'}{r} + \frac{1}{r^2}\left(\frac{3z'^2 - r'^2}{2}\right) + O\left(\left(\frac{r'}{r}\right)^3\right)\right) d^3r'$$

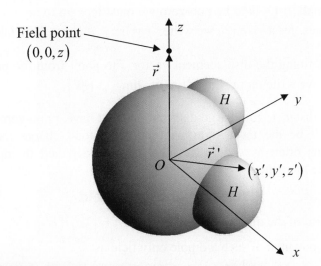

Field point
$(0,0,z)$

Figure 8-2: Coordinate system setup for calculating the moments of an arbitrary charge distribution in which we monitor the field along the z-axis only. Because a definite distribution of some type is needed for the drawing, the schematic image of the highly polar water molecule is shown.

Equation 8.3 can be written as:

8.4
$$\phi(r) = \frac{Q_0}{r} + \frac{Q_1}{r^2} + \frac{Q_2}{2r^3} + O(r^{-3})$$

$$Q_0 = \frac{1}{r}\int \rho(\vec{r}\,')d^3r', \quad Q_1 = \int z'\rho(\vec{r}\,')d^3r', \quad Q_2 = \int (3z'^2 - r'^2)\rho(\vec{r}\,')d^3r'$$

Where, Q_0 is called the monopole moment, Q_1 the dipole moment and Q_2 the quadrupole moment…

Consider the simple examples using point charges given below:

$$Q_0 \neq 0 \qquad Q_0 = 0 \qquad Q_0 = 0$$
$$Q_1 = 0 \qquad Q_1 \neq 0 \qquad Q_1 = 0$$
$$Q_2 = 0 \qquad Q_2 = 0 \qquad Q_2 \neq 0$$

If one wants a pure dipole, obviously all terms but Q_1 must be zero. Of course, because we've restricted ourselves to analyzing the plana along the z-axis, the only dipole that matters is one stretched out along it (why?). If we just had two point particles of charge q each placed a distance $s/2$ on either side of the origin, we see, by substitution into the definition for Q_1 given in equation 8.4, that the dipole moment is $p \equiv Q_1 = qs$. However, in order for this to be true as close to the source as we like, we must let s go to zero, so that the condition $\dfrac{r}{s} \gg 1$ which is required for the validity of the expansion given in 8.4 allows r to be arbitrarily small. If Q_1 is to be non-zero we must let s go to zero in such a way that qs remains finite. Thus, for a given p, we write: $p \equiv \lim_{\substack{s \to 0 \\ q \to \infty}} (qs)$. And, taking into account all three axes and defining the displacement vector $\vec{s}$ to point from the positive charge to the negative charge, we write more generally: $\vec{p} \equiv \lim_{\substack{s \to 0 \\ q \to \infty}} (q\vec{s})$.

Clearly, at a distance, for charge distributions that have a non-zero dipole moment, the dipole term will be the most significant after the total charge itself. Now, since molecules and atoms often have zero net charge, to understand them, it is crucial to understand the dipole moment.

The General Dipole Field

Given the form of the dipole moment above, $Q_1 = \int z' \rho(\vec{r}\,') d^3r'$, we can write the general form for the potential of such a dipole only field as:

8.5
$$\phi(\vec{r}) = \frac{1}{r^2} \int \hat{r} \cdot \vec{r}' \rho(\vec{r}\,') d^3r'$$

Here we have used the fact that z' just represents the projection of the vector to the source point onto the direction to the viewpoint. This expression for the potential can, in turn, be re-written generally as:

8.6
$$\phi(\vec{r}) = \frac{\hat{r} \cdot \vec{p}}{r^2}$$

Which can be written in spherical coordinates, using $\vec{p} = p\hat{z}$, as: $\phi(\vec{r}) = \dfrac{p \cos \theta}{r^2}$. This field is shown in Figure 8-3a.

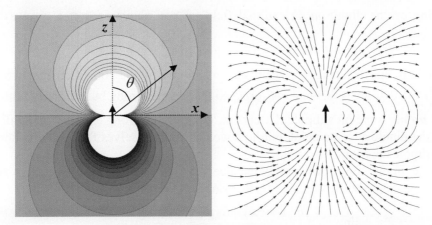

Figure 8-3: a. Contour plot of an ideal dipole ϕ-field in z-x plane (given by equation 8.6). **b.** *E*-field of a dipole in z-x plane (given in equation 8.8). To picture the full three dimensional field, imagine rotating the plot about the axis defined by the dipole (which is shown as a short arrow in the middle of each plot).

In a problem, you are asked, using Cartesian coordinates, to show that this yields the following electric field, in mixed spherical/Cartesian coordinates:

8.7 $$\vec{E} = \frac{p}{r^3}\left(3\left(\hat{p}\cdot\hat{r}\right)\hat{r} - \hat{p}\right)$$

Expressed completely in spherical coordinates, using $\vec{p} = p\hat{z}$, we have:

8.8 $$\vec{E} = \frac{p}{r^3}\left(2\cos\theta\hat{r} + \sin\theta\,\hat{\theta}\right)$$

Separating the two components, we can write:

8.9 $$E_r = \frac{2p\cos\theta}{r^3}, \ E_\theta = \frac{p\sin\theta}{r^3}, \ E_\phi = 0$$

This field is shown in Figure 8-3b.

Actions on the Dipole by External Fields

Of course, as we saw in the case of water being bent, dipole distributions that create the field described by equation 8.9 can also be acted on by an external field. Using our knowledge of basic mechanics, we can now explain this action, including giving the appropriate equations.

In particular, an electric dipole in an external field will, except in special cases, feel a force and a torque. The figures below show the dipole modeled as two point charges stuck together by an arbitrarily short and narrow bar. Since forces are prerequisites to torques, we begin with calculation of the force.

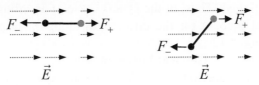

The simplest force case is shown above on the left where a dipole is shown as a dumbbell with a positive charge at one end and a negative charge at the other. The net force acting on the dipole in an external field (shown above as dotted arrows) is easily shown to be:

8.10 $\qquad F_{net} = F_- + F_+ = -qE(x) + qE(x + \Delta x)$

Note that it is the difference between the magnitude of the forces on the two charges, $|F_+|$ and $|F_-|$, that gives rise to a *net* force, so that a very strong but perfectly uniform field can exert **no** net force on a dipole. Of course, in reality, a strong enough field will result in breaking the dipole at some point, but here we consider ourselves far from this limit so that our ideal dipole approximation holds.

Now, in the limit in which the distance, Δx, between the charges goes to zero, with $p = q\Delta x$ held constant, we get, using $q = p / \Delta x$:

8.11 $\qquad F_{net} = \lim_{s \to 0} p \dfrac{E(x + \Delta x) - E(x)}{\Delta x} = p \dfrac{\partial E}{\partial x}$

Extra for Advanced Students

You should note that the E-field will not always be simply along the x-axis. However, since the other components of $\vec{E}$ cannot contribute to the force in the x-direction, we can make the substitution $\vec{E} \to E_x$ for the case of an arbitrary E-field. Furthermore, in general, the dipole will not be aligned with any one axis but will have a projection along each axis, thus, we make the substitution $\vec{p} \to p_x$ into the equation, and note the resulting term $p_x \dfrac{\partial E_x}{\partial x}$ is only one piece of the total force on the dipole in the x-direction. In particular, if the E-field varies along the two other axes, each axis will contribute a term just like it. The term in the y-direction, for example, accounts for the differing E_x, and thus differing F_x, between the two charges of the dipole, as one moves off in the y-direction. Now, if these were only the terms, we would have the case illustrated on the left.

Here the dipole can be broken up into the two perpendicular dipole components shown on the right. That is, it can be thought of as two dipoles: one of magnitude p_x which is acted on by the differing F_x in the x-direction *and* one of magnitude p_y which is acted on by the differing F_x in the y-direction, so that the net force in the x-direction is the sum of these two. We can consider these two terms separately because the "corrections" that involve both Δx and Δy are second order and higher in the arbitrarily small size of the dipole (namely $\Delta x \Delta y$ and higher order multiples of these), and we only divide out one such quantity so that those mixed terms all go to zero in the limit (see end of chapter problem for formal verification of this). Thus, in the general case of an arbitrarily short dipole in an arbitrary field, one must account for the differences along each axis, and the force on a dipole is given by equation 8.12.

General Force on Dipole

It was shown in the previous "extra's" sub-subsection that equation 8.11 can be generalized, for arbitrary E-field and arbitrary dipole moment, to:

8.12 $\qquad F_i = p_x \dfrac{\partial E_i}{\partial x} + p_y \dfrac{\partial E_i}{\partial y} + p_z \dfrac{\partial E_i}{\partial z}$, where i is x, y or z.

In our mathematical shorthand, we have:

8.13 $\qquad F_i = \vec{p} \cdot \vec{\nabla} E_i$, where i is x, y or z.

These forces explain how the water and paper are attracted by the charged comb. Again, it is the change in the field over a distance that matters in terms of giving the paper and the water impetus to move.

Torque

It is clear that these forces will usually cause the dipole to rotate as seen in the figure below in which we revert, for simplicity, back to a uniform E-field:

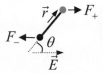

It is not hard to see that the torque on the dipole is given by:

8.14 $\qquad \vec{\tau} = \vec{r} \times \vec{F}_+ + \vec{r} \times \vec{F}_- = qsE\sin\theta = pE\sin\theta = \vec{p} \times \vec{E}$

The energy given by the field to the dipole in rotating it to the point of lowest potential energy is easily calculated from this equation. It releases, in moving from maximum potential energy, U, at $\theta = 0$ to some greater angle θ_0:

8.15 $\qquad U(\theta_0) = \int_0^{\theta_0} \tau d\theta = pE \int_0^{\theta_0} \sin\theta d\theta = pE(1 - \cos\theta_0)$

Thus, the maximum energy given to the dipole is $U(\theta = \pi) = 2pE$.

These torques are, of course, responsible for rotating permanent dipole moment molecules such as water, so that they are attracted to a charged static object such as the comb. Note well that if the atom or molecule is rotated 180 degrees so that the negative charge is closest to the comb, it will be repelled as the stronger force acts on the charge closest to the source. Hence, in the case of the comb, if the negative charge is closest, a stronger force is acting to push the close negative charge away than is acting to pull the more distant positive charge toward it.

But, how can atoms and molecules have constant dipole distributions such as we have been assuming? After all aren't they composed of charges in constant motion? The answer is yes, but the motions are so quick (described by quantum mechanics they are on the order of 10^{-16} seconds) that we can, by definition, ignore them and more in our static assumption. Indeed, we can average on such a tiny time scale that one can still study changes that happen very fast on the macroscopic scale without compromising the validity of the approximation. Because of this fact, we will often retreat into the abstraction that leaves out this motion. That is, we retreat to the model of a fixed charged distribution, and this is indeed helpful, as you may already gather, since an effective analysis of those issues requires incorporating material outside the level of abstraction of classical E&M itself.

Dipole Moments of Atoms and Molecules
Induced Dipoles

We first consider the type of dipole that is induced by an applied electric field. In particular, if we take an atom, such as the hydrogen atom shown below and put it in a uniform field, we will stretch the charge distribution. The left shows the spherically symmetric ground state of a hydrogen atom's single electron drawn as a static charge distribution, or more accurately as a probability distribution. The electron fills the first level of the so-called 1s-shell. That shell is illustrated below on the left by a gray scale shading which indicates the probability of finding the electron in a given region; the shading is thus an indication of the effective charge density as well. On the right, we illustrate the effect of an external electric field, which is to pull negative charge upward. Here white represents the maximum negative charge density and black zero charge density, while red is the positively charged proton.

 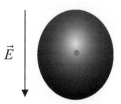

Hence, since we have broken the spherical symmetry of the charge distribution effectively creating two poles of charge, we say we have induced a dipole moment onto the atom. We have exaggerated the effect as we can see immediately by estimating the field inside the hydrogen atom. Given the size of approximately 1A, we get (in *SI*):

$$E = \frac{1}{4\pi\varepsilon_0} \frac{e}{\left(10^{-10}m\right)^2} \sim \frac{9\times10^9}{10^{-20}}\left(1.6\times10^{-19}\right)\frac{V}{m} \sim 1.4\times10^{11}\frac{V}{m},$$ which is about 100 billion

volts per meter! Thus, it is clear that the picture on the right is exaggerated, since ordinary external fields are many orders of magnitude smaller than this. We can estimate the size of the induced dipole moment in the following way.

Since the external field, E_{ext}, is much less than the internal field, E_{int}, we can approximate the fractional change, to lowest order, as proportional to $\dfrac{E_{ext}}{E_{int}}$. This is fundamentally the statement that, at some level, the linear approximation to a curve is sufficient, so that doubling E_{ext} doubles the difference in position of the charge centers, Δx. Hence, given the characteristic length, r_0, we can write:

8.16
$$\frac{E_{ext}}{E_{int}} \sim \frac{\Delta x}{r_0}$$

Using $E_{int} = \dfrac{e}{r_0^2}$, and $p = e\Delta x$ gives:

8.17
$$p \sim r_0^3 E$$

The ratio $\alpha \equiv \dfrac{p}{E}$, is called the atomic polarizability. For *H*, we get $\alpha = 10^{-24} cm^3$, which is within better than a factor of two of the measured value. A list of atomic polarizabilities are given in the table below.

Halogens $(\times 10^{-24}\,cm^3)$		Noble gases $(\times 10^{-24}\,cm^3)$		Alkali metals $(\times 10^{-24}\,cm^3)$	
		He	.20	Li^+	.029
F^-	1.07	Be	.40	Na^+	.23
Cl^-	3.6	Ar	1.65	K^+	.90
Br^-	4.45	Kr	2.42	Rb^+	1.68
I^-	6.92	Xe	3.93	Cs^+	2.47

Table 8-1: List of atomic polarizabilities by group.[1]

Though we will not estimate it, clearly a similar effect can and does happen in molecules.

Polar Molecules (*Permanent Dipoles*)

We next consider permanent dipoles. A molecule that has a significant dipole moment is called a *polar* molecule. For example, water is a strongly polar molecule with $p \sim 1.8 \times 10^{-18}\,esu \cdot cm$.[2] Non-polar molecules, such as those of kerosene, have no significant dipole moments. Being polar or non-polar is an important distinguishing feature of molecules, which can be seen, for instance, in how radically it affects their ability to mix. Polar (non-polar) molecules generally mix into solutions easily with each other, whereas they do not with non-polar (polar) molecules. For example, water dissolves sugar (a polar substance) well but water doesn't dissolve oil (which is non-polar). However, kerosene can break up (dissolve) oil, which is also non-polar.

$$p = 1.8 \times 10^{-18}\,esu \cdot cm$$

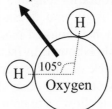

Water has a considerable permanent dipole moment.

Kerosene, which consists of varying length chains of carbon and hydrogen, is non-polar.

Most fundamentally, whether a molecule is polar or not determines how strongly it reacts to an external field. Paper thus should not be nearly so strongly attracted by your charged comb as water. From this, one may glean that, for our context, a molecule will be considered "polar" if its permanent dipole moment is significantly larger than that induced by the (external) applied field, and non-polar when the reverse is true. In particular, we will be most interested in this text in the effect of dielectrics in capacitor-like configurations. Before getting to the capacitor and more general analysis, we need to do one more brief study.

---

[1] Taken from "Dielectrics in Electric Fields" by Gorur G. Raju, 2003 Marcel Decker Inc, NY, NY. One can convert from Fm^2 (the units he uses) to cm^3 by multiplying by: $.899 \times 10^{12}\,cm\,/\,F\,\left(10^2\,cm\,/\,m\right)^2 = 8.99 \times 10^{15}\,cm^3\,/\,\left(Fm^2\right)$

[2] If we multiply this by $\sim 3.3 \times 10^{-12}$, which is the conversion factor to *mks (SI)*, we get: $p \sim 6 \times 10^{-30}\,Cm$

Uniformly Polarized Material

To better understand polarized bodies, i.e. bodies with their polarized parts aligned to some degree, we look at the fields of two important cases: a uniformly polarized slab and a uniformly polarized sphere.

Slab

We first consider a slab that has a uniform polarization that is fixed into the body so we can manipulate and experiment with it freely. For simplicity, we draw a piece of a macroscopic slab as if all the dipole parts align, and we show the parts (atoms) as much bigger than they are.

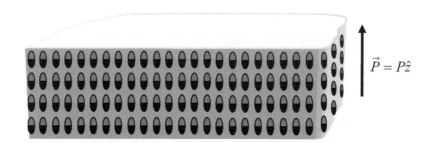

$$\vec{P} = P\hat{z}$$

Our full slab, of which the above is only a cut-out, is taken to be arbitrarily long and wide with some definite thickness. The polar nature of the given part is illustrated using an ellipse with one half colored red to indicate positive and the other half colored black to indicate negative. Note how each such dipole stacks above the dipole below it. In fact, the most essential aspect of this slab can be illustrated by a single stack of dipoles such as shown below. While we have shown the stack with a circular cross section, any other shape would do just as well. Putting enough of such stacks side-by-side will allow us to create an arbitrarily large slab of the thickness of the stack.

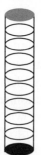

Note how the dipoles in the center are white, indicating no *net* charge. This is because the charge from the lower dipole, which is negative, just cancels the positive charge from the upper dipole. Only the top and bottom layers of charge are left uncanceled. Hence, the effect for an observer who is sufficiently far away, is as the same as if the slab simply had a sheet of positive charge, Q, on the upper surface and a sheet of negative charge, -Q, on the lower surface.

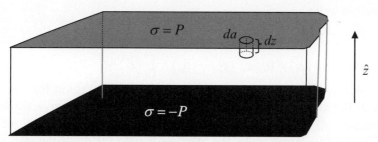

Now, to say more, we need to specify something more about the strength of the dipoles. This is given in terms of polarization per unit volume, P, so that:

8.18
$$d\vec{p} = \vec{P}dV$$

If we take the z-axis to be along the thickness of the slab and da to be the differential surface area as shown above, then we can write: $dp = P\,da\,dz$ for the polarization of a part of the body near the top surface, so that the effective dipole moment of length dz has surface charge of Pda. This in turn means the total charge on the surface of the slab is: $Q = PA$, where A is the total area of the surface, or in terms of the surface charge density, σ, we can write:

8.19
$$\sigma = P$$

Now, because the effective charge distribution of the uniformly polarized slab shown above is the same as that of the two parallel plate surfaces (with a vacuum between them) shown below, the *external* ϕ-field, and hence the external E-field, is the same for both.

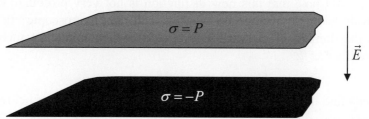

What about the inside? Is the E-field inside the two plates the same as that inside the slab? The answer is: yes in one way and no in another. Clearly, the detailed E-field, the so-called *microscopic E-field*, is not the same. Inside the *plates*, there can be no strong field. Indeed, the field must be uniform for the ideal parallel-plate capacitor, *while inside the slab* there will be very high fields near the nucleus of any given atom of the material.

However, the situation is different if one considers the average field over a small volume, which we call the *macroscopic* field, inside the slab, which we can write mathematically as:

8.20
$$\left\langle \vec{E} \right\rangle = \frac{\int_V \vec{E}dV}{V}$$

Also, we can write:

8.21
$$\left\langle \phi \right\rangle = \frac{\int_V \phi dV}{V}$$

So, we can therefore write:

8.22 $\left\langle \vec{E} \right\rangle = -\vec{\nabla}\left\langle \phi \right\rangle$,

This, in turn, implies $\nabla \times \left\langle \vec{E} \right\rangle = 0$ and $\left\langle \phi_b \right\rangle - \left\langle \phi_a \right\rangle = \int_a^b \left\langle \vec{E} \right\rangle \cdot d\vec{l}$

In particular, we have to average over a large enough region that the effect of the discrete nature of the atoms is smoothed over. Moving to the larger scale in this way is therefore a method of leaving out (abstracting out) the discreteness of the charge distribution. Thus, on this larger scale, the average E-field inside of the slab should be the same throughout the slab because each piece of a uniform slab is, by definition, the same as every other piece. Furthermore, the potential difference between two parallel plates with a surface charge density P and a slab of polarization density P must be the same since the *external* ϕ-field is the same for both. This means the average potential drop, $\left\langle \Delta\phi \right\rangle$, per unit distance, Δx, i.e. the average E-field, is the same for both systems. Or, more precisely, since the E-field inside of two infinite plates (with only plana between them) is $\vec{E} = -4\pi\vec{P}$, the average field inside the infinite slab can be expressed by the relation: $\left\langle \vec{E} \right\rangle = -4\pi\vec{P}$.

Note that, when dealing with dielectric materials in this text, we will often simply write $\vec{E}$ *without* the angle brackets for the macroscopic field, since the context clearly will indicate when we mean macroscopic field and when not. This context sensitive use of the symbol $\vec{E}$ will allow us to write an extended simple form of Maxwell's equations that uses the macroscopic field inside of dielectrics and the microscopic field outside of them. However, you should note that this new distinction is not very profound, since we have already effectively taken an average for what we call the "microscopic" field in order to remain within the domain of classical E&M itself (which implicitly includes classical mechanics)--see Chapter 2.

Sphere

Now, let's find the field for a uniformly polarized sphere shown below.

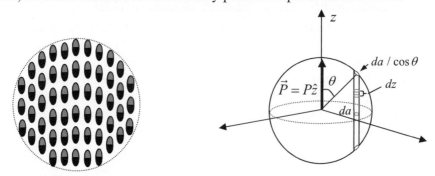

Figure 8-4: a. (*left*) Cross section of a sphere that is uniformly polarized. **b.** (*right*) one column of a sphere that is uniformly polarized with polarization density $\vec{P}$. The differential area of a cross-section of the column is shown as da. We mark a small length dz as one of the dipoles making up the column. Though the polarization vector applies to all regions within the sphere, only one arrow is shown at the center to avoid cluttering the diagram.

If the sphere has uniform polarization density P, then by the same stacked dipole argument used above for the slab, the charge of each stacked dipole, treating the interior ones which are cylindrical, is $dq = Pdadz$. This charge is then distributed over the top of the stack which has an area of $da / \cos\theta$ as illustrated in Figure 8-4. Thus, the surface charge density for the top surface of the given stack is:

8.23
$$\sigma = \frac{Pdadz}{da / \cos\theta} = P\cos\theta$$

Hence, following the analogy of the slab further, the polarized sphere is like a spherical shell with surface charge density given by equation 8.23. It can be shown that this is, in turn, like two uniformly *charged* sphere's of charge of opposite sign whose centers are slightly separated. The charges of each sphere that are not canceled by nearby charges from the other sphere end up on the outer portions of the combined spheres making a pattern that looks in cross section like the following.

Figure 8-5: a. (*left*) Two uniformly and oppositely charged spheres whose centers are slightly displaced in the vertical direction yield the net charged distribution shown. All but the two facing crescent moon portions of the charged distributions cancel. **b.** (*right*) The ϕ-field outside the approximately  spherical region enclosed by the outer boundary of the two crescent "moons," is the same as that caused by all the charge in each sphere being concentrated into the points shown.

You can see that this pattern, which we have exaggerated to make it more evident, appears like two crescent moons of different charges facing each other. This distribution of charge can be shown to closely approximate the cosine distribution of charge given in equation 8.23 as long as the thickness of the crescents can be neglected. Note that the outer boundary of the surface of the two closely overlapping spheres can be made arbitrarily close to that of a sphere, and thus the overlapping spheres can indeed be used to model the polarized sphere.

Hence, we can now easily determine the ϕ-field. Namely, since *each* of the two closely overlapping spheres of charge Q, produces the same external field as a point particle of charge Q located at its center would, we can calculate the field outside the polarized sphere by using the separated charges shown in Figure 8-5b. This is clearly just a simple dipole, so we have:

8.24
$$\phi = \frac{p_T \cos\theta}{r^2}, \qquad \text{where, the total polarization is: } p_T = Qd$$

We now need to express p_T in terms of P. To do this, take each sphere to be composed of n particles each of charge q per unit volume, so that $Q = nqV$, where $V = \frac{4}{3}\pi R^3$ is the volume of the sphere of radius R. Since the spheres are separated by a distance d, the polarized sphere that they model can be thought of as consisting of a uniform distribution of dipoles of magnitude $p = qd$. This means that the polarization density is $P = np = nqd$; hence, we have:

8.25
$$p_T = Qd = nqVd = \frac{4}{3}\pi R^3 P$$

Of course, if we were not trying to illustrate the utility of the overlapping sphere analogy, we could simply integrate the constant P over the volume of the sphere to get the same result.

Thus, using equation 8.6, we have the potential outside the polarized sphere in terms of P:

8.26
$$\phi = \frac{\frac{4}{3}\pi R^3 P \cos\theta}{r^2}, \quad \text{(outside)}$$

from which the E-field quickly follows. Note that this expression is valid (in our averaged macroscopic sense) right up to the surface of the polarized sphere because the field of the overlapping spheres is equivalent to that of an arbitrarily small dipole at the center.

But, what about the field inside the polarized sphere? We begin by noting that *on the surface* of the sphere equation 8.26 implies (why?) that:

8.27
$$\phi = \frac{4}{3}\pi P z \quad \text{(on the surface)}$$

Now, because there is only one possible solution for a given potential specified on a boundary, if we find one solution, we have *the* solution (see Chapter 2). Now, equation 8.27 is a solution to Laplace's equation everywhere inside:

8.28
$$\nabla^2 \langle\phi\rangle = -4\pi \langle\rho\rangle = 0$$

Hence, therefore, we can write (leaving out the angle brackets):

8.29
$$\phi = \frac{4}{3}\pi P z \quad \text{(inside)}$$

From, which we get:

8.30
$$\vec{E} = -\frac{4}{3}\pi P \hat{z} = -\frac{4\pi\vec{P}}{3} \quad \text{(inside)}$$

Of course, this cannot apply outside since there are net charges on the surface.

The E-field in a cross section (with azimuthal angle $\phi = 0$) of the polarized sphere is illustrated in the figure below:

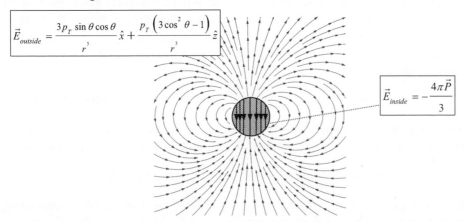

$$\vec{E}_{outside} = \frac{3p_T \sin\theta \cos\theta}{r^3}\hat{x} + \frac{p_T\left(3\cos^2\theta - 1\right)}{r^3}\hat{z}$$

$$\vec{E}_{inside} = -\frac{4\pi\vec{P}}{3}$$

After a careful look at the figure, you will notice a discontinuity in the E-field at the boundary of the polarized sphere. For example, at the north pole of the polarized sphere, the discontinuity in the perpendicular component of the E-field is:

$\Delta E_\perp \left(r = R, \theta = 0 \right) = E_{out} - E_{in} = \dfrac{2 p_T}{R^3} - \left(-\dfrac{p_T}{R^3} \right) = 4\pi P$. In fact, using Gauss's law and the fact that there is charge on the surface, one can show that there must be such a discontinuity. Indeed, $\Delta E_\perp = 4\pi P$ everywhere on the surface, as can be verified by fairly simple calculations. However, since $\nabla \times \vec{E} = 0$ in any static field even where there is charge, there is no discontinuity in the parallel component anywhere on the surface of our polarized sphere. It is left as an exercise for you to verify both these points.

Such "boundary conditions" on E, i.e. conditions on the components of E that are parallel and perpendicular to a boundary, will be discussed in an end of chapter problem which makes use of the dielectric constant, our chief parameter for describing a dielectric material.

Before turning to the dielectric constant, we emphasize one further point. Recall that E-fields add linearly. That is, plana qualified by a preexisting field receives the action of a new source in such a way that the new source is able to increase the strength of the preexisting field by the same amount as it would if there were no preexisting field. Mathematically, we simply say a second E-field adds vectorially to the old field to get the new field. Because of this, we can find the field that results from placing a uniformly polarized sphere into an already existing field by simple vectorial addition. So, for example, if we put the sphere into a uniform field and that field did not modify the sphere's polarization, the resulting field would be obtained by vectorially adding the field of the polarized sphere to that of the uniform field. Of course, this would remain true if the uniform field caused the given polarization of the sphere.

The Dielectric Constant

Simple Parallel Plate Capacitor

The dielectric constant, ϵ, of a material is a measure of how susceptible that material is to macroscopic polarization. In particular, consider the parallel plate capacitor shown below. In the top diagram, we show a capacitor with no massive body between the plates. In the bottom diagram, we show the same capacitor with a dielectric material filling the region between the plates. And, we define q_{free} as the charge that is pumped by the battery onto the capacitor plates, and q_{bound} is the effective charge that appears on the top surface of the dielectric as a result of the dielectric's polarization. The sum of these two charges is q_0, the same as in the vacuum case. This clearly must be true since once the effect of the dielectric is accounted for by the "bound" surface charges just described, we can simply treat the interior as (macroscopically) the same as a vacuum, as we saw in previous sections.

The fact that the total surface charge is the same in the dielectric-filled case as it is in the vacuum case can also be seen in the following way. Since the potential difference, V, is kept the same and since the medium is uniform in both cases, the electric field, i.e. the drop in potential per unit length, must be the same for both. Thus, the uniform planar charge distribution near the top (and bottom) surface, $q_0 = q_{free} + q_{bound}$ (and $-q_0$ for the bottom), that causes the interior fields must be, as illustrated below, the same.

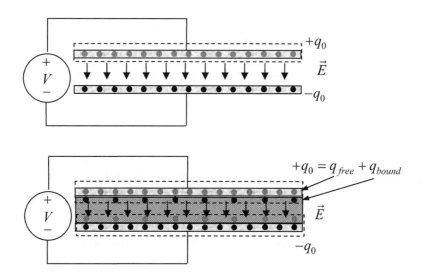

With the charge definitions given in the above figure, *the dielectric constant* is defined as the following ratio:

8.31 $$\epsilon \equiv \frac{q_{free}}{q_0}$$

This means that the voltage source needs to supply $q_{free} = \epsilon q_0$ of charge, which is $q_{free} - q_0 = (\epsilon - 1)q_0$ more charge than without the dielectric. Since we define the capacitance as: $C \equiv \frac{q_{free}}{V}$, the capacitance of a dielectric-filled capacitor is ϵ times bigger than the vacuum equivalent:

8.32 $$C_{dielectric} = \epsilon C_{vacuum}.$$

Now, for a moment, take off the voltage source and just consider a vacuum capacitor with a charge q on the top plate and $-q$ on the bottom. The voltage across the capacitor is then $V = q/C$. Now consider putting $q_{free} = \epsilon q$ on the same capacitor, this results in a voltage of ϵV. Hence, the electric field in the first case is: $E = eV/d$ and in the second case, we have:

8.33 $$\vec{E}_{free} = \epsilon \vec{E}.$$

Now, if we put a material with dielectric constant ϵ in a vacuum capacitor that has q_{free} charge on it, we get a resulting field, which we call simply E, that is the sum of the field already existing due to q_{free} plus the field that results from the *effect* of that field on the atoms in the dielectric: $\vec{E} = \vec{E}_{free} + \vec{E}_{bound}$. The bound field, as we saw in the polarization section, is given by $\vec{E}_{bound} = -4\pi\vec{P}$. Thus, we have $\vec{E} = \epsilon\vec{E} - 4\pi\vec{P}$, so that we can write:

8.34 $$\chi_e = \frac{P}{E} = \frac{\epsilon - 1}{4\pi}$$

χ_e is called the *electric susceptibility* and is useful because the macroscopic polarization induced by the application of an electric field to a dielectric material is, in many cases, simply proportional to the *E*-field over a significant range. The susceptibility is another way of speaking about the dielectric constant.

The dielectric constants of some interesting materials are shown in the table below. It should be no surprise that the dielectric constant of a non-polar substance is often of order one, in contrast to that of polar substances which can be quite big. Water, for example, has a dielectric constant of about 80.

Material	Phase	Temp.(C)/ Pressure	Dielectric constant [3] (ϵ)
Water	liquid	25C	78.54*
Water vapor (steam)	gas	140C/1atm	1.00785*
Air (dry)	gas	25C/1 atm	1.000536**
Alumina (china)	solid		3.1 - 3.9
CO_2	gas	0C/1atm	1.00098*
Clay	solid		1.8 - 2.8
Coal (fine powder)	solid		2 - 4
Ethyl alcohol	liquid	25C	24.3
Ferrous Oxide	solid	15C	14.2*
Freon 11	liquid	21.1C	3.1
Kerosene	liquid	21.1C	1.8
Octane	liquid	20C	1.96*
Oil, petroleum	liquid	25C	2.1
Paper	solid		2
Paraffin	solid		1.9 - 2.5
Quartz	solid		4.2
Rubber	solid		3
Silicon	solid		11.0 - 12.0*
Germanium	solid		15.8**
Gallium Arsenide	solid		13.13**

Notice that the dielectric constant of silicon is fairly high so that charge can separate easier in silicon than, for example, in air (or analogically in a vacuum through pair creation) because as an electron moves away from a nucleus, it and the nucleus's field can rearrange (polarize) the intervening silicon so as to cancel some of the effect of their fields. This is very important in silicon semiconductor electronics. A look at the table reveals that this same reasoning applies to Germanium and GaAs, which are also semiconductors that are commonly used in electronics.

Dielectric Sphere in a Parallel Plate Capacitor

We would also like to be able to deal with electric fields in and around dielectrics with other shapes. For example, consider the case previously mentioned of putting a dielectric sphere in a uniform field such as found inside of a parallel plate capacitor. This case is illustrated in the figure below. The field caused by the parallel plates will polarize the dielectric sphere causing a new field, which is now a result of the charges on the plates *as well as* the charges within the sphere. As we will see, the new field will be non-uniform near the sphere.

---

[3] Values not marked are taken from http://www.flowmeterdirectory.com/dielectric_constant_03.html. Values marked with * are taken from 37th edition of the "Handbook of Chemistry and Physics" put out by Chemical Rubber Publishing Company (often called the CRC). Values marked with ** are taken from "American Institute of Physics Handbook," 3rd edition.

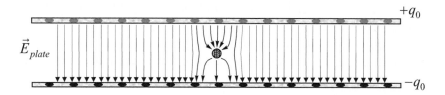

We start by assuming that the uniform E-field from the plates induces a uniform polarization in the sphere. We will verify the reasonableness of this assumption by showing that it leads to a self-consistent solution. With that assumption and using the formula for the internal field given in the section titled "Uniformly Polarized Material: Sphere", the total field inside the sphere, symbolized by E_{inside}, is then:

8.35 $$\vec{E}_{inside} = \vec{E}_{plate} + \vec{E}_{bound} = \vec{E}_{plate} - \frac{4\pi}{3}\vec{P}$$

We also have:

8.36 $$\vec{P} = \chi_e \vec{E}_{inside} = \frac{\epsilon - 1}{4\pi}\vec{E}_{inside}.$$

Note that here we have written $\vec{P}$ is proportional to $\vec{E}_{inside}$ rather than $\vec{E}_{plate}$, because the relevant field is the field that finally exists in the sphere after all the interaction has ceased, for it is this final field that determines how much macroscopic polarization the sphere will have, though it is $\vec{E}_{plate}$ that starts the process that is worked out as the interplay between the charges in the sphere and $\vec{E}_{plate}$. We thus get:

8.37 $$\vec{E}_{inside} = \frac{3}{2 + \epsilon}\vec{E}_{plate}$$

Clearly, there is a solution for $\vec{P}$ with this value of $\vec{E}$, thus showing our assumption is self-consistent. A similar analysis yields the E-field outside the sphere.

Brief Peek at "Microscopic" Physics

We now turn to give a very general look at the microscopic reason that macroscopic polarization occurs in the presence of an applied E-field. Since this problem properly belongs to statistical and quantum mechanics, we will only take a cursory look.

The two types of microscopic polarization that we discussed were: induced and permanent polarization, or non-polar and polar. In both cases, it is the sum total of the alignment of all the particles that causes the macroscopic polarization. We here focus on the polar molecules. The problem of calculating the susceptibility of a solid or a liquid is in general very hard because each molecule affects the next. Indeed, when the interaction is strong between neighboring atoms, one cannot even uniquely isolate one particular group of atoms as a dipole that lines up to make P. Still some general principles apply.

These principles are most evident in the case of a gas in which there is minimal interaction between the constituent atoms or molecules. For concreteness, take water vapor in a uniform E-field. As shown in the figure below, as we saw earlier in this chapter, the polar water molecule experiences a torque that tries to rotate the molecule so that its positive side is closest to the negative plate. One might expect, therefore, that all molecules would line up facing in that way. This would result in huge (see end of chapter problem) macroscopic polarizations. This, however, is not what happens because the molecules have various random motions. That is, they have impetus rotating their parts. As the molecules bump each other, different impeti are activated causing different motions. Hence, they do not all line up at any one moment, but they do tend to line up and thus create some effective, or on-the-average, line up. Because one is only interested in average effects, one needs to do a statistical analysis that takes into account the degrees of freedom available. Thus, as we mentioned from the start, these effects need to be studied using methods of statistical physics as well as general information from quantum mechanics.

Figure 8-6: Water vapor in a uniform electric field. Clearly the positive ends of the polar water molecules are attracted to the negative plates and vice-versa, tending to align the molecules in a particular orientation. However, because the water vapor molecules already undergo various rotational motions associated with their temperature, the electric field is only partially successful in orienting a given molecule at a particular time. The net result is an "on the average" orientation in the preferred direction.

Maxwell's Equations Extended to Include Dielectrics

Our previous macroscopic analysis of dielectrics can apply to any shape dielectric in the presence of any shaped conductors. This is best seen by writing down the promised extended version of Maxwell's equations that involves dielectric materials.

We begin with Maxwell's first equation and use the distinction between free and bound charge. By definition, the field produced by free charge is ϵE; hence, we can write this formally as:

8.38 $$\nabla \cdot \vec{D} = 4\pi\rho_{free}, \quad \text{(Maxwell's extended 1}^{st}\text{ equation)}$$

where we define the *displacement field* as: $\vec{D} \equiv \epsilon\vec{E}$

Recalling $\nabla \cdot \vec{E} = 4\pi\rho$ and noting that $\rho = \rho_{bound} + \rho_{free}$, we have:

8.39 $$\nabla \cdot \left((\epsilon - 1)\vec{E} \right) = -4\pi\rho_{bound}$$

Using the general form of equation 8.36, we also have:

8.40 $$\nabla \cdot \vec{P} = -\rho_{bound}$$

Maxwell's second and third equations, the source-less equations, $\nabla \cdot \vec{B} = 0$ and $\nabla \times \vec{E} = -\dfrac{1}{c}\dfrac{\partial \vec{B}}{\partial t}$, are left unchanged. However, to get the fourth equation, we need to incorporate changing fields into our understanding and our formalism. In particular, we must recognize that changing fields will cause changing dipole moments. As an aside, note that induced dipole moments will be able to change quicker than polar ones because the later involve more rotational inertia. Both, however, (ignoring interaction between the molecules, for example, as one can in gases at STP) can generally respond relatively unimpeded into the MHz range.

When the microscopic dipole moment, p, is changing, we know that charge is moving, i.e. there is impetus activated charge, and we know that impetus activated-charge generates an A-field and a B-field (somewhere). Indeed, taking the displacement between the charges of the dipole to be $\vec{l}$, we can write this current as a "bound" current in the following way:

8.41 $$\vec{J}_{bound} = nq\frac{d\vec{l}}{dt} = \frac{d(n\vec{p})}{dt} = \frac{d\vec{P}}{dt}$$

Thus, the fourth Maxwell's equation can be written in the following intermediate form:

8.42 $$\nabla \times \vec{B} = \frac{4\pi}{c}\vec{J} + \frac{1}{c}\frac{\partial\vec{E}}{\partial t} = \frac{4\pi}{c}\left(\vec{J}_{free} + \frac{d\vec{P}}{dt}\right) + \frac{1}{c}\frac{\partial\vec{E}}{\partial t}$$

Now, recalling that $\vec{E} = \epsilon\vec{E} - 4\pi\vec{P}$ and $\vec{D} \equiv \epsilon\vec{E}$, we write $\vec{D} = \vec{E} + 4\pi\vec{P}$. Hence, we can write the final form of the dielectric-extended Maxwell's equation as:[4]

8.43 $$\nabla \times \vec{B} = \frac{4\pi}{c}\vec{J}_{free} + \frac{1}{c}\frac{\partial\vec{D}}{\partial t}$$ (Maxwell's extended 4th equation)

Note that *sometimes* the split between free and bound charge is a matter of calculational convenience, there being, in those cases, no significant physical principle of immediate concern behind our choice. However, often there is a significant distinction in reality that is responsible for the convenience. For example, in the case of the capacitor example given earlier, the free charge moves between the battery and the plates, whereas the bound charge is trapped in the dielectric.

---

[4] Upon seeing this, you might feel some of the same temptation Maxwell might have felt to assume the plana was simply a species dielectric material, i.e. a polarizable massive body.

Summary

Dielectrics are massive bodies that react and/or respond to an *E*-field even when they have no net charge. This *E*-field can be of either the primal type $(-\nabla\phi)$ or the secondary type $(\dot{A})$, but since we in this chapter deal mostly with static cases, we typically refer to the first type. Dielectrics can be thought of as composed of dipoles.

Roughly speaking, a *dipole*, as the name (*di*-pole) implies, is any system with positive and negative poles of charge. An ideal dipole consists of two point particles of charge q separated by an arbitrarily small displacement, $\vec{s}$, such that $\vec{p} = q\vec{s}$, the dipole moment, remains constant; formally: $\vec{p} = \lim\limits_{s\to 0, q\to\infty} q\vec{s}$, where $\vec{s}$ points from negative to positive as shown here. $\bullet\overset{\vec{p}}{\longrightarrow}\bullet$

There are *two types of dielectrics*: (1) *polar*, those with molecules that have permanent dipole moments, i.e., dipole moments that exist without the imposition of an external electric field and (2) *nonpolar*, those with molecules that have a dipole induced by the application of a field.

Far from any finite sized charge distribution, the potential, i.e. the ϕ-field, can be expanded in *multipole moments*. Formally, in terms of the distance r from the distribution, if we *limit our view to along the z-axis*, we can write:

$$\phi(\vec{r}) = \int \frac{\rho(\vec{r}\,')}{|\vec{r} - \vec{r}\,'|} d^3 r' = \frac{Q_0}{r} + \frac{Q_1}{r^2} + \frac{Q_2}{2r^3} + O\left(r^{-3}\right)$$

Where Q_n is the nth moment of the charge distribution, $\rho(\vec{r})$. In particular, we have:

$$Q_0 = \frac{1}{r}\int \rho(\vec{r}\,')d^3 r', \quad Q_1 = \int z'\rho(\vec{r}\,')d^3 r', \quad Q_2 = \int \left(3z'^2 - r'^2\right)\rho(\vec{r}\,')d^3 r'$$

Q_0 is the monopole moment; Q_1 is the dipole moment; Q_2 is the quadrupole moment.[5] In general, Q_1 is a vector written as:

$$\vec{p} = \int \vec{r}\rho(\vec{r}\,')d^3 r'.$$ Ideal forms of each of these can be represented as shown to the right using point charges.

$\bullet\, q$ $\quad\quad$ $\displaystyle\Big|$ $\quad\quad$ $\displaystyle{\Big|}2q$

$Q_0 \neq 0 \quad Q_0 = 0 \quad Q_0 = 0$
$Q_1 = 0 \quad Q_1 \neq 0 \quad Q_1 = 0$
$Q_2 = 0 \quad Q_2 = 0 \quad Q_2 \neq 0$

For a *general dipole:*

The ϕ-field is:

$$\phi(\vec{r}) = \frac{1}{r^2}\int \hat{r}\cdot\vec{r}\,'\rho(\vec{r}\,')d^3 r' = \frac{\hat{r}\cdot\vec{p}}{r^2}$$

In spherical coordinates, using $\vec{p} = p\hat{z}$, this is:

$$\phi(\vec{r}) = \frac{p\cos\theta}{r^2}$$

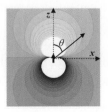

---

[5] In general, the quadrupole moment is a matrix (tensor) and Q_2 is the *zz* component written: Q_{zz}:

The E-field is:

$$\vec{E} = \frac{p}{r^3}\left(3(\hat{p}\cdot\hat{r})\hat{r} - \hat{p}\right), \vec{E} = \frac{p}{r^3}\left(2\cos\theta\hat{r} + \sin\theta\hat{\theta}\right)$$

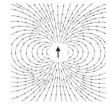

An external E-field[6] that varies in the x-direction exerts a *net force on an ideal dipole*, which is given by: $F_{net} = p\dfrac{\partial E}{\partial x}$. If the x-component of the E-field, E_x,

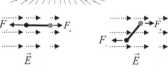

also varies along the y and z direction, terms corresponding to those two variations also contribute to the force in the x-direction, F_x, in a similar manner. As a result, for the ith - component of the force, F_i, we have, with p_x, p_y, p_z representing the dipole moments along each axis:

$$F_i = p_x\frac{\partial E_i}{\partial x} + p_y\frac{\partial E_i}{\partial y} + p_z\frac{\partial E_i}{\partial z}, \text{ or } F_i = \vec{p}\cdot\vec{\nabla}E_i \qquad \text{where } i \text{ is } x, y \text{ or } z.$$

For dipole orientations not parallel to the x-direction, a *torque is also exerted on the dipole*. This torque is responsible for the alignment of the molecular dipoles of *polar* materials in an applied field. The torque, $\vec{\tau}$, exerted by an external field on an ideal dipole is given by: $\vec{\tau} = \vec{p}\times\vec{E}$. The potential energy, U, of a dipole at a given angle is given by: $U(\theta_0) = pE(1 - \cos\theta_0)$.

In *non-polar* substances, the applied field causes the dipole moment. For example, an electric field applied to a ground state H atom distorts the spherically symmetric 1s state, i.e. makes it less likely, relative to an isolated H atom, that the electron will be found in the direction in which the field

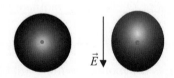

points. The ratio of the displacement of the negative charge center from the positive charge center, Δx, to the radius, r_0, of the atom (or molecule) is proportional to the ratio of the external E-field trying to stretch the atom to the internal E-field trying to hold the atom together. This gives: $p \sim r_0^3 E$

The ratio $\alpha \equiv \dfrac{p}{E}$, is called the atomic polarizability, a measure of the degree to which an external field can polarize a substance.

A molecule that has a significant dipole moment is called a *polar* molecule. Water is a strongly polar molecule with $p \sim 1.8\times10^{-18}\,esu\cdot cm$. A molecule will be considered "polar" if its permanent dipole moment is significantly larger than that induced by the

---

[6] By external, we mean any field that is not generated by the charges under consideration, in this case the charges of the ideal dipole.

(external) applied field, and non-polar when the reverse is true. The distinction between polar and non-polar molecules is important. For example, polar (non-polar) molecules generally mix into solutions easily with each other, whereas they do not with non-polar (polar) molecules.

We consider two simple types of uniformly polarized bodies, a slab and a sphere, which consist of microscopic dipoles that lineup.

A *slab* consisting of a dielectric that fills the region between two planes, which is shown below to the left, generates the same (macroscopic) E-field as occurs inside and outside of the vacuum region bounded by the same two planes but with surface charge density $\sigma = P$ shown in the figure below to the right. The charge in the interior region of the dielectric is zero if one does an average over many atoms because, on this scale, the positive end of one dipole cancels with the negative end of the one above it, leaving only an effective charge on the top and bottom boundaries, so that $\vec{E}_{inside} = -4\pi\vec{P}$.

A ball or *sphere* consisting of dielectric material bounded by a spherical surface, which is shown below, generates the same (macroscopic) E-field as occurs inside and outside of the vacuum region bounded by the same spherical surface but with surface charge density $\sigma = P\cos\theta$. This, in turn, is the same as two slightly offset (say by distance d) spheres of charge Q and $-Q$, where $Q = Nq$ and N is the number of atoms (molecules) with a dipole moment which have paired charges q and $-q$ as shown here on the left. The field outside the two spheres is equivalent to that of point particles of charge Q and $-Q$

separated by distance d located at the original sphere's center (see above right). That is, the ϕ-field *outside* the uniformly polarized sphere of radius R, is: $\phi = \dfrac{\frac{4}{3}\pi R^3 P\cos\theta}{r^2}$, while

inside the field is $\phi = \dfrac{4}{3}\pi P z$, which means there is a uniform E-field inside

given by: $\vec{E} = -\dfrac{4\pi\vec{P}}{3}$. The E-field of the uniformly polarized sphere is

shown on the right. The discontinuity in the E-field at the boundary of the sphere signals the presence of the surface charge just discussed.

The dielectric constant, ϵ, which in *cgs* is greater than one, is a measure of how susceptible a material is to macroscopic polarization. Said in terms of the simple parallel plate capacitor, though it applies generally, the dielectric constant is the ratio of the charge, q_{free}, that one needs to apply to one of the plates of a *dielectric*-filled capacitor to produce a voltage V to the charge, q_0, that needs to be applied to a *vacuum*-filled capacitor to produce that same

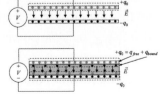

voltage. Formally: $\epsilon \equiv \dfrac{q_{free}}{q_0}$ or $q_{free} = \epsilon q_0$ (which gives $C_{dielectric} = \epsilon C_{vacuum}$). The charged plate, by polarizing the dielectric material, i.e. lining up the permanent dipole moments, effectively pulls oppositely charged particles near it nullifying part of its charge, thus requiring more charge to be dumped to get the same voltage. This is valid for any system with a spatially uniform dielectric constant, not just the parallel plate capacitor. The total field in a dielectric is given by: $\vec{E} = \vec{E}_{free} + \vec{E}_{bound}$, where $\vec{E}_{free} = \epsilon \vec{E}$ and for the parallel-plate capacitor $\vec{E}_{bound} = -4\pi \vec{P}$. Hence, we get the electric susceptibility $\chi_e = \dfrac{P}{E} = \dfrac{\epsilon - 1}{4\pi}$, which is another way of speaking about the dielectric constant. Note well that the E in the above equations is the E that results *after* the field of the bound charge has been taken into account.

The field inside of a spherical dielectric after being placed in a uniform field, $\vec{E}_a$, is given by: $\vec{E}_{inside} = \dfrac{3}{2 + \epsilon} \vec{E}_a$.

In an actual physical situation, the atoms (or molecules) undergo random motions associated with their temperature so that there is only some alignment of the dipoles in an applied field. Indeed, the orientation of any given atom is constantly changing, but on the average, there is an alignment. To understand these macroscopic systems in terms of the microscopic variations would require statistical mechanics as well as input from quantum mechanics that are beyond the scope of this text.

A set of extended Maxwell's equations that treats the interior of dielectric material as averaged over an appropriate scale is derived. The second and third Maxwell's equations do not need to be changed. The extended first equation can be obtained by noting that the field produced by free charge is $\epsilon \vec{E}$, while the second equation is obtained by taking account of the bound charge currents that can occur in dielectrics. Thus, we have:

$$\nabla \cdot \vec{D} = 4\pi \rho_{free}, \quad \text{(Maxwell's extended 1st equation)}$$

$$\nabla \times \vec{B} = \frac{4\pi}{c} \vec{J}_{free} + \frac{1}{c} \frac{\partial \vec{D}}{\partial t} \quad \text{(Maxwell's extended 4th equation)}$$

where $\vec{D} \equiv \epsilon \vec{E}$ is called the displacement field.

The total and bound charge densities are respectively given by:

$$\nabla \cdot \vec{E} = 4\pi \rho_{total}, \nabla \cdot \vec{P} = -\rho_{bound}$$

Helpful Hints

Note that the electric susceptibility, $\chi \equiv P / E$, differs from the polarizability, $\alpha \equiv p / E$, in that the χ tells *the polarization per unit volume* produced by a field of strength E, while α tells the *polarization for a given molecule* (or atom) of the material. Thus, though χ is dimensionless, α has dimensions of cubic length.

Problems

1. a) Calculate the capacitance of each of the capacitors shown to the right. Each is half filled with a material of dielectric constant ϵ in the manner shown; their plates have  area A and are separated by a distance d. b) What is the E-field in each region of each capacitor?

2. a) Calculate the monopole, dipole and quadrupole moments q, p_z, Q_2 for the configuration shown here. b) The quadrupole moment actually has nine components (five of which are independent) in general. One such component is: $Q_{xz} \equiv \int (3x'z') \rho(\vec{r}') d^3 r'$, calculate this also.

3. To get a very rough idea of the size of the dipole moment of water, assume the configuration shown to the right. Also, assume the effective charge separation between each hydrogen atom and the oxygen atom, l, to be about the Bohr radius of the H atom, $r_b = .5A$. Take the black point (the O atom) to have charge $2e$ and each of the end points have charge e, where $e = 4.8 \times 10^{-10} esu$ is the charge of an electron. After calculating the dipole moment for water, compare your calculated value to the measured dipole moment of water which is $1.8D$ (where D stands for a "debye," with $1D = 1 \times 10^{-18} esu \cdot cm$) and comment on any discrepancies.

4. a) Using Maxwell's media equations, derive an equation for the field inside of a parallel plate capacitor filled with a dielectric. Express the field in terms of the charge q and area A of the plates and the dielectric constant ϵ. What happens to the field when $\epsilon \to \infty$? For finite ϵ, what is the charge density on the surface of the dielectric next to each plate? What happens when $\epsilon \to \infty$? b) Replace the dielectric with a perfect conductor. What is the charge density on the surface of the conductor next to each plate? Comment on the relationship between the three cases (i.e., finite ϵ, $\epsilon \to \infty$ and a perfect conductor).

5. Given a horizontal plane with a charge surface density of σ that has a vacuum above it and a material with dielectric constant ϵ everywhere below it, give expressions for the E-field above and below the plane. Verify your answer for the cases $\epsilon \to 1$ and $\epsilon \to \infty$. Explain the meaning of these two extremes.

6. a) Consider a planar surface of area A that has somehow been charged to a surface charge density σ. Suppose the small face of a piece of paper (which also has area A) that is much thicker than it is wide or long is brought arbitrarily close parallel to this surface. Very roughly, making the simplest possible assumptions (such as treating the paper as a conductor), what force does the surface exert on the paper? Give an equation; then, give a numerical answer for the force (in newtons) by assuming there is a charge $q \sim 10^{-7} C$ over an area $A \sim 4mm^2$ of the plane. b) Now, instead assume a point particle of charge q is put at a distance x from a square piece of paper of length l and thickness, w, so that the normal

at the center of the plane of the paper points at the particle. Approximately what force does the charge exert on the paper? Using the same values as above with $w \sim .1mm$ and $x = 1mm$, give the force in newtons. c) Which of these makes a better approximation to the force that a charged comb exerts on a small piece of notebook paper of area A? Which numerical answer seems more reasonable; convert to pounds to help in answering this part of the question.

7. Show that the following two boundary conditions hold for any surface: $\Delta D_\perp = 4\pi\sigma_s$ and $\Delta E_\parallel = 0$, where ΔE_i represents the change in E_i across the boundary.

8. Prove that equation 8.9 follows from equation 8.7.

9. Prove the two dimensional equation for the force on a dipole discussed in the text, i.e.

$$F_x = p_x \frac{\partial E_x}{\partial x} + p_y \frac{\partial E_x}{\partial y} \text{ and } F_y = p_x \frac{\partial E_y}{\partial x} + p_y \frac{\partial E_y}{\partial y}$$ where we consider only the x-y plane.

Make sure to show why the higher order dependences on Δx and Δy (e.g. $\Delta x \Delta y$ or Δx^2) do not come into play.

10. Suppose that (linearly polarized) microwave frequency electromagnetic radiation of frequency f impinges on water in a 1 kW microwave oven. Take the dipole moment of a water molecule as: $p \sim 2 \times 10^{-18} esu \cdot cm$. a) Making the (grossly inaccurate) assumption that one molecule does not influence the next, write down the differential equation for the angular motion of a given water molecule. Using dimensional analysis and this equation, what is the characteristic frequency of rotation in terms of p, E_0 (the maximum strength of the field of the radiation) and physical parameters of the water molecule (use $r_{hydrogen} \sim .5A$)? b) Give a numerical value for this frequency, making the rough approximation that the X-ray tube radiates a beam of area $A \sim 100cm^2$ directly at the water and that all of it is absorbed by the water. c) Assuming that this characteristic frequency is the minimum possible value of the resonant frequency and given that the microwave oven has an operating frequency of about 3GHz, explain generally how a microwave oven works.

11. How long would a parallel-plate capacitor of thickness .1 mm have to be to have a capacitance of 1 farad? Solve for: a) the case with vacuum between the plates and b) the case with a material of dielectric constant $\epsilon = 100$.

12. Adding a few facts from quantum[7] and statistical mechanics, derive a rough equation for the electric susceptibility, χ_e, of an ideal gas. That is, use the fact that the rotational motions of each molecule have an average energy, kT, associated with the temperature, T, of the gas. Also, use the fact that the fraction of the particles in the gas that are aligned with the field is approximately proportional to the ratio of the rotational potential energy (due to

---

[7] The number of degrees of freedom is different from what one would expect via a purely classical analysis of the atoms.

the E-field) to the rotational kinetic energy (due to the temperature). Neglect all other forces and write the equation in terms of: n, the number of molecules per unit volume, p, the dipole moment of a each molecule of the gas, and T.

13. Suppose, using a charged plastic rod, one dumps $1\mu C$ into the homemade capacitors used in the electrostatic motor discussed in an end of chapter problem in Chapter 2. Assume the capacitor is a cylinder made of plastic ($\epsilon \sim 3$) and has dimensions $h \sim 8cm, r_{inside} \sim 2cm, r_{outside} \sim 2.2cm$, and that it is effectively lined inside and out with a perfect conductor (of course, the inside is not shorted to the outside). How much voltage will the capacitor have? How many times will we have to charge the rod for the capacitor's voltage to reach 10kV?

14. Show that one cannot, in general, ignore the mutual interaction of the constituent particles when calculating the polarizability, α, of a solid or liquid material.

15. Imagine spreading 1 gram of water into a relatively thin slab. Supposing that the dipole moments of all the molecules were to exactly align, give an approximate value for the strength of the E-field that would be produced inside the water. The dipole moment of water is: $p \sim 2 \times 10^{-18} esu \cdot cm$. Give your answer in SI units. Comment on the magnitude of your answer relative to fields one encounters in ordinary life, i.e. with the things you touch.

Chapter IX

Magnetics:
Static Magnetic Fields Inside Massive Bodies

Introduction

We have seen that magnetic fields are intimately related to electric fields, so it will be no real surprise to find more parallel's appearing as we apply these principles to the particular case of dielectrics and magnetics. However, we also expect differences, as impetus activated charge is different though related to charge itself.

We are all familiar with the strong attraction or repulsion of magnets for or against each other, with their ability to attract iron, and even with their amazing ability to act through some materials, through glass or even the surface of your desk if it is thin enough.

However, few are familiar with the weaker attraction magnets have for some materials and with the even weaker repulsion they have for nearly all materials (even when they are overwhelmed by the former.)

We have thus three distinct types of induced behaviors to explain:

(1)   **Ferromagnetic**, such as a magnet's ability to attract iron relatively strongly
(2)   **Paramagnetism**, weak ability of a magnet to attract some materials
(3)   **Diamagnetism**, weak ability of a magnet to repel nearly all materials

Once we have explained these three, we will have the primary principles to explain the attraction (and repulsion) of magnets for each other, and their ability to act through some materials. We will see in a moment that, as with the electric case, the dipole moment is the second order approximation to any source of magnetic field. However, because the source of magnetism is impetus activated charge, not a special type of positive (say, north) and negative charge (south), unlike the electric case, there is no monopole term and the dipole term is the first term that can exist.

Dipole and Other Moments

In parallel to the dielectric case of the last chapter in which we started with the source of the ϕ-field (i.e. charge), to study magnetics, we begin with the source of the A-field, i.e. impetus activated charge (recall that we treat even electron spin in this way at our level of our abstraction). As we know, the mathematical treatment of the generation of an A-field by a current distribution can be formulated in a parallel way to that of the generation of a ϕ-field by a charge distribution. Namely, recall that

9.1 $$\vec{A} = \frac{1}{c} \int \frac{\vec{J}(\vec{r}')}{|\vec{r} - \vec{r}'|} d^3 r'$$

For simplicity of formal calculation, we, without a true loss of generality, as we did in the electric case, choose a field point on the z-axis: $\vec{r} = z\hat{z}$. Borrowing the expansion from the electric case, substituting $\vec{J}$ for ρ and inserting the $1/c$ factor, we get the expansion for the A-field far from the current sources:

9.2 $$\vec{A}(\vec{r}) = \frac{1}{cr} \int \vec{J}(\vec{r}') d^3 r' + \frac{1}{cr^2} \int (\hat{r} \cdot \vec{r}') \vec{J}(\vec{r}') d^3 r' + O(r^{-3})$$

It can be shown that the first integral is zero for any current distribution confined to a finite region that is conserved (so that $\nabla \cdot \vec{J} = 0$) and that the second term can be written as follows:

9.3 $$\vec{A}(\vec{r}) = \frac{\vec{m} \times \hat{r}}{r^2}, \text{ where } \vec{m} = \frac{1}{2c} \int \vec{r}' \times \vec{J}(\vec{r}') d^3 r',$$

This is called the *magnetic dipole* term. Since the first term in the expansion given in equation 9.2 is identically zero, we see, as mentioned earlier, that the magnetic dipole term is always the first non-zero term.

General Magnetic Dipole Field

For easier access to the principles, we return to the simplest type of dipole, i.e. the small circular current loop which we already analyzed at the end of Chapter 3. You can see in an end of the chapter problem how any shaped looped will look arbitrarily close to a simple dipole, and thus the circular loop, from far enough away. For convenience, such a dipole and its *A*-field in the *x-y* plane is reproduced below.

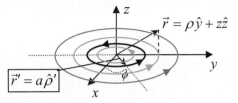

Recall also from Chapter 3 that we obtained:

9.4
$$\vec{A}(\rho,\phi,z) = \frac{2I}{c}\hat{\phi}\int_{\pi/2}^{3\pi/2} \frac{\left(a\sin\phi' \, d\phi'\right)}{\sqrt{\left(a\cos\phi'\right)^2 + \left(\rho - a\sin\phi'\right)^2 + z^2}}$$

Using $r = \sqrt{\rho^2 + z^2}$ and $r' = a$, we can write:

$$\vec{A}(\rho,\phi,z) = \frac{2I}{c}\hat{\phi}\int_{\pi/2}^{3\pi/2} \frac{\left(a\sin\phi' \, d\phi'\right)}{\sqrt{a^2 - 2\rho a\sin\phi' + r^2}} = \frac{2I}{rc}\hat{\phi}\int_{\pi/2}^{3\pi/2} \frac{\left(a\sin\phi' \, d\phi'\right)}{\sqrt{1 - 2\frac{\rho}{r}\frac{a}{r}\sin\phi' + \frac{a^2}{r^2}}}$$

If we now take the viewpoint, $\vec{r}$ to be very far away compared to the ring size, i.e. $r \gg r'$, which means $\sqrt{\rho^2 + z^2} \gg a$, we can write, using $\rho/r = \sin\theta$:

9.5
$$\vec{A}(\rho,\phi,z) \approx \frac{2I}{rc}\hat{\phi}\int_{\pi/2}^{3\pi/2} a\sin\phi' \, d\phi' \left(1 + \frac{\rho}{r}\frac{a}{r}\sin\phi'\right) = \frac{I\pi a^2 \sin\theta}{cr^2}\hat{\phi}$$

Thus, we have:

9.6
$$A_\phi = \frac{IA}{cr^2}\sin\theta = \frac{m}{r^2}\sin\theta, \quad \text{where } A \text{ is the area of the loop.}$$

We note here that we take the magnetic moment as

9.7
$$\vec{m} = \frac{I\vec{A}}{c} = \frac{IA}{c}\hat{z},$$

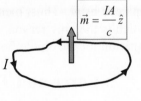

thus reproducing the result given in equation 9.3. This remains true when the loop is any "smooth" shape as long as we take A as the area of that loop.

We, thus, can define the *ideal magnetic dipole moment* as the limit in which the area of the loop, $\vec{A}$, gets arbitrarily small while $\vec{m}$ remains constant thus requiring that the current I gets infinitely large. In our mathematical formalism, we write: $\vec{m} \equiv \frac{1}{c}\lim_{A\to 0, I\to\infty} I\vec{A}$.

Re-expressing equation 9.6 in terms of Cartesian coordinates, we calculate the *B*-field of an ideal dipole as follows. Using $\hat{\phi} = -\sin\phi\hat{x} + \cos\phi\hat{y}$, we can easily project onto the three Cartesian axes and write the result in terms of x, y and z by recalling $\cos\phi = x/\sqrt{x^2+y^2}$, $\sin\phi = y/\sqrt{x^2+y^2}$ and $\sin\theta = \dfrac{\sqrt{x^2+y^2}}{\left(x^2+y^2+z^2\right)^{1/2}}$. We get:

9.8
$$A_x = \frac{m}{\left(x^2+y^2+z^2\right)}\frac{\sqrt{x^2+y^2}}{\left(x^2+y^2+z^2\right)^{1/2}}\frac{-y}{\sqrt{x^2+y^2}} = \frac{-m\,y}{\left(x^2+y^2+z^2\right)^{3/2}}$$

$$A_y = \frac{m\,x}{\left(x^2+y^2+z^2\right)^{3/2}}$$

$$A_z = 0$$

Thus, we get, using $\vec{B} = \nabla \times \vec{A}$:

9.9a,b,c
$$B_x = \frac{\partial A_z}{\partial y} - \frac{\partial A_y}{\partial z} = \frac{3mxz}{\left(x^2+y^2+z^2\right)^{5/2}} = \frac{3m}{r^3}\sin\theta\cos\phi\cos\theta$$

$$B_y = \frac{\partial A_x}{\partial z} - \frac{\partial A_z}{\partial x} = \frac{3myz}{\left(x^2+y^2+z^2\right)^{5/2}} = \frac{3m}{r^3}\sin\theta\sin\phi\cos\theta$$

$$B_z = \frac{\partial A_y}{\partial x} - \frac{\partial A_x}{\partial y} = \frac{m\left(3z^2 - \left(x^2+y^2+z^2\right)\right)}{\left(x^2+y^2+z^2\right)^{5/2}} = \frac{m\left(3\cos^2\theta - 1\right)}{r^3}$$

Note that in the last equality in each line, we have converted to spherical coordinates. These formulas can, in turn, be written (see end of chapter problem) compactly as:

9.10ab
$$\vec{B} = \frac{m\left(3\left(\hat{m}\cdot\hat{r}\right)\hat{r} - \hat{m}\right)}{r^3}$$

$$= \frac{m\left(3\cos\theta\,\hat{r} - \hat{z}\right)}{r^3} \qquad \text{where we took: } \vec{m} = m\hat{z}$$

Notice that this is exactly the same formula as that for the electric dipole field with $\vec{p}$ replaced by $\vec{m}$. Thus, memorizing the relatively simple formula will give you two for one, which is the only reason we have taken the trouble to write this more compact form of the equation for the dipole fields.

Force and Torque on Dipoles due to External Fields

In explaining the three types of repulsion and attraction mentioned above, we will need to understand how a dipole is acted on by an external field, such as that of a magnet. We thus calculate the forces on a dipole and then consider what torques those forces might cause.

Force

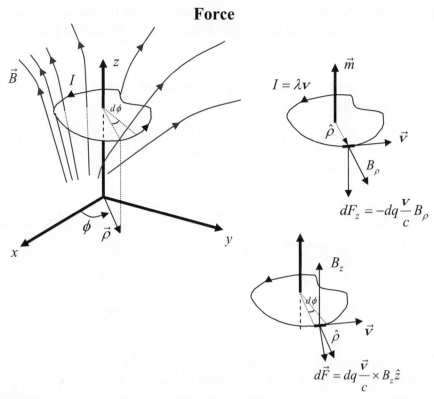

Figure 9-1: a. (*left*) A small arbitrarily-shaped current loop in an arbitrary *B*-field. **b.** (*top right*) Force exerted by the *radial* component of the *B*-field ($\vec{F} = q\dfrac{\vec{v}}{c} \times \vec{B}$) on a differential piece, *dl*, of the loop. The differential piece, which is shown thicker than the rest of the loop, has a charge $dq = \lambda\,dl$, where λ is the linear charge density. **c.** (*bottom right*) Force exerted by the *z-component* of the *B*-field on *dl*. Note that this force is in the *x-y* plane but is *not* necessarily in the radial direction.

We need a finite source, not an ideal dipole, to do our force calculation. The most straightforward choice is an arbitrarily small circular loop of current. However, to help bring home the point that, for any given *B*-field, a small enough loop will look arbitrarily close to an ideal dipole and thus its shape will not matter, we will work out the force due to *the radial component of the field* for an arbitrarily-shaped loop in an arbitrary *B*-field. The physical configuration is illustrated in Figure 9-1a.

In Figure 9-1b, we show how the radial component of *B*, B_ρ, acts on a differential piece of the loop, *dl*, causing a force in the negative *z*-direction, thus indicating that a small bar magnet put into the field (at the given point) oriented with its south pole downward will be pulled downward. Figure 9-1c shows the effect of the *z*-component of the field, B_z, on the *dl*. To avoid further complexity, as already implied, we will address this component (B_z) and the ϕ-component of *B*, B_ϕ, only for the case of a circular loop in a symmetrical field, only calculating B_ρ for an arbitrarily-shaped loop in an arbitrary field.

In the figure below, we focus on the force caused by the radial component. In particular, the left drawing shows the various angles in play. Using that figure, we write the force in the z-direction as: $dF_z \hat{z} = dq\frac{\vec{v}}{c} \times B_\rho \hat{\rho} = -\hat{z}\, dq\frac{v}{c} B_\rho \sin\beta$. Using the right drawing below, we have the charge of the differential line element: $dq = \lambda dl = \lambda \rho d\phi / \sin\beta$, where λ is the linear charge density. Thus, we have:

9.11
$$|F_z| = \int_0^{2\pi} \lambda \frac{v}{c} B_\rho\left(\rho,\phi,z\right)\sin\beta\,\frac{\rho(\phi)d\phi}{\sin\beta} = \frac{I}{c}\int_0^{2\pi} B_\rho\left(z,\rho,\phi\right)\rho(\phi)d\phi$$

where we have used $I = \lambda v$ in the last equality.

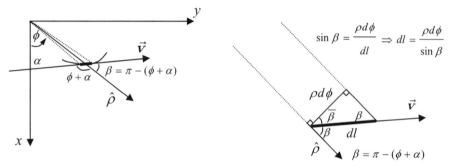

In order to get a particularly convenient form of the force, we now rewrite B_ρ in terms of the z component of B. Must B_ρ and B_z be related? Yes, remember no flux lines of B-field can end, since there are no sources of B. The formal way of writing this differentially is $\nabla \cdot \vec{B} = 0$. We use the integral form, $\oiint \vec{B} \cdot d\vec{A} = 0$,

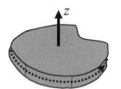

by drawing an arbitrarily short box with top and bottom surfaces exactly the same size and shape as the loop as shown to the right.

Using a simple circular loop to illustrate the key principles, we can see the relationship between B_ρ and B_z and its importance in the following way. Picking one differential element of the loop, we note first that if the radial strength of B, B_ρ, is the same on the opposite side of the loop, there will be no net force, since the forces on each side will be equal and opposite. If, however, B_ρ is different on the other side of the loop, say as shown below, then to conserve the flux of B, there must be a decrease in B_z. Here's why. For simplicity, assume that B_ρ only changes across the diameter[1] currently under consideration. If we imagine a container such as shown above on the right, then there is more B_ρ coming in from the left than from the right so that there must be a compensating difference between the B piercing the top and bottom as shown. Note, by the way, that a change in the remaining component of B, B_ϕ, cannot have any effect since all of it "remains" in the container.

---

[1] Of course, by this we mean an arbitrarily small angle of diameters around the one we are considering, for only in the completed limit (being of reason) is it an infinitely thin line. And, the intersection of that line with the infinitely thin current loop gives the two relevant points on the current loop.

$$B_z(R,\pi,z+\Delta z)$$

$$B_\rho(R,\pi,z) \longleftrightarrow B_\rho(R,0,z)$$

$$B_z(R,\pi,z)$$

> **Note on Coordinates**: cylindrical coordinates can be confusing if one is not careful. Note for example, if $\vec{B} = B_\rho \hat{\rho}$ with B_ρ constant does not mean the B-field is constant. It means it has a constant magnitude in the radial direction. But the radial direction *depends* on the value of ϕ. Nothing similar happens in Cartesian coordinates, because they do not mix the dimensions.

We now return to our arbitrary shaped loop and make this analysis more precise. In this case shown in Figure 9-1a in which a net B_ρ leaves the sides, it is appropriate to express conservation of flux as follows. The flux leaving from the top minus that coming in from the bottom must equal that coming in from the sides. Keeping in mind the convention that outward flux is positive, we write this as:

9.12
$$\int \left(B_z(\rho,\phi,z+\Delta z) - B_z(\rho,\phi,z)\right) \frac{\rho(\phi)^2}{2} d\phi = -\Delta z \int B_\rho(\rho,\phi,z)\rho(\phi)d\phi$$

Dividing each side by Δz and taking the limit to arbitrarily small Δz gives a derivative in the integrand on the left hand side. Namely, equation 9.12 becomes:

9.13
$$-\int \frac{\partial B_z}{\partial z}(\rho,\phi,z) \frac{\rho(\phi)^2}{2} d\phi = \int B_\rho(\rho,\phi,z)\rho(\phi)d\phi$$

Making our assumption of an arbitrarily small loop more precise we require that the $\dfrac{\partial B_z}{\partial z}$ changes arbitrarily little over the loop, so that we can take that derivative as constant in the integral on the left hand side. Thus, we have $\int_0^{2\pi} B_\rho(z,\rho,\phi)\rho(\phi)d\phi = -\dfrac{\partial B_z}{\partial z} A_{loop}$, where A_{loop} is the area of the loop. Thus we can write equation 9.11 for the total force on the loop in the z-direction as:

9.14
$$F_z = \frac{IA_z}{c}\frac{\partial B_z}{\partial z} = m\frac{\partial B_z}{\partial z}$$

Now, as according to our plan, we take the loop to be circular and the B-field to be symmetrical around that loop so that there is no ϕ-dependence in B when the origin of the coordinate system is placed at the center of the loop. Clearly, in this case, the velocity of the charge in the loop is parallel to ρ-component of B, so there is no force anywhere on the loop due to it. Also, it is easy to see, referencing Figure 9-1c, that for a circular loop, $dF_\rho = \lambda\rho d\phi B_z \dfrac{v}{c}$ on each differential element of the loop. Each such force is balanced by one diametrically across from it, so that these radial forces yield no net force.

It can be shown that, in general, the force on a magnetic dipole is:

9.15
$$\vec{F} = \nabla\left(\vec{m}\cdot\vec{B}\right)$$

When $\nabla \times \vec{B} = 0$, this formula reduces to the same formula as that obtained for the electric dipole with $\vec{p}$ replaced by $\vec{m}$:

9.16
$$\vec{F} = \left(\vec{m} \cdot \vec{\nabla} \right) \vec{B}$$

Or explicitly in terms of components we can write:

9.17a,b,c
$$F_x = m_x \frac{\partial B_x}{\partial x} + m_y \frac{\partial B_x}{\partial y} + m_z \frac{\partial B_x}{\partial z}$$

$$F_y = m_x \frac{\partial B_y}{\partial x} + m_y \frac{\partial B_y}{\partial y} + m_z \frac{\partial B_y}{\partial z}$$

$$F_z = m_x \frac{\partial B_z}{\partial x} + m_y \frac{\partial B_z}{\partial y} + m_z \frac{\partial B_z}{\partial z}$$

It can be shown that when $\nabla \times \vec{B} \neq 0$, one must subtract $\vec{m} \times \left(\nabla \times \vec{B} \right)$ from the general formula 9.15, i.e. $\vec{F} = \nabla \left(\vec{m} \cdot \vec{B} \right)$, to get equation 9.16, i.e. $\vec{F} = \left(\vec{m} \cdot \vec{\nabla} \right) \vec{B}$ (see end of chapter problem).

Torque

In a circular loop in the center of and aligned with a cylindrically symmetrical B-field, there will thus be a force, but there will be no torque trying to twist the loop. We say the loop is in its lowest rotational energy condition, though not its lowest translational energy condition. This is simply another way of stating that there are no torques that "want" to change the orientation of the loop, but there are forces that will give impetus to the loop as a whole, thus causing center of mass motion. In this subsection, we discuss only spin, i.e. rotation about the center of mass, leaving orbital motion, i.e. rotation about other points, aside.

The simplest way to explain how a torque arises is to consider a rectangular loop in a constant field. Figure 9-2 below shows the forces that lead to a torque around the x-axis. Mathematically, the torque due to the force on the short sides is:

$$\vec{\tau} = \sum \vec{r} \times \vec{F} = \left(-r_y \hat{y} + r_z \hat{z} \right) \times \left(-\lambda a \frac{v_x \hat{x}}{c} \times \left(B_x \hat{x} + B_y \hat{y} + B_z \hat{z} \right) \right)$$

$$+ \left(r_y \hat{y} - r_z \hat{z} \right) \times \left(\lambda a \frac{v_x \hat{x}}{c} \times \left(B_x \hat{x} + B_y \hat{y} + B_z \hat{z} \right) \right)$$

$$\vec{\tau} = \frac{2Ia}{c} \left(-r_y \hat{y} + r_z \hat{z} \right) \times \left(-B_y \hat{z} + B_z \hat{y} \right)$$

Which yields:

9.18
$$\vec{\tau}_{short} = \frac{2Ia}{c} \left(r_y B_y - r_z B_z \right) \hat{x}$$

$$= m \left(B_y \cos \theta - B_z \sin \theta \right) \hat{x}$$

In the last equality we have used: $r_y = \dfrac{b}{2}\cos\theta$ and $r_z = \dfrac{b}{2}\sin\theta$, and $m = \dfrac{IA_{loop}}{c}$. We now seek a simple way to express this relation. Using Figure 9-2a, we see that $\vec{m} = -m\left(\cos\theta\hat{z} + \sin\theta\hat{y}\right)$. Thus, we can rewrite 9.18 as:

9.19 $\qquad \vec{\tau}_{short} = \left(m_y B_z - m_z B_y\right)\hat{x}$

And, since for the short side only B_y and B_z are relevant, we write: $\vec{B} = B_y\hat{y} + B_z\hat{z}$. Thus noting, the form of cross product result seen in equation 9.19, we can write $\vec{\tau}_{short} = \vec{m} \times \vec{B}$.

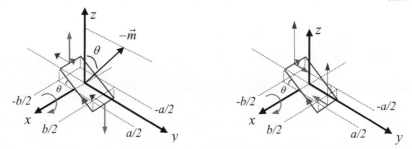

Figure 9-2: a. (*left*) Rectangular current loop (brown), of width a and length b tilted at an angle θ from the x-y plane, experiences a torque around the x-axis due to the action of the y-component of the B-field, B_y, acting on the short sides of the loop. The B-field is shown in blue, and the forces are shown in green. The negative of the magnetic dipole moment vector, $\vec{m}$, is also shown. **b.** (*right*) The same loop also experiences a torque around the x-axis due from B_z.

Using the same argumentation, a similar formula can be obtained for the long sides, where this time only B_x and B_z are relevant and $\vec{m}$ has only m_x and m_y. In this way, we show that $\vec{\tau}_{long} = \vec{m} \times \vec{B} = \left(m_z B_x - m_x B_z\right)\hat{y}$ confining ourselves to the relevant components of B. We thus have the x and y components of the torque. The z-component can be handled, for example, by moving our loop into the y-z plane and tilting it outward from that plane by an angle θ similar to what we have already done. In this way, we show generally that the torque acting on an ideal dipole (a loop for example that is small enough that its shape does not matter) is:

9.20 $\qquad \vec{\tau} = \vec{m} \times \vec{B}$

The torque produced by the B-field and quantified by this equation is responsible, as we will see, for both paramagnetism and ferromagnetism. For now, however, we turn to the cause of diamagnetism which is fundamentally different. Paramagnetism and ferromagnetism rely on something much closer to the idealized dipole than does diamagnetism.

Diamagnetism

For diamagnetism, as well as for para and ferromagnetism, we must return to the atomic structure of ordinary substances, for we will see that it is atomic electrons that are responsible for these effects. Remember all physical substances have extension (quantity in the first sense of the word), parts one outside the next, and in every ordinary substance, an essential such part is the atom. We can understand the essential cause of diamagnetic behavior by assuming a given electron follows a circular orbit around the nucleus. Of course, it does not, but its motion around the nucleus in this simple fashion allows us to get to the heart of the matter.

So, consider the electron moving quickly in a stable circular orbit around a fixed singly charged nucleus under the influence of the electric field generated by it as shown below in Figure 9-3a. As you will recall, the centrifugal Coulomb force causes the changing momentum, and this relation, represented simply in terms of cylindrical coordinates, is:

9.21 $$\frac{e^2}{R^2} = \left|\frac{d\vec{p}}{dt}\right| = \left|\frac{d\left(m_e v \hat{\phi}\right)}{dt}\right| = \frac{m_e v^2}{R}$$

Where e, R, v and m_e are, respectively, the electron's charge, orbital radius, speed and mass.

If we now average over a long enough time period, so that the electron looks smeared over its orbit, rather like the spokes of a quickly moving bike, we can treat its motion as a current. Since the electron carries a charge e around once in a time T, the current can be written: $I = \frac{e}{T} = e\frac{v}{2\pi R}$. Thus, treating the atom in the dipole moment approximation described earlier, its magnetic moment can be written as:

9.22 $$\vec{m} = \frac{IA}{c}\hat{z} = \frac{evR}{2c}\hat{z}$$

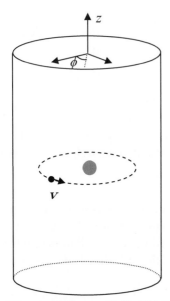 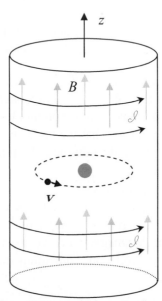

Figure 9-3: a. (*left*) Electron in a circular orbit around a singly charged nucleus before and after **b.** (*right*) the application of an *A*-field, which specifies the uniform *B*-field as shown. The field is generated by applying a uniform current to the surface of the long cylinder. The current density on the surface is $\mathscr{I}$, i.e., I *esu / s* per *cm* in the *z*-direction.

When the current is switched on in the cylinder around the atom, a circular A-field is activated in the plana where the electron rotates. Given the applied current is in the same direction as the motion of electron, that plana near the electron "wants" to resist the increasing A-field by increasing the impetus of the electron. That is, since the electron is negatively charged, increasing its impetus means the electron tries to generate an A-field in the direction opposed to its rotation, thus opposing the increasing A-field from the cylinder. In terms of the B-field, the increasing electron impetus, opposes the increasing B-field caused by the cylinder.

The magnitude of these effects can be approached in the following way using the A-field. From Chapter 3, we have:

9.23
$$\vec{A} = \frac{2\pi\rho\mathscr{I}}{c} \hat{\phi}$$

Because we will need to express the results in terms of B, we note that $\vec{B} = B_0\hat{z}$, with

$B_0 = \dfrac{4\pi\mathscr{I}}{c}$, which gives:

9.24
$$\vec{A} = \frac{B_0}{2} \rho \hat{\phi}$$

Thus, recalling that $\dfrac{e}{c}\Delta A = \Delta p = m_e\Delta v$, we see that, taking the atom to be aligned in the center of the cylinder as pictured in Figure 9-3, the speed of the electron increases by:

9.25
$$\Delta v = \frac{eB_0R}{2m_ec}$$

Hence, incorporating the opposing nature of the magnetic field and using equation 9.22, the change in the magnetic moment is:

9.26
$$\Delta\vec{m} = -\frac{e^2B_0R^2}{4m_ec^2}\hat{z}$$

Now, you should be worried about the implicit assumption made above that the radius of the electron's orbit is unchanged by turning on the field. We can easily verify that it does not change by showing that the increase in the centrifugal force that results because of the presence of the $\vec{v}\times\vec{B}$ force is exactly "compensated by" the increased rate of change of linear momentum (see end of chapter problem).

Now, how does this explain diamagnetism? Diamagnetic substances can be modeled as composed of electrons circulating around nuclei. In the absence of an external B-field, there is generally no force that gives preference to one direction of rotation over another. Thus, as with the dielectric case, given the thermal activity, the *average* net dipole moment of the whole with no applied field is zero; $\langle\vec{m}\rangle = 0$. However, once a field is applied, from the above analysis, we see that a dipole is imposed. If, for simplicity and to get a rough approximation, we assume every atom in a given substance contributes a dipole-like contribution, and we ignore the factor that comes about from averaging over the angles, we can get the total diamagnetic force by simply adding the force acting on each of the induced magnetic dipole moments. This force is given by:

9.27 $$F = \Delta m \frac{\partial B}{\partial z}$$

So, the total force is just N, the total number of dipoles, times F.

This gives the fundamental quantitative account. We can qualitatively summarize the diamagnetic effect by noting that the induced dipole will always oppose the direction of the change of the applied field, no matter what direction the electron rotates relative to the field, so that the body is always moved in the direction of decreasing field strength. Thus, when one approaches a diamagnetic substance with a bar magnet, for example, that substance will be repelled, though perhaps imperceptibly.

To connect our analysis to some experimental results, assuming that the induced dipole is proportional to the applied field, we write Δm, in analogy to what we did for the electrical case, in terms of a magnetizability as: $\Delta m = \bar{\alpha}_m B$, (where we use an overbar to indicate that this will not be our formal definition of magnetizability (or the related susceptibility $\bar{\chi}_m = \vec{M} / B$)) and using the equation 9.26 we write:

9.28 $$\bar{\alpha}_m = \frac{e^2 \langle r \rangle^2}{4 m_e c^2} = \frac{e^2 a_0^{\;2}}{4 m_e c^2} \approx 2 \times 10^{-30} \, cm^3$$

Where we take the Bohr radius $a_0 = .5 \times 10^{-8} cm$, $e = 4.8 \times 10^{-10} esu$,

$m_e = 9.11 \times 10^{-28} gm$ and $c = 3 \times 10^{10} cm / s$. The calculation is simplified by noting

that the classical electron radius $r_e = \dfrac{e^2}{m_e c^2} \approx 2.8 \times 10^{-13} cm$.

Comparison of this with typical electric polarizability given in Chapter 8, i.e. in the range 10^{-24} to 10^{-23}, shows how weak this effect is. However, some materials, such as graphite, allow for electron orbits larger than a single atom and thus have much higher diamagnetic magnetizability, but still not approaching the level of the polarizability of dielectrics. Still, the sensitivity of the magnetizability to the average radius, $\bar{\alpha}_m \propto \langle r \rangle^2$, is interesting in its own way. Thus, for example, the diamagnetic magnetizability of graphite at cold temperature is an order of magnitude larger than a typical material. The relative size of the magnetizabilities of some diamagnetic materials can be gleaned from Table 9-1.

Diamagnetism is common, but it is weak. Paramagnetic materials have to overcome innate diamagnetism before they can exhibit their paramagnetic effects. However, the effects of diamagnetism can have visible consequences. Consider the levitating frog in the figure to the right.[2] It is actually being stably suspended by a magnetic field via its diamagnetic nature. Now, water is

---

[2] This was done by scientists at Radboud University in Nijmegen the Netherlands. You can watch the video of the frog at:
scientistshttp://web.archive.org/web/20070211113825/http://www.HFML.RU.nl/pics/Movies/frog.mpg

virtually present in the frog. Indeed, in terms of percentage of the total mass, water is dominant, so that we can, for our purposes, take the frog to be water of the same mass. With this model of the relevant part of the frog's nature and with the diamagnetic value for water given in Table 9-1, you can calculate what magnitude B-field and gradient of B is needed to accomplish this feat (see end of chapter problem).

Magnetized Material

Before we discuss the remaining two phenomena of para- and ferromagnetism, it is helpful to understand the working of the ordinary magnets. To do this, we, in turn, begin with a study of the general nature of magnetized material.

Uniformly Magnetized Material

As with the dielectric case, we start with an ideal slab of uniformly magnetized material that we take to be composed of perfectly aligned ideal magnetic dipoles as illustrated below.

Magnetized Slab

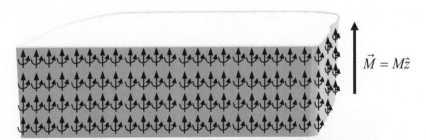

Also as in the dielectric case, this image is only a portion of an arbitrarily long and wide slab of some fixed thickness. The direction of the dipole moments is illustrated by an arrow, while the idea that each of the dipoles arises from a small current loop is illustrated by a directed curve around that arrow. Whereas in the dielectric case we noted a cancellation within a vertical slice of the slab, here we can see one in a horizontal slice.

In particular, take a one-deep horizontal slice of the slab. Then peel off any one row of the dipoles, say the top row pictured above. Note that on the left hand side of any dipole within the row (i.e. not including the edges) the current circulates out of the page, while, at that same spot, the dipole on its left circulates the same current inward. This can easily be seen in the below left figure, which shows three dipoles in a side view. The middle figure shows the top view, and the far right figure shows the dipoles as due to square shaped currents to illustrate the cancellation effect. Notice how the red arrows representing the current from the first dipole cancel with the blue arrows from the second. Since these arrows represent equal and opposite currents in nearly the same place, the net field they cause is effectively canceled out.

Hence, all the internal currents cancel, and one ends with only a band of current, so that the row can be replaced with a band of current as shown here on the right.

This same analysis applies to the sequence of such rows that make up any horizontal slice of the slab. Each row is represented by a band, and for any slice, we can draw the figure shown below on the left, which illustrates three such bands, and how they can be reduced to a single effective band.

Of course, for simplicity, I have drawn the boundaries as straight lines, but the argument is not changed by using any closed curve for the boundary. Furthermore, since an ideal dipole can be approximated by a small loop of *any shape*, our analysis goes through as long as we stay sufficiently far from the atomic scale.

Thus the whole slab can, at least as far as the external field is concerned, be replaced by a band. That is, in direct analogy to the dielectric case, we have the following "empty slab" equivalent source, where the surface current has a (measured) density written as $\mathscr{I}$.

Note well that none of our arguments require our magnetized slab to be thin, so our conclusions also hold, for example, for a long uniformly magnetized cylinder.

Current Band

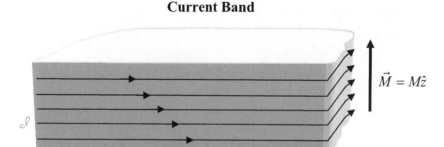

Now, to deal with the effect of magnetic dipoles in a massive body, we introduce $\vec{M}$ shown above, which gives the magnetization per unit volume of the given material and points in the direction of the net magnetic dipole. If we take each one of the dipoles imaged above as having a surface area da and height dz, we would say the strength[3] of their dipole moment is given by:

9.29 $m = M \, da \, dz$

Each cell is like a baby version (of the given dimensions) of the whole slab, and like the whole slab, it has a current running uniformly around its surface. Again, in this picture, our whole slab is composed of rectangular solids of height dz and surface area da, which, as you are aware from calculus, if made small enough can arbitrarily approach filling the volume of any reasonable surface. As implied earlier, the cells that make up the slab must be much larger than the atomic (or molecular) structure responsible for the dipole.

---

[3] Recall our shorthand, whereby we sometimes use the word strength to mean the number that results by comparing the strength of interest with a standard unit of strength.

To convert M to a current density on the surface in charge per unit length, we recall $m = \dfrac{Ida}{c}$, so that we can write using 9.29:

9.30 $\mathscr{I} = Mc$ where, as usual: $\mathscr{I} \equiv I$ per unit length along $\hat{z}$

Hence, the field *outside* (and far enough away from atomic structure) of a slab of magnetization density M can be modeled by a band of current of surface charge density Mc flowing in the appropriate direction on the surface of what would be the slab.

What about inside? Again, an argument similar to, but not the same as, the dielectric case applies. In the dielectric case, we used the fact, that, for the static configurations, $\nabla \times \vec{E} = 0$. Here in the magnetic case, we need the, as far as current science knows, always true relation: $\nabla \cdot \vec{B} = 0$.

In the electric case, $\nabla \times \vec{E} = 0$ along with $\nabla \cdot \vec{E} = 4\pi\rho$ means that we do not have to worry about impetus-activated charge causing an E-field, since *only* charge causes E-field. Said another way, the E-field can be known by specifying only a scalar potential, i.e. the quality specified in the plana that we call the ϕ-field, which is, in turn, determined by the charge distribution. Averaging over volumes sufficiently large compared to the atomic scale allows us to ignore the detailed structure inside of a dielectric and thus have an effective cancellation of the interior charge separation. This leaves $\langle\phi\rangle$ to be determined solely by the effective charge distribution on the top and bottom surfaces of the slab.

For the magnetic case, $\nabla \cdot \vec{B} = 0$, while $\nabla \times \vec{B} = \dfrac{4\pi\vec{J}}{c}$, means that *only* impetus activated charge causes B-field. And, said another way, the B-field is a particular disposition of the plana qualified by the A-field, $\vec{B} = \nabla \times \vec{A}$, which is only determined by the impetus activated charge. As above, averaging over volumes sufficiently large compared to the atomic scale allows us to ignore the detailed structure inside and thus have an effective cancellation of the interior currents. This leaves $\langle A \rangle$ to be determined by the effective surface charge distribution on the sides of the slab. Hence, as in the dielectric case, if interior averages are considered, the field *inside and outside* of a slab of uniformly magnetized material is the same as that of a sheet of current traversing around the boundaries of a surface of the same shape and size as the slab.

To further explore the formalism and its implication, we also look at the question of the B-field inside of a magnetic material using $\nabla \cdot \vec{B} = 0$ more directly. That is, we use the fact that the net B-flux out of any closed surface is zero. First, recall that we showed before the above analysis that the B-field outside of the slab is the same as that outside of sheet of current of the same size as the slab's boundary. Then, consider a cylindrically shaped uniformly magnetized body, i.e. a particular type of magnet, with a Gaussian surface drawn such as shown below in Figure 9-4a. Also, draw, as shown in Figure 9-4b, the same surface on a cylindrical current shell of the same dimensions as the magnetized body.

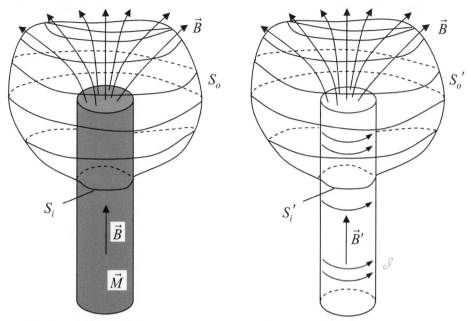

Figure 9-4: a. (*left*) A uniformly magnetized cylindrical body (a magnet) with a Gaussian surface around it **b.** (*right*) A uniform cylindrical current band of the same size as the boundary of the cylindrical body (say, a finely wound single-layer coil) on the left. It also has the exact same Gaussian surface. Since (1) the net flux into such a closed surface made up of the outside surface, S_0, and inside surface, S_i, is zero and (2) the outside field is the same in both case a and case b and (3) the outside B-field along the middle of the long cylinder is negligible, we can prove that the average B-field inside of the solid cylinder is the same as the average B inside of the cylindrical shell on the right, i.e. $\left\langle \vec{B}_{inside} \right\rangle = \left\langle \vec{B}'_{inside} \right\rangle$

Since the flux through the outside part of the Gaussian surface, S_0, is the same for both the magnet and the cylindrical current shell, in order for there to be no net flux in each case, the flux through S_i and S'_i must be the same in both cases. That is, mathematically, we write:

9.31 $\int_{S_i} \vec{B} \cdot dS = \int_{S'_i} \vec{B}' \cdot dS$

This is the heart of the argument that we will now make proving that the average B-field inside of a uniformly magnetized cylinder, i.e. the magnet, is the same as the average field inside of the cylindrical shell of current. Taking the z-axis to be the axis of the cylinder, the average field inside the magnet is:

9.32 $\left\langle \vec{B} \right\rangle = \dfrac{\int_V \vec{B}\, dV}{V} = \dfrac{\sum_i \int_{S_i} \vec{B} \cdot d\vec{S}\, dz_i}{V}$

In the second equality, we use a series of closely spaced, i.e. a distance dz_i apart, parallel surfaces $\{S_1, S_2, S_3...\}$ to define slices of the volume V of height l within the magnet. (Note here that the index "i" now does double duty, standing for both the "inside surface" *and* indicating which of a series of surfaces one is referencing.) Using the same reasoning, we can write the average B-field inside the cylindrical current shell as:

$$9.33 \qquad \left\langle \vec{B}' \right\rangle = \frac{\int_V \vec{B}' \, dV}{V} = \frac{\sum_i \int_{S_i'} \vec{B}' \cdot d\vec{S} \, dz_i}{V}$$

We now use the fact that the field outside the cylindrical shell is negligible for a long cylinder. Hence, when we move the surface S_i, although the surface S_0 changes, it changes in a region of negligible field, so the flux through it does not change. And, since the flux through the new S_0 does not change, the value of the flux through the new S_i cannot change because it must exactly cancel with the latter. Formally, this simply means that each of the surface integral terms given in equation 9.33 must be equal. And, since the exterior field of the magnet is the same as that of the cylindrical shell, each of the surface integral terms in equation 9.32 must likewise be equal. Thus, given equation 9.31, the right hand side of equation 9.32 equals that of 9.33, which tells us that the average field inside of a uniformly magnetized cylinder is the same as the average field inside of a cylindrical current shell, i.e.: $\left\langle \vec{B} \right\rangle = \left\langle \vec{B}' \right\rangle$. Since no larger[4] scale average is necessary in the cylindrical shell, we can drop the average symbol and write: $\left\langle \vec{B} \right\rangle = \vec{B}'$. And, moreover, following the convention adopted in our study of dielectrics, we can just let the physical context signal when averages are being taken, and simply write, for the fields inside and outside of the magnet and the cylindrical shell, $\vec{B} = \vec{B}'$.

Comparison of a Permanent Magnet to a Permanent Dielectric

Figure 9-5a and b show, respectively, the field of a uniformly magnetized rod and the field of a uniformly (electrically) polarized rod. Observe Figure 9-5a closely and you will see that there is a discontinuity in slope of the B-field along the top and bottom surfaces where there is an effective surface charge density. By contrast, Figure 9-5b reveals that there is no such discontinuity on those surfaces, whereas there is along the left and right surfaces where there is an effective bound surface charge distribution.

Note also that the (electric) polarization vector $\vec{P}$ points opposite to the internal field, whereas the magnetization vector $\vec{M}$ points in the same direction as its internal field. This is because when the electric dipoles line up they tend to cancel the effect of any applied surface charge, whereas when the magnetic dipoles line up, being composed of currents, they tend to reinforce any applied surface currents. In short the difference arises from the fundamental fact that the electric field is caused by charge and the magnetic field is caused by impetus activated charge.

---

[4] Remember that even for the case of free plana (i.e. "vacuum") we take a smaller scale average to avoid small scale issues that are beyond the domain (analogical level of abstraction) of classical E&M.

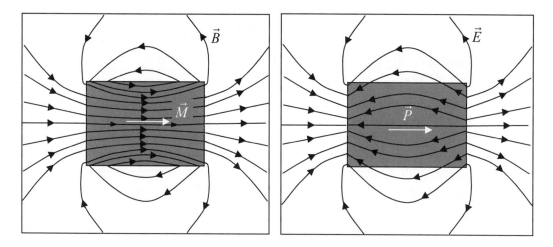

Figure 9-5: a. (*left*) A cross section of the B-field due to magnetized bar of uniform magnetization density $\vec{M}$, where we have shown the average or macroscopic field inside. Note the discontinuity in the direction of B on the top and bottom surfaces due to the effective surface currents there. **b.** (*right*) E-field due to polarized bar of uniform polarization density $\vec{P}$, where we again have shown the average or macroscopic field inside. Note the discontinuity in the slope off E is due to the effective charge distribution on the left and right surfaces. Note that $\vec{P}$, unlike $\vec{M}$, points in the opposite direction to the internal fields, since the "bound" charges, unlike the bound currents tend to act against any applied field.

Non-Uniformly Magnetized Material

Lastly, we consider the case of a material in which the magnetization density, M, varies from place to place. Indeed, the uniformly magnetized material in free plana discussed above is an example of a varying M, if one treats M as describing the whole of space. In that case, there is a sharp change in M at the boundary of the massive body.

Given what we have learned above, we can quickly obtain a relationship between the M and the current distribution that would be needed to cause the same field. Drawing the side and top views of the various dipoles again, but this time allowing the current in each cell to differ from the one before it so that there is a net current in the x-direction, we get Figure 9-6a and b below.

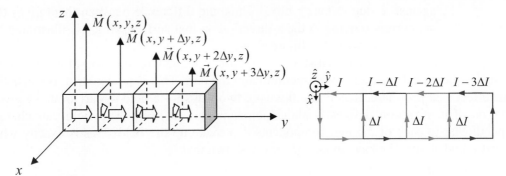

Figure 9-6: a. (*left*) Side view of part of a row of dipoles in a nonuniformly magnetized body. **b.** (*right*) Top view of "a", showing how currents do not completely cancel on the sides but leave a residue of ΔI because of the linear fall off of M that we have chosen in this example.

The current caused in the x-direction is thus, taking that current as going through an area $\Delta y \Delta z$, and using equation 9.30:

$$9.34 \qquad J_x(x,y) = \frac{\Delta I_x(x,y)}{\Delta y \Delta z} = \frac{c\big(M_z(x,y+\Delta y,z) - M_z(x,y,z)\big)\Delta z}{\Delta y \Delta z} \sim c\frac{\partial M_z}{\partial y}$$

Now, this relation is only valid if there is no change in M_y in the z-direction, for this also will produce excess current in the x-direction. An a non-zero M_y means that there is a current circulating in the x-z plane, and picturing a sequence of circulating currents that change as one progresses along the z-axis, we can see how a current deficit or excess can arise along the x-direction.

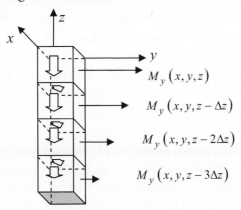

Mathematically, we write:

$$9.35 \qquad J_x(x,y) = \frac{\Delta I_x(x,y)}{\Delta y \Delta z} = \frac{c\big(M_y(x,y,z-\Delta z) - M_y(x,y,z)\big)\Delta y}{\Delta y \Delta z} \sim -c\frac{\partial M_y}{\partial z}$$

Note that, because the direction of $\vec{M}$ defines the axis around which the bound current circulates, the first term in the numerator (i.e. the term proportional to $M_y(x,y,z-\Delta z)$) tells, for the surface between the "$z-\Delta z$ box" and the "z box," how much current is going along $\hat{x}$ due to the "$z-\Delta z$ box," whereas the second term tells how much current is going against $\hat{x}$ due to the "z box." Thus, the difference between them gives the net current moving in the x-direction on that surface, which is illustrated by the small arrow in the figure above.

The top view drawings such as that shown in Figure 9-6b, may remind you of the drawings we made when we discussed Stokes theorem and the curl. This is no accident, because the same effect is at play here. Recall that there are two ways for a field to curl around a given direction. The direction defines a plane perpendicular to itself. The field can change perpendicular to each of the two components that make up the field. This is exactly what we just found above. We have in fact generally shown that:

9.36 $\vec{J} = c\nabla \times \vec{M}$

The *H*-field and Magnetic Permeability (μ)

As we mentioned earlier in this chapter, there is a large class of substances that, like dielectric materials, respond in proportion to the strength of an applied B-field. That is, the dipole moment per unit volume induced is proportional to field applied, $\vec{M} \propto \vec{B}$. If for example, we double the field, we double the magnetization. As we have also noted, we will not write this fact in that simple way. Though a principled analysis of the causes of magnetization would lead us to write the magnetization as $\vec{M} = \bar{\chi}_m \vec{B}$, if one is of a more practical bent, one might also like to bring in the fact that *we almost always control free currents directly, not the B-field*. We now proceed to find a way to incorporate this fact into our formalism, keeping in mind that this is more of a purely practical path than we usually follow.

In the electric case, we saw that we can most easily control the voltage, not the free charge[5], and hence the $\vec{D}$-field, which is a construct related to free charge, was not very useful. For example, in a parallel plate capacitor, our iconic configuration for the electric case, once we set the voltage (i.e. the ϕ-field drop) on the plates, we know what $\vec{E}$ is inside, independent of the dielectric used. This is because the bound and free charge arrange in such a way that the potential drop (and hence the E-field) is what our controlling circuitry is regulating it to be. Hence, $\vec{E}$ itself is often fairly directly accessible. By contrast, we will see that the analog to $\vec{D}$ in magnetism, $\vec{H}$, will be quite useful because of our direct access to currents.

To find an expression for $\vec{H}$, which will be our analog to $\vec{D}$, we must develop part of the extension of Maxwell's 4th equation. In the section on magnetized material, we discussed magnets that have built-in or "bound" currents, currents that we cannot easily directly control because they are closely related to the nature of the materials involved. We just saw that the bound currents can be written as:

9.36' $\vec{J}_{bound} = c\nabla \times \vec{M}$.

Clearly, these currents are real currents, for they are just those currents within the material whose effect is not canceled out by nearby currents. Since they are real, we can write:

$\nabla \times \vec{B} = \dfrac{4\pi \vec{J}_{bound}}{c}$. Hence, adding the free currents, we get:

---

[5] This ultimately derives from a generic aspect of the nature of atoms.

9.37
$$\nabla \times \vec{B} = \frac{4\pi}{c}\left(\vec{J}_{bound} + \vec{J}_{free}\right)$$

Using equation 9.36', we can now *define* $\vec{H}$ by writing:

9.38
$$\boxed{\nabla \times \vec{H} = \frac{4\pi}{c}\vec{J}_{free} \qquad \text{with } \vec{H} \equiv \vec{B} - 4\pi\vec{M}}$$

This definition of $\vec{H}$ now allows us to explain how taking $\vec{M} \propto \vec{H}$ is functionally the same as the principled definition $\vec{M} \propto \vec{B}$. In particular, we define the *magnetic susceptibility*, χ_m, as:

9.39
$$\vec{M} = \chi_m \vec{H} \qquad \text{(standard definition)}$$

Note that χ_m, which is *called the volume magnetic susceptibility*, is dimensionless because H has the same dimensions as B. To see this, note that B has units (according to the dipole field formula, $B \sim m/r^3$) of dipole moment per unit volume, i.e. the units of M.

This means, using the definition of $\vec{H}$ in equation 9.38, we have:

9.40
$$\boxed{\vec{B} \equiv \mu\vec{H} \qquad \text{where } \mu \equiv 1 + 4\pi\chi_m}$$

which provides the definition of *magnetic permeability*, μ.

This, in turn, implies: $\vec{M} = \dfrac{\chi_m}{\mu}\vec{B}$, so that $\bar{\chi}_m = \chi_m / \mu$. This contrasts with the more

straightforward electrical definitions: $\vec{D} = \epsilon\vec{E}$ and $\vec{P} = \chi_e\vec{E}$.

Note again that $\vec{H}$ is not a measure of the direction and strength of any real quality of the plana (or anything else we know of), but only a convenient construct for isolating and keeping track of the effect of the free currents. We start with the admittedly rather convoluted proportionality of equation 9.39, which few, if any, theoretical physicists would now choose, but which physicists do, in fact, use because such a starting equation is a matter of longstanding usage and because $\vec{H}$ does have practical advantages.

To understand how to use $\vec{H}$ more generally, we turn to the iconic physical configuration for magnetics, the cylindrical current sheet with a current per unit length ϑ, such as shown in Figure 9-4b. It is the iconic case because it is the simplest closed current configuration[6] that can generate a uniform B-field in a region, just as the parallel plate capacitor is the simplest voltage driven configuration that can generate a uniform E-field.

We draw a rectangular loop perpendicular to the cylindrical surface as shown. We then use the line integral form of equation 9.38, which is obtained by integrating that equation over the surface area of the loop and applying Stokes' theorem. The line integral form, which applies to any loop encircling a current I, is:

9.41
$$\oint \vec{H} \cdot d\vec{l} = \frac{4\pi I}{c}$$

---

[6] For a closed current, i.e. one for which $\nabla \cdot \vec{J} = 0$, we must mix at least two dimensions to complete the circuit. It is simpler to leave out the third. Now, the simplest curve that mixes two dimensions is a circle, since the "amount of mixing" near each point is the same.

If we let the perpendicular dimension of the loop become arbitrarily small in standard fashion, we then get, recalling that the outside field is arbitrarily close to zero taking the z-axis as the axis of the cylinder: $\vec{H}_{inside} = \dfrac{4\pi \mathcal{I}}{c}\hat{z}$. Note that the calculational *procedure* is just the same as it was for $\vec{B}$ when we knew nothing about magnetic materials. And, as it must, the result reduces to the vacuum case when we eliminate the magnetic material by setting $\mu = 1$.

> *Note on units nomenclature*: the *B*-field is measured in gauss in *cgs*, and because μ is dimensionless, we take *H* also to be measured in gauss. However, some give the name *orested*, which are again dimensionally the same as a gauss, to the units of the *H*-field. If one chooses to accept the bookkeeping complication this introduces, one can think of the unit *orested* as a way of distinguishing the *H*-field, which is just a construct, from the *B*-field.

Materials have a wide range of magnetic permeabilities as can be deduced from Table 9-1 below, which gives the volume susceptibilities (and some μ values) for various materials. Note there are three main ways of giving magnetic susceptibility: 1) volume susceptibility, 2) molar susceptibility and 3) mass susceptibility, also called specific susceptibility; various references chose different definitions.[7]

Material[8]	Temp (C)/ Pressure (*atm*)	Volume Susceptibility, $\chi_m \cdot$ (cgs)
Diamagnetic		
Water[9]	20/1	-7.190×10^{-7}
Bismuth[10]	20/1	-1.32×10^{-5}
Diamond[11]	~20/1	-1.7×10^{-6}
Graphite[12] $\chi_{\parallel}$	~20/1	-4.9×10^{-5}
Graphite[12] $\chi_{\parallel}$	-173/1	-6.6×10^{-5}
He[13]	20/1	-7.84×10^{-11}
Xe[13]	20/1	-1.89×10^{-9}
N$_2$[13]	20/0.781	-4.03×10^{-10}
Paramagnetic		
Al		1.75×10^{-6}

---

[7] To convert from various references, it is helpful to know that: $\chi_m(volume) = \chi_m(molar)\rho/M$, $\chi_m(molar) = \chi_m(specific)\cdot M$ and $\chi_m(volume) = \rho\chi_m(specific)$, where ρ is the mass density and M is the mass per mole (all in *cgs* units). Note also to convert to *SI* or what we sometimes call *mks*, one has: $\chi_m^{SI}(volume) = 4\pi\chi_m^{cgs}(volume)$

[8] http://en.wikipedia.org/wiki/Magnetic_susceptibility.

[9] G. P. Arrighini, M. Maestro, and R. Moccia (1968). "Magnetic Properties of Polyatomic Molecules: Magnetic Susceptibility of H$_2$O, NH$_3$, CH$_4$, H$_2$O$_2$". *J. Chem. Phys.* **49**: 882–889.

[10] S. Otake, M. Momiuchi and N. Matsuno (1980). "Temperature Dependence of the Magnetic Susceptibility of Bismuth". *J. Phys. Soc. Jap.* **49** (5): 1824–1828.

[11] J. Heremans, C. H. Olk and D. T. Morelli (1994). "Magnetic Susceptibility of Carbon Structures". *Phys. Rev. B* **49** (21): 15122–15125.

[12] N. Ganguli and K.S. Krishnan (1941). "The Magnetic and Other Properties of the Free Electrons in Graphite". *Proc. R. Soc. London* **177** (969): 168–182.

[13] R. E. Glick (1961). "On the Diamagnetic Susceptibility of Gases". *J. Phys. Chem.* **65** (9): 1552–1555.

Manganese[*]	18	7.36×10^{-5}
O_2[13]	20/0.209	2.97×10^{-8}
O_2[*]	-196	3×10^{-4}
Ferromagnetic		
Nickel (99% pure)[14]		~50 (μ=600)
Ferrite M33[14]		~60 (μ=750)
Iron (99.8% pure)[14]		~400 (μ=5000)
Mu-metal[15] 75% Ni, 15% Fe, plus Cu and Mo		1.6-4,000 (μ=20-50,000)
* Values marked with * are taken from 37th edition of the "Handbook of Chemistry and Physics"		

Table 9-1: Volume Magnetic Susceptibility, χ_m, of various substances.

The negative indicates that the induced magnetization is in the opposite direction to the applied field. The direction of the response is responsible for a magnet attracting para and ferro-magnetic substances while repelling diamagnetic ones.

The above tabulated values give a handle on the degree to which various substances can augment or diminish an applied B-field. As indicated above and previously described, there are three major causes of this B-field induced effect: dia-, para- and ferro-magnetism. We summarize, in a rough way, the relative sizes of these effects, as well as the linear relation between H and B, and hence M, by the graph given in **Figure 9-7**. The graph shows the B-field that results, for example, inside of a cylinder of magnetic material of permeability μ when a uniform sheet of current is applied to its surface. Since we control free currents, which are directly related to H, and those free currents are ultimately responsible for the B-field (being responsible for both the applied field and the resulting induced magnetization), we talk of B resulting from H and put H on the independent axis, the x-axis.

Note that to make understanding the plot conceptually simple, we consider a long cylinder or a toroid of the given material (which means there is no divergence of M nearby) that is tightly wrapped with a current carrying wire, so that the H-field represents the B-field that *would be* present if there were no magnetic material. We also do not consider current reversals, which, as we will see when we study ferromagnetism, make the plot more complex.

Figure 9-7: Very rough graph of B versus H, showing the magnetic permeability for ferro-, para- and dia-magnetism (μ_f, μ_p, μ_d), the various general types of induced magnetic effects.

Clearly, the scale is *not* consistent in order to show the various effects which are so different in magnitude. One can roughly think of the two parts of the above graph as views of a graph with a consistent scale as seen with two different magnifying lenses, one for the ferromagnetic line and the other for the para and diamagnetic lines near the $\mu = 1$ line. As we will see, the ferromagnetic curve

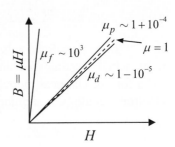

actually bends and indeed depends on the past history of the application of the B-field to the material.

---

[14] http://info.ee.surrey.ac.uk/Workshop/advice/coils/mu/

[15] http://en.wikipedia.org/wiki/Permeability_%28electromagnetism%29; ferrites are a ceramic with iron(III) oxide Fe_2O_3 as their principal components

With the *H* construct in place, and with an overview of the magnitudes of the three magnetic effects, we can now move to study paramagnetism and ferromagnetism. We have already seen how diamagnetism arises by inducing a change in the rotation of electrons in the atoms of the given material. Para and ferromagnetism arise from something more intrinsic to the electron, a property that was not even remotely understood until the advent of quantum mechanics and still today remains somewhat of a mystery, though much empiriometric illumination has been given by the great advances in quantum field theory in the last century.

Paramagnetism

Spin

This new property is called spin. To get a sense of how strange it is, note that quantum mechanics successfully predicts that if one rotates *the state of an electron* through 360 degrees it will not return to the same state, as for example you *would* if you rotated through that same angle. However, this mystery belongs to quantum mechanics, and we will leave it there. We are here interested in the electron's spin and magnetic moment. It has angular momentum in some analogical sense and thus, in some analogical sense, we can say it is rotating; note carefully the qualifications here. It also has a magnetic field and is like a tiny ideal dipole in some ways.

We will treat it as if it were simply, though we know it is not, a tiny rotating ball with uniformly distributed negative charge throughout. Like the orbiting electrons we studied in the diamagnetic case, this impetus activated charge creates a field that looks like an ideal dipole at distances far from it. Also like it, the dipole moment vector points in the opposite direction to the angular momentum vector.

Because quantum mechanics has shown us that the electron cannot simply be a rotating charge, we introduce further formal distinctions to incorporate some of these differences. We define, for example, the gyromagnetic ratio:

9.42 $$\gamma = \frac{|\vec{m}|}{\vec{L}}$$

Classically, for a uniform ring of radius *a*, and mass, *m*, and charge *e*, rotating at angular speed ω, we get:

9.43 $$\gamma = \frac{m}{L} = \frac{\frac{IA}{c}}{ma^2\omega} = \frac{e\frac{\omega}{2\pi}\pi a^2}{ma^2\omega} = \frac{e}{2m}$$

Since a uniform sphere of rotating charge is simply a sum of such rings of various radii, which becomes an integral in the limit of arbitrarily small rings, this ratio is also valid for our classical model of the electron. The actual quantum mechanical value for the electron's analogical angular momentum (spin) and moment is:

9.44 $$\gamma_e = g_s\gamma_{classical} \quad \text{where } g_s \sim 2\left(1 + \frac{\alpha}{2\pi} + ...\right) \text{ is called the } \textit{electron g-factor.}$$

Dirac's relativistic quantum mechanics predicts the factor of 2, rather than the classical value of unity, while the corrections to that factor of two are derived from quantum field theory. The correction terms are known analytically to third order in

$\alpha = \dfrac{e^2}{\hbar c}$,   which   is   called   the   fine   structure   constant.   Experimentally,

$g_e \sim 2.0023193043617(15)$. Note that, by contrast to the case of spin, the g-factor for an orbiting electron is the same as the classical case, i.e., $g = 1$.

These distinctions further bring out the unique nature of electron spin, but, what does spin have to do with paramagnetism? You probably already see the answer. Though electrons have certain important differences with classical dipoles, such as in relationship between its spin and the magnitude of its moment, they will experience a torque under the action of an applied field that will tend to align them with the field. This, in turn, means that if the applied field has a gradient, there will be a net force on the body.

Some Microphysics

Let's slow down and look at paramagnetism a little more carefully at the electron level.

First, *does every electron contribute*? No. Unlike with diamagnetism, most electrons do not contribute! Quantum mechanics indicates that the preferred state of the electrons in the lower energy states, those that tend to hang out closest to the nucleus (the inner shells), is in pairs. Electrons like to pair up. They pair up in such a way that their dipole moments anti-align so that they are effectively like a single particle with no dipole moment. Only those that do not pair can contribute to a paramagnetic effect. In turns out that this only happens in rare cases. Oxygen (atomic number 8), for example, is paramagnetic even though it has an even number of electrons because the two outer electrons do not pair up. They do not pair up because of facts peculiar to the nature of oxygen that have to do with the way its quantum mechanical orbitals are filled.

Second, *do those electrons that do participate in paramagnetism line up perfectly*, like soldiers marching along a road? No. Like the dielectric case, the molecules are jostled by thermal forces and only align "statistically." That is, at any one moment a given electron might align to some degree with the applied field, but a moment later it might anti-align. What matters is that the electrons that participate spend more time near alignment than near anti-alignment. The result is that the net dipole moment, when averaged over a time longer than the jostling, increases in proportion to the strength of the applied B-field. This is the core of the paramagnetic effect. Now, because, the electrons do align, the B-field activated plana acts to pull the material toward a magnet. This can be seen from equation 9.14. Taking the magnet to be at the origin and taking B and m along the z-axis, the gradient of the magnet's field is negative indicating a negative force, a force back toward the magnet.

An important aspect common to the dia- and para-magnetism is that the field must dispose the material before it can cause impetus in it.[16] If, for example, in the case of

---

[16] William Gilbert (1544–1603 AD) the great codifier of medieval thought on electricity and magnetism, especially of Peter Peregrinus's pivotal work, who himself had some significant insights and was a great advocate of experiment, pointed out the importance of this generic aspect of magnetic phenomena. In commenting on St. Thomas Aquinas's explaining this point, Gilbert wrote: "Thomas Aquinas, writing briefly on the loadstone in Chapter (sic) VII of his *Physica*, touches not amiss on its nature, and with his divine and clear intellect would have published much more, had he been conversant with magnetick (sic) experiments." William Gilbert, *On the Magnet*, Chiswick Press, London (1900AD), pg. 3, translated from the original Latin. Reading the relevant part of Chapter 7 of Aquinas's commentary on Aristotle's *Physics* reveals that the results of magnetic experiments were only partially known by St. Thomas and that some had

paramagnetism, the orientation of the electron's magnetic moment was not changed by the field, then there could be no net effect on the whole body, because each atom would be in some random position and as likely to repel as attract. Ferromagnetism shares this aspect as well, but in other ways is fundamentally different than even paramagnetism.

Ferromagnetism

Ferromagnetism, as a quick glance at Table 9-1 reveals, is a much stronger effect than the other two. It, for example, can arise when whole groups of electrons spontaneously align. In fact, three effects are generally at play: (1) the tendency of electrons to align in groups, (2) the preferred axes of alignment, and (3) domain building by application of a field.

The alignment-in-groups effect is still another important quantum mechanical phenomenon. In ferromagnetic substances, the atomic structure is such that outer shell electrons tend to want to align their dipole moments. From a classical point of view in which we leave out of consideration such effects, two aligned electrons should be higher energy and want to anti-align, as can be seen if you throw two bar magnets on a table next to each other.

We can understand this funny alignment behavior semi-classically if we recall that the electrons also have a negative charge and thus, from this perspective, rather stay away from each other as much as possible. It turns out that electrons follow a rule called the Pauli exclusion principle which requires, among other things, electrons to have opposite spin if they are otherwise in the same state. Thus, two electrons that have the same spin cannot have the same spatial state, which means they are on the average more separated from each other than if they were in the opposing spin states and thus able to be in the same spatial state. Hence, the aligned state keeps the electrons further apart, thus satisfying the charges' mutual desire to push away from each other. Now, the magnetic interaction between electrons is much weaker than the electrostatic effect; so it is not, of itself, important at this level, but the spin, which is deeply related to that magnetic moment, is dominant in the sense just described. Hence, the result is that the outer shell electrons tend to align. Why then don't *all* the electrons that are able to align just line up and spontaneously make every piece of ferromagnetic material a permanent magnet?

This brings us to the *second* effect. For concreteness, consider iron, which has a body-centered-cubic crystal structure, one cell of which is shown here. Consider this cell to be in the middle of a large chunk of single crystalline iron. Electrons in the corner of the cube in such a structure, for example, will find it easier to align along the directions defined by the edges of the cube.

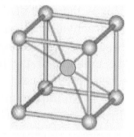

Hence, electrons that happen to be aligned with one axis will not easily be pushed into another alignment with another axis while those aligned in-between the axes will tend more easily to fall into one of the six preferred directions. Electrons find it hard to all line up once they are in a certain orientation. However, the tendency towards mutual alignment

been incorrectly passed along to him, which, as implied by Gilbert obviously hindered the value of his analysis.

just discussed can operate more successfully when the spin structure is first set. For example, we could start with molten iron and let it cool, or we could simply start with an iron crystal that is above its Curie point. The Curie point is the temperature above which the thermal jostling is so great that no macroscopic alignment or magnetic ordering as it is called, occurs. Beyond the Curie point, the material (iron) is paramagnetic. As our sample cools at some point, its temperature approaches the Curie point. As the temperature approaches very near the Curie point, the electrons tend to align more and more. A spin alignment of one orientation or another begins to dominate in various portions of the crystal, creating regions of the crystal called spin domains. The figure here illustrates such domains in a schematic of the cross section of a crystal with arrows to indicate which direction all the many, many electrons in the given domain point. Note the random directions of the domains.

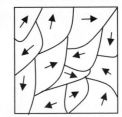

What determines that a given spin direction, not another, dominates in a given region? It is more or less determined by the direction that the spins in the given region happen to have at the time at which the forces become strong enough to set the alignment. This fundamentally means that the alignment is set by the initial conditions at the time of the creation of the sample and external conditions at the moment or in the past, not by an innate property of the iron itself, such as the tendency to spontaneously align. In this way, some parts of the crystal form in this direction others in that, until the crystal is filled with different spin domains. Note well that spin domains form even in single crystals such as we are now considering. Polycrystalline materials clearly will have their own natural boundaries that will also come into play.

Now, such a material filled with randomly oriented spin domains has no net magnetization. Note that if all of the spins did spontaneously line up making one large magnet, this would require sucking a large amount of energy from the environment, something that is highly unlikely. In other words, on a large enough scale, the classical tendency of the dipoles to anti-align takes over.

The *third* and last piece needed to understand the ferromagnetic effect is that these domains can grow and diminish according to the application of an external field. As the applied field is increased, those domains aligned with it are joined by neighbors in domains that are in an alignment least resistant to joining the field alignment. The wall, which is many atoms wide, doesn't so much come down as move gradually, taking over more and more of the given domain. In this way, there is a linear increase in the total magnetization of the material.

Of course, this implies that we reach a limit at which the whole iron is one domain, and this is indeed the case. This is called the saturation point. Beyond this point, the magnetic material, in this case the iron chunk, cannot further reinforce the field caused by the free currents, and the only increase in field intensity is that caused by the free currents directly.

It is informative to explore this a little further. Consider the following setup that will allow us to generate a plot of B versus H.

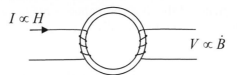

We drive the right hand side of the iron-core coil with a current source of gradually increasing current $I(t)$. Knowing this free current means that we, in principle, know H and thus have the x-component of our desired plot. Furthermore, the voltage induced in the secondary is proportional to the rate of change of the B-field inside the core, so that if we start from zero current and thus $B = 0$ [17] and integrate over time from that moment, we can determine B itself. In this way, we can get the data to make a plot that would look roughly like that shown in Figure 9-8.

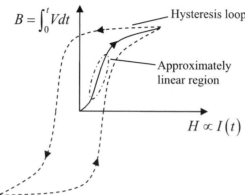

Figure 9-8: Plot of B versus H for the donut-shaped coil configuration above. As more domains align with each other along the applied field, the net B-field is increased. At some point, the applied field has finished aligning the easy domains and must begin to align the spins that are in harder-to-move orientations, and this is responsible for the steepening slope near the top of the solid curve. Finally, one begins to exhaust the number of domains left to align, and the curve approaches saturation. If we continue to cycle the current, we will generate the full hysteresis loop shown.

The solid curve in the figure shows how the magnetic field increases sharply and linearly in a certain region as domains line up as the applied current is increased. During this rising part of the curve, the applied field is reinforced by the aligning of the magnetic domains in the way described in the figure caption until it reaches the saturation point. The dashed continuation of the curve in the figure shows what happens if, after reaching near saturation, we cease increasing the current and begin to turn it down. The curve does not retrace the path it made coming upward. This so called "hysteresis" arises because the newly spin-aligned chunk of iron needs something to force those spins out of their present alignment. This only begins to happen when the current goes negative, i.e. *reverses* direction. Notice that the current does not have to go as far negative as it went positive to accomplish the alignment, but, nonetheless, a significant current is required to reestablish a non-magnetized bar.

Before leaving this topic, we point out that the word "ferromagnetic" is used in two different senses that you may encounter. The broad sense is, as we have used it and will continue to use it here, any material that has spontaneous magnetic ordering that is affected by applied magnetic fields. However, technically it is sometimes also used to distinguish

---

[17] This, of course, assumes that the core is not initially permanently magnetized.

ferromagnetism from ferrimagnetism, which involves local moments that have a tendency to anti-align.

Magnetism and its Connection back to Ordinary Life

We started our discussion of magnetism in Chapter 1 and Chapter 3 with ordinary direct experience with magnetism. This, as you know well by now, is the necessary starting point for all of our knowledge, for we start with what we see, hear, touch, smell or taste. We often left this beginning point as we investigated more deeply the nature of magnetic causes and their effects. It is thus natural to return to explain what, after all, we started out trying to explain!

What causes magnets to attract one another? Why is iron attracted to a magnet? How can magnets act through certain materials? How can one magnet suspend a paper clip that then can, in turn, suspend another paper clip in a series such as shown in the picture in Chapter 3? Why are there permanent magnets?

We start with the last, because we just saw the answer in the previous section. Ferromagnetic materials spontaneously magnetize and if they are in the presence of a magnetic field when they cool below their Curie temperature, we get a natural magnet, such a lodestone.

Magnets attract (or repel) each other because they have parts that have impetus-activated charge, or spin that is in some way like impetus-activated charge, that cause *B*-fields that can act one on the other. Iron is attracted by a magnet because it has domains of magnetic dipoles (the electrons' spin magnetic moments) that can align with the field of the magnet and thus be attracted like ideal dipoles are in a field with a gradient.

Since the effects of diamagnetism (which affects about everything) and paramagnetism are ordinarily too weak to be seen, *B*-fields are only affected by ferromagnetic material in a noticeable way. For example, bringing a largely diamagnetic material like your hand into a region with an already existing ordinary strength *B*-field, say of a typical permanent magnet, will not affect the state of that field enough for you to notice any differences. As we have seen, since your hand is diamagnetic, it will indeed have the effect of reducing the *B*-field near it. When your hand moves into a region of *B*-field, that field causes a change in the orbiting of the electrons of your hand which, in turn, causes a contrary tendency that works against the ambient *B*-field. However, it is a very small affect. Hence, *B*-fields more or less "go through" such materials, and this explains how your magnet can operate through a thin wood (a diamagnetic material) desk.

The series of paper clips suspended from a single magnet is explained by ferromagnetism. Since a paper clip is ferromagnetic, it reinforces the magnet's *B*-field along its length, so that the magnet is able, acting through the clip, to hold the clip touching it. That clip, in turn, "channels' the field to the next until the field is no longer strong enough to support further paper clips.

Now, we are ready to give the formal (empiriometric) summary of our dielectric and magnet study that incorporates our understanding into a revised Maxwell formalism.

Maxwell's Equations Extended to Include Magnetics and Dielectrics

We already have the first three of Maxwell's equations. The first comes from the last chapter:

9.45 $\nabla \cdot \vec{D} = 4\pi \rho_{free}$, where $\vec{D} = \vec{E} + 4\pi \vec{P}$.

The second and third are unchanged:

9.46a,b $\nabla \cdot \vec{B} = 0$, $\nabla \times \vec{E} = -\dfrac{1}{c}\dfrac{\partial \vec{B}}{\partial t}$.

The last equation is not finished yet. In the previous chapter we developed the following form:

9.47 $\nabla \times \vec{B} = \dfrac{4\pi}{c}\left(\vec{J}_{free} + \vec{J}_{non-dielectric\,bound}\right) + \dfrac{1}{c}\dfrac{\partial \vec{D}}{\partial t}$

Where we break out the current density used in that chapter really into a sum, i.e., the actually free currents plus any other currents besides the dielectric bound currents. Since there are no other bound currents besides the ones associated with magnetism, we can write: $\vec{J}_{non-dielectric\,bound} = \vec{J}_{bound\,of\,magnetized\,material}$.

Now, using the definition of H given in equation 9.38, $\vec{B} = \vec{H} + 4\pi \vec{M}$, we get:

$$\nabla \times \left(\vec{H} + 4\pi\vec{M}\right) = \nabla \times \vec{H} + 4\pi\nabla \times \vec{M} = \dfrac{4\pi}{c}\left(\vec{J}_{free} + \vec{J}_{bound\,of\,magnetized\,material}\right) + \dfrac{1}{c}\dfrac{\partial \vec{D}}{\partial t}$$

and using the relationship between $\vec{M}$ and $\vec{J}_{bound\,of\,magnetized\,material}$ given in equation 9.36',

$\vec{J}_{bound} = c\nabla \times \vec{M}$, we get the new form of the 4th Maxwell equation:

9.48 $$\nabla \times \vec{H} = \dfrac{4\pi}{c}\vec{J}_{free} + \dfrac{1}{c}\dfrac{\partial \vec{D}}{\partial t}$$

Thus, we have an extended set of equations for use in a magnetic and dielectric media, which are listed together in the "Summary" section.

Upon inspection of these equations, you might wonder what else we can say about H; for example, what is its divergence? To answer this, consider again the case of the uniformly magnetized bar discussed in Figure 9-5a. Yes, H is even useful when there are no free currents.

If we were to calculate the H-field for our uniformly magnetized material, using the definition in equation 9.38 we would see that the H-field has exactly the same form as the E-field of our permanent dielectric. This points to a deep analogy between E and H. We can see this analogy clearly in the static, no free current, no free charge case in which Maxwell's vacuum and media equations are:

9.49a-d $\nabla \cdot \vec{E} = 4\pi\rho_b$, $\nabla \cdot \vec{H} = 4\pi\bar{\rho}_b$

$\nabla \times \vec{E} = 0$ $\nabla \times \vec{H} = 0$

Note the exact symmetry between E and H. The key is in how this arises. The first equation follows from $\nabla \cdot \vec{E} = 4\pi(\rho_{bound} + \rho_{free})$, and the third and fourth follow even more trivially from equations 9.46b and equation 9.48 respectively. The second equation is the interesting one.

The non-zero divergence of H arises from the divergence of M; namely, $\nabla \cdot \vec{H} = \nabla \cdot \left(\vec{B} - 4\pi \vec{M} \right) = -4\pi \nabla \cdot \vec{M}$, so that we *define* $\bar{\rho}_b \equiv -\nabla \cdot \vec{M}$. More intuitively, one can see from looking at a simple bar magnet, such as shown in Figure 9-5, that there is a discontinuity in the magnetization density, i.e. the M-field, at the right and left boundaries of the magnet; therefore, there is a divergence in M. We can thus, solely for purposes of calculation, interpret this as a layer of magnetic "charge," and thus make an identification between the treatment of E and H. That is, we have bound magnetic charge at the boundary in the Figure 9-5a and bound electric charge at the boundary in Figure 9-5b.

Some interesting constraints on the boundary conditions of the various fields follow from the full media Maxwell's equations (see end of chapter problem). Some of these conditions are evident at the boundaries of the uniformly polarized and uniformly magnetized bars shown in Figure 9-5.

Speed of Waves in Dielectric/Magnetic Media

We have yet to say anything about electromagnetic radiation in a media. This is a vast topic, but we do want to introduce it very briefly. With the four media Maxwell equations in hand, we can write down the equation and the plane wave solution for electromagnetic waves in a linear media, i.e., one in which: $\vec{B} = \mu \vec{H}, \vec{D} = \epsilon \vec{E}$.

To begin, we set all free currents to zero and using the linear media relations to substitute for H and B equations 9.48 and 9.46b become:

9.50a,b $$\nabla \times \vec{B} = \frac{\mu \epsilon}{c} \frac{\partial \vec{E}}{\partial t}, \quad \nabla \times \vec{E} = -\frac{1}{c} \frac{\partial \vec{B}}{\partial t}$$

Next, we cross the first equation on both sides gives with ∇, and use the identity $\nabla \times \left(\nabla \times \vec{B} \right) = \vec{\nabla} \left(\nabla \cdot \vec{B} \right) - \nabla^2 B$, (which again you are not expected to be able to memorize or prove). We then apply $\nabla \cdot \vec{B} = 0$ and substitute for $\nabla \times \vec{E}$ using 9.50b to get the sought after "wave equation:"

9.51 $$\nabla^2 \vec{B} = \frac{\mu \epsilon}{c^2} \frac{\partial^2 \vec{B}}{\partial t^2}$$

A similar line of reasoning gives the same equation for $\vec{E}$.

A standard solution to this equation is the plane wave traveling in the positive z-direction given by:

9.52 $$E_x = E_0 \sin \left(kx - \omega t \right), \quad B_y = B_0 \sin \left(kx - \omega t \right) \quad \text{(all other components are zero)}$$

where the speed of the wave is given by $c' = \dfrac{\omega}{k} = \dfrac{c}{\sqrt{\mu \varepsilon}}$.

One can easily verify that this is a solution by plugging into equation 9.50 and its E-field pair. You can check directly that equations 9.52 do indeed solve all of Maxwell's media equations.

Notice that the effective speed of travel is down by a factor of $\dfrac{1}{\sqrt{\mu \varepsilon}}$, since both ϵ and μ are greater than unity. What causes the difference in speed?

To answer, keep in mind that in both the dielectric case and the magnetic case, we have prescinded from two different types of "E-field"; we've treated the E-field as one agent. In fact, as we have seen, our empiriometric definition of E consists of two agents; one associated with the "$\dot{A}$ effect," i.e. a magnetic effect, and the other with ϕ. Also, in our analysis of the response of magnetic materials to external current sources, we have not distinguished between the $\dot{A}$ effect and the proper B-field effects. Lastly, we have in equation 9.48 treated the real charge involved in dielectric and magnetic effects as part of "space." The last is the key to sorting out how the slower speed arises.

In fact, we have not abrogated our fundamental Maxwell equations but only extended them. Hence, it is clear, classically, that the A and ϕ-fields propagate through both free and virtual plana traveling at speed c. However, the effective speed is different in a media because there are sources all along the route that are activated altering the original fields as they propagate. Thus, the effective speed, that is the distance within the material that the field can reach and act on something in a unit of time, is now reduced by a factor called the index of refraction, $n = \sqrt{\epsilon\mu}$.

In *SI* units, μ and ϵ for free plana have dimensions and are called, respectively, the permeability and permittivity of free space. However, keep in mind that these are not measures of strengths of new properties of the plana, but only reflect the way we have chosen to measure the properties already discussed. This is most evident in our formalism through the fact that $\mu = 1$ and $\epsilon = 1$ for free plana (in the *cgs* system of units).

Summary

There are three different types of magnetic behavior:
(1) **Ferromagnetic**, such as a magnet's ability to attract iron relatively strongly
(2) **Paramagnetism**, weak ability of a magnet to attract some materials
(3) **Diamagnetism**, weak ability of a magnet to repel nearly all materials

Similar to the way that for dielectrics we expand the solution for ϕ in terms of charge, for magnetics, we expand the solution for A in terms of current to get the following *multipole moment expansion* for a viewpoint on the z-axis as:

$$\vec{A} = \frac{1}{c}\int \frac{\vec{J}(\vec{r}')}{|\vec{r}-\vec{r}'|}d^3r' = \frac{1}{cr}\int \vec{J}(\vec{r}')d^3r' + \frac{1}{cr^2}\int (\hat{r}\cdot\vec{r}')\vec{J}(\vec{r}')d^3r' + O(r^{-3})$$

It can be shown that the first integral is zero for any current distribution confined to a finite region that is conserved (so that $\nabla\cdot\vec{J}=0$) and that the second term, called the *magnetic dipole* term can be written, in general, as:

9.3 $\vec{A}(\vec{r}) = \dfrac{\vec{m}\times\hat{r}}{r^2}$, where $\vec{m} = \dfrac{1}{2c}\int \vec{r}'\times\vec{J}(\vec{r}')d^3r'$.

At an arbitrarily large distance, the magnetic dipole moment $\vec{m}$ of an arbitrarily shaped (but "smooth") planar loop of current is given by $\vec{m} = \dfrac{IA}{c}\hat{z}$, where A is the area of the loop and $\hat{z}$ is the normal to the plane of the loop. This can be shown in a straightforward manner for a circular loop. We define the *ideal magnetic dipole moment* as the limit in which the area of the loop, $\vec{A}$, gets arbitrarily small while $\vec{m}$ remains constant thus requiring that the current I gets infinitely large. In our mathematical formalism, we write: $\vec{m} \equiv \dfrac{1}{c}\lim_{A\to 0, I\to\infty} I\vec{A}$.

The *B-field of an ideal magnetic dipole is*: $\vec{B} = \dfrac{m\left(3(\hat{m}\cdot\hat{r})\hat{r}-\hat{m}\right)}{r^3} = \dfrac{m\left(3\cos\theta\,\hat{r}-\hat{z}\right)}{r^3}$. The latter case applies when $\vec{m}=m\hat{z}$. The B-field of the magnetic dipole is the same as that for the E-field of an electric dipole with $\vec{p}$ replaced by $\vec{m}$.

Force on a dipole
A B-field with a gradient can cause a net *force on a magnetic dipole* (such as shown here) the magnitude and direction of which can be calculated by integrating over the "$\vec{v}\times\vec{B}$" forces on each differential element of the loop. For the radial component, B_ρ, in cylindrical coordinates, we get (where β is the angle between $\vec{v}$ and $\vec{B}$ and $\sin\beta = \rho\,d\phi/dl$):

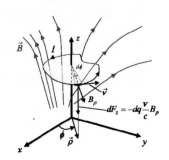

$$\left|F_z\right| = \int_0^{2\pi} \lambda \frac{v}{c} B_\rho\left(\rho,\phi,z\right)\sin\beta \frac{\rho(\phi)d\phi}{\sin\beta} = \frac{I}{c}\int_0^{2\pi} B_\rho\left(z,\rho,\phi\right)\rho(\phi)d\phi$$

Because of B-flux conservation ($\nabla\cdot\vec{B}=0$), the net flux of B_ρ into the sides must be balanced by the net flux exiting out the top and bottom, i.e., along the z-direction. In the magnetic dipole limit of an arbitrarily small loop, this can be expressed as: $\int_0^{2\pi} B_\rho\left(z,\rho,\phi\right)\rho(\phi)d\phi = -\frac{\partial B_z}{\partial z}A_{loop}$, so that we can write the net force along the z-direction as: $F_z = m\frac{\partial B_z}{\partial z}$, where $m = \frac{IA}{c}$. This can be generalized to:[18]

9.15 $$\boxed{F_i = \nabla_i\left(\vec{m}\cdot\vec{B}\right)}$$ where $m_i = \frac{IA_i}{c}$ and $i\in\{x,y,z\}$ (works for $\nabla\times\vec{B}\neq 0$)

For the special case in which $\nabla\times\vec{B}=0$, we have:

9.16 $$\boxed{F_i = \left(\vec{m}\cdot\vec{\nabla}\right)B_i}$$

Torque on a dipole
As shown in the figures to the right, the *torque* around the x-axis on a rectangular loop of current (brown) arises from differential forces (green) that result from the action of B_y (blue in the left figure) *and* from B_z (blue in the right figure). This leads to the general formula for the torque:

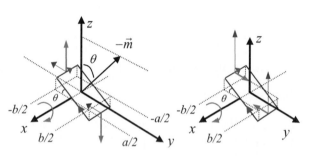

9.20 $$\boxed{\vec{\tau} = \vec{m}\times\vec{B}}$$ The torque produced by the B-field and described by this equation is responsible for both paramagnetism and ferromagnetism.

We can understand the *essential cause of diamagnetic behavior* by assuming a given electron follows a circular orbit around the nucleus. If we average over long enough times, we can think of the electron path as a current of magnitude: $I = e/T = ev/2\pi R$ and magnetic moment $\vec{m} = \frac{IA}{c}\hat{z} = \frac{evR}{2c}\hat{z}$. When a rotational A-field, $\vec{A} = A_\phi\hat{\phi}$, that corresponds to a uniform B (such as that inside a cylindrical sheet of current) is switched on, the virtual plana of the atom resists the change in A-field by attempting to activate further impetus in the electron, resulting in an increased speed but unchanged radius, that results in an increase in the magnetic moment corresponding to $\Delta\vec{m} = -\frac{e^2 B_0 R^2}{4m_e c^2}\hat{z}$. (Note that this increase in momentum of the electron corresponds precisely to the decrease in potential

---

[18] $\nabla_i\vec{m}=0$.

momentum[19] (for a negatively charged particle) in the given direction due to the presence of the *A*-field). The radius of the electron's path does not increase (even though the electron's increased impetus means it has greater ability to move in its present Cartesian direction of motion) because the increase in centripetal force due to the presence of the *B*-field is exactly enough to balance the effect of the increase in momentum.

Thus, though with no applied *B*-field the average *net* dipole moment of the material is zero, an applied field induces aligned dipole moments in the diamagnetic material that the field then acts on in the way described in equation 9.15, i.e., according to: $F = \Delta m \dfrac{\partial B}{\partial z}$.

Because $\Delta m < 0$, this equation shows that the applied *B*-field gives impetus to the diamagnetic material in the direction of decreasing field. That is, a diamagnetic material is pushed away from a magnet.

We give intermediate non-standard but more sensible definitions of magnetization and magnetic susceptibility which we define and call, respectively, magnetization-bar, $\Delta m = \bar{\alpha}_m B$ and magnetic susceptibility-bar, $\bar{\chi}_m = \vec{M} / B$. From our analysis above, we get

$$\bar{\alpha}_m = \frac{e^2 a_0{}^2}{4 m_e c^2} \approx 2 \times 10^{-30} \, cm^3,$$ which is much smaller than the ballpark electrical susceptibility $\alpha_e \sim 10^{-24}$. Though weak, diamagnetism is common. Paramagnetic materials have to overcome innate diamagnetism before they can exhibit their paramagnetic effects.

The *uniformly magnetized slab* (which is an idealized example of a magnet) can be modeled as individual magnetic dipoles (atoms or molecules) lined up in a given direction such as shown to the right. Each dipole is shown as a small current loop whose circulation is oriented upward. The slab is characterized

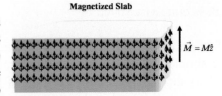

by $\vec{M}$, which specifies the measure of its direction and magnetization per unit volume. The external *B*-field of the slab is equivalent to that of a band of current around the outside. This can be seen by considering three dipoles in a row such as shown here. In the top view, we see how the currents in adjacent cells, being equal but opposite, cancel each other just leaving the currents on the boundary. The same effect happens between rows, leaving just the outer boundary. That is, the external *B*-field caused by the uniformly magnetized slab is equivalent to that caused by a band of current (of thickness equal to the thickness of the slab) traveling along the surface of the slab (with slab itself removed) in a right handed sense around the vertical axis (*z*-

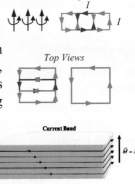

---

[19] One starts with no *A*-field, so no potential momentum in the $\hat{\phi}$ direction. With the *A*-field on, one has a negative potential momentum (for the electron) in that direction, i.e., there is a decrease in potential momentum.

axis). The current on the surface per unit length along the z-direction is given by: $\mathscr{I} = Mc$.

We can show that the *field inside of the current band is the same as the average field inside the uniformly magnetized bar*. For example, if we average over the appropriate length scale, only the impetus activated charge on the surface remains to cause any A-field either inside or outside of the surface, the effects of the interior currents having, *on the average*, canceled each other. We can also see the equivalence by analyzing the case of an arbitrarily long cylinder (for which all the arguments in the previous paragraph also hold). Making use of the conservation of the flux of B, ($\nabla \cdot \vec{B} = 0$) as well as the negligibly small exterior field far from the ends of the cylinder, we show that the B-flux piercing an interior surface whose outer boundary is defined by the cylinder is independent of any translation of that surface along the axis of the cylinder. Then, since the flux integrals of these translated surfaces must separately equal each other in the uniformly magnetized cylinder and in the cylindrical shell of current *and* since the flux integral of any one such surface in the magnet must equal any one in the shell, they must all equal each other. And, because these surfaces can be used to build a volume integral of B within each physical system, these volume integrals must be equal. Hence, since we chose the volumes to be equal, the average interior B-fields are equal.

As in the dielectric case, because of the need for larger than atomic-scale averaging, our analysis breaks down if, for instance, we try to examine the field less than an atom's radius away from the surface of the magnet.

And, following the convention adopted in our study of dielectrics, we let the physical context signal when averages are being taken, and simply write, for the fields inside and outside of the magnet and the cylindrical shell, $\vec{B} = \vec{B}'$. We use this convention in general unless otherwise stated.

The iconic physical configuration for magnetics is the cylindrical current sheet because it is the simplest closed current configuration that can generate a uniform B-field in a region, just as the parallel plate capacitor is the simplest voltage driven configuration that can generate a uniform E-field.

Comparison between polarized and magnetized rod
The external field *patterns* of a uniformly magnetized rod and a uniformly polarized rod are the same, as shown. By contrast, the internal fields are different because the source of electric field is charge, while the source of magnetic field is impetus activated charge (not magnetic charge).

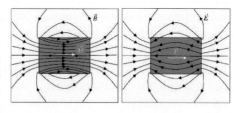

There are discontinuities in the slope of the B-field (left figure) at the top and bottom surfaces because of the effective currents on those surfaces. There are discontinuities in the slope of the E-field (right figure) at the left and right surfaces because of the effective charge on those surfaces. (Note that, inside the bodies, $\vec{M}$ points with the direction of the B-field, while $\vec{P}$ points opposite to the direction of the E-field, which, again, arises because of the different nature of the sources of B and E.)

An analysis of a *non-uniformly magnetized body* using the techniques described reveals that there is no longer perfect cancellation between the cells. Using $\mathscr{I} = Mc$, one instead gets the following formula for "bound current" in the x-direction due to changes in M_z in the y-direction:

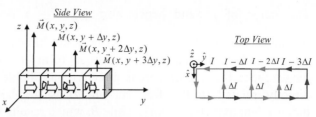

$$J_x(x,y) = \frac{\Delta I_x(x,y)}{\Delta y \Delta z} = \frac{c\big(M_z(x,y+\Delta y,z) - M_z(x,y,z)\big)\Delta z}{\Delta y \Delta z} \sim c\frac{\partial M_z}{\partial y}.$$ Variations in M_y

along the z-direction also imply a contribution to J_x. This analysis gives the general formula for the relation between the "bound current," i.e., currents associated with the atomic electrons within the material, and the magnetization vector:

9.36'
$$\vec{J}_{bound} = c\nabla \times \vec{M}$$

The bound current here specifies only those currents whose effects are not on the average canceled by neighboring currents. We already saw such bound currents in the uniform magnetized slab at the boundaries between the slab and the vacuum where $\vec{M}$ changes perpendicular to itself,[20] i.e,. has a curl.

$\vec{H}$ and μ

Because it is generally much easier to control the voltage (i.e., the ϕ-field within an additive constant) than the free charge, it is most convenient, in dielectrics, to work with E directly. Specifying the voltage immediately determines E. However, we do not directly control the A-field, but rather control the free currents. Hence, in working with magnetic materials it is most convenient to work with another quantity, $\vec{H}$, in our equations, a quantity analogous to $\vec{D}$. For the special cases in which there is no (net) divergence of $\vec{M}$ nearby, the $\vec{H}$ construct is most closely analogous with B and corresponds to the measure of the B-field that *would* obtain if there were no magnetic material present. More generally by making use of our formalism, we can write an extended form of the static form of the last Maxwell's equation to give the definition of $\vec{H}$ in the following way:

9.38
$$\nabla \times \vec{H} = \frac{4\pi}{c}\vec{J}_{free} \qquad \text{with } \vec{H} \equiv \vec{B} - 4\pi\vec{M}$$

Hence, following this practical motivation, we define the magnetic (volume) susceptibility as:
$$\vec{M} = \chi_m \vec{H}$$

We can see that taking $\vec{M} \propto \vec{H}$ in this way is functionally the same as the more principled definition $\vec{M} \propto \vec{B}$ by writing:

9.40
$$\vec{B} \equiv \mu\vec{H} \qquad \text{with } \mu \equiv 1 + 4\pi\chi_m$$

This contrasts with the more straightforward electrical definitions: $\vec{D} = \epsilon\vec{E}$ and $\vec{P} = \chi_e\vec{E}$. These magnetic (and electrical) relations are useful because many materials have a range in which these quantities are approximately linearly proportional.

---

[20] That is, as discussed in Chapter 5, it has a *net* "change perpendicular to itself."

The range of μ (and hence also $\chi_m = \dfrac{\mu-1}{4\pi}$) is wide, as illustrated here in the figure which does not have a consistent scale in order to include the many orders of magnitude of that range. Since the free current is directly controllable and H is directly related to the free current, H is graphed on the independent axis (the *x*-axis), while B, which results from both the free current and the bound currents that it induces, is graphed on the dependent axis.

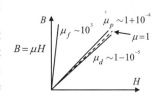

The *spin of the electron* is the key physical fact behind para- and ferro-magnetism. One needs quantum mechanics to describe spin of an electron. The *gyromagnetic ratio* is defined as: $\gamma \equiv \dfrac{|\vec{m}|}{\vec{L}}$, and the classical value of γ for a rotating spherical particle that has a uniformly distributed charge e and mass m_p is: $\gamma_{classical} = \dfrac{e}{2m_p}$. For the actual spin of an electron, the gyromagnetic ratio is: $\gamma_e = g_s \gamma_{classical}$ where we define the *g*-factor, which in this case, is $g_e \sim 2.0023...$ By contrast, the *g*-factor for an orbiting electron is the same as the classical case, i.e. $g = 1$.

Electrons are an essential part of the substances that we touch everyday. We say they are virtually present, not substantially present; for example, the electron is part of the copper not copper itself.[21] The spin nature of this particular part called the electron is critical in the existence of ordinary substances. *The Pauli Exclusion Principle* describes the aspect of spin that requires that no two electrons (which are called spin ½ particles) can be in the same state. Without this, all electrons would fall into the ground state, and none of the ordinary substances could exist as we know them.

Paramagnetism arises when an external *B*-field acts on the electrons of a certain substances. Because of the electron's dipole nature, the field exerts a torque, turning the dipole moments of the electrons to align with the external field. Once aligned with the field the dipole is attracted to the region of stronger field, as can be seen using the force formula equation 9.14. That is, paramagnetic materials are attracted toward a magnet. Note that diamagnetic materials are repelled because the induced dipole is anti-aligned, while paramagnetic ones are attracted because the field aligns the dipole with itself.

Due to thermal agitation, the atoms do not exactly align with the *B*-field, but each only aligns more or less at any given moment, making a statistical treatment necessary to further describe paramagnetism.[22]

Unlike in diamagnetism, most of the electrons in a paramagnetic material do *not* contribute to the paramagnetic effect. This is because all but the outer electrons form pairs in which one member of the pair anti-aligns with the second, so that for these inner electrons there is no net dipole moment.

---

[21] This use of "virtual" should not be confused with a later use of "virtual" that will appear in quantum field theory, which has a related but not identical meaning.

[22] The result for the magnetic susceptibility of paramagnetic materials (ignoring their diamagnetic component) is analogous to that of electrical susceptibility (see problem on χ_e at end of Chapter 8).

Note that in both dia- and para-magnetism the field must dispose the material, i.e. induce or align dipoles in it, before it can cause impetus in it. This is also true of the attraction of a non-magnetized ferromagnetic substance by a magnet.

Ferromagnetism is much stronger than diamagnetism and even paramagnetism. Three effects are at play: (1) the tendency of electrons to align in groups, (2) the preferred axes of alignment, and (3) domain building by application of a *B*-field.

In ferromagnetic substances, the atomic structure is such that outer shell electrons tend to want to align their spins (and hence their dipole moments) in order to respond to the tendency of particles with the same type of charge to repel each other. However, they do not all align spontaneously because there are preferred directions defined by the crystal structure of the material. Furthermore, complete alignment implies a substantial magnetic field 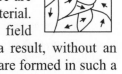 operating that would tend to turn some of the dipoles around. As a result, without an applied field, (microscopic) domains all with similar dipole alignment are formed in such a way on the average the net dipole of the ferromagnetic material is zero, as shown in the drawing above.

However, if a ferromagnetic material at a temperature above its *Curie temperature* (where it is paramagnetic) is cooled below that point in the presence of a *B*-field, magnetic domains can become "locked" in position, resulting in a permanent magnet. It is believed that lodestone (also called magnetite) is formed in this way.

If we apply a current to a toroidal coil filled with a ferromagnetic material, the material behaves as represented here in a graph of *B* versus *H*. Starting from a material with no permanent field ($\vec{M}=0$) and no applied current (thus, $H=0$),[23] increasing the applied current expands the dipole domains that align more with the direction of the 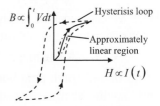 applied field at the expense of those that are less aligned. At first, because of the weakness of the applied field, relatively few dipoles can respond. As the current is increased further, one passes a threshold in which many dipoles can be made to change domain alliances. In this region, the response is roughly linear. At a later point, one begins to run out of dipoles to "join our side," and this is the onset of saturation. At the *saturation point*, the slope of the curve bends close to zero, so that an increase in applied current causes relatively little change in *B* because it no longer gets a boost from the ferromagnetic material.

If one now decreases the current, the domains resist realigning, and, thus, the decreasing *H* curve does not retrace the increasing *H* curve. This is called *hysteresis*. Indeed, one must reverse the current to force the ferromagnetic material back to its *B*=0 state.

Diamagnetism, paramagnetism, and ferromagnetism explain all of our ordinary-life (and beyond) experience[24] with magnetism.

---

[23] If there is no applied current and no bound current (i.e., $\vec{M}=0$), i.e., no impetus activated charge, then $\vec{B}=0$; hence, $\vec{H} \equiv \vec{B} - 4\pi\vec{M} = 0$.

[24] By ordinary experience, we here mean those experiences which are accessible to anyone who has access to natural substances, for example, lodestone, water and iron. Thus, some ancient peoples had the ordinary encounter with magnetism of which we speak. As more technical devices enter common usage, people will confront more of the "non-ordinary" phenomenon.

Incorporating our definitions of D and H, we can write a full set of Maxwell's Equations for fields within media:

Maxwell's equations in a medium

$$\nabla \cdot \vec{D} = 4\pi \rho_{free} \qquad\qquad \text{where: } \vec{D} = \vec{E} + 4\pi \vec{P} \ \ (\text{also } \vec{D} = \epsilon \vec{E}).$$

$$\nabla \cdot \vec{B} = 0$$

$$\nabla \times \vec{E} = -\frac{1}{c}\frac{\partial \vec{B}}{\partial t}$$

$$\nabla \times \vec{H} = \frac{4\pi}{c}\vec{J}_{free} + \frac{1}{c}\frac{\partial \vec{D}}{\partial t} \qquad \text{where: } \vec{H} \equiv \vec{B} - 4\pi \vec{M} \ \ (\text{also } \vec{B} = \mu \vec{H}).$$

Note also that the last equation incorporates the bound charge by including it in $\vec{J}$ and then subtracting it off both sides of the last Maxwell's equation in its vacuum (i.e. fundamental) form, making use of equation 9.36' and the definition of H.

In the static, no free current, no free charge case, Maxwell's vacuum and media equations yield a *set of equations symmetrical in E and H*:

9.49a-d $\qquad \nabla \cdot \vec{E} = 4\pi \rho_b , \ \nabla \cdot \vec{H} = 4\pi \bar{\rho}_b$

$$\nabla \times \vec{E} = 0 \qquad \nabla \times \vec{H} = 0$$

Here we can speak of magnetic charge. Of course, this magnetic charge is not physical, but a construct to aid in solving certain problems. The magnetic charge density is defined as $\bar{\rho}_b \equiv -\nabla \cdot \vec{M}$. A bar magnet, such as discussed above, obviously has a divergence in $\vec{M}$ at its north and south pole boundaries that, externally, can be treated as if they were "north" and "south" charge. In addition, the H-field inside the magnet, unlike the B-field inside, *also* looks as if it were caused by such charges. Note that unlike for B, $\nabla \times \vec{H} = 0$ does not imply $H = 0$, because in general, as seen above, $\nabla \cdot \vec{H} \neq 0$; hence, H remains useful even when there are no free currents.

Keeping in mind this difference, we can summarize, as we did for charge in the dielectric case, the equations for the various currents as follows: $\vec{J}_{total} = \frac{c}{4\pi}\nabla \times \vec{B}$,

$$\vec{J}_{free} = \frac{c}{4\pi}\nabla \times \vec{H}, \ \vec{J}_{bound} = \frac{c}{4\pi}\nabla \times (\vec{B} - \vec{H}) = c\nabla \times \vec{M}.$$

Using the Maxwell's equation for a media (a massive body that has both dielectric and magnetic properties), we get a *modified wave equation* ($\nabla^2 \vec{B} = \frac{\mu\epsilon}{c^2}\frac{\partial^2 \vec{B}}{\partial t^2}$ and its E counterpart) that describes something of the complex behavior of an electromagnetic wave traveling through, for example, the virtual plana of a massive substance, taking into account its interactions with the media's charged massive parts as it travels.

This equation implies that the speed of travel of electromagnetic radiation in a media is reduced from its vacuum speed by a factor known as the *index of refraction*; i.e., we have:

$c' = \dfrac{c}{n}$, where $n = \sqrt{\mu\varepsilon}$. Note that for vacuum, $n=1$, since $\mu=1$ and $\epsilon=1$.

Problems

1. Assuming a disk shaped magnet of 2 *cm* radius generates a 100 gauss field at its front surface, how much effective bound current, in amps, flows around the edges of the disk.

2. Consider, as illustrated here, the boundary surface between two uniform materials that have, as indicated, ϵ_1, μ_1 and ϵ_2, μ_2. a) Using the static Maxwell's equations for a media, derive the following boundary condition equations: $\Delta D_\perp = 4\pi\sigma_s$, where σ_s is the free surface charge density on the boundary between the materials, $\Delta E_\parallel = 0$, $\Delta B_\perp = 0$, $\Delta H_\parallel = -\dfrac{4\pi}{c}\mathcal{I}$, where $\mathcal{I}$ is the free current per unit length along the boundary on the boundary surface. Note that $\Delta V = V_2 - V_1$, where V represents a component of *E, D, B* or *H* that is parallel or perpendicular to the surface and the subscript (i.e., "1" or "2") indicates the material in which the field exists. ΔV is the difference between the value of the *V* immediately on each side of the two surfaces. b) Use these equations to write the relationship: (i) between the perpendicular components of *E* on each side of the boundary and (ii) between the parallel components of *H* on each side of the boundary.

3. In Chapter 8, we saw that the *E*-field inside of a uniformly polarized slab of arbitrary length and width is given by: $\vec{E} = -4\pi\vec{P}$. What is the *B*-field inside of an arbitrarily long uniformly magnetized cylinder of magnetization $\vec{M}$? Explain how you can derive this result from Maxwell's media and vacuum equations. Also, explain how it is that *H* is, in this case, the *B*-field due to free charge alone?

4. Show how to derive the equations for the *B*-field given in equations 9.10a,b from the Cartesian equations given in equations 9.9a,b,c.

5. Recall our calculation of the force on a current loop in the text. Explain why we only needed to take into account the external field. After all, the field at any given point of the plana is not identical with the external field, for the currents in the loop itself also contribute to the field. That is, there are contributions to the field from the current in the loop as well as from the currents that are responsible for the external field. Are there not? If so, how can we ignore them?

6. $\vec{F}_c = \left(\vec{m} \cdot \vec{\nabla}\right)\vec{B}$ (equation 9.16) only describes the force on an ideal magnetic dipole in the static case for which $\vec{\nabla} \times \vec{B} = 0$. This equation is also true for electric dipoles with $\vec{m} \to \vec{p}$. Hence, in such cases, we can think of the magnetic dipole as composed of magnetic charges instead of a loop of current; this is the reason we use the subscript "c" above. However, when there are currents in the region of interest so that: $\vec{\nabla} \times \vec{B} \neq 0$, we must return to current loops and the correct equation for the force on a magnetic dipole is: $\vec{F}_l = \vec{\nabla}\left(\vec{m} \cdot \vec{B}\right)$. Using $\vec{a} \times \left(\vec{b} \times \vec{c}\right) = \left(\vec{c} \cdot \vec{a}\right)\vec{b} - \left(\vec{b} \cdot \vec{a}\right)\vec{c}$, show that $\vec{F}_c = \vec{F}_l - \vec{m} \times \left(\nabla \times \vec{B}\right)$. This

difference in forces was used in the 1950's to show that the neutron magnetic moment must be modeled by a current loop, not magnetic charge.

7. In this problem, we investigate the physical aspects of the previous problem. a) Using an arbitrarily small *circular* loop whose axis points along the z-direction, show that the x-component of the force is given by: $F_x = m_z \dfrac{\partial B_z}{\partial x}$. Noting that this is not formally the same as equation 9.16: $\left(\vec{F}_c \right)_x = \left(\left(\vec{m} \cdot \vec{\nabla} \right) \vec{B} \right)_x = \left(m_z \dfrac{\partial}{\partial z} \right) B_x$, show that it does reduce to it if $\vec{\nabla} \times \vec{B} = 0$. b) Consider a region that *has a current distribution within it* that causes a field given by: $\vec{B} = kx\hat{z}$. Calculate the force on a magnetic dipole lying on the z-axis with $\vec{m} = m\hat{z}$ in the following two ways: i) Assume the magnetic dipole is a circular loop of current and take, keeping m constant in the usual way, the ideal dipole limit of an arbitrarily small loop. ii) Assume the magnetic dipole consists of a north magnetic monopole separated by a distance from a south magnetic monopole and take, keeping p constant in the usual way, the ideal dipole limit of an arbitrarily small separation. Explain your results physically.

8. Calculate the potential energy of two identical dipoles of magnetic moment, m, separated by a distance d when a) they are aligned parallel to each other and b) when they are anti-aligned, thus showing why, in ferromagnetic materials, all the domains don't prefer to align. *Hint:* thinking about bringing in the dipoles from infinity along a given line, L, in the following orientation. One dipole is aligned with L, the other is perpendicular to L. Explain, using the proper force equation, why this is helpful.

9. Consider a disk-shaped magnet of 2 *cm* radius that has 300G field close to its front surface. Treat the effective current circulating around the edge of the disk as a ring of current of the same radius. a) What is $m \equiv \dfrac{IA}{c}$? b) Approximately what is the maximum force along the axis of symmetry that the magnet can exert on a cube of steel ($\mu \sim 100$) with length .2 *cm*? Give an equation and a numerical answer in dynes. c) To get a better feel for the strength of this force, give the amount of massive material (in grams) that would be needed to cause the same force in weight (on the Earth). Compare your answer with the mass of the cube of steel.

10. a) Taking a standard compass needle as an ideal dipole, calculate its dipole moment in the following way. Align two identical compasses one directly above the other next to a vertical ruler as shown. Move the top compass near the bottom one until the fields of each compass interfere in a significant way, so that we can say the fields of each needle at that distance are about equal to the field of the Earth $(\sim .5G)$. We find this distance to be about 3.5 *cm*. b) What is the approximate maximum torque, in $dyne - cm$, that the Earth's field exerts on a standard compass needle? If this torque were generated by a perpendicular force on the end of the needle

(approximate needed size, for example, by using the picture of the woman's hand above), what would the magnitude of the force need to be? Also express your answer in grams of material that would have the same force pulling it gravitationally to the Earth.

11. Consider a compass that has little damping (usually the cheaper ones), but which has the value of the magnetic dipole moment, m, calculated in the previous problem. a) Use the torque to calculate the small angle resonant frequency of the needle, treating it like a pendulum. Note that you will have to estimate the mass of the needle. b) Given the fact that it takes about $\tau \sim 3s$ for the amplitude of the needle to drop to one third of its initial value, give a very rough estimate of the torque needed to get the compass moving. Give the torque in $dyne-cm$ and in $g-cm$, where the "g" refers to grams of material that would be needed to cause the same force in weight (on the Earth). Given the maximum field on Mars is $200nT$, will the compass likely work (i.e., roughly point along the field lines) on Mars?

12. Qualitatively describe, in terms of the A-field, how two aligned magnet dipoles, one directly in front of the other, act.

13. In this problem, we consider the fundamentals of levitating an animal such as the frog shown in the chapter. a) What condition must the B-field satisfy for it to be able to hold up a diamagnetic material against gravity? Address only the magnitude of the force and its first derivative; ignore the stability requirement (note in passing that *without* the quantum mechanical realities behind diamagnetism, stable levitation would not be possible).[25] b) What is this condition for water? For graphite? Express your answer in *cgs* and in *SI* units. c) How much current would be needed for a 2 *cm* ring of current to hold a very tiny animal along the axis of the ring? Use the diamagnetism of water for the animal. d) Approximately how much current is needed for an arbitrarily long cylinder (along the axis of the cylinder)? Comment, without derivation, on the potential for using a solid cylinder of current for holding up an animal.

14. Show why, in the diamagnetic effect described in the text, the radius of the electron orbit in our model does not change.

15. If on our way to Mars (or on Mars), we end up using an electric or magnetic shield for radiation, we may need something to, in turn, shield our experiments from the electric or magnetic fields. In seeking a solution, we might look to the following. Consider a closed shell of dielectric material with constant ϵ and a closed shell of magnetic material with constant μ. a) Describe and explain what happens inside the dielectric shell when it is placed in a uniform electric field. Note what happens when $\epsilon \to \infty$. b) Describe and explain what happens inside the magnetic shell when it is placed in a uniform magnetic field. Note what happens when $\mu \to \infty$. *Hint*: Use a shell with the orientation and relative dimensions shown here

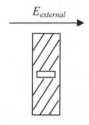

---

[25] It had already been shown in the mid 1800's that it is impossible to statically stably maintain a body in the air by magnetism

and take the body to be arbitrarily long in the dimension out of the page. Analyze the electric case first and then use the formal analogy between E and H. Also answer using your physical understanding of dielectric and magnetic materials. Do not worry about explaining the general case, although the results obtained here can be generalized.

16. Explain how, as mentioned in Chapter 4, a magnetic material with a high μ facilitates transformer operation.

17. a) Explain why, for a diamagnetic, paramagnetic or ferromagnetic body in a B-field, if you *double* the field, within a certain field range, you will get *quadrupole* the force on the body. Also, explain why if the applied field is above a certain strength, you only observe double the force. b) Calculate the magnitude of the B-field for which iron will magnetically saturate by assuming that two of the outer electrons for each atom are involved. Give your answer in gauss and tesla.

18. Explain and draw the H-field inside the uniformly magnetized bar magnet, thus completing the analogy with the uniformly polarized bar.

Chapter X

A Radio Frequency Transmitter

Introduction

The whole of classical electricity and magnetism is summarized in Maxwell's equations, and if you have followed along, you now have a fundamental understanding of those equations and their physical meaning. We are thus ready to apply those principles to our practical theme, understanding and, if you chose, building a radio frequency (RF) transmitter.

To make our transmitter easy to test, we can pick either an AM or an FM band transmission frequency, since you almost certainly have or can easily get a radio that picks up both these bands. Now, these names, AM and FM are highly misleading since, in this often-used context, they refer to frequency ranges, whereas AM means amplitude modulation and FM means frequency modulation, which refer to the way the voice information is encoded in the circuit. The association between the encoding technique and the frequency band arises because the FCC (Federal Communications Commission) assigned the frequencies in the range 520-1710 kHz for standard AM usage, and 87.5-108.0 MHz for FM usage.[1] There are no fundamental physical reasons that prevent transmitting in the first frequency range using frequency modulation *or* in the second range using amplitude modulation, though there are certain trade-offs.

---

[1] The AM band is within the frequency range more technically called the Medium Frequency (MF) band, (300-3000kHz), while the FM band is in the Very High Frequency (VHF) band (30-300MHz) that also encompasses the traditional TV bands.

How do we decide what method of modulation to use and what frequency range? Well, high frequencies, such as the FM band, tend to get absorbed quicker (the very low frequency band (VLF) can even go through sea water), but they bend less well around obstacles. Whereas, AM band waves bounce off the atmosphere; for example, you may have noticed that AM stations at night can come in from very far away. AM band electromagnetic waves can travel along the surface in a mode called ground waves. Whereas, FM broadcasts tend to stay local, since FM band waves do not bounce (refract) well off the ionosphere or travel well along the ground. These are only some of the considerations that arise. Clearly, we cannot attack the problem of what frequency to use in general. Let's consider a radio for use on Mars confining ourselves to the AM and FM bands. We would probably want line-of-sight communications for astronauts working at a given location on the surface, because line-of-sight is the most efficient and because bending can cause the same transmitted wave to break up into pieces and then arrive at the same spot in different phases creating hot and cold spots of reception. Also, we want line of sight to orbiting spacecraft for communication back to base and back to earth as shown on the cover. Furthermore, we cannot rely on signals bounced off the atmosphere of Mars because there is no ionosphere off of which AM band transmissions could bounce.

Hence, we will go with FM transmission band and will use frequency modulation largely because of the convenience already mentioned, but it, of course, does have real intrinsic advantages as well.

To understand our transmitter and how it is designed, we need a little more understanding of components and circuits. We will also eventually need to answer what AM and FM encoding is.

Components

We have already introduced some of the components that we need, i.e., resistors, capacitors, and inductors. However, our capacitor and inductor calculations need to be modified in the light of what we learned in the last two chapters.

Furthermore, we do not, at this point, know any of the physics of two key components in modern electronics, bipolar diodes and transistors. After addressing the capacitors and inductors, we will analyze them, beginning with the diode as a step toward the transistor. Since our focus remains *Electricity and Magnetism*, though from a more practical, what-we-can-do-with-it, point of view, we will avoid long excursions into crystal structure, quantum mechanics and detailed specific calculations around them and give only a rudimentary explanation of each device. Still, the exposition can serve as a solid starting point for further investigation into these fascinating topics.

Capacitors and Inductors

Our capacitors can now, as they nearly always do in actual circuits, have dielectric material between their plates. Given a parallel plate capacitor of area A and thickness d, we use equation $\nabla \cdot \vec{D} = \rho / \epsilon_0$, which is our first media Maxwell equation in *SI* (*mks*) units. We use *SI* units because they are the standard in the electronic industry. Applying Gauss's theorem, we have $\epsilon_0 EA = q$, which implies $V = Ed = \dfrac{q}{\epsilon \epsilon_0} \dfrac{d}{A}$, so that we get:

10.1 $C \equiv \dfrac{q}{V} = \epsilon \epsilon_0 \dfrac{A}{d}$, where $\epsilon_0 = 8.85 \times 10^{-12} F / m$

Where F, a farad, is the standard *SI* unit for inductance, and ϵ is the so-called relative dielectric constant (permittivity), which equals the value of ϵ in *cgs*.

Similarly, inductors can have ferromagnetic cores. The formula for a long, finely-wound coil of radius a and length l with N turns is calculated as follows. Using the 3rd Maxwell equation in *SI* units, $\nabla \times \vec{B} = \mu \mu_0 \vec{J}$, we get: $B_z = \dfrac{\mu \mu_0 N I(t)}{l}$. Then using

$\nabla \times \vec{E} = -\dfrac{\partial \vec{B}}{\partial t}$, also in *SI* units, to write Faraday's law; $V = -\dfrac{\partial \sum\limits_{i=1}^{N} \oiint\limits_{S_i} \vec{B} \cdot d\vec{a}}{\partial t}$, we get:

$V = -\dfrac{\partial}{\partial t}\left(N B_z \pi a^2 \right)$, which, defining V, I and the coil's winding direction in the proper way, gives:

10.2
$$L \equiv \frac{V}{\dfrac{dI}{dt}} = N^2 \mu \mu_0 \frac{\pi a^2}{l}, \quad \text{with } \mu_0 = 4\pi \times 10^{-7} H/m$$

Where H, a henry, is the standard *SI* unit for inductance, and (analogously to the dielectric case above) μ is the so-called relative magnetic permeability, which equals the value of μ in *cgs*.

Diode

The diode is a device that allows current to flow one direction but not the other. If we think of charge carrier flow as like water flow, the diode is like a one way valve. To understand how that valve works, we need to investigate the "p-n junction," which was discovered in 1939 by Russell Ohl of Bell Laboratories, and was the heart of the condensed matter electronics revolution that is still with us today. Before this revolution, electronics tended to focus on vacuum tubes, which means on manipulating electrons in a region that is free of massive materials, i.e., in a vacuum.

To understand the p-n junction, consider a piece of silicon. Silicon is a type of material called a semiconductor, because it does not conduct freely like a metal, but yet does conduct more than an insulator (such as plastic). The conduction of silicon can be increased by adding atoms of a certain type during growth of the crystal. Such atoms can also be added by diffusion or an ion implant processes. This process of adding atoms of a different type is called doping. If doping silicon (or other semiconductor) with a given material results in potentially mobile electrons, then we call that material a donor material. And, a semiconductor doped with a donor atom is said to be n-type, because it has free negative carriers. If instead the doping material leaves an unused bonding site, we call it an acceptor material and say it creates holes. A material doped with acceptor atoms is said to be p-type because these "holes" can travel like free positive carriers.

A silicon crystal looks schematically like the below figure. Note that all of silicon's valence electrons, which are shown as black dots, are tied to particular places in the crystal structure. This is because silicon's atoms like to be in the state illustrated to the right of the crystal structure diagram, namely with 4 electrons most readily available for bonding.

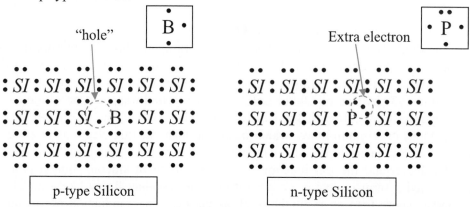

Now, consider silicon doped with boron. Boron, by contrast to silicon, only has three valence electrons available for bonding to silicon (all its other electrons are bound up). We draw this configuration as shown on the top of the left drawing below. A boron atom's presence in place of a silicon atom leaves a hole in the electron structure of the silicon crystal. This is shown circled in red in the left diagram below. It turns out that, through interactions spread out over the valence electrons of many atoms, this "absence of an electron" behaves, at some level, in the same way as a positively charged particle of the same magnitude of charge as the electron. We will treat it as such and, because of the empty spot that it leaves call each such effective carrier a "hole," keeping in mind that this is just a construct for the operation of *many* electrons throughout the local region. These holes are the majority carriers of current when a field is applied, and we thus speak of the material as a p-type material.

If we dope silicon with phosphorus, we get yet n-type material. Phosphor has 5 electrons available for bonding with silicon. Phosphor's electron arrangement is shown on top of the right figure above. The p-doped silicon crystal structure is the main illustration on the right. It shows the extra electron circled in red. This, as in the p-type case, does not mean there is extra charge in this region. It means that the crystal structure has no ready place for the electron that is *needed* in the region to keep it electrically neutral. This, in turn, means that the electron is free to act as a current carrier. That is, when an *E*-field is applied to the material, this electron can get impetus and move away from the P, leaving a fixed positively charged ion. This is precisely what we need for a diode to work.

If we dope half of a rectangular piece of silicon with boron and half with phosphor, we have half p-type and half n-type, and in the middle we have a p-n junction, the heart of the diode, which works as follows. If we apply no voltage, we get the follow situation.

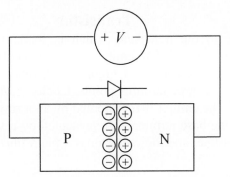

Figure 10-1: Voltage applied to a diode, which is shown as a rectangular piece of silicon, half is doped p-type and the other half is doped n-type, making a p-n junction where they meet. The circled plus and minus signs indicate the depletion region formed near the junction. Above the rectangle representing the physical diode is the schematic symbol for the diode. The direction of the arrow in the symbol tells the direction that current is allowed to flow.

There are more free electrons in the n-type region on the right than there are on the left. Therefore, as a pinch of sugar dropped in a cup of coffee soon ends up in the whole cup, so too the electrons diffuse to the left to "fill" the whole diode. Similarly, since there are an excess of holes on the left, they diffuse to the right. Complete "mixing" is however inhibited because these "particles" carry charge. As mentioned earlier, holes leaving from the left leave fixed negative ions, while electrons leaving from the right leave fixed positive ions. This means that an electric field is developed in a region depleted of carriers called the *depletion region*. The field operates against diffusion, so that at some point, an equilibrium between diffusion (which is fundamentally a thermal process) and the electric field built up by successful diffusion is reached. The potential drop is, of course, dependent on the degree of dopant on each side of the junction. The alert reader may now ask how, if there is a voltage drop across the junction, the voltage across the whole diode, which includes the bulk p and n-type materials, can be zero? The answer is that there are also voltage drops across the metal semiconductor gap that exist on each side of the diode.

If we now apply a negative voltage, thus trying to force current against the "one-way valve" direction, which is indicated by the direction of the arrow in the schematic symbol in Figure 10-1, we can see why the diode is indeed a one-way valve. As we apply more negative voltage, we aid diffusion and thus allow further build up of the depletion region, this means in steady state, that holes in the p region are repelled more strongly so that the more voltage that is applied the less current that can flow. At some point, we will reach an electrical breakdown of the material, but we do not consider that here.

By contrast, if we apply a positive voltage, we decrease the depletion region and holes can more easily flow across the junction. In fact, because a so-called Boltzmann distribution (that is derived from statistical mechanics) is involved, there is an exponential increase of current flow with voltage. The formula, which can be calculated from simple assumptions, is:

10.3 $I = I_s \left(1 - e^{-\frac{eV}{kT}} \right)$, where e is the charge on an electron, T is the temperature in K.

And, $k = 1.38 \times 10^{-16} erg / K$ in cgs and $k = 1.38 \times 10^{-23} J / K$ in SI units.

Transistor

We now move to the transistor, which is fundamentally a current amplifying device. It has three terminals called an emitter, base and a collector. The schematic symbol, which is shown on the right middle of Figure 10-2, represents these three terminals as follows. In the symbol, the three terminal points, labeled *E*, *B* and *C*, connect via lines to a thicker horizontal line. The emitter is represented by an arrow angled out from the left side of the horizontal line, the base terminal is represented by a vertical line connecting to the center of the horizontal line, and the collector is a plain line segment angled out from the right.

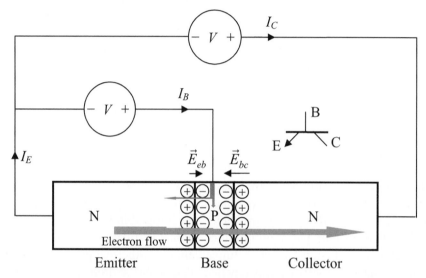

Figure 10-2: Transistor shown biased for active operation, in which a small current, I_B, into the base results in a large current in the collector, I_c. The gray arrow shows *electron* flow. The small red arrows show *hole* flow into and from the base. Though not indicated directly in the drawing, the electron current is much greater than the hole current.

Figure 10-2 also shows a graphic of the physical regions of an npn transistor, such as we will use in our transmitter. It shows the transistor with a voltage source applied from the collector to the emitter. We say that this positive voltage *forward biases* the base-emitter junction and *reverse biases* the collector-base junction. Electrons from the wire are given impetus to move to the right through the n-region by the electric field caused by the voltage source. When they reach the b-e (base-emitter) junction, they can cross it because the depletion field that would normally act against the flow of electrons into the base has been somewhat decreased by the application of the base emitter voltage. These electrons "emitted" into the base from the "*emitter*" find it relatively easy to cross the base region. What is the major danger facing electrons crossing the p-type base region? Recombination, i.e. falling into one of the holes (open sites) in the p-type base region. An electron recombining with a hole means that a negative charge has been deposited in that region of the base. Each such negative charge increases the barrier against further electrons moving across the base. Now, because the base region is thin only a small fraction of the current is used up in recombination, but it can nonetheless, if unchecked, completely stop current flow across the base. The base current fed by the voltage source that establishes the b-e bias carries these electrons out, making a positive current inward into the base. Note well that because the recombination rate is a small fraction of the current flowing across the b-e junction, the required base current is a small fraction of the emitter current.[2]

Having made it across the base, the electrons are now "collected" at the "*collector*" depletion region. The reverse bias on the base-collector junction insures that there is a strong *E*-field that acts like a strong vacuum cleaner sucking up all the electrons that make it near it.

Now, here is the key result. Because only a small fraction of the emitter current recombines in the base, only a small base current is needed to prevent charge build up in the base, and thus a small increase in the base current means there will be much larger increase in the emitter current (there being nothing left to stop it). In short, a small change in the base current makes a big change in the emitter current. In this sense, we can say that the transistor's operation is controlled by the base current; one could say that its "base" of operations is the "*base.*" A small base current regulates a much bigger emitter, and hence collector current. It is this amplifying property of the transistor that is the key to its usefulness in both analog and digital circuits.

Some Circuit Fundamentals

It may now seem that we have everything we need to discuss our transmitter circuit. However, a basic understanding of the components is not enough, we also have to consider how to analyze circuits that have any kind of voltage (or current) applied to them. Why so? Well, generally speaking, practical circuitry is about manipulating voltages and currents. This may be best seen by looking at a particular practical case, so we introduce our schematic in Figure 10-4, which is several pages ahead in the next section.

Our design process starts with you wanting your speech to be transmitted to your radio so it can be heard over it at a distance. This happens by means of a microphone such as that shown schematically to the far left of Figure 10-4. When someone speaks, his vocal chords, voice box (or larynx), and the various parts of his mouth form certain pressure waves in the air. These pressure waves can reach and move your eardrum, thus transmitting to you, via the air, something of the nature of the voice-generating parts of the speaker's body. In our case, these pressure waves activate impetus in the diaphragm of the microphone, which is analogous to the ear drum. The microphone, in turn, changes the state of the current flowing through it. Ideally, the microphone causes a change in current proportional to the strength of the pressure wave at a given moment. Now, although speech is already a very complicated sound wave, you will probably want your transmitter to carry any kind of sound you can hear. So, this first section of our circuit should be able to handle a wide variety of waves. This first section is, not surprisingly, called the "audio section,"

---

[2] Note the base is designed so that electrons can survive a long time in it; we say the base has a long lifetime for minority carriers. We should further note that there are also some holes injected into the emitter region from the base, but that rate of injection is a small fraction of the rate of electrons injected into the base, because we much more lightly dope the base than the emitter. Hence, the base current can easily take care of both recombination in the emitter and the base and still remain a small fraction of the emitter current. Said another way, some of the base current holes recombine with the emitter current electrons, reducing that flow by only a very little, and the rest of the base current holes recombine with the emitter current that makes it into the base region, again reducing that flow through the transistor by only a little more.

and is outlined in red, and after "converting" the sound to current, it amplifies those current changes to a level that can be used in the "RF section," which is outlined in green.

But, how do we begin to analyze a current that changes in some arbitrary way, the graph of which is some nearly arbitrary shape? We do it in analogy to the way we figured out the area of an arbitrarily shaped surface by breaking it into simple rectangular shapes that we understand. Clearly, in this case, we are dealing with a curve rather than an area, but there is a more important difference. In this case, we are dealing with a curve that represents the measure of the quickness of a motion, e.g. the quickness of the flow of a current, at various times. Therefore, breaking it into pieces to think of it as a series of line segments might tend to lose what we are interested in, which is the pattern over a period of time. We need a simple way to decompose it that will capture something more fundamental about the whole wave form.

Fourier Analysis

If you know anything about music, the idea of notes and harmonics might come to you. Music has fundamental tones out of which in some analogical sense it is composed. It has parts that play off each other, analogically like extensive parts (in the first category of quantity) of an animal, such as the head, body and legs of a rabbit, to make a single unity. Your vocal cords are analogically like strings on, for example, a violin. Such strings vibrate in a primary mode that includes a simple sinusoid as its base. Of course, a sinusoid, or a "sine wave," is a curve that can be written as a linear combination of sines and cosines:

10.4 $\qquad I(t) = A\cos\omega t + B\sin\omega t$

Said another way, we call any such wave a "sine wave" because all such waves are sine waves with different phases; $I(t) = K\sin(\omega t + \phi)$. For example, a cosine curve is a sine curve shifted to the left by $\pi/2$ radians. Still, as we will see, it is important to distinguish between sine and cosine, since such shifts are real and the above decomposition witnesses to their complimentarily.

From the point of view of extension, i.e. the primary meaning of quantity, the decomposition given by equation 10.4 makes sense, because the sine and the cosine are fundamental descriptions related to a circle. And, motion in a circle, as noted in *PFR-M*, is the simplest type of bound motion. In sound generation, like other types of generation, we are dealing with the activity of parts confined to a finite region; hence, it is reasonable to expect that this simplest bound activity will play an important compositional role. Since the sine and cosine waves are just the plots of the activity of each dimension of such bound motion, it is natural to divide any plot of sound intensity over time into such waves.[3] Indeed, it can be shown that any curve can be decomposed into an infinite series of such

---

[3] Of course, this does not mean that the wave is *actually* composed of sine waves. For example, in the case of electromagnetic radiation, it only means that the given field activated in the plana is *as if* it were generated by sine wave sources of the requisite frequencies and amplitudes at the given location--for the given field is sufficiently close to one that is actually generated that way. Note that this statement implicitly requires that the nature of the source and its action in the plana be such that the field that any given source generates in the plana adds linearly with the field due to any number of other such sources (remember we called this complex of properties, the property (extended sense) of linear superposition). Similarly, returning to the primary meaning of decomposition, i.e. breaking into parts, which is quantitative, it should be clear that a curve is not *actually* a sum of a finite number (still less an infinite number) of sine waves (see also end of chapter problem).

waves. This is called a Fourier decomposition of a curve and is named after the man who discovered the usefulness of such decomposition in solving the heat equation. The simplest form of such decomposition divides any wave that has a period of repetition T into a harmonic series. For example, a current, $I(t)$ that starts at zero, i.e. $I(0)=0$, and repeats every T seconds, can be decomposed into an infinite series of sine waves written as:

10.5
$$I(t)= \sum_{m=1}^{m=\infty} I_m \sin 2\pi m \frac{t}{T}$$

Thus, if we can understand how our components respond to a sine (or cosine) wave current, we can in principle calculate the response to any shaped current input, which, in turn, means to any pressure wave that our microphone can translate linearly to a current variation.

Impedance

Having shown the *fundamental* importance of sine waves, which, in electrical terms, are also called alternating currents, AC, and contrasted with direct currents, DC, we now introduce the simplest way to treat the response of an electrical "component" to a voltage of given frequency, phase and amplitude.

Since we are generally concerned with the relationship between voltage and current in a circuit, we look, in analogy to the resistor formula, i.e. $V = IR$, at the general equation: $V = IZ$. Z is called the impedance of the given circuit element. We treat the circuit element as a "black box" with two ports as shown below. A "black box" is a circuit element which we ignore the interior workings of, focusing only on what happens at the external terminals; it is as if it's pitch black inside, and we cannot see what is in the "box."

In analogy to resistance, the impedance can be thought of as the current, I, that results upon the application of a voltage, V, that varies in a sine wave fashion with time. We could, for example, apply a sine wave varying voltage and measure the resulting current, and then do the same experiment with a cosine wave. Now, if these two waves have the same amplitude, they correspond to the y and x components of a circle. Hence, we can treat the waves as the two components of a complex number in the following way. Recall complex numbers are graphed with their imaginary part on the y-axis and their real part on the x-axis. Symbolically, we write the applied voltages as: $V = V_0 \cos \omega t + i V_0 \sin \omega t = V_0 e^{i\omega t}$. This formalism is a very powerful tool that has many facets to it. Using it, we define the impedance as the response of a circuit element to sine and cosine waves.

We start by finding the impedance for a capacitor. The definition of capacitance recalled in equation 10.1 gives, after taking the time derivative:

10.6
$$I_c = C \frac{dV_c}{dt}$$

With $V_c = V_0 e^{i\omega t}$, we get $I_c = C V_0 i\omega e^{i\omega t} = i\omega C V_c$. Hence, the complex impedance, the analogical extension of resistance for AC input, for a capacitor is:

10.7
$$Z_c \equiv \frac{1}{i\omega C}$$

The inductor definition in equation 10.2, gives $V_L = L\dfrac{dI_L}{dt} = Li\omega I_L$, so that the impedance of the inductor is:

10.8 $Z_L = i\omega L$

Adding a resistor to these two impedances gives us the ability to do a fairly complete "lumped circuit element analysis." Lumped circuit elements are just elements that, at some level, can be treated in isolation from each other, e.g. as having no induction between them.

An Electronic Harmonic Oscillator

The most fundamental circuit we can make out of these three components is a damped harmonic oscillator. The voltage (and current) of such a circuit can go through a complete cycle like a mechanical harmonic oscillator. For example, recall from *PFR-M* that a child's swing can be an example of a harmonic oscillator. It makes a complete cycle as it goes from its most extended forward (positive) position through its equilibrium and then to its most extended backward (negative) position then back to its equilibrium position and back to its most forward position. Note how this motion is very like the *x*-component of the motion of a body moving in a circle, which is a sine wave (general sense). In fact, at a low enough amplitude, as we saw in *PFR-M*, the swing motion does take on such a sine wave character. Of course, because of friction, i.e. damping, the oscillation finally dies out.

The fundamental feature of every oscillator is its ability to store energy, i.e., its ability to "store" energy in part of a cycle and use it, i.e., activate actual impetus, in the next part. In the swing example, climbing higher in the gravitational field stores up energy. The impetus of the swing, which changes under the action of the force of gravity and its suspension wires, moves it upward along an arc of a circle until all of its impetus is deactivated. At that time, gravity, including its action through the suspension wires, increasingly gives the swing impetus in such a way that it moves more and more quickly downward along part of an arc of a circle.[4] A similar exchange occurs with voltage (and current) in an electrical harmonic oscillator.

The circuit shown in Figure 10-3a is a schematic for such a harmonic oscillator, called a tank circuit. To illustrate simply the primary storage/release aspect of the oscillator, we show no resistance, which would, in a real circuit, act to gradually damp out the oscillation, like friction does for a swing.

---

[4] Of course, ignoring damping, at any point during the cycle, the total (potential plus kinetic) energy (rate of transfer of impetus) is the same.

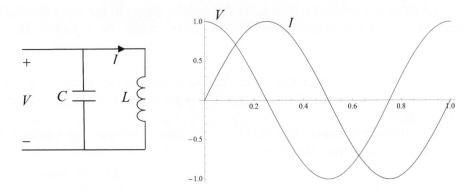

Figure 10-3: a. (*left*) harmonic oscillator circuit. **b.** (*right*) In the idealized case, starting with no current, $I = 0$, but with a voltage on the capacitor, with no resistance the circuit on the left will oscillate indefinitely. The red curve (cosine wave) shows the voltage, V, as a function of time. The blue curve (sine wave) shows the current, I, as a function of time.

The current and voltage curves for an ideal isolator are shown in Figure 10-3b. To explain the oscillators operation, we need to include a tiny resistance to avoid obscuring its causal structure. For simplicity, we take the inductor to be the tightly wound arbitrarily-long cylindrical coil analyzed in an earlier chapter.

We start by assuming that the capacitor has a finite voltage, V_0, across it and that it is then hooked to the inductor, so that initially there is no current, I. Assuming a tiny resistance in the coil and a much tinier one in the connection wires, we see that the charges in the capacitor (which cause a potential drop V) set up an E-field in the metal of the inductor and the connecting wires, causing current flow according to Ohm's law. As current flows through the inductor, an A-field is activated in the plana of the inductor. Note, through the very design of an inductor, the A-field generated within it is able to couple to the other parts of it, but not to wires and components outside of it, thus allowing a "lumped" component analysis. The (virtual) plana inside the coil wire resists the change of its state by attempting to deactivate the impetus of the mobile charges in the coil, thus competing against the ϕ-field caused by the charges on the capacitor. This $\dot{A}$ effect allows the current to rise only so quickly until it cancels out all of the EMF applied by the capacitor except for a small bit which maintains the current flow. The limiting of the current by the $\dot{A}$ effect, which is not the same as the electric field caused by the static charges on the capacitor, is a key point here. If the coil were a simple resistor, a current of V / R would flow *instantly* in the circuit. By contrast, because the rate of current increase is slowed down by the $\dot{A}$ effect, the current rises at some rate as shown in the blue curve in Figure 10-3b above. As charge leaves the capacitor, the voltage across it drops so that the field driving the inductor current decreases and thus the current does not increase as fast. At some point, all the initial charge on the capacitor is used up so that the driving voltage goes to zero; this is illustrated in the above figure by the first zero crossing of the red curve. Note that, at this point, the current (blue line) is at its maximum positive value. With no voltage (ϕ-field) to drive it, the current starts dropping, however, the plana does not like the change in the state of A that results, so it responds by trying to sustain the current, i.e.

the $\dot{A}$ effect (which is a force) now acts on the mobile charge in the same direction as the existing current, and thus begins to establish a voltage across the inductor. This voltage gets bigger as the current falls more rapidly. When the capacitor is completely charged in the other direction, it overwhelms the strength of the inductive ($\dot{A}$) effect and begins to drive the current back the other way as in the first quarter cycle. In this way, the whole cycle plays out and continues until the energy is dissipated into heat via the tiny resistances that we have allowed.

Using the empiriometric formalism that we have developed, we can now develop an expression for the current that takes into account the major effects at play. Recalling Ohm's law in terms of the field, $\vec{J} = \sigma \vec{E}$, we can write the current in the inductor as:

10.9 $\qquad I = JA = \sigma A E ,$

where σ is the conductivity of the coil wire and A is its cross section.

The electric field, E, consists of two parts: the part driven by the capacitor voltage, V_c, which equals $V_c / (N 2\pi r_0)$, where N is the number of times the wire is wound around the coil and r_0 is the radius of the coil and the $\dot{A}$ effect part which equals:[5] $\dfrac{-N\mu_0 \dot{I} r_0}{2l}$, where l is the length of the inductor coil. Inserting the sum of these two parts in for E in equation 10.9, taking a time derivative, using $\dot{V}_c = -I / C$ and rearranging gives:

10.10 $\qquad \ddot{I} + \dot{I}\left(\dfrac{2l}{\mu_0 N r_0}\right)\dfrac{1}{\sigma A} + \dfrac{I}{C}\left(\dfrac{l}{\mu_0 N^2 \pi r_0^2}\right) = 0$

The last expression in parenthesis is the inverse of the coil's inductance, i.e., $1/L$.

Lastly, note that for our assumption of arbitrarily small resistance the conductivity σ is taken to be arbitrarily large so that only the first and the last term on the left hand side will contribute, leaving only:

10.11 $\qquad \ddot{I} = -\omega_0^2 I$, where $\omega_0 = 1/\sqrt{LC}$

This is the wave equation, which has a sine wave solution of frequency ω_0 as can be quickly shown by substituting a trial solution of the form: $A \sin \omega_0 t$.

This harmonic oscillator is fundamental to the operation of our radio frequency (RF) generator. With the understanding of this new effect, we are finally ready to analyze our circuit.

Understanding the Transmitter

Audio Section

As we have already explained, the input to the audio circuit is the microphone which converts pressure waves of sound into changes of current. The particular type of microphone we use is symbolized by a circle with a capacitor within it (Figure 10-4). This is because it is a so called "condenser" microphone.[6] It originally worked by a mechanical connection of the diaphragm to one of the plates of a capacitor so that as the pressure

---

[5] This can be found by using the equation for the B-field inside the coil ($B_0 = N\mu_0 I / l$) and applying Stokes theorem to $\vec{B} = \nabla \times \vec{A}$ to get A.

[6] Invented at Bell labs in 1916 by E. C. Wente.

waves caused the diaphragm to move, it changed the capacitance of the device. The newer version of the condenser microphone, which we use here, uses an electret and is called an electret microphone. An *electret* is the electrical equivalent of a magnet, i.e. it is a permanently[7] polarized piece of material such as we discussed in Chapter 8. Using an electret, among other advantages, avoids the need for a biasing supply; however, power is still generally needed for the preamplifiers built into them.

The current changes caused by the microphone as sound waves impact it are amplified by the two transistors, *Q1* and *Q2*. Now, for a transistor to act as linear amplifiers, e.g., doubling their output when you double the input, it needs a certain voltage across each of its junctions. This applied voltage (or current) is called bias. As we already saw above, generally, the emitter needs to be forward biased and the collector needs to be reversed biased. More specifically, a DC bias is needed to set the operating point of the transistor in the active (i.e. linear) region. This DC bias is provided by the resistors R2 and R3. It is a simple divider, splitting the power supply voltage in half. Since the voltage drop across the base-emitter junction of a transistor in the active regime is about .6 v, this means there is about .56*ma* flowing in the emitter wire and thus about the same flowing in the collector if we can show that it is reversed biased.

---

[7] They can last centuries.

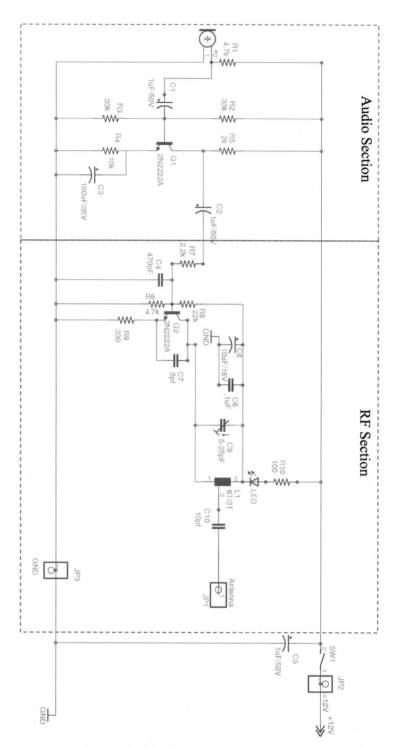

Figure 10-4: Schematic of RF transmitter.

Now, we distinguish between large signal and small signal operation of the transistor, because, in the active region with which we are concerned, the transistor does behave linearly for small signals, whereas it does not for large signals. A small signal is, as the name suggest, simply a voltage (or current input) that only perturbs the transistor (or other circuit), not moving it far from its DC bias operating point. Once the large signal analysis has been used to fix our DC bias (the detailed analysis of which will take us too far astray from our central topic but which we encourage you to investigate (see end of chapter problem)), we no longer consider it, but instead model the circuit in the small signal limit. Engineers make a small signal equivalent circuit and work with it. A typical equivalent circuit for a transistor is as shown below in Figure 10-5b. Figure 10-5c is a more complicated equivalent that incorporates capacitances within the device.

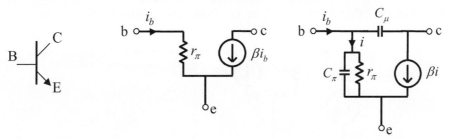

Figure 10-5: a. (*left*) schematic symbol for bipolar transistor **b.** (*middle*) small signal model for transistor without capacitance. Notice how a small current, i_b into the base results in a current β times larger flowing into the collector **c.** (*right*) small signal model with capacitance.

The capacitances arise from the depletion regions around the two junctions. β, the current gain already discussed, is order of magnitude 100, and for $I_E \sim .5ma$, $r_\pi \sim 5k\Omega$, using:

10.12 $$r_\pi = \left.\frac{\partial v_{be}}{\partial i_{be}}\right|_{I_B} = \frac{kT}{qI_B}$$

Note that standard notation uses capitol letters for large signal analysis and lowercase for small signal analysis. Since the 2N2222A is a fast transistor, its capacitances are small and can be neglected at audio frequencies ($20Hz - 20kHz$). Thus, we will only need the simpler equivalent circuit (middle figure above) for our analysis of the audio section.

The audio circuit can, using Figure 10-4 and Figure 10-5b and making appropriate approximations, be drawn as shown below in Figure 10-6. In this circuit, we replace the microphone circuit with a voltage source (using analysis, which is beyond the scope of this text, of the metal-oxide semiconductor field effect transistor (MOSFET) that is inside the electret microphone package) and neglect the relatively large values of R2 and R3. The circuit in Figure 10-6 is easily solved after we eliminate the two capacitors C1 and C3 in the following way. C3 effectively shorts out R4 above $f = \dfrac{1}{2\pi RC} \approx .16Hz$ as can be seen by noticing that the magnitude of its impedance equals that of the resistor at this frequency. Hence, for the audio frequencies we are interested in, this is an effective short. Similar arguments (see end of chapter problem) can be made for C1, which is called a coupling capacitor because it couples only AC signals, thus leaves the DC bias of the transistor unaffected. Applying Kirchhoff's voltage and current laws to the various loops, we get the voltage gain of the amplifier as: $Gain \equiv G = \dfrac{V_0}{V_i} = -\beta\dfrac{R_L}{r_\pi}$. Using equation 10.12, we then get the gain of the amplifier to be about 45 in the following way (where the voltages used in the second line are discussed in an end of chapter problem):

10.13

$$|G| = \beta \frac{R_L}{r_\pi} \approx \beta \frac{R_L}{\frac{kT}{qI_E}(\beta+1)}$$

$$\approx R_L \frac{I_E(ma)}{25mV} \approx \frac{R_L}{R_E} \frac{(6V-.6V)}{25mV} \approx 225\frac{R_L}{R_E} \approx 45$$

Note that the emitter resistor $R_E \equiv R_4$ shows up because it provides bias.

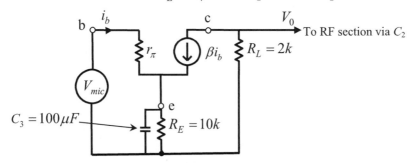

Figure 10-6: Model for audio section of transmitter circuit.

As we have said this gain only applies to frequencies above a certain range. It also only applies to frequencies below a certain range. That is, sine waves with frequencies outside the bandwidth of the amplifier will not have this gain. Again, this statement has meaning for any time development of the voltage (current) through understanding the wave as if it were made from a superposition of sine waves. In this spirit, we say there is a high frequency cutoff for the amplifier. Because this transistor has a gain-bandwidth product[8] of 350 MHz, it cuts off at about 8 MHz. Since this is much more bandwidth than we need, you might want to add a capacitor across $R_L = R_5$ to make the cut-off frequency closer to the 20 kHz end of audio frequency band. What value capacitor will do this?

Radio Frequency Section

Before discussing the RF section, we have to understand frequency and amplitude modulation, FM and AM. We want our circuit to transmit electromagnetic radiation to a receiver so that sounds on this end can be recreated on the other end. We thus must somehow encode the intensity variation of the pressure waves of sound onto the electromagnetic radiation. The most obvious and simplest way to do this would be to have the intensity of the electromagnetic wave vary in the same way as the sound waves. This is amplitude modulation. However, the very naturalness of this encoding makes it subject to interference, for as you may imagine, there are many electromagnetic waves in our environment that change in amplitude at audio rates and thus interfere with AM signals; two examples are: lightning and fluorescent lights. Frequency modulation (FM) changes the (instantaneous) frequency of the base electromagnetic radiation (called the carrier) at the frequency of the sound pressure wave, and the amplitude of this change is proportioned to the amplitude of the sound wave. Hence, for example, if we want to transmit silence, we

---

[8] This can be found from a data sheet available, for example, on the web or in a semiconductor manufacturer data book.

send a pure carrier wave, say an 88 MHz electromagnetic sine wave (which is in the radio frequency (RF) range). If we want to send a 440 Hz sine wave (which is the fundamental tone of A above middle C), we change the frequency of 88 MHz at a rate of 440 Hz with an amplitude of some chosen (instantaneous) frequency excursion Δf. Volume is determined by the amplitude of this frequency deviation, Δf; the lower the deviation, the lower the volume. The electromagnetic carrier is provided by our RF oscillator to which we now return.

Capacitor C2 marks the divide between the audio frequency (AF) and radio frequency (RF) sections of our circuit. C2 is a coupling capacitor that "filters out" DC, protecting the bias of Q3 and also (in conjunction with the effective output impedance of the audio amplifier and the effective input impedance of the RF oscillator circuit) filters out high frequencies (see end of chapter problem for details).

The RF section is basically a fancy harmonic oscillator whose frequency is changed by the changing voltage applied by the output of the audio amplifier. It works as follows. The heart of the oscillator is the tank circuit consisting *partially* of C9 and L1. C9 is a *variable* capacitor, so that we can adjust the oscillation frequency to match an open frequency on our FM radio. However, and this is key to understanding the circuit, C9 is effectively in parallel with the collector base capacitance of the transistor $Q2$, so that the tank circuit consists of $C_9 + C_{bc}$ and $L1$. The changing voltage applied by the audio amplifier changes C_{bc}, and hence changes the oscillation frequency.

The oscillator works by feedback. You have likely heard a feedback oscillation before. It occurs in public announcement systems when a microphone is, for example, placed too close to a speaker. When this happens the tiniest noise can be amplified through the microphone and outputted from the speaker and then re-amplified by the microphone, i.e. feeding back into the amplifier. This sets up an oscillation which will continue until the feedback is eliminated by, for example, moving the microphone or turning down the amplifier gain. Rather than doing a detailed analysis of the RF section (see end of chapter problem for some such analysis), we will briefly explain feedback and feedback oscillation, which is a general way of analyzing oscillators, including ours.

To understand feedback oscillation, we start with the general block diagram of a feedback system shown in Figure 10-7. Each functional element of such a diagram is defined by its frequency response, which is given the technical name "transfer function." In our case, the "feed forward" element has a gain as a function of frequency that is represented by $A(f)$, and the transfer function for the "feed back" portion of the signal is called $F(f)$. From the block diagram, we see: $V_0 = A(V_{in} - FV_0)$, and basic algebra yields transfer function for the entire system:

10.14
$$\frac{V_0}{V_i} = \frac{A}{1 + AF}$$

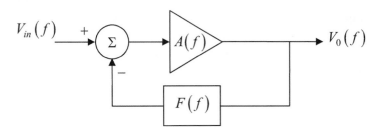

Figure 10-7: General block diagram for a feedback system.

The key point for us to note is that when F is negative and $AF = 1$, denominator goes to infinity so that a small input yields a huge output. This is the empiriometric condition for oscillation. More physically, this means there must be positive feedback with an amplification of unity, so that what comes out gets fed back in to keep the cycle going. Our oscillator is designed so this happens at only one frequency which is set, as we have said, by the tank circuit. Note that we need do an active element, such as transistor $Q2$ in our transmitter, to compensate for the damping introduced by the power dissipated in the resistive elements which are a part of any such circuit.

The last element in the RF section is the antenna. As we have discussed in previous chapters, the changing impetus of the mobile carriers in the metal cause electromagnetic waves in the plana that consist of changing A-fields that result in a combination of electric (analogical type of electric field due to $\dot{A}$ effect) and magnetic fields. These fields can cause impetus (i.e., exert a force) along the direction of motion in, for example, a receiving antenna. And, what is of interest here, it causes charge motion in such an antenna.

We have drawn a capacitor in series with the connection to the antenna to help in experimenting to optimize broadcasting. We will further discuss the issues involved in transmission including coupling the RF oscillator to the antenna in the "testing" section, but first we have to build the device.

At the receiving end, your AM/FM radio has to "undo" what we have done in the transmitter. It must collect the modulated electromagnet wave that we broadcast, amplify it if necessary, then separate out (demodulate) the "encoded" signal and, lastly convert it to pressure waves in the air.

Making the Device

The first step is obviously to acquire the needed parts, which includes the components and the printed circuit board. Table 10-1 and Table 10-2 show where you can purchase the needed components. The printed circuit board (PCB), which is basically a board with all the "wiring" built into it, has already been designed for you and can be purchased at PCB (design number 45120 at batchpcb.com).[9] Alternately, you may want to purchase a PCB kit, which includes acid to etch your own board. To keep expenses down, to have help and just to make it more fun, I recommend that you do this project with a group of your fellow students.

---

[9] In particular, go to: http://www.batchpcb.com/index.php/Products/46614 (for changes in the design number and/or this link; see iapweb.org/pfrem/errata)

Jameco® Catalog Number	Description	Quantity Needed*
216283	BAT HOLDER,2-AAA,WIRES	1
216452	9v BAT SNAP,SAFETY,6",26AWG	1
93761	CAP,RADIAL,100μF,25V,20%	1
332612	CAP,MONO,470PF,100V,10%,X7R	1
332304	CAP,CERM,DISC,5pF,1kV,20%	1
94212	CAP,RADIAL,10μF,25V	1
330431	CAP,RADIAL,1μF,25V	3
544921	CAP,CER,RADIAL,0.1μF	1
81543	CAP,MONO,10pF,50V,20% (10)	1
2077554	CAP,CERM,TRIM,8mm,5-25pF	1
691420	RES,CF,220K OHM, 1/4 WATT,5%	1
103166	RESISTOR Assortment,1/4W, 5%	This package has all the other resistors needed
178511	TRANSISTOR,PN2222A, NPN	2
333851	LED,RED,660nm,T-1	1
320179	MIC CART,1.5-12V,2PIN LEADS	1
* Most of these parts have minimum buy numbers		

Table 10-1: List of parts needed that can be purchased from Jameco

Radio Shack® Catalog Number	Description	Quantity Needed
275-327	6 piece slide switch kit (This is the size switch that fits in the PC board below.)	1
278-1345	(for inductor L1) This package includes three roles of wire; recommend using gold wire (22 gauge) because its stiffness helps the coil keep its shape.	< 3 inches
64-2051	15 W Soldering Iron, with small holder	1

Table 10-2: List of parts needed that can be purchased from Radio Shack

Figure 10-8: Picture of completed transmitter circuit

Once you have the parts, you need to know how to identify the type and value of each component. For identifying components by their shape and approximate relative size the completed circuit is shown in Figure 10-8. There are two types of capacitors: electrolytic, which have a polarity that needs to be respected and non-electrolytic, which do not. Electrolytic capacitors tend to be tall cylinders whereas the non-electrolytic ones we are using tend to be somewhat flat with a round cross section and a yellow tint of color. The capacitor values are written on the case; often for ceramic capacitors the following code is used: first digit, second digit followed by the number of zeros to add, so that 471 means 470 pF. Note that capacitors are usually measured in microfarads ($10^{-6}F$) or picofarads ($10^{-12}F$).

The transistors have a half-moon cross section and are about as tall as they are wide. The resistors are cylindrical shaped with their connection leads coming out parallel to their axis. They also have color bands on them to indicate their value. These bands can be interpreted using the code illustrated in Figure 10-9. The resistor illustrated tells us the first digit is red, which is 2, the second digit is violet which is 7 and the last band tells us to multiply by 10^5 since green is 5; hence, the resistor is $2.7M\Omega$. The figure also shows a mnemonic for the color code: "Big Boy ROY G BV Goes West for Gold and Silver." Since the color code has within it the whole spectrum except indigo, we borrow the common mnemonic "Roy G. Biv" for the spectrum and exclude the "i," since few people distinguish indigo in ordinary color identification.

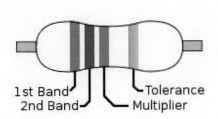

Color	Number	Tolerance
Black	0	-
Brown	1	-
Red	2	-
Orange	3	-
Yellow	4	-
Green	5	-
Blue	6	-
Violet	7	-
Gray	8	-
White	9	-
Gold	-	5%
Silver	-	10%

Big **B**oy **R** **O** **Y** **G**. **B** **V** **G**oes **W**est for **G**old and **S**ilver

Figure 10-9: Resistors have 4 color bands that symbolize the value and tolerance of the resistor using the code illustrated to the right of the resistor. The resistor illustrated has a value of $2.7M\Omega$ and a tolerance of 5%. The last line shows the mnemonic, "Big Boy ROY G BV Goes West for Gold and Silver" for the color code.

The PCB board comes marked with component values and names (e.g., R1 or C1), so you only have to solder them into place. Soldering is not hard and can be learned by reading a few online tutorials and then doing a little practice before hand. You do get to make one component, the inductor.

The inductor is made by closely wrapping 6 turns of about 22 gauge wire around a 0.5cm (~3/16 inch) diameter screw driver; start next to the screw driver's handle and work your way outward. These wraps make the primary part of our inductor. Then, wrap 3 turns on top of this very close to one end; this will be the end that connects to the collector of Q3. This second set of wraps becomes a secondary coil for coupling to the antenna. After wrapping, you will need to strip the coating off the ends (with, for example, sandpaper) before soldering them into the printed circuit board.

As you may recall, we have already very roughly estimated the value of such an inductor using the special case of formula for inductance given equation 10.2 in Chapter 4.

First, assemble the audio section only, so that you may test your assembly one section at a time. Test it by hooking a probe from your oscilloscope from ground to either side of C2. Once you are sure it works, then assemble the RF section.

Testing

With the board complete except for the antenna, you now can put your device next to an FM radio receiver. With the transmitter off, find a station that has no existing radio broadcasting. In my area, the bottom part of the dial (about 88MHz) is most open; your area may be different.

Now, turn on the switch and adjust the variable capacitor C9 until you hear feedback or your voice or both. If you hear a load feedback sound your transmitter is working. Note that when you put your hand near C9 you will probably change the capacitance to ground that your oscillator sees and this will change its oscillation frequency, so you have to compensate for this effect in your adjustments. Once you have a signal you can play as you please; however, the question of an antenna probably comes to mind as you want to be able to transmit further. The circuit board provides for a connection to an antenna.

The coupling capacitor, C10, which we mentioned above, minimizes the effect of the antenna-to-ground capacitance (why? And what is the worry here?). You will also find a pair of pads associated with a resistor labeled $R13$. These pads are only meant to serve as a slot to solder any extra component that, after experimenting, you decide might be helpful in optimizing transmission and reception. Keep these accessories in mind as you think through the next paragraphs and while you are experimenting.

Remember that transmitting means that the field generated is such that it can escape the confines of the circuit. In short, we do not want the $\dot{A}$ effect to cause voltages in our circuit but in the receiver. This means, among other things, it is generally preferable to have a tuned antenna.

What type of antenna you want will depend on how you want to transmit. If you want to transmit to one location, a directional antenna is best. Why so?

An antenna that transmitted in a spherically symmetric way,[10] that is the same in each direction, would activate each new region of the plana from a spherical surface. Thus, (as we already saw, for example, for the electric field) the area over which the field must activate new parts of the plana increases with the square of the radius and thus the power per unit area decreases in this same way. This, in turn, means that at a given fixed location the power decreases as the inverse square of the distance.

We can do much better than this if we do not waste power by sending it irrelevant directions. Clearly if you want to be a radio DJ, then you may want everyone in all directions to hear you and then there are no irrelevant directions. However, for point to point communications clearly a lot can be gained by directing the rays.

As with all the interesting things we discuss, one can break off the main trail to find many interesting side trails. To keep on the main trail, it is probably best to focus on the simplest antenna we can start with and then move to a simple antenna that does do some "ray directing."

The simplest antenna is just a wire that one moves until the best signal is obtained. Once you have the best signal, you should move the receiver away from the radio to eliminate feedback (making whatever antenna adjustments may help keep the signal strong) and then talk into the mike until you find an optimum distance at which your speech is not distorted (note that a 1kHz tone generator and an oscilloscope monitoring the audio output may be helpful here).

---

[10] Note that it is actually impossible to have a coherent *isotropic* electromagnetic radiation (though an incoherent one is possible). Still, electrical engineers reference antenna patterns to an isotropic antenna because, in terms of the first property of quantity (extension), it embodies the generic reality of being distributed equally in all angular directions around a small region which is idealized to a point. This latter is also the ultimate reason behind our use of the isotropic antenna here. In particular, we are only manifesting the sharp contrast between electromagnetic radiation that is directed at a chosen target and radiation that is directed so as to spend some fraction of its energy in irrelevant directions.

The next simplest antenna is the dipole antenna that we mentioned in Chapter 6, that uses the Earth as a kind of mirror to generate the effect of the other half of the dipole. For 100MHz, this is $\lambda / 4 = \dfrac{c}{4v} \approx .75m$. The next simplest is a full dipole antenna which has this length on two sides like shown below. Both these broadcast their waves in a pattern shown below in Figure 10-10. In experimenting with one version of this transmitter, we found that we could transmit a clear signal 1/5 mile away, using a near-resonant dipole made out of flimsy wires, through houses and trees. Perhaps you can do even better.

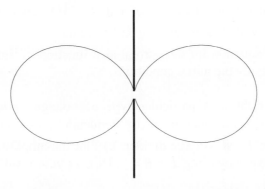

Figure 10-10: Cross section of the pattern of emission of electromagnetic radiation from a dipole antenna. The dark vertical lines represent the two elements of the dipole antenna, which could just be two stretched wires. The driving wires are connected to capacitor $C10$ and to the circuit ground.

You are encouraged to fiddle around and learn as much as you can about the operation of your transmitter, including electromagnetic wave propagation.

Problems

1. Given the circuit to the right with $L = 1\mu H$ and $R = 1k\Omega$ that starts with a current I_0, how long will it take for it to fall to about $I_0/3$? To answer, derive the general differential equation for the current as a function of time, then solve that equation and apply the initial conditions.

2. Given the circuit to the right with $C = 1\mu F$ and $R = 1k\Omega$ that starts with a voltage V_0, how long will it take for it to fall to about $V_0/3$? Start by deriving the differential equation for the current as a function of time, then solve that equation and apply the initial conditions.

3. The circuit to the right is a particular type of voltage divider. a) Calculate the ratio V_o/V_i as a function of the frequency, thus giving the transfer function for this $L-R$ voltage divider. b) Plot the amplitude and phase as a function of frequency using $L = R = 1$. Discuss your results.

4. a) Using the small signal model in Figure 10-6, calculate that effective input resistance of the audio section. That is, what small signal resistance, r_i, appears to a voltage source driving the base of Q1? We can model the interface between the voltage on the microphone and the voltage that appears on the base of the transistor in the small single model using the circuit on the right. b) Use this model along with the impedance of a capacitor to explicitly derive the formula for the base voltage (i.e. the potential at point b) as a function of the frequency of the voltage source. c) Show that the cutoff frequency, i.e. the frequency at which the voltage drops by $1/\sqrt{2}$, is $\omega = 1/(RC)$ by roughly plotting the magnitude of v_b/v as a function of frequency. Calculate the time constant, i.e. $R \cdot C$, associated with the coupling done by capacitor C1 and thus show that C1 can be treated as roughly a short in our small signal model for the frequencies of interest.

5. In the text, we discussed Fourier decomposition, which adds sine waves to create new waves. a) Explain this starting from the first property of physical things, extension (quantity). In particular, explain in what sense the curve associated with $\sin x$ and the curve associated with $\sin(-x)$ yield the straight line: $y = 0$. Use a drawing of the one cycle of each of the curves in your explanation. Then, explain in what sense this analogically applies to two sources that would generate certain fields, call them F-fields, that have spatial patterns given by: $F^1_x = \sin x$, $F^2_x = \sin(-x)$.

b) Consider one cycle of a square wave of amplitude $\pi/4$ and period 2π, such as shown here. Draw the first three terms of the Fourier expansion of the wave

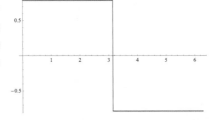

which are: $S(x) \approx \sin x + \frac{1}{3}\sin 3x + \frac{1}{5}\sin 5x$. Sum the first two terms and explain the result using a plot.

6. In this problem, we establish the DC biases of the audio amplifier portion of our transmitter circuit. Using the fact that very little current goes into the base of Q1, calculate the approximate DC bias voltage on the base and the collector of Q1 as well as the DC current flowing through the collector of Q1. *Hint*: when a bipolar transistor such as the 2N2222A is used in the linear active region, its base-emitter voltage is approximately $.6V$.

7. a) Using an approach similar to that used in the above problem, calculate the DC voltage on the base and the emitter of Q2 and its approximate collector current. You will need the potential drop across the series combination of the LED and R_{10} which is about $2.1V$. b) Using the small signal model given in Figure 10-6, neglecting phase and taking C_3 as a short at audio frequencies, give the audio frequency *input impedance of the RF oscillator circuit*, i.e. specify: $\Delta v / \Delta i$, where Δv is the small change in base voltage that results from a small change in current, Δi, injected into the base of Q2. Also using that small signal model, calculate the audio frequency *output impedance of the audio amplifier* (i.e. as seen "looking across" the collector of Q1 and ground) c) Using these results, calculate the frequency at which the voltage gain (from the effective voltage driving the output impedance of the AF amplifier to the base of Q2) drops by $1/\sqrt{2}$, i.e. the low frequency cut-off frequency. Before doing your calculation draw a schematic of your reduced small signal model. d) Calculate the high frequency cut-off. e) What are the purposes of these two cut-offs?

8. Draw the small signal model of the RF circuit, including a base-collector capacitor, C_{cb} of about 4pf. Treating all components that do not have a significant value near RF (i.e., in the FM band in our case) as shorts, make a rough estimate of the frequency of oscillation. Use our long cylinder approximation for the inductor (L_1), then reduce it by a factor of 2 to compensate for the poor coupling between the various wires of the coil.

9. Suppose one wants to use a variable capacitor that one takes from an old transistor radio. How would one determine its range of values so that the right inductor can be wound to pair with it in our transmitter circuit? *Hint*: consider using a fairly low frequency capacitor divider network.

10. Mars or the Earth (especially Mars because of its ferric content) can block radio waves. Of course, it is best to not have to transmit through the planet. a) Give a general equation for how high, h, one needs to be above the Earth or Mars of radius r_0 to avoid it. The answer should be in terms of the distance d over the surface that one would like to transmit. b) Give an approximation for distances that are small compared to the circumference of the planet. c) Give the numerical relation for h in feet for the Earth ($r_E = 6.38 \times 10^3 km$) and for

Mars ($r_M = 3.40 \times 10^3 \, km$) with d in miles. d) Using this formula, how high of a tower does one need to transmit $10 \, miles$ on Earth? on Mars?

11. Is Kirchhoff's voltage law (KVL) always valid? If it is not, give an example of when it is not and explain why. Is KVL fundamental? To answer, use the definition of voltage drop as how much energy per unit charge a particle is given by the E-field. Also, specify the E-field everywhere in a simple circuit that violates KVL. How is this discussion relevant to our circuit? *Hint:* Remember the analogically general meaning of E includes the $\dot{A}$ effect.

12. One can write the amplitude modulated (AM) carrier wave of frequency f_c as: $A_m(t)\cos(2\pi f_c t)$, with $A_m(t) = A_{m0}\cos(2\pi f_m t)$, where we show a modulation tone of frequency f_m. One can write, assuming the same modulation tone, the frequency modulated (FM) carrier as: $A_{m0}\cos(2\pi f_c t + \phi(t))$, where $\phi(t) = \dfrac{\delta f}{f_m}\cos 2\pi f_m t$. Show that the instantaneous frequency, i.e. the best approximation to the frequency of the wave at a given moment in time, tracks the modulation tone.

13. Explain qualitatively why a good conductor can block the signal of your transmitter.

Appendix I
The Concepts of Form and Abstraction

On the Use of the Word Form

Recall that the definition of a physical substance is one that can change, can be other than it is. We say that every physical thing is something but can be something else. We call what it is, its "form" and what it can become its "matter." This definition of the word "form" can bring up two questions that can lead us to think more deeply about these two "components" of every physical thing. The first is: why do we use the word form, whose first dictionary meaning is shape? The second is: What do we mean, in the context of modern physics, when we say (and it is said fairly often), "formally," which is clearly derived from the word "form."

Genesis of the Word Form to Mean What a Thing Actually is

The word *form* probably originally comes from a change of shape. To understand this, let's look at the four causes involved in the making of a statue. Let's say you make a life-size statue of Einstein out of copper, which starts as some large amorphous shape. Copper is obviously, the "material cause" of the statue, since that is the material out of which Einstein's shape is made. Clearly the new shape, the new *form*, the new actuality into which it is made, which we call the "formal cause" is Einstein's shape. Since you, the sculptor, are responsible for the change, you are the "efficient cause." The reason that you make the statue, i.e., the "final cause" of the statue, is your desire to have a reminder of that great physicist. Note this is clearly an extrinsic final cause, not belonging intrinsically to the copper but external to it. Indeed, there is no intrinsic final cause of Einstein's shape in the copper. This is so because it is not essential, but merely accidental, for copper to take the *form* (shape) of Einstein. Though this use of "form" is first in terms of etymology of the usage, this use is not first in principle. This use involves a property (shape) the thing has, rather than what the thing is primarily in its substance. What a thing is in its substance is *form's* first and most important meaning. The reason the substantial use of the word "form" is first in principle is that something has to exist (substance) before it can have its properties.

Use of the Word Form in Modern Physics

In modern physics, the use of the word "form" might seem to contradict the primary principled usage, i.e. what a physical thing actually is. This is because, in modern physics, "form" usually refers not to the physical reality itself, but to the theoretical framework, i.e., the system of rules *especially as represented by the equations*, that is used to make predictions about physical reality. For example, we might say after solving an equation: "formally the answer is." Thus, there are two radically different types of forms at play, the form of the theory and the form of the reality that the theory describes.

To understand this more completely, we have to enter the empiriometric mindset with, roughly said, its goal of reaching the simplest most elegant set of equations that correctly predict the measured outcome of experiment. Equations in modern physical theories serve to carry something of the nature of the physical realities around us. The equations and their attending system of conceptual understandings that relate them to experiment, what we call the theory, are the substitute (or "model") for the full realities that we are trying to reach. We manipulate the equations to give predictions and understanding about the realities they represent. While we are doing so, they represent those realities and are often the only vehicle for contact with them that we have. In so doing, we, at some level, look through the equations to the reality they describe. The physical *theory* and its particular equations are signs of a physical reality; they are things (actualities) that refer to other things (actualities); this is what we mean by a sign. Of course, because we know the symbolic forms and their attending rules (which are themselves a different type of form) are not the reality themselves, when we come to focus on them, as opposed to the realities they describe, we see that the equations themselves are simply mental constructs (beings of reason), existing only in our minds. Yet, the equations and rules are the actualities ("forms") that we directly think about and, through the symbols on the paper,[1] see and manipulate. Hence, when we note aspects in them that are *purely* formal, we are noting that those aspects are forms (actualities) that are *not* themselves the thing we are interested in *and* do *not* refer to the external reality that we are interested in. That is, we are noting those aspects are forms *that are not signs*; hence we call them "purely formal." Again, in the purely formal cases, they no longer refer (at least as directly) to realities other than themselves. We are denying their referential status, denying that they refer to anything outside themselves, but affirming their reality.

Of course, our goal is to understand external physical reality, so we have to be careful not to confuse the system of equations with all its predictive power with the physical reality which it can reveal. Using our key word form, we can say that, in the last analysis, we are seeking not the form of the equations but the form of the physical things which they "describe." That is, we want to know something of the nature of physical things, i.e. something of what they are, which is what we mean by form in the primary principled sense. The very denial of the referential status of an equation which is implied in saying "purely formal" is indeed an implicit acknowledgement of this goal, for if equations were never referential, we would never have to make the distinction between aspects which are and are not referential.

We can also talk about various "forms" of an equation (or system of equations), where again we are still using the word "form" to mean an actuality, namely, a particular actual state of an equation. Now, because we know that some forms of our equations are closer to the physical situation under consideration than others, we can say things such as: "this result is formally correct but less physical than this other." For example, we might say that $m\ddot{x} = F = 0$ is more formal than $x = vt + x_0$, which is the solution for the motion. A result that is *purely* formal might, for example, derive from a manipulation of symbols that contains no significant new physical content and thus draws one back to the symbols and system of equations itself, not to the exterior physical reality. Such a *purely formal result* might come about in many ways. For example, it could arise from (1) a mathematical or logical necessity of the theory that does not have a reflection in the physical reality or (2) a

---

[1] This is a *second* level of depth. The letters and symbol of our equations are actually ink on the page, but for us they refer to our theory which exists in our minds, not in the ink on the page; that theory, in turn, refers to the physical world.

manipulation of symbols that relates only trivial information or information that a less formal result already expressed (for example, given $Z = XY$, $X = Z / Y$ is, at one level, a purely formal result) or (3) merely the definitions of the symbols themselves or the like (for example, given the definition of $\sum$, then $3 \times \sum\limits_{i=1}^{3} x_i = 3\left(x_1 + x_2 + x_3\right)$ is a purely formal result).

In summary, the equations (including their attending theoretical interpretation) are themselves *actual* ideas in our minds, but they *only* exist *in our minds* (though they have symbolic forms on paper); so that when we say a result is simply formal, we are recalling this obvious fact, and denying the result any referential status. However, as mentioned, such a statement would be a waste of words, if the equations did not normally refer to something beyond themselves, i.e. to actualities in the physical world. Now, because of this fact, using the word form **_only_** in such a context, e.g., when we say a "purely formal" mathematical result, can be dangerous, because such a use does derive from the *primal* use, i.e. what a *physical* thing actually is. If we use "form" *only* in the *secondary* context, we can build a tendency to think that mathematical theories are the first things that we know or even that they are the only things that exist. Both of which, of course, we know are not true. For the first thing we know is not a mathematical theory, but physical things (though we only know them, at the first level, in a general and vague way). Indeed, the symbols themselves are physical things on the page before they are incorporated into the structure of a theory in our mind. And, the theories themselves *are built out of* the basic understandings of physical things that we get through the senses, and so could not exist without them.[2] (Moreover, only by reflecting on the physical things that we know directly through the senses, do we even realize that we have ideas). From this perspective, the very use of the word "form" in the *secondary* way should make us recall that the things we know first are not things in our mind, not our ideas, but things in the external world, and thus the *first* meaning (for us) of actuality, and thus form, is what *physical* things are.

On Abstraction

The first in principle, and thus most important, meaning of the word abstraction comes from our intellectual ability to abstract, or pull out, a general understanding from sensorial particulars. For example, when you see (with your senses) a particular pie cut into two pieces your mind (your intellectual power) naturally can pull out the general idea of two. The *general* idea of two, as a *generality*, is not floating around the pie as general; it is only general after your mind leaves out all else but the "two-ness" of the pie. From there, we can then talk about 2 of anything we'd like, specifying that general idea of two in many

---

[2] Note that distinctions (parallel in many ways to what we have made here for physics) must be made in modern mathematics. There we have symbols and rules of manipulation which also only represent actual quantitative aspects of reality, but are not those realities themselves. Those same rules and symbols are, of course, themselves used in modern physics and by analogy are applied to physical realities. The symbols and rules we make in modern mathematics that constitutes "its formalism", allows us to move around the realities we have capture in that formalism. Again, the distinction between sign and thing referenced by the sign needs to be made. For example, the commutative law represented by $a \times b = b \times a$ or taking a specific case $3 \times 2 = 2 \times 3$, i.e. ability to switch places without affecting the result, is not identical with the fact about numbers that 3 twos is the same as 2 threes, but only a handy manipulation that allows us to take account of that fact in our formalism.

different ways. For example, we can understand what is meant when we say, "a bicycle has *two* wheels," only after we have understood the general idea of two. Thus, in going from sense knowledge to intellectual knowledge, we then *are* moving from the particular to the general. However, as long as we remain in the sensory realm, we have not yet done any science, for science is about understanding the world from its principles, and thus is an intellectual act. So, as long as we do not confuse sensorial and intellectual knowledge, we see that we must always go from general to specific; that is, we get the general understanding and add to it more specifics, gradually making our way back to the particular, which after all is the only way things exist in the external world. The mind does this naturally, as we saw with above example of two or in the example of the need to get the general idea of a triangle before we can know the specific idea of an equilateral triangle.

Now, in modern science, through the use of the empiriometric method, we try to copy the natural *abstractive* activity of the mind by creating a theory which is centered around a system of equations. The theory can serve to bring out and organize certain concepts that are only *analogically* like those obtained by natural abstraction. We start by gathering enough particular information to infer a theory. Experiment will then affirm or verify this theory in its domain. Our first theories will of necessity leave much out of consideration, i.e. be more general, and thus have limited domains of validity. Later theories will have to incorporate more of those aspects of reality that we had previously left out; it is in this principled sense that we thus say that we are further specifying our understanding of reality. The theory has captured more of the specifics of reality and thus, in that sense, is closer to it. If we use a practical, rather than a principled definition of "general," i.e. if we mean simply the theory can be used in more cases, then, of course, the newer theory is more general. However, in physics, our *first* interest is in the physical principles, not the practical ease and benefit that they give; indeed the main role of probing a theory's experimental predictions and their scope is to guide us as to how correct the theory is in principle and in what domain those principles are valid.

With this priority of principle understood, we can speak of a more advanced theory, such as general relativity compared to special relativity, as being more general in the sense of covering more cases without inadvertently attacking the fundamental principle that our *understanding* (as opposed to our sensory knowledge) of the world moves from generic to specific.

.

Appendix II

What it Means for the Active and Passive Charges to be Proportional to one Another

To answer, we must look at the field and its action more deeply from a first principles point of view. The strength of a power only exists in relation (though perhaps merely in potency, not in act, at any given moment) to the ability of some substance to receive its action. Action and reception are correlative; that is, they are like top and bottom, each can only exist by being complemented by the other. Action and reception are relational properties, not intrinsic properties. Of course, the power itself, being a quality, is an intrinsic property, but, in so far as the power is ordered to act outside of the substance, that action, by its very notion, implies a relation to another substance which receives that action.

A charged body acts to cause a power, a field, in the plana with a certain strength, i.e. with a certain ability to cause impetus at a certain rate in a massive body receptive to the action of the field. The field has a strength that is proportioned to activate impetus in some second body. Suppose that we know the least receptive body to such action that exists or can exist (in the entire natural contingency of the universe). Thus, the receptivity of this body in a real way conditions or defines the limits of any given field's power to cause impetus in a given time. This least receptive body in this way defines the unit or standard of receptivity, so that all receptivity to the electric field can then be measured relative to this least receptive body. Thus, we can assign this body's *receptivity* to be unity: $q_{least\_passive}^{P} \equiv 1$. Similarly, if we can find the least active body, we can, assign its *activity* to be unity: $q_{least\_active}^{A} \equiv 1$. All others will be compared to (i.e. measured by) these. Note, this still leaves the actual effect unspecified; that is, the intensity of impetus that results in the receiving body in a given time is still undetermined. This bears investigation. By bringing out a few distinctions that have been left fuzzy, we can lay out a clear understanding of all the specifications that are needed.

Upon thought, we can see that specifying only the strength of the power of a body that activates the field in the plana does not even completely specify *the field caused in the plana*. The actual field caused in the plana also depends on the *receptivity of the plana*, though we usually neglect it because we take it to be the same everywhere for "free" and otherwise unqualified plana. Neglecting it does not make it not exist; it must be specified. Furthermore, even given a field specified in the plana, if a force[3] is to act, the plana must

---

[3] Here we have an ambiguity in the use of the word force that we should try to resolve. On the one hand, we often use force to mean only acting powers, powers actually causing impetus. But sometimes we speak of the force as the power itself, whether acting or not, just as we speak of the power to think, even when we are not thinking. Later in the Newtonian course and, in this E&M course practically from the beginning, the term field is used for the power to act and the term force is used only when there is a receiving body upon which

act on the receiving body, so that both this and the previous action/reception *pairs* must be specified. Typically, even though we are confronted with multiple properties, we naturally lump some of them into one measure, and thus one number, as we have been implicitly doing here up till now. In particular, the following are lumped into one number given by q^A: (1) the strength of the power of the first body to act *on the plana*, (2) the amount of receptivity of the plana to the action, and (3) part[4] of the strength of the power *of the plana*, i.e., the magnitude of the electric field, at a given plana part; thus we write: $E = \dfrac{q^A}{r^2}$, where we are lumping the remaining measure of the strength (namely, the decrease in that strength as the length of plana, across which the source must support a field, increases) into

---

the field is acting, either causing impetus or being impeded from causing impetus by another force on the receiving body. In thinking about the language ambiguity around the word "force," we ought to bear in mind this latter nuance; namely, we still use the word "force" for powers that act without causing impetus, but which just resist other powers that can cause impetus.

[4]The measure of the strength of the action of the charge on the plana (described in (1) and (2)) and the propagation of the field by one part of the plana acting on the next are not clearly divided between q and the inverse square law. The reason is that, in classical E&M, we never directly measure the action of the charge on the plana separately. Plana near a larger particle, for example, might have a different receptivity than a smaller one; after all, the plana has a different shape and its minimal parts are likely different as well, possibly indicating such a difference. Such a difference would, because the radius is measured from the charge center, make the receptivity depend on radius and tie it up with the fall off due to propagation. On the other hand, plana could have the same receptivity independent of the size and shape it's forced into, thus decoupling these aspects from the fall off. Also, the receptivity could depend on the strength of the field to be actuated in the given plana cell. This gets complicated because a similar effect might occur in the propagation of the field in the plana, because part of the reason the field strength falls off the way it does may be due to a field strength dependence of the plana receptivity (to plana activated field which might be different than charge activated field). We also could argue that we typically measure the field at larger scales, not right where the field is created, so we are always measuring an electric field that has been propagated; the charge causes the field in the plana and then it propagates, then we measure its effect; however, we should chose large enough particles and plana cells so that this is not an issue because, by definition, we are considering only what can be investigated by classical E&M.

We can, from simplicity of assumption arguments, take all of the first action (item (1) in text) to be measured by q^A, leaving the second (and the action of the plana on the charge) for the inverse square law. And, though it is not clear that this is the case, the evidence is highly suggestive that it is. Further analysis is needed to determine what exactly can be proven even within the context of current understanding of E&M (cf. footnote 10, also 7). The following formula is a mathematical summary of the force law with its terms written explicitly: $F = \tilde{q}_a R_{q \to p} A^{(1)}_{p \to p} R^{(2)}_{p \to p} ... A^{(n-1)}_{p \to p} R^{(n)}_{p \to p} A^{(n)}_{p \to q} \tilde{q}_p$; where $\tilde{q}_a$ and $\tilde{q}_p$ are the strength of the charge acting on the plana and the receptivity of the charge to the action of the plana, $R_{q \to p}$ is the receptivity of the plana to the action of the charge causing a field in it, $A^{(n-1)}_{p \to p}$ is the strength of the power of $(n-1)^{th}$ plana cell acting to cause an electric field in the nth plana cell, $R^{(n)}_{p \to p}$ is the receptivity of the nth plana cell to the action of that previous cell (we can take this to be the same in all cells), $A^{(n)}_{p \to q}$ is the strength of the power of the nth plana cell to cause impetus in the charge. The simplicity assumption at the beginning of this paragraph leads to: $q^A = \tilde{q}_a R_{q \to p}$ and from the main text of article: $q^P = \tilde{q}_p$.

our inverse square fall-off.[5,6] The action of the plana on the charge is also contained in $\dfrac{q^A}{r^2}$, while the measure of level of receptivity of the second body to the action of that field (i.e. rate of activation of impetus in second massive body) is q^P, so that we write: $F = q^P E$. However, it is clear that accounting for these more subtle distinctions is still not sufficient to specify the force. More precisely, even given (at our level of analogical abstraction) (1) the linear nature of the passive and active charges, i.e. increased receptivity or activity does not change the way the force acts[7] and (2) the inverse square law fall-off of the field's strength, more information is needed to specify what force results from the field caused by the minimal active charge acting via the plana on the minimal receptive charge. With the given definition of charge units, this extra information can be specified in Coulomb's law

---

[5] For the case of an electric field activated in a plana cell immediately in contact with a charge (assuming the chosen cell size is not smaller than the domain of applicability of classical E&M), the field would, by definition, still decrease according to this law. The field in that cell and the field felt by the receiving charge would still be determined by the distances involved. However, one would see, if one tries to assign a single value to the cell, a noticeable discretization error. This is because the cell structure is created assuming that one will only be considering a large enough number of cells that the effect of the discretization will be insignificant. (The cell structure is created to illustrate the fact (though we don't know the detail) that the field must propagate from one part of the plana to another, not from one point to another). Again, if the cells are large enough and if (at this level) we avoid assigning a single value to each cell, clearly part of the strength of the field seen to act on the charge will still be approximately represented by the inverse distance portion of the relation.

Of course, as one moves beyond the limit of the domain of classical E&M, Coulomb's law will fail, revealing the need for further (analogical) specification. For example, bringing small charged particles into close proximity requires high energies where other effects come into play. On the other hand, macroscopic particles in close proximity require one to include in one's analysis the compositional nature of such particles, including the nature of the particle's parts and their distances from the parts of the receiving particle.

[6] These details were not mentioned in *PFR: Mechanics* because the focus there was on the interaction of massive bodies generally, not fields as it is, largely, here.

[7] In one sense the linear form of the charges in Coulomb's law is definitional. Namely, one defines the strength of the field by how much impetus it can generate in a given charge in a given unit of time, and one defines the strength of the power to create that field strength, as q^A. By this we mean precisely that the strength of the power to create the field in the sense parsed above (which breaks out its action-reception-action nature) is proportioned to the strength of the field created. However, because there is a separate meaning of charge intensity that can be found in reality, namely the discrete charges, this meaning of charge might in principle give us a separate meaning for charge. In particular, starting with one unit of active charge, say that of an electron, one might talk about doubling the charge by putting two electrons together. Now, the ability to cause field might be, for instance, compromised (perhaps by one interfering with another) by the two coming together. Hence, we might think of writing the numerator of Coulomb's law as: $q_1{}^A q_2{}^P = f\left(\overline{q}_1{}^A\right) g\left(\overline{q}_2{}^P\right)$, where the over-bar indicates this second definition of charge based on the number of discrete charges assembled in a region. However, all experimental evidence suggests that this numerator equals $\overline{q}_1{}^A \overline{q}_2{}^P$. Now, it is easily shown, (by assuming that each function is analytic and expanding each in a power series and equating terms), that the only way this can be true (within a multiplicative constant that can be absorbed in our definition of units) is for $\overline{q}_i{}^A = q_i{}^A$. Another way to define charge might appear if we had a handle on the field other than through the amount of impetus it activates in another massive body. If so, the strength revealed in that new way might reveal, for example, an effect on the plana of the field not before recognized. Something like this happens in the case of general relativity with respect to Newton's law of gravity.

by the inclusion of a constant k: $\quad F = k\dfrac{q^A q^P}{r^2}$.[8] In less fundamental but more practical

units, such as *mks*, this constant will also incorporate the redefinition of charge and distance.[9] This constant represents the determination of how much impetus is caused in a given unit of time by the specification of the full complex of powers, thus revealing something about the natures of these powers and through that of the substance of which they are properties.[10]

So, what then does the *universal* proportionality between the active charge of a given body and its passive charge mean (see equation 2.1)? It means that all bodies of a given active charge will act on all other bodies of the same active charge in the same way. Thus, for example, the previously mentioned body with least active charge will cause (via the plana) the same impetus per unit time in all bodies with that least active charge. Similarly, the same obtains for bodies with active charge of 10 or a hundred units of this least active charge. More completely, this means that a field of a given intensity created by *any* charge in free (and otherwise unqualified) plana is proportioned to be received in the same way by every other charged body, dependent only on that receiving body's active charge, because its active charge is proportionate to the receptive charge. Now, it is <u>not</u> obvious, *a priori*, that this must be the case.

Indeed, one could imagine bodies that could activate fields but could not be acted on by them at all, yielding an infinite ratio of active to passive charge in some bodies. As shown in the figure below, one could also imagine one body (call it a type A body) being *twice as receptive* but with the *same activity* as another; call it type B. In such a case, a type B body acting (via the plana) on another type B body would yield a certain impetus per unit time in each other. However, type A (or type B) acting on type A would yield twice that amount, though, of course, the type A/type B interaction would violate Newton's third law.

---

[8] Or, we could redefine charge to absorb this constant as previously mentioned. One can also write the law in

fundamental units such as: $\quad \dfrac{m_e c^2}{l} \dfrac{F}{\dfrac{m_e c^2}{l}} = \dfrac{\left(\dfrac{q_1}{e}\right)\left(\dfrac{q_2}{e}\right)}{\left(\dfrac{r}{l}\right)^2} \dfrac{e^2}{l^2} \Rightarrow \dfrac{m_e c^2}{l} f = \dfrac{m \cdot n}{j^2} \dfrac{e^2}{l^2}, \; f = \dfrac{m \cdot n}{j^2}\left(\dfrac{\dfrac{e^2}{l^2}}{\dfrac{m_e c^2}{l}}\right)$, where m,

n and j are integers, the first two specifying how many units of charge and the last how many units of distance separate the charges. m_e is the mass of an electron, l is the fundamental length say the Planck length, e is the charge on an electron and f is the number of fundamental units of force.

[9] We have not addressed the fundamentals of force which should be in terms of momentum (measured intensity of impetus relative to a fundamental standard unit of impetus) per fundamental unit time. The units chosen for impetus and time, of course, also end up affecting the value of k.

[10] Note there is at least one other issue that needs to be treated, which we will *not* treat here; broadly, we can call it the issue of superposition of fields. For example, how does the already existing field of the receiving charge affect the receptivity of the plana to the action of the electric field on plana and a charge on plana.

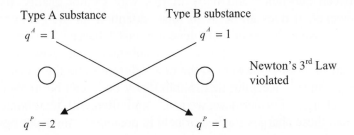

Type A substance   Type B substance
$q^A = 1$ $q^A = 1$

Newton's 3rd Law
violated

$q^P = 2$ $q^P = 1$

Now, if we try to go beyond saying "the ratio of active to passive charge is the same for every charged body" by trying to assign a value other than one to this ratio, we run into problems. Suppose instead of assigning, as we normally do, $q^A = q^P$, we take $q^A = 10q^P$. If we continue to assign a value of unity to the receptivity of the least receptive body previously mentioned ($q_{least\_passive}^P = 1$), the activity of the least active body can no longer be assigned a value of unity, but must be greater than or equal to ten ($q_{least\_active}^A \geq 10$), for if it were assigned any less, it would imply the contradiction that it had less receptivity than the least receptive body. But, assigning 10 or more to the least active body would be merely a construct (and not a particularly useful one), because, as we have seen, the least active body should properly be the standard by which we measure active charge. Any such non-unitary definition is also equivalent to redefining the constant *k*.

At our level of analogical abstraction, the fundamental point is that the active charge of a body is a power ordered to cause a field in the plana and thus is ordered (final cause) to cause impetus via the plana in another body. In this way, the activity of the activating charge is *proportioned* to the receptivity of the receiving charge. Now, as we have said, experiment reveals that the active and passive charges are always proportioned in the same way in every particle. Of course, one particle can still have a larger active and passive charge than another particle, as long as they are proportioned in the same way as any other particle. To account for this possibility, we need a measure of the overall charge of a body. Since action is notionally prior to reception, we take this overall charge, q, to be equal to the active charge: $q = q^A$. Hence, once we specify the active charge and that the action and reception are proportioned in only one way for a given active charge, we then only need to specify the level of the effect that that particular proportion causes. Namely, as we have also said earlier, we need to specify the amount of impetus per unit time that results for a given activity and a given receptivity; in the Coulomb's equation this is, again, *k*.

In actuality, of course, we have found things like protons and electrons (cf. footnote 7) that have exactly the same magnitude of charge (though of different directional type or mode represented by a difference in sign) from whose interaction substances are made that radically incorporate them. Since the charge of an electron in a new substance acts, at some analogically generic level, in the same way as it does when part of a group of electrons, we can "abstract" from whether one deals with charge in a substance or group of substances or even part of a substance—you will recall we made this analogical abstraction in Newtonian physics and used the term "body" to represent that meaning. This is a crucial point and is essential to building the empiriometric theory of E&M. As usual, leaving out the distinction between substance and body in this way cannot change the reality of this distinction. However, it does allow us to focus attention on the qualities of the electron called active and passive charge or more generally just "charge." We can say the electron is virtually present, i.e. present by its powers, in a substance. The powers of the electron (in this case, its charge), became powers of the new substance and the electron ceased to exist as a separate substance.[11] Keeping this in mind, any body can be treated as composed of many individual charges. Furthermore, we can,[12] and often will, treat bodies as if they were nothing more than these charges stacked or held in position in some configuration.

---

[11] One may claim that the free electron is a property of the plana by saying that further research into the empiriological grounding and context (and/or further empiriometric research) might reveal evidence pointing in this direction. However, at our current level of understanding, all the evidence supports or is consistent with the view that a free electron is a substance, as tenuous a one as it may be.

[12] At one level of analogical abstraction.

Appendix III

Mathematics of Conservative Forces[13]

If we have a force, how do we know if it is conservative? We can graph it and look to see if a closed path integral is always zero. But, we can also look at the curl $\vec{\nabla} \times \vec{F}$ of the force field, $\vec{F}$, which can be written in the following way:

(1)
$$\vec{\nabla} \times \vec{F} = \left(\hat{x}\frac{\partial}{\partial x} + \hat{y}\frac{\partial}{\partial y} + \hat{z}\frac{\partial}{\partial y} \right) \times \left(F_x\hat{x} + F_y\hat{y} + F_z\hat{z} \right)$$

The curl tells us how much the force "curls" around a given point arbitrarily close to that point. We may have guessed this by recalling that $\vec{A} \times \vec{B}$ introduces the notion of rotation that implicitly brings in 3-D through the concept of cross product (see Chapter 2, *PFR-M*). We can show explicitly how $\vec{\nabla} \times \vec{F}$ yields amount of curling.

Take the small loop shown in Figure III-1 below with some new notation to facilitate our present task.

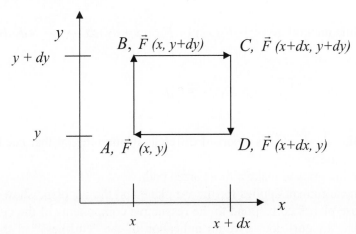

Figure III-1: Small rectangular path of side length dx and dy around which we calculate the work.

Note lines *AB* and *CD* each extend dy while lines *BC* and *DA* extend dx,

---

[13] The pages in this appendix are reproduced from Chapter 5 of PFR-M (pages 246 to middle of 247 just after Fig. 5-17 of first edition) with minor emendations to prevent referencing to figures or equations not reproduced here or to simplify their numbering.

Consider first the AB and CD portion of the line integral: $\oint \vec{F} \cdot d\vec{r}$. Using F_y to mean the y component of the Force, $\vec{F} = F_x \hat{x} + F_y \hat{y} + F_z \hat{z}$, and using the average of the forces at the two end points to calculate the integral along each segment, we get:

AB: $\left(\dfrac{F_y(x,y) + F_y(x, y+dy)}{2} \right) dy \approx F_y(x,y) dy$ because the second term differs from the first only in second order in dy.

CD: $-\left(\dfrac{F_y(x+dx, y+dy) + F_y(x+dx, y)}{2} \right) dy \approx -F_y(x+dx, y) dy$ because the first term differs from the second only in second order in dy.

So, the $AB + CD$ term of line integral gives: $\dfrac{\left(F_y(x,y) - F_y(x+dx) \right)}{dx} dx\, dy \rightarrow \dfrac{\partial F_y}{\partial x} dx\, dy$

Similarly for the two horizontal components, BC and DA of the line integral

BC: $\left(\dfrac{F_x(x, y+dy) + F_x(x+dx, y+dy)}{2} \right) dx \approx F_x(x, y+dy) dx$

DA: $-\left(\dfrac{F_x(x+dx, y) + F_x(x,y)}{2} \right) dx \approx -F_x(x,y) dx$

So, $BC + DA$ term of line integral gives: $\left(-F_x(x,y) + F_x(x, y+dy) \right) dx \rightarrow -\dfrac{\partial F_x}{\partial y} dy\, dx$

Giving:

(2) $\displaystyle\oint_{ABCD} \vec{F} \cdot d\vec{r} = \dfrac{\partial F_y}{\partial x} dy\, dx - \dfrac{\partial F_x}{\partial y} dy\, dx = (\vec{\nabla} \times \vec{F})_z\, dx\, dy$.

Recall this line integral, $\displaystyle\oint_{ABCD} \vec{F} \cdot d\vec{r}$, is a sort of differential line integral that needs to be summed with other line integrals to make a finite sized path.

Exactly the same argument applies for the x-z plane and the y-z plane showing that the line integrals in those planes correspond to the respective components of the curl of F. This, in turn, shows that the curl does give an indication of the "curliness" of the field, because, near a point, this is precisely what one means by curliness, i.e. that the line integral is non-zero. So, if curl is zero, the line integral around the region arbitrarily close to the point is zero.

To move to a finite loop integral, we complete the previously mentioned sum. The path of a finite line integral is shown in blue in Figure III-2; the smaller black loops approximate the larger loop better and better as one approaches infinitesimally small loops. In short, in the limit, these small line integrals sum up to the line integral around the area because where the regions touch, i.e. in the middle, leads to a net integral of zero because each integral path in the interior is paired with one of the opposite sign. In the diagram, only those with only one arrow on a line segment contribute to the sum of all the differential line integrals.

It is intuitively clear that this argument will work for any Euclidean surface. Combining this with *Equation (2)* , we, thus, obtain the finite version of that equation, which is also called ***Stokes Theorem:***

(3)
$$\oint_{Path} \vec{F} \cdot d\vec{x} = \oiint_{Surface\ defined\ by\ Path} \left(\vec{\nabla} \times \vec{F} \right) \cdot \hat{n}\, dA$$

Exercise: re-derive this in detail without referencing the textbook

Stokes theorem, in turn, implies that when the curl of a force field is zero, its line integral around any path is zero and, thus, that force field is path independent and thus conservative.

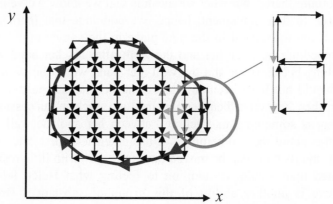

Figure III-2: We add up the line integrals of all the small square loops that approximately cover the *interior surface bounded by the blue line*. In the limit of infinitesimally small loops, this sum equals the integral along the blue curve because the integral along *each* little line segment in the interior has an opposite sign pair that cancels its contribution to the sum. This cancellation is represented in the figure by arrows of opposing directions on a given segment. The upper right shows a close up of two such small path integrals with the cancellation at the shared boundaries shown. The two orange arrows in this magnified view represent contributions from the neighboring loops not visible in the magnification.

Appendix IV

Why do we need plana?

Remember everything we know comes by way of what we feel, see, hear, smell and taste. To begin to answer our title question, consider a blind and deaf person, such as Helen Keller, in a room with no wind. She reaches her hand out and feels her arm extending *without encountering noticeable resistance* until it reaches the wall, at which point she can push very hard, yet her hand will move no further. Though she doesn't feel the substance(s) through which she moved her hand, she *does* feel her arm being extended,[14] i.e. feels it moving forward, which means it is going from one place to the next in an environment, i.e. in a something.[15] You and I, who have the benefit of sight, might quickly say, "there's nothing there," but what we mean is that we know we've moved, and we have felt no resistance to that movement; hence, we conclude that there is *nothing that can noticeably resist* my movement to the new place in the something. Again, Helen realizes that there is something between her and the wall; after all, her hand went forward and if there were nothing between her and the wall, her hand would be stopped immediately by the wall. She would not have been able to extend her hand, would not have been able to move her hand then her wrist then her forearm, etc. each in succession to a given place. She knows there is something intervening between her and the wall; however, she only knows two things about the "something." It has extension (first category of property of a physical thing), and its parts can be moved out of the way with little resistance.

A sighted man would, in addition to feeling what Helen felt, *see* the wall and realize that there is another ability of this group of substances (free plana, nitrogen, oxygen, etc.) which we call air; namely, it has the power to carry information about the wall from the wall to him. If he (or Helen), in a quick motion, swatted the air, he would feel the something moving between his fingers and realize that it is less different from the things around him than he might first have guessed. This air seems to be a very "thin" version of a liquid, i.e. it's very easy to move through. Thus, in ordinary life, Helen's experiment never happens, because we *can* feel air.

However, our sighted man (or Helen) might *imagine* further. He might imagine making the air arbitrarily thin, i.e. arbitrarily easy to move through, getting to what we call a vacuum. He may realize that air gets thinner as one climbs higher above the Earth's surface, leading eventually to the vacuum of space. This, however, would not be "nothing." A vacuum is not "nothing." Nothing is no-thing; one must be very careful with this word.

---

[14] Unlike your hair, for example, you can feel the position of your arm. Someone might tell a friend that her hair is wind blown and tangled, but he does not have to tell her that her arm has bent in a funny way and should be rearranged.

[15] One might invent a new concept of motion in which one imagines stretching the size of the universe, which would then not be "motion" properly speaking at all but some kind of growth. For example, one could imagine the universe *growing* in some part or all of its extension. Growth, insofar as what we know directly, would have to be growth into something, but if a line of argument leads one to the necessity of a growth of the entire universe, which by definition cannot grow into any environment, one would have to posit this extended type of growth. (In fact, until such a proof from what we *actually* know is established, one could not even know whether such an extended understanding of growth makes any sense.) However, this is not all the line of discussion here, which is about what we have direct knowledge of.

If taken literally, it can often lead to nonsense. In loose speech, it sometimes means literally nothing and sometimes means the absence of some particular thing or class of thing. If clarity were reached, we would realize we usually want to say the latter. "There's nothing in the closet" does not mean that there are no shelves or shelf paper or air. It means there is no food or clothes or whatever is called for by the context of the sentence.

In the "vacuum" of space, Helen Keller, this time in her space suit, could repeat her above experiment, this time reaching out to touch the outer hull of her spacecraft. She would come to the same fundamental conclusions; there is something there that has extension and can be moved out of the way. Again, her sighted companion would add to this that this substance, which we call plana, has a property (or properties) that allow information about the hull to be transmitted through it. This is what we call light. Light, as we have discussed in the text, is a property of the plana.

Thus, Helen knows that there is something between her and the spacecraft. When we pump the air (i.e., the massive substances: the nitrogen, oxygen, etc.) out of a vacuum chamber, the walls of the chamber remain separated though the massive components are gone. It is clear there is still something in the chamber. For example, because there is, in fact, an interior to the chamber, it is clear that each set of opposing walls of the chamber is separated, i.e. there is an extension between them; this is, after all, what we mean by an interior. Now, "nothing" is not extended -- only something is. Hence, the question is not "Is there nothing in the chamber?", but "What is the nature of the something in the chamber?"

All we do with the word plana is give that "something" a name. We only know very general things about the something, such as what Helen and the sighted man understood at first. Our attention then naturally turns to learning more about the nature of this unusual something. Whether it is called "vacuum," "space" or "plana," the issue remains; we either handle it directly or let it be handled indirectly through our equations. This latter method is very useful and even critical when we wish to focus our attention on other things, but clearly it is no help when we want to focus on the nature of the plana itself or just to recognize that it is there. Remember, we simply call it "plana" after the feminine of the Latin word for "field", because its most obvious specifying qualities are its receptivities to various fields. The word plana carries an extremely general connotation[16], leaving open the question of its more specific nature.

Finally, remember whether you are walking across the street or space-walking outside the shuttle, even if your goal is not the wall or anything massive, even if your goal is the middle of the room or the middle of space (free plana), you did not walk through nothing and you did not end up nowhere!

---

[16] This contrasts markedly with already greatly specified notions like that of the classical ether.

Appendix V: Mars Fact Sheet

Mars/Earth Comparison

Bulk parameters

(Mars/Earth)	Mars	Earth	Ratio
Mass (10^{24} kg)	0.64185	5.9736	0.107
Volume (10^{10} km^3)	16.318	108.321	0.151
Equatorial radius (km)	3397	6378.1	0.533
Polar radius (km)	3375	6356.8	0.531
Volumetric mean radius (km)	3390	6371.0	0.532
Core radius (km)	1700	3485	0.488
Ellipticity (Flattening)	0.00648	0.00335	1.93
Mean density (kg/m^3)	3933	5515	0.713
Surface gravity (m/s^2)	3.71	9.80	0.379
Surface acceleration (m/s^2)	3.69	9.78	0.377
Escape velocity (km/s)	5.03	11.19	0.450
GM (x 10^6 km^3/s^2)	0.04283	0.3986	0.107
Bond albedo	0.250	0.306	0.817
Visual geometric albedo	0.150	0.367	0.409
Visual magnitude V(1,0)	-1.52	-3.86	-
Solar irradiance (W/m^2)	589.2	1367.6	0.431
Black-body temperature (K)	210.1	254.3	0.826
Topographic range (km)	30	20	1.500
Moment of inertia (I/MR^2)	0.366	0.3308	1.106
J$_2$ (x 10^{-6})	1960.45	1082.63	1.811
Number of natural satellites	2	1	
Planetary ring system	No	No	

Orbital parameters

(Mars/Earth)	Mars	Earth	Ratio
Semimajor axis (10^6 km)	227.92	149.60	1.524
Sidereal orbit period (days)	686.980	365.256	1.881
Tropical orbit period (days)	686.973	365.242	1.881
Perihelion (10^6 km)	206.62	147.09	1.405
Aphelion (10^6 km)	249.23	152.10	1.639
Synodic period (days)	779.94	-	-
Mean orbital velocity (km/s)	24.13	29.78	0.810
Max. orbital velocity (km/s)	26.50	30.29	0.875
Min. orbital velocity (km/s)	21.97	29.29	0.750
Orbit inclination (deg)	1.850	0.000	-
Orbit eccentricity	0.0935	0.0167	5.599
Sidereal rotation period (hrs)	24.6229	23.9345	1.029
Length of day (hrs)	24.6597	24.0000	1.027
Obliquity to orbit (deg) *(tilt of axis)*	25.19	23.45	1.074

Mars Observational Parameters

```
Discoverer:      Unknown
Discovery Date:  Prehistoric
```

```
Distance from Earth
        Minimum (10^6 km)                        55.7
        Maximum (10^6 km)                       401.3
Apparent diameter from Earth
        Maximum (seconds of arc)                 25.1
        Minimum (seconds of arc)                  3.5
Mean values at opposition from Earth
        Distance from Earth (10^6 km)            78.39
        Apparent diameter (seconds of arc)       17.9
        Apparent visual magnitude                -2.0
Maximum apparent visual magnitude                -2.91
```

Mars Mean Orbital Elements (J2000)

```
Semimajor axis (AU)                      1.52366231
Orbital eccentricity                     0.09341233
Orbital inclination (deg)                1.85061
Longitude of ascending node (deg)       49.57854
Longitude of perihelion (deg)          336.04084
Mean Longitude (deg)                   355.45332
```

North Pole of Rotation

```
Right Ascension: 317.681 - 0.108T
Declination    :  52.886 - 0.061T
Reference Date : 12:00 UT 1 Jan 2000 (JD 2451545.0)
T = Julian centuries from reference date
```

Martian Atmosphere

```
Surface pressure:  6.36 mb at mean radius (variable from 4.0 to 8.7 mb
depending on season)
                [6.9 mb to 9 mb (Viking 1 Lander site)]
Surface density: ~0.020 kg/m^3
Scale height:  11.1 km
Total mass of atmosphere: ~2.5 x 10^16 kg
Average temperature:  ~210 K (-63 C)
Diurnal temperature range: 184 K to 242 K (-89 to -31 C) (Viking 1 Lander
site)
Wind speeds:  2-7 m/s (summer), 5-10 m/s (fall), 17-30 m/s (dust storm)
(Viking Lander sites)
Mean molecular weight: 43.34 g/mole
Atmospheric composition (by volume):
    Major       : Carbon Dioxide (CO_2) - 95.32% ; Nitrogen (N_2) - 2.7%
                  Argon (Ar) - 1.6%; Oxygen (O_2) - 0.13%;
                  Carbon Monoxide (CO) - 0.08%
    Minor(ppm) : Water (H_2O) - 210; Nitrogen Oxide (NO) - 100;
                  Neon (Ne) - 2.5;
                  Hydrogen-Deuterium-Oxygen (HDO) - 0.85;
                  Krypton (Kr) - 0.3; Xenon (Xe) - 0.08
```

Satellites of Mars

	Phobos	Deimos
Semi-major axis* (km)	9378	23459
Sidereal orbit period (days)	0.31891	1.26244
Sidereal rotation period (days)	0.31891	1.26244
Orbital inclination (deg)	1.08	1.79
Orbital eccentricity	0.0151	0.0005
Major axis radius (km)	13.4	7.5
Median axis radius (km)	11.2	6.1
Minor axis radius (km)	9.2	5.2
Mass (10^{15} kg)	10.6	2.4
Mean density (kg/m^3)	1900	1750
Geometric albedo	0.07	0.08
Visual magnitude V(1,0)	+11.8	+12.89
Apparent visual magnitude (V_0)	11.3	12.40

*Mean orbital distance from the center of Mars.

If no sub- or superscripts appear on this page - for example, if the "Mass" is given in units of "(1024 kg)" - you may want to check the notes on the sub- and superscripts at http://www.spds.nasa.gov/planetary/factsheet/fact_notes.html.

Procession of Mars Equinox 175,000 earth years (cf. wikipedia)
Procession of Earth Equinox 26,000 (25,800) earth years
Tilt of earth's axis: 23 degrees 27 arcminutes See above Obliquity.

Highest point on surface Olympus Mons (about 24 km above surrounding lava plains)

Atmospheric components 95% carbon dioxide, 3% nitrogen, 1.6% argon Surface materials basaltic rock and altered materials[17]

---

[17] Most of information culled from NASA's Dr. David R. Williams' information sheet (dave.williams@gsfc.nasa.gov, Greenbelt, MD 20771 +1-301-286-1258).

Appendix VI

Common Physical Constants

		cgs	*SI* (and others)
Speed of light	c	$3.00 \times 10^{10} \, cm/s$	$3.00 \times 10^8 \, m/s$
Elementary charge	e	$4.803 \times 10^{-10} \, esu$	$1.60 \times 10^{-19} \, C$
Electron mass	m_e	$9.11 \times 10^{-28} \, g$	$9.11 \times 10^{-31} \, kg$
Mass-energy relation	c^2	$8.99 \times 10^{20} \, erg/g$	$8.99 \times 10^{16} \, J/kg$ (931.5 *MeV/u*)
Proton mass	m_p	$1.67 \times 10^{-24} \, g$	$1.67 \times 10^{-27} \, kg$
Bohr radius	r_B	$.529 \times 10^{-8} \, cm$	$.529 \times 10^{-10} \, m$
Avogadro constant	N_A		$6.02 \times 10^{23} \, mol^{-1}$
Boltzmann constant	k	$1.38 \times 10^{-16} \, erg/K$	$1.38 \times 10^{-23} \, J/K$ ($8.62 \times 10^{-5} \, V/K$)
Gravitational constant	G	$6.67 \times 10^{-8} \, dyne \cdot cm^2/g^2$	$6.67 \times 10^{-11} \, N \cdot m^2/kg^2$
Planck constant	h	$6.63 \times 10^{-27} \, erg \cdot s$	$6.63 \times 10^{-34} \, J \cdot s$ ($4.14 \times 10^{-15} \, eV \, s$)
Bohr Magneton	μ_B	$9.27 \times 10^{-21} erg/G \, {(e\hbar/(2m_e c))}$	$9.27 \times 10^{-24} \, J/T \, {(e\hbar/2m_e)}$
Coulomb's law constant	k_e	1	$8.99 \times 10^9 \, Nm^2/C^2$
Permittivity of free space		1	$\epsilon_0 = 8.85 \times 10^{-12} \, F/m$
Permeability of free space		1	$\mu_0 = 4\pi \times 10^{-7} \, H/m$
Electron Volt (eV)		$1.60 \times 10^{-12} erg$	$1.60 \times 10^{-19} \, J$

Common Physical Properties

Air (dry, at 20°C and 1 atm)

Density	$1.21 \, kg/m^3$
Specific heat molar at constant pressure	$1010 \, J/kg \cdot K$
Ratio of molar specific heats	1.40
Speed of sound	$343 \, m/s$
Electrical breakdown strength	$3 \times 10^6 \, V/m$
Effective molar mass	$0.0289 \, kg/mol$

Water

Density	$1000 \, kg/m^3$
Speed of sound	$1460 \, m/s$
Specific heat at constant pressure	$4190 \, J/kg \cdot K$
Heat of fusion (0°C)	$333 \, kJ/kg$
Heat of vaporization (100°C)	$2260 \, kJ/kg$
Molar mass	$0.0180 \, kg/mol$

Earth

Mass	$5.98 \times 10^{24} \, kg$
Mean radius	$6.37 \times 10^6 \, m$
Free-fall acceleration at the Earth's surface	$9.81 \, m/s^2$
Standard atmosphere	$1.01 \times 10^5 \, Pa$
Period of satellite at 100-km altitude	$86.3 \, min$
Radius of the geosynchronous orbit	$42,200 \, km$
Escape speed	$11.2 \, km/s$
Magnetic dipole moment	$8.0 \times 10^{22} \, A \cdot m^2$
Mean electric field at surface	$150 \, V/m$, *down*

Astronomical Distances

From Earth to:

Moon	$3.82 \times 10^8 \; m$
Sun	$1.50 \times 10^{11} \; m$
Nearest star	$4.04 \times 10^{16} \; m$
Galactic center	$2.2 \times 10^{20} \; m$
Andromeda galaxy	$2.1 \times 10^{22} \; m$
Edge of the observable universe	$\sim 10^{26} \; m$

Common Conversion Factors

Mass and Density

$1 \; kg = 1000 \; g = 6.02 \times 10^{26} \; u$

$1 \; slug = 14.6 \; kg$

$1 \; u = 1.66 \times 10^{-27} \; kg$

$1 \; kg/m^3 = 10^{-3} \; g/cm^3$

Length and Volume

$1 \; m = 100 \; cm = 39.4 \; in. = 3.28 \; ft$

$1 \; mi = 1.61 \; km = 5280 \; ft$

$1 \; in. = 2.54 \; cm$

$1 \; nm = 10^{-9} m = 10 Å$

$1 \; light\text{-}year = 9.46 \times 10^{15} \; m$

$1 \; m^3 = 1000 \; L = 35.3 \; ft^3 = 264 \; gal$

Time

$1 \; d = 86,400 \; s$

$1 \; y = 365 \; ¼ \; d = 3.16 \times 10^7 s$

Angular Measure

$1 \; rad = 57.3° = 0.159 \; rev$

$\pi \; rad = 180° = ½ \; rev$

Speed

$1 \; m/s = 3.28 \; ft/s = 2.24 \; mi/h$

$1 \; km/h = 0.621 \; mi/h = 0.278 \; m/s$

Force and Pressure

$1 \; N = 10^5 \; dyne = 0.225 \; lb$

$1 \; lb = 4.45 \; N$

$1 \; Pa = 1 \; N/m2 = 10 \; dyne/cm^2$
$\qquad = 1.45 \times 10^{-4} \; lb/in.^2$

$1 \; atm = 1.01 \times 10^5 \; Pa = 14.7 \; lb/in.^2$
$\qquad = 76 \; cm\text{-}Hg$

Energy and Power

$1 \; J = 10^7 \; erg = 0.239 \; cal = 0.738 \; ft\text{-}lb$

$1 \; kW\text{-}h = 3.6 \times 10^6 \; J$

$1 \; cal = 4.19 \; J$

$1 \; eV = 1.60 \times 10^{-19} \; J$

$1 \; horsepower = 746 \; W = 550 \; ft \cdot lb/s$

Some Relationships between *SI* and *cgs* Systems of Units

	Symbol	*SI (mksA)*	*cgs*
length	$l, d, r...$	$1\ m$	$10^2\ cm$
mass	m	$1\ kg$	$10^3 g$
momentum	$p = mv$	$1\ buridan\ (B),\ (kg{\cdot}m/s)$	$10^5\ phor\ (g{\cdot}cm/s)$
force	$F = dp/dt$	$1\ B/s\ (kg{\cdot}m/s^2)$	$10^5\ dyne\ (g{\cdot}cm/s^2)$
Energy	E	$1\ joule\ (J),\ (kg{\cdot}m^2/s^2)$	$10^7\ erg\ (g{\cdot}cm^2/s^2)$
charge	q	$1\ coulomb\ (C)$	$3 \times 10^9\ esu$
charge density	ρ	$1\ C/m^3$	$3 \times 10^3\ esu/cm^3$
current	I	$1\ C/s$	$3 \times 10^9\ esu/s$
current density	$\vec{J} = \rho \vec{v}$	$(C/s)/m^2$	$3 \times 10^5\ (esu/s)/cm^2$
scalar potential (ϕ-field) (potential energy per unit charge)	ϕ, V	$1\ volt\ (V),\ (J/C)$ $$\phi = \frac{1}{4\pi\epsilon_0}\frac{q}{R}$$	$1/300\ esu/cm,\ (statvolt),$ (erg/esu) $$\phi = \frac{q}{R}$$
vector potential (*A*-field) (potential momentum per unit charge)	A	$1\ B/C,\quad (T{\cdot}m)$ $$d\vec{A} = \frac{\mu_0}{4\pi}\frac{Id\vec{l}}{R}$$	$10^6 esu/cm,\ (G{\cdot}cm)$ $$d\vec{A} = \frac{Id\vec{l}}{cR}$$
E-field	E	$1\ V/m$ $$\vec{E} = \frac{1}{4\pi\epsilon_0}\frac{q}{R^2},\ \vec{E} = -\nabla\phi - \frac{\partial \vec{A}}{\partial t}$$	$3.3 \times 10^{-5} statvolt/cm\ (esu/cm^2)$ $$\vec{E} = \frac{q}{R^2},\ \vec{E} = -\nabla\phi - \frac{1}{c}\frac{\partial \vec{A}}{\partial t}$$
B-field	B $(\vec{B} = \nabla \times \vec{A})$	$1\ tesla\ (T)$ $$d\vec{B} = \frac{\mu_0}{4\pi}\frac{Id\vec{l} \times \hat{R}}{R^2}$$	$10^4\ gauss\ (G),\ (esu/cm^2)$ $$d\vec{B} = \frac{Id\vec{l} \times \hat{R}}{cR^2}$$
force due to *E*-field	F	$\vec{F} = q\vec{E}$	$\vec{F} = q\vec{E}$
force due to *B*-field	F *point particle,* *2 line currents*	$\vec{F} = q\vec{v} \times \vec{B},$ $$\frac{F}{l} = \frac{\mu_0}{2\pi}\frac{I_1 I_2}{r}$$	$\vec{F} = q\dfrac{\vec{v}}{c} \times \vec{B},$ $$\frac{F}{l} = \frac{1}{c^2}\frac{2I_1 I_2}{r}$$
permittivity of free space	ϵ	*dimensionless*	$\epsilon_0 = F/m$
permeability of free space	μ	*dimensionless*	$\mu_0 = H/m\ or\ N/A^2$
resistance	R	$1\ ohm\ (\Omega)$	$1.113 \times 10^{-12}\ s/cm$
capacitance	C	$1\ farad\ (F)$	$9 \times 10^{11} cm\ (1\ cm \sim 1\mu F)$
Inductance	L	$1\ henry\ (H)$	$1.113 \times 10^{-12} s^2/cm$

Some other helpful relations:

cgs: $\sqrt{dyne} = esu/cm$, SI: B-field tesla=*buridan/m*

$4.4N \sim 1lbs$; $1N \sim$ force of gravity on a 100 gram body.

SI: $\vec{S} = \dfrac{1}{\mu_0}\vec{E} \times \vec{B}$, $dU_E = \dfrac{\epsilon_0}{2}\vec{E} \cdot \vec{E}\ dV$, $dU_B = \dfrac{1}{2\mu_0}\vec{B} \cdot \vec{B}\ dV$

cgs: $\vec{S} = \dfrac{c}{4\pi}\vec{E} \times \vec{B}$, $dU_E = \dfrac{\vec{E} \cdot \vec{E}}{8\pi}\ dV$, $dU_B = \dfrac{\vec{B} \cdot \vec{B}}{8\pi}\ dV$

Maxwell's Equations *(SI)*

I. $\nabla \cdot \vec{E} = \rho/\epsilon_0$ II. $\nabla \cdot \vec{B} = 0$

III. $\nabla \times \vec{E} = -\dfrac{\partial \vec{B}}{\partial t}$ IV. $\nabla \times \vec{B} = \mu_0 \vec{J} + \dfrac{1}{c^2}\dfrac{\partial \vec{E}}{\partial t}$

Maxwell's Equations *(cgs)*

I. $\nabla \cdot \vec{E} = 4\pi\rho$ II. $\nabla \cdot \vec{B} = 0$

III. $\nabla \times \vec{E} = -\dfrac{1}{c}\dfrac{\partial \vec{B}}{\partial t}$ IV. $\nabla \times \vec{B} = \dfrac{4\pi \vec{J}}{c} + \dfrac{1}{c}\dfrac{\partial \vec{E}}{\partial t}$

(See next page)

Maxwell's Equations *(SI)*

I. $\nabla \cdot \vec{E} = \rho / \epsilon_0$ II. $\nabla \cdot \vec{B} = 0$

III. $\nabla \times \vec{E} = -\dfrac{\partial \vec{B}}{\partial t}$ IV. $\nabla \times \vec{B} = \mu_0 \vec{J} + \dfrac{1}{c^2} \dfrac{\partial \vec{E}}{\partial t}$

Maxwell's Equations *(cgs)*

I. $\nabla \cdot \vec{E} = 4\pi\rho$ II. $\nabla \cdot \vec{B} = 0$

III. $\nabla \times \vec{E} = -\dfrac{1}{c} \dfrac{\partial \vec{B}}{\partial t}$ IV. $\nabla \times \vec{B} = \dfrac{4\pi \vec{J}}{c} + \dfrac{1}{c} \dfrac{\partial \vec{E}}{\partial t}$

Maxwell's Media Equations *(SI)*

I. $\nabla \cdot \vec{D} = \rho_{free}$ II. $\nabla \cdot \vec{B} = 0$

III. $\nabla \times \vec{E} = -\dfrac{\partial \vec{B}}{\partial t}$ IV. $\nabla \times \vec{H} = \vec{J}_{free} + \dfrac{\partial \vec{D}}{\partial t}$

Maxwell's Media Equations *(cgs)*

I. $\nabla \cdot \vec{E} = 4\pi\rho_{free}$ II. $\nabla \cdot \vec{B} = 0$

III. $\nabla \times \vec{E} = -\dfrac{1}{c} \dfrac{\partial \vec{B}}{\partial t}$ IV. $\nabla \times \vec{H} = \dfrac{4\pi \vec{J}_{free}}{c} + \dfrac{1}{c} \dfrac{\partial \vec{D}}{\partial t}$

$$\vec{D} \equiv \epsilon \vec{E}, \vec{B} = \mu \vec{H}$$

cgs:
$$\vec{D} = \epsilon \vec{E} = \vec{E} + 4\pi \vec{P} = \left(1 + 4\pi\chi_e\right)\vec{E}$$
$$\vec{B} = \mu \vec{H} = \vec{H} + 4\pi \vec{M} = \left(1 + 4\pi\chi_m\right)\vec{H}$$

NB: top boxes repeated for convenience

Appendix VII
Grad, Div and Curl and Vector Relations

Grad, Div, Curl and Laplacian

Cartesian:

$$\nabla \Phi = \frac{\partial \Phi}{\partial x}\hat{x} + \frac{\partial \Phi}{\partial y}\hat{y} + \frac{\partial \Phi}{\partial z}\hat{z}$$

$$\vec{\nabla} \cdot \vec{V} = \frac{\partial V_x}{\partial x} + \frac{\partial V_y}{\partial y} + \frac{\partial V_z}{\partial z}$$

$$\vec{\nabla} \times \vec{V} = \left(\frac{\partial V_z}{\partial y} - \frac{\partial V_y}{\partial z} \right)\hat{x} + \left(\frac{\partial V_x}{\partial z} - \frac{\partial V_z}{\partial x} \right)\hat{y} + \left(\frac{\partial V_y}{\partial x} - \frac{\partial V_x}{\partial y} \right)\hat{z}$$

$$\nabla^2 \Phi = \frac{\partial^2 \Phi}{\partial x^2} + \frac{\partial^2 \Phi}{\partial y^2} + \frac{\partial^2 \Phi}{\partial z^2}$$

Cylindrical:

$$\vec{\nabla}\Phi = \frac{\partial \Phi}{\partial \rho}\hat{\rho} + \frac{1}{\rho}\frac{\partial \Phi}{\partial \phi}\hat{\phi} + \frac{\partial \Phi}{\partial z}\hat{z}$$

$$\vec{\nabla} \cdot \vec{V} = \frac{1}{\rho}\frac{\partial}{\partial \rho}\left(\rho V_\rho \right) + \frac{1}{\rho}\frac{\partial V_\phi}{\partial \phi} + \frac{\partial V_z}{\partial z}$$

$$\vec{\nabla} \times \vec{V} = \left(\frac{1}{\rho}\frac{\partial V_z}{\partial \phi} - \frac{\partial V_\phi}{\partial z} \right)\hat{\rho} + \left(\frac{\partial V_\rho}{\partial z} - \frac{\partial V_z}{\partial \rho} \right)\hat{\phi} + \frac{1}{\rho}\left(\frac{\partial}{\partial \rho}\left(\rho V_\phi \right) - \frac{\partial V_\rho}{\partial \phi} \right)\hat{z}$$

$$\nabla^2 \Phi = \frac{1}{\rho}\frac{\partial}{\partial \rho}\left(\rho \frac{\partial \Phi}{\partial \rho} \right) + \frac{1}{\rho^2}\frac{\partial^2 \Phi}{\partial \phi^2} + \frac{\partial^2 \Phi}{\partial z^2}$$

Spherical:

$$\vec{\nabla}\Phi = \frac{\partial \Phi}{\partial r}\hat{r} + \frac{1}{r}\frac{\partial \Phi}{\partial \theta}\hat{\theta} + \frac{1}{r\sin\theta}\frac{\partial \Phi}{\partial \phi}\hat{\phi}$$

$$\vec{\nabla} \cdot \vec{V} = \frac{1}{r^2}\frac{\partial}{\partial r}\left(r^2 V_r \right) + \frac{1}{r\sin\theta}\frac{\partial}{\partial \theta}\left(\sin\theta V_\theta \right) + \frac{1}{r\sin\theta}\frac{\partial V_\phi}{\partial \phi}$$

$$\vec{\nabla} \times \vec{V} = \frac{1}{r\sin\theta}\left[\frac{\partial}{\partial \theta}\left(\sin\theta V_\phi \right) - \frac{\partial V_\theta}{\partial \phi} \right]\hat{r} + \left[\frac{1}{r\sin\theta}\frac{\partial V_r}{\partial \phi} - \frac{1}{r}\frac{\partial}{\partial r}\left(rV_\phi \right) \right]\hat{\theta} + \frac{1}{r}\left[\frac{\partial}{\partial r}\left(rV_\theta \right) - \frac{\partial V_r}{\partial \theta} \right]\hat{\phi}$$

$$\nabla^2 \Phi = \frac{1}{r^2}\frac{\partial}{\partial r}\left(r^2 \frac{\partial \Phi}{\partial r} \right) + \frac{1}{r^2 \sin\theta}\frac{\partial}{\partial \theta}\left(\sin\theta \frac{\partial \Phi}{\partial \theta} \right) + \frac{1}{r^2 \sin^2\theta}\frac{\partial^2 \Phi}{\partial \phi^2}$$

We also can write : $\dfrac{1}{r^2}\dfrac{\partial}{\partial r}\left(r^2 \dfrac{\partial \Phi}{\partial r} \right) = \dfrac{1}{r}\dfrac{\partial^2}{\partial r^2}\left(r\Phi \right)$

Vector Relations

Particular relations:

$$\nabla \times \vec{r} = 0, \ \nabla \times \hat{r} = 0, \ \nabla \cdot \vec{r} = 3, \ \nabla \cdot \hat{r} = \frac{2}{r}, \quad (\vec{a} \cdot \nabla)\hat{r} = \frac{1}{r}\left[\vec{a} - \hat{r}(\vec{a} \cdot \hat{r})\right] \equiv \frac{a_\perp}{r}$$

General relations:

Between ordinary vectors:

$$\vec{a} \cdot \left(\vec{b} \times \vec{c}\right) = \vec{b} \cdot \left(\vec{c} \times \vec{a}\right) = \vec{c} \cdot \left(\vec{a} \times \vec{b}\right)$$

$$\vec{a} \times \left(\vec{b} \times \vec{c}\right) = (\vec{a} \cdot \vec{c})\vec{b} - \left(\vec{a} \cdot \vec{b}\right)\vec{c}$$

$$\left(\vec{a} \times \vec{b}\right) \cdot \left(\vec{c} \times \vec{d}\right) = (\vec{a} \cdot \vec{c})\left(\vec{b} \cdot \vec{d}\right) - \left(\vec{a} \cdot \vec{d}\right)\left(\vec{b} \cdot \vec{c}\right)$$

With del-operator:

$$\nabla \times \nabla \phi = 0$$

$$\nabla \cdot \left(\nabla \times \vec{a}\right) = 0$$

$$\nabla \times \left(\nabla \times \vec{a}\right) = \nabla\left(\nabla \cdot \vec{a}\right) - \nabla^2 \vec{a}$$

$$\nabla \cdot \left(\phi\vec{a}\right) = \vec{a} \cdot \nabla\phi + \phi\nabla \cdot \vec{a}$$

$$\nabla \times \left(\phi\vec{a}\right) = \nabla\phi \times \vec{a} + \phi\nabla \times \vec{a}$$

$$\nabla\left(\vec{a} \cdot \vec{b}\right) = \left(\vec{a} \cdot \vec{\nabla}\right)\vec{b} + \left(\vec{b} \cdot \vec{\nabla}\right)\vec{a} + \vec{a} \times \left(\nabla \times \vec{b}\right) + \vec{b} \times \left(\nabla \times \vec{a}\right)$$

$$\nabla \cdot \left(\vec{a} \times \vec{b}\right) = \vec{b} \cdot \left(\nabla \times \vec{a}\right) - \vec{a} \cdot \left(\nabla \times \vec{b}\right)$$

$$\nabla \times \left(\vec{a} \times \vec{b}\right) = \vec{a}\left(\vec{\nabla} \cdot \vec{b}\right) - \vec{b}\left(\vec{\nabla} \cdot \vec{a}\right) + \left(\vec{b} \cdot \vec{\nabla}\right)\vec{a} - \left(\vec{a} \cdot \vec{\nabla}\right)\vec{b}$$

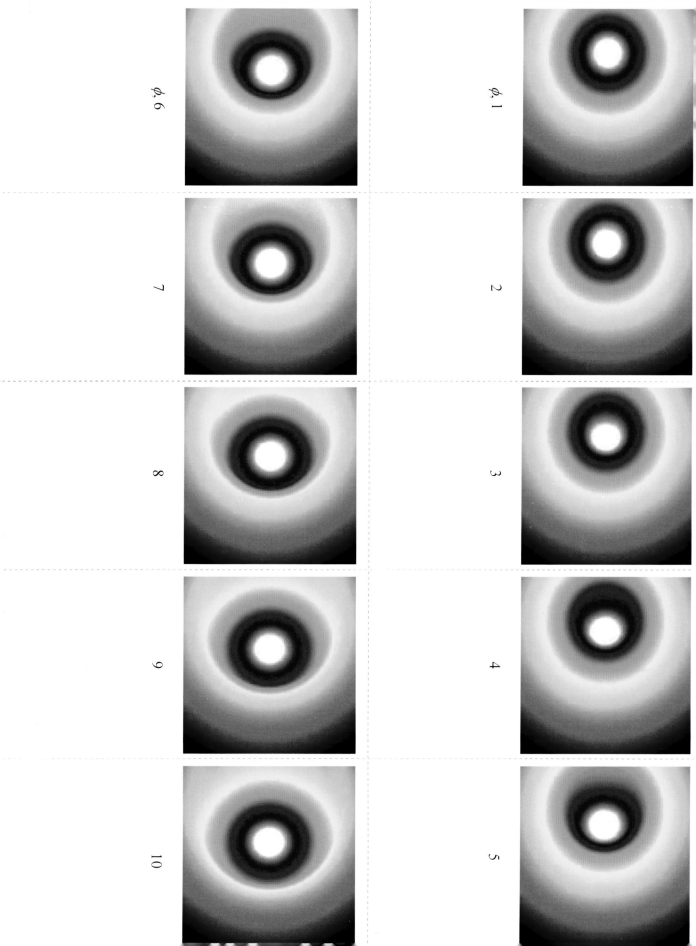

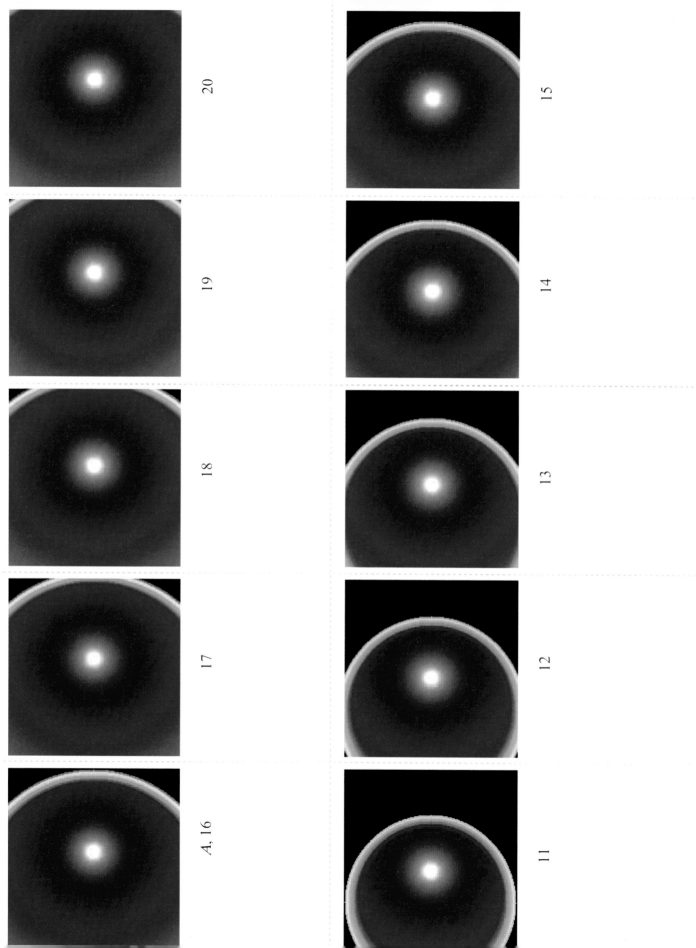

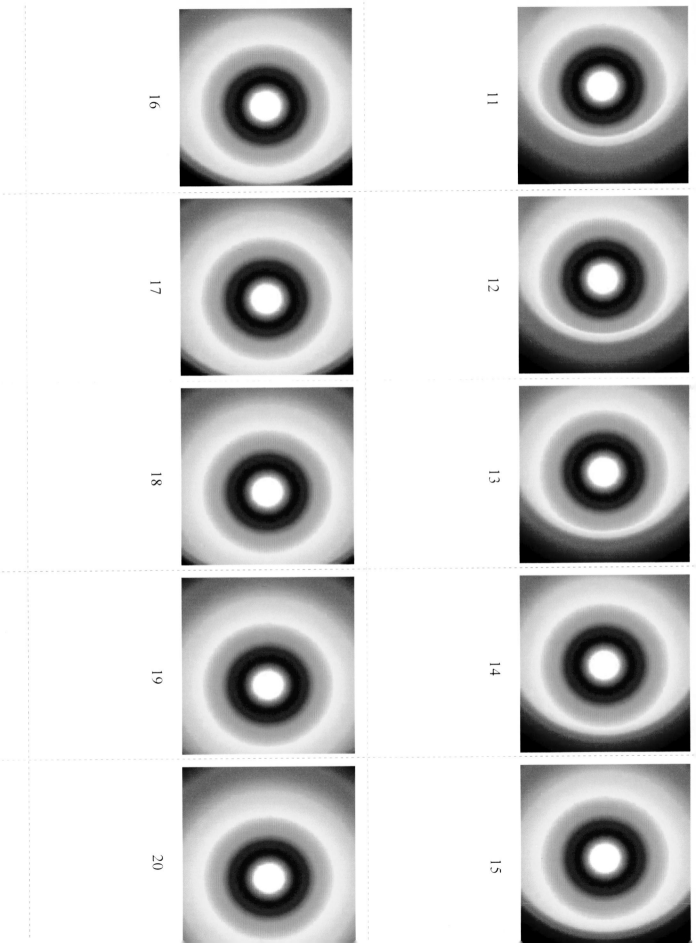

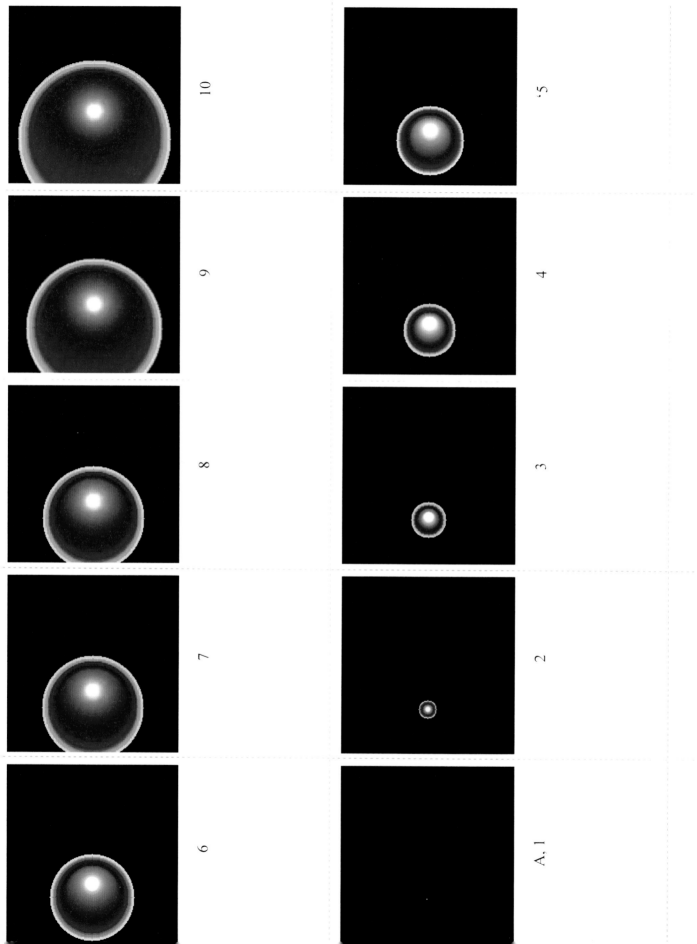

Time Sequence Images for: ϕ, $\vec{A}$, $\dfrac{d\phi}{dz}$, $\dot{\vec{A}}$, and $\dfrac{d\phi}{dx}$

(see Chapter 6)

Flip Book Instructions

How to use the previous two sheets to create flip books for ϕ and $\vec{A}$

i) Remove the two perforated sheets from the book.
ii) Tear out each card along its perforations. It often helps to fold back and forth along the perforation before tearing.
iii) Arrange the cards in order with ϕ, 1 first and ϕ, 20 at the end.
iv) Put a couple staples parallel to the width about ¾ of the way up the empty area of the bottom of the card.
v) Flip with your thumb to view the ϕ movie.
vi) Turn the book over and flip to see the A movie.

Description of Colors and Parameters used in Simulation Images

Parameters for ϕ, $\vec{A}$, $\dfrac{d\phi}{dz}$, $\dot{\vec{A}}$ and $\dfrac{d\phi}{dx}$ simulations

All images show the state of the plana in an *x-z* plane slice of the plana through the line of motion of the point charge. The image represents a width that is 2 light-seconds (which we abbreviate *ls*) across. Note: a light second is the distance light goes in 1 second (approximately: $3 \times 10^{10}\, m$). The particle sits at rest for a long enough time for the plana to be activated with a static ϕ-field. The first frame shows the plana and the particle just up to the moment it moves. At this moment, it begins to undergo a uniform acceleration: $a = 2ls / s^2$. It does this for .2 seconds, moving .04 *ls* during that time and ending with a speed of .4c. The 20 frames show evenly spaced time for 2 seconds so that each frame is separated by .1 second; however, the second frame has been removed from the beginning to allow one more frame of the wave propagation to be kept at the end.

Color scale:
 i) ϕ *images*
From least magnitude (closest to zero) to the most positive magnitude using a logarithmic scale:

> ### black red orange yellow green blue indigo violet and white

(in shorthand we write: black, ROYGBIV, white)

Thus, black is the least magnitude and white is the greatest: black= $\log(\min \phi)$, white = $\log(\max \phi)$

ii) *All other images*
Logarithmic scale from *most negative* (red) to *close to zero* (black) to *most positive* (purple):

> ### red orange yellow **black** green blue purple

$d\phi / dz$

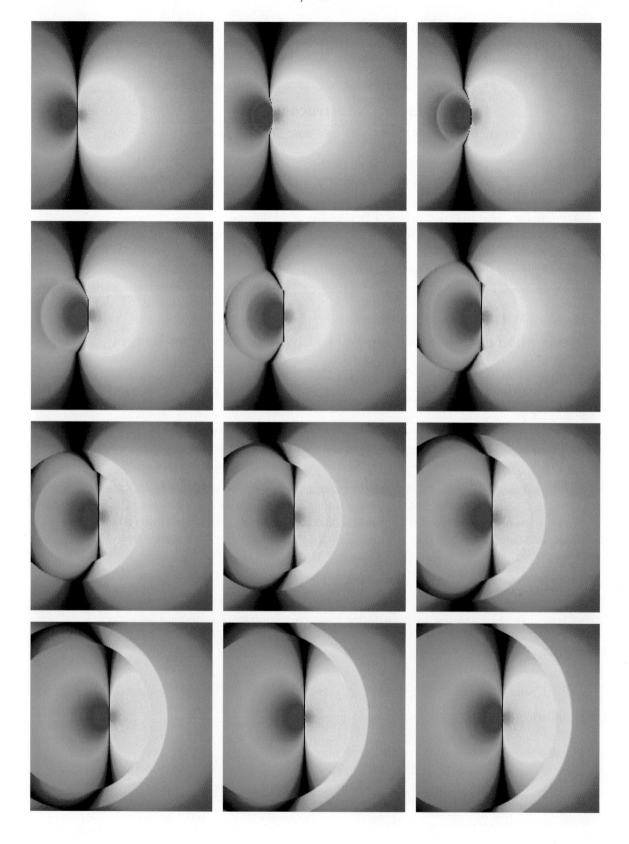

$d\phi / dz$

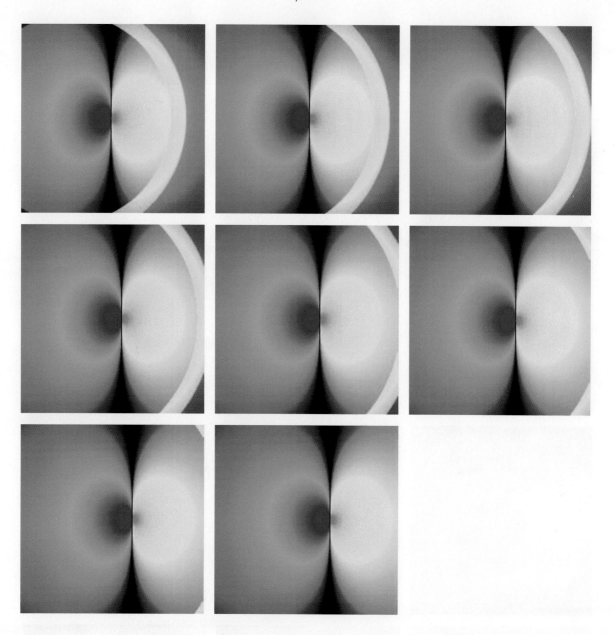

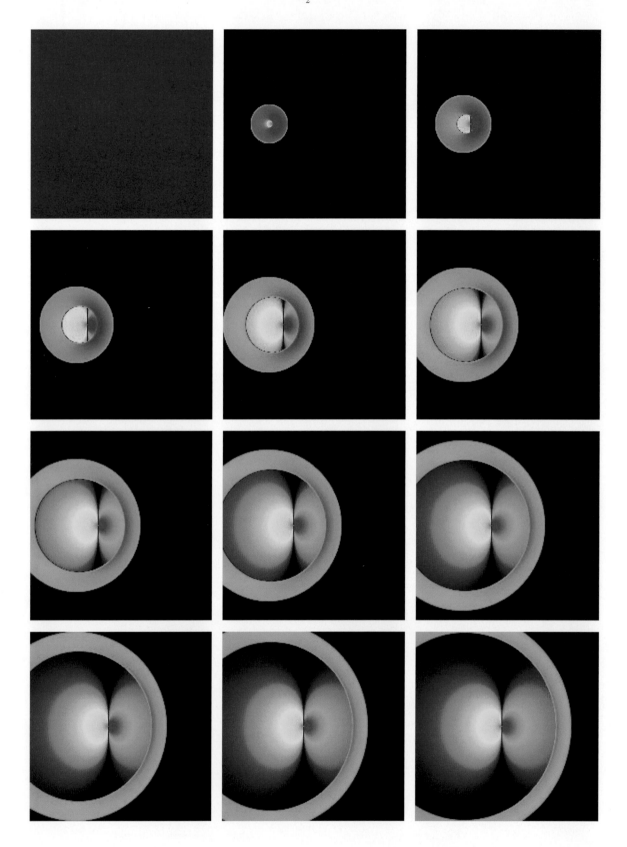

$\dot{A}_z$

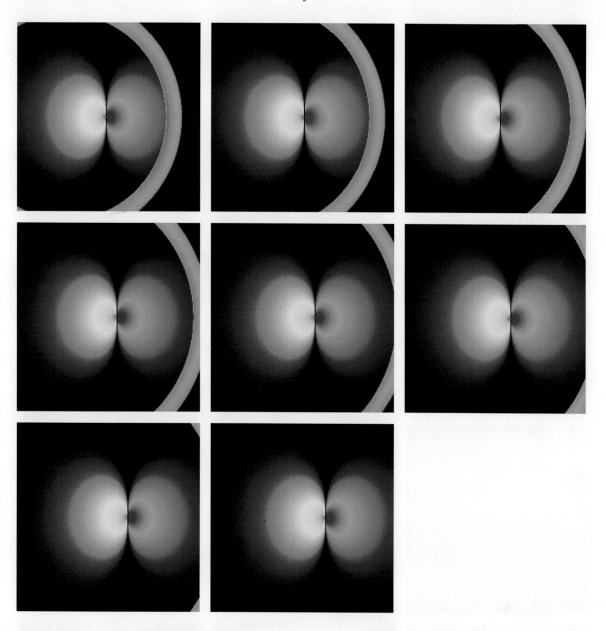

$d\phi \, / \, dx$

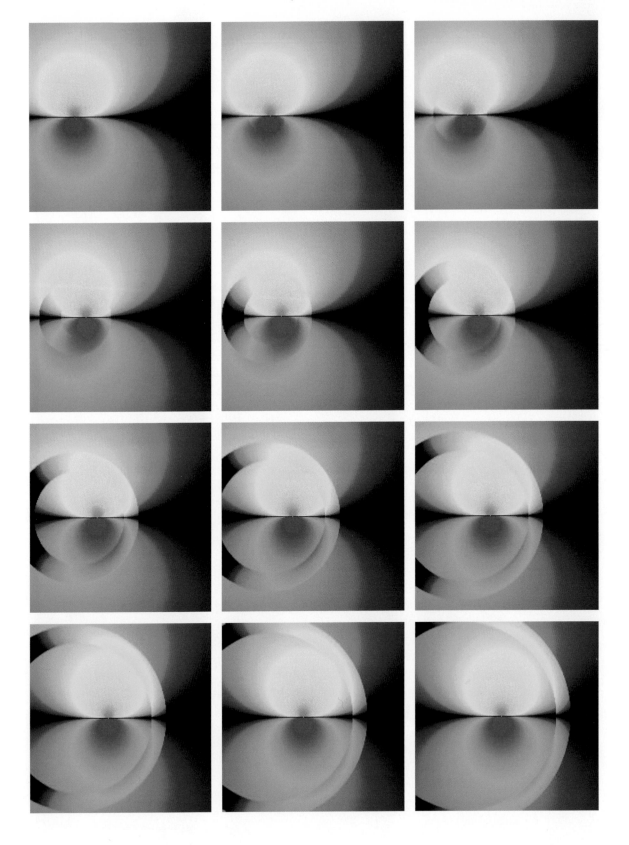

$d\phi / dx$

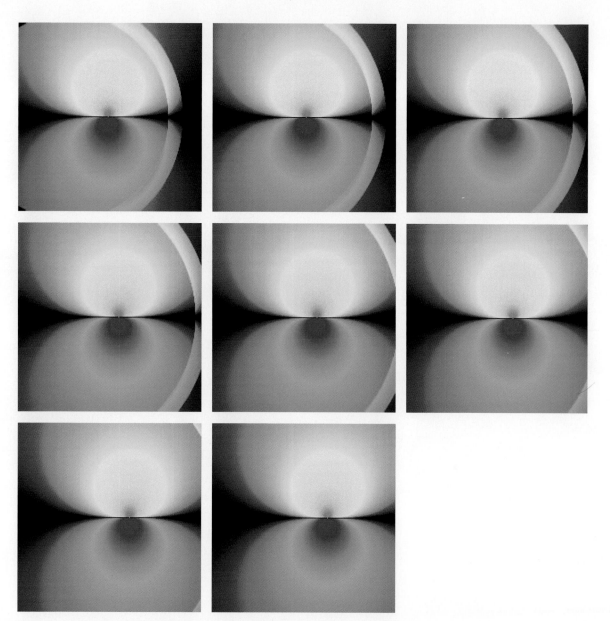

Index